Mathematics

for the IB Middle Years Programme

MYP Year 2

SERIES EDITOR: IBRAHIM WAZIR

KASSANTHRA COYNE, ROBERT EUELL, DIANE OLIVER,
KATHERINE PATE, MICHELLE SHAW

Published by Pearson Education Limited, 80 Strand, London, WC2R 0RL.

www.pearson.com/international-schools

Text © Pearson Education Limited 2021
Development edited by Julie Bond, Keith Gallick and Eric Pradel
Edited by Linnet Bruce
Indexed by Georgina Bowden
Designed by Pearson Education Limited
Typeset by Tech-Set Limited
Picture research by SPi Global
Original illustrations © Pearson Education Limited 2021
Cover design © Pearson Education Limited 2021

The right of Kassanthra Coyne, Robert Euell, Diane Oliver, Katherine Pate, Michelle Shaw and Ibrahim Wazir to be identified as the authors of this work has been asserted by them in accordance with the Copyright, Designs and Patents Act 1988.

First published 2023

24
10 9 8 7 6 5

British Library Cataloguing in Publication Data
A catalogue record for this book is available from the British Library

ISBN 978 1 292 36741 5

Printed and bound by CPI Group (UK) Ltd, Croydon, CR0 4YY

Acknowledgements

Cover: Shutterstock: Who is Danny/Shutterstock

Photos Acknowledgements:

123RF: yuriz/123RF 1, 2; bloodua/123RF 24; andreykuzmin/123RF 40; Andriy Popov/Shutterstock 71; Georg Henrik Lehnerer/123RF 98; Weerapat Kiatdumrong/123RF 99; ALLA ORDATII/123RF 124; Sergejus Bertasius/123RF 144; Oksana Tkachuk/123RF 152; Ioana Davies/123RF 161, 162; normaals/123RF 165; wutlufaipy/123rf.com 263; Brian Kinney/123RF 277; Sally Wallis/123RF 312; Piotr Pawinski/123RF 327; daisydaisy/123RF 360; Robert Nyholm/123RF 362; Tatiana Popova/123RF 371; Igor Zakowski/123RF 374; Tatiana Popova/123RF 377; Esther Derksen/123RF 376; Siegfried Damm/123rf.com 385; **Alamy Stock Photo:** Jui-Chi Chan/Alamy Stock Photo 250; christopher jones/Alamy Stock Photo 301; **Shutterstock:** Vrezh Gyozalyan/Shutterstock 12; Klever LeveL/Shutterstock 19; Maksim Shchur/Shutterstock 23; Taras Vyshnya/Shutterstock 24; Robert J Daveant/Shutterstock 24; Budimir Jevtic/Shutterstock 57, 58; Photobac/Shutterstock 59; Zerbor/Shutterstock 66; Elena Veselova/Shutterstock 79; CHULKOVA NINA/Shutterstock 88; PhuShutter/Shutterstock 89; pingebat/Shutterstock 90; Shawn Hempel/Shutterstock 95, 96; Maridav/Shutterstock 97; Polina Katritch/Shutterstock 98; Maridav/Shutterstock 100; Foto-Ruhrgebiet/Shutterstock 104; Billion Photos/Shutterstock 129, 130; Sashkin/Shutterstock 131; Flamingo Images/Shutterstock 132; Oleg Golovnev/Shutterstock 150; Sashkin/Shutterstock 151; VaLiza/Shutterstock 155; Sergei Drozd/Shutterstock 163, 179; Soleil Nordic/Shutterstock 165; A G Baxter/Shutterstock 191, 192; Garsya/Shutterstock 194; szefei/Shutterstock 206; Maxim Vetrov/Shutterstock 207; vichie81/Shutterstock 217; Shahjehan/Shutterstock 239, 240; Puwadol Jaturawutthichai/Shutterstock 243; NAS CREATIVES/Shutterstock 246; DarwelShots/Shutterstock 247; Greir/Shutterstock 248; ffolas/Shutterstock 251; Andre Boukreev/Shutterstock 251; Globe Guide Media Inc/Shutterstock 255; Iakov Filimonov/Shutterstock 257; LacoKozyna/Shutterstock 260; Aleks Kend/Shutterstock 261; sportpoint/Shutterstock 265; MeeRok/Shutterstock 269, 270; Bertl123/Shutterstock 277; elfinadesign/Shutterstock 279; Kjarra/Shutterstock 279; Morphart Creation/Shutterstock 283; George Dolgikh/Shutterstock 283; ronstik/Shutterstock 285, 286; stocksolutions/Shutterstock 289; northlight/Shutterstock 289; SSS55/Shutterstock 293, 294; Joseph Sohm/Shutterstock 299; Vicky Jirayu/Shutterstock 300; NaughtyNut/Shutterstock 301; Rocksweeper/Shutterstock 301; 99Art/Shutterstock 302; joserpizarro/Shutterstock 305; Charlie Goodall/Shutterstock 312; deryabinka/Shutterstock 319; Julia Weinrichter/Shutterstock 321; PaulSat/Shutterstock 325; Who is Danny/Shutterstock 333, 334; Bakhtiar Zein/Shutterstock 340; Margaret M Stewart/Shutterstock 341; Bakhtiar Zein/Shutterstock 343; ESB Professional/Shutterstock 346; SmartPhotoLab/Shutterstock 356; Aleksander Krsmanovic/Shutterstock 358; stockphotofan1/Shutterstock 369, 370; ronstik/Shutterstock 372; Ian Dikhtiar/Shutterstock 377; **Getty Images:** maradaisy/Getty Images 42; Brett Taylor/iStock/Getty Images Plus/Getty Images 303; **Studio 8:** Studio 8/Pearson Education Ltd 373; **Sanjay Charadva:** Sanjay Charadva/Pearson India Education Services Pvt. Ltd 336.

Text Acknowledgements:

Central Statistics Office: Population and Migration Estimates - CSO - Central Statistics Office. CSO, 20 Aug. 2020 41; **Chalk Dust Company:** How to Calculate Population Density by Antony Caruso. Chalk Dust Company. 87; **Insider Intelligence Inc.:** He, Amy. US Adults Are Spending More Time on Mobile Than They Do Watching TV. Insider Intelligence, Insider Intelligence, 4 June 2019 179; **National Geographic Society:** National Geographic Society. Using a Grid With a Zoo Map. National Geographic Society, 7 Aug. 164; **Stockingblue:** European Union States by Population Density. (FEB 12, 2018). Stocking Blue RSS. 89; **The Federal Geographic Data Committee:** District of Columbia Prototype USNG Street Atlas. USNG Page Number 18S UJ 2400, Page Number 2. The Federal Geographic Data Committee. 165.

All other images © Pearson Education

Contents

Course Structure

Mathematics can be fun!

'I can't do it!'

Have you ever thought or exclaimed these words when stumped by a mathematics problem? I bet every one of us has said these words to themselves at least once in their lifetime. And not just because of a mathematics problem! In order to be engaged as a learner, regardless of age, we like to experience things in a fun and interactive way. Not only that, learning only happens when we leave our comfort zone.

This series is dedicated to the idea that mathematics can be (and is) fun.

Our mission is to accompany you, dear learner, out of your comfort zone and towards the joy of mathematics. Do you accept this challenge?

Bearing in mind the latest research about learning mathematics, the driving ideas behind this series are the following:

- We believe that everyone can do mathematics. Of course there are a few that find it 'easier' than others, *but mathematics learning*, done right, *is for all*. We believe in you, the learner.
- *The essence of mathematics is solving problems*. We will work together to help you become a better problem solver. Problem solvers make mistakes. Plenty of them. With perseverance, they end up solving their problems. You can too. Making mistakes and learning from them is part of our education. Our approach is backed by research on *growth mindsets* and follows in the steps of George Pólya, the father of problem solving.
- Other than some special inventions, most of our societies' development is done by groups. That is why we will, with the help of your teachers, support you to achieve your goals within a group environment.
- Mathematicians' work, no matter how 'advanced' the result, starts with an exploration. Ideas do not magically materialise to a mathematician's mind by superpowers. Mathematicians work hard, and while working, discoveries are made. Whoever discovered gravitational forces did not sit back and then all of a sudden come out with the idea. It was observation first.
- Once you have an idea, you can investigate it to develop your understanding in more depth. We have included many opportunities for you to expand your knowledge further.

Fact

The legend is that Isaac Newton discovered gravity when he saw a falling apple while thinking about the forces of nature. Whatever really happened, Newton realised that some force must be acting on falling objects like apples because otherwise they would not start moving from rest.

It is also claimed that Indian mathematician and astronomer Brahmagupta-II (598–670) discovered the law of gravity over 1000 years before Newton (1642–1727) did. Others claim that Galileo discovered it 100 years before Newton.

How to use this book

No one can teach you unless you want to learn. We believe that, through this partnership with you, we can achieve our goals.

In this book, we have introduced each concept with an Explore. First and foremost, when you start a new concept, try to do the Explore. Have courage to make guesses but try to justify your guesses. Remember, it is ok to make mistakes. Work with others to analyse your mistakes.

Explore 9.4

Look back at the data for rolling a dice 36 times from Explore 9.3.
How would you represent this data to make it easy to read?

Throughout this book, you will find worked examples. When you are given a worked example, do not jump immediately to the solution offered. Try it yourself first. When you do look at the solution offered, be critical and ask yourself: could I have done it differently?

Worked example 8.3

Marta wants to cut 1 metre of ribbon into three equal lengths.

How long should each length be?

Give your answer to a suitable degree of accuracy.

Solution

$100 \div 3 = 33.333...$ cm

It is not possible to measure 33.333... cm accurately.

Each length is 33.3 cm (to the nearest millimetre).

At the end of any activity, we encourage you to reflect on what you have done. There is always a chance to extend what you have learned to new ideas or different perspectives. Not only in studying mathematics, but in any task you should always take the opportunity to reflect on what you did. You will either feel that the task is completed, or you may find that you need to improve on some parts of it. This is true whether you are a student, a teacher, a parent, an engineer, or a business leader, to mention a few. You will find reflection boxes throughout the book to help you with this.

Reflect

In Worked example 8.3, how did it help to convert the measurements to cm?

Can you round 5.26 m to the nearest 10 cm without converting to cm first?

At the end of each section of the book, you will find practice questions. It is recommended that you do these, and more, until you feel confident that you have mastered the concept at hand.

Practice questions 8.2

1 Write 7.517 metres as ___ m ___ cm ____ mm

2 Write 3 m 24 cm 5 mm:

a in centimetres, to the nearest mm

b in metres, to the nearest mm.

Instead of summarising each chapter for you, we have you review what you learned from the chapter in a self-assessment. These self-assessments are checklists. Look at them, and if you feel you missed something, revisit the section covering it.

Self assessment

- I can identify natural numbers, integers and real numbers.
- I can identify and use the place value of digits in natural numbers up to hundreds of millions.
- I can identify and use the place value of digits in decimals.

Finally, at the end of every chapter, it is good practice to look back at the chapter as a whole and see whether you can solve problems. Each chapter contains check your knowledge questions for this purpose.

? Check your knowledge questions

1 Complete these measurement conversions.

a $420\,\text{m} = \square\,\text{cm}$ b $530\,\text{cm} = \square\,\text{m}$

c $0.4\,\text{km} = \square\,\text{mm}$ d $546\,\text{mm} = \square\,\text{cm}$

e $2450\,\text{m} = \square\,\text{km}$ f $3.4\,\text{m} = \square\,\text{mm}$

2 How many books of width 18 mm will fit on a shelf 1 m long?

During your course, your teacher will help you work in groups. In group work, ask for help and help others when asked. The best way of understanding an idea is when you explain it to someone else.

Remember, mathematics is not a bunch of calculations. Mathematical concepts must be communicated clearly to others. Whenever you are performing a task, justify your work and communicate it clearly.

Additional features

Matched to the latest MYP Mathematics Subject Guide

Key concepts, related concepts and global contexts

Each chapter covers one key concept and one or more related concepts in addition to being set within a global context to help you understand how mathematics is applied in our daily lives.

KEY CONCEPT

Relationships

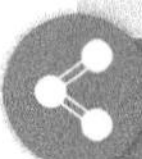

RELATED CONCEPTS

Patterns, Quantity, Representation, Systems

GLOBAL CONTEXT

Globalisation and sustainability

Statement of inquiry and inquiry questions

Each chapter has a statement of inquiry and inquiry questions that lead to the exploration of concepts. The inquiry questions are categorised as factual, conceptual and debatable.

Statement of inquiry

Using number systems allows us to understand relationships that describe our climate, so we are able to acknowledge human impact on global climate change.

Approaches to learning tags

We have identified activities and questions that have a strong link to specific approaches to learning to help you understand where you are using particular skills.

Thinking skills

Do you recall?

At the start of each chapter, you will find do you recall questions to remind you of the relevant prior learning before you start a new chapter. Answers to the do you recall questions can be found in the answers section at the back of the book.

Do you recall?

1 What are directed numbers?

2 What are the number operations?

3 What mental methods do you know for adding two 2-digit numbers?

4 What mental methods do you know for subtracting from a 2-digit number?

Investigations

Throughout the book, you will find investigation boxes. These investigations will encourage you to seek knowledge and develop your skills. They will often provide an opportunity for you to work with others.

Investigation 1.1

Collect magazine, newspaper or online articles that use global temperatures, sea levels and carbon dioxide emissions. Explain in each case what the data tells you.

Research temperatures and the amount of rainfall for five different locations on the same day or month each year for 20 years. What does your data show you?

Fact boxes

Fact boxes introduce historical or background information for interest and context.

Fact

A pescatarian is someone who eats fish, but does not eat any other meat.

Hint boxes

Hint boxes provide tips and suggestions for how to answer a question.

Hint Q9

Note the different units.

Reminders

These boxes are used to recap previous concepts or ideas in case you need a refresher.

Reminder

Always state how you have rounded the measurement in the final answer.

Connections

These boxes highlight connections to other areas of mathematics, or even other subjects.

Connections

You learned how to measure and draw angles accurately in Chapter 4.

Challenge Q12

Challenge tags

We have identified challenging questions that will help you stretch your understanding.

This series has been written with inquiry and exploration at its heart. We aim to inspire your imagination and see the power of mathematics through your eyes.

We wish you courage and determination in your quest to solve problems along your your MYP mathematics journey. Challenge accepted.

Ibrahim Wazir, Series Editor

A note for teachers

Alongside the textbook series, we have also created digital Teacher Guides. These Guides include, amongst other things, ideas for group work, suggestions for organising class discussion using the Explores and detailed, customisable unit plans.

1

Year 1 review

1 Year 1 review

KEY CONCEPT

Relationships

RELATED CONCEPTS

Quantity, Representation

GLOBAL CONTEXT

Orientation in space and time

Statement of inquiry

Throughout history, mathematics has provided a way to represent quantities and the relationships between them.

Factual questions:

- How do we classify real numbers?

Conceptual questions:

- Why do we need an order of operations?

Debatable questions:

- Do we need multiple forms to represent the same number?

Do you recall?

1 What are the different number sets and what notation is used to represent each set?
2 Using the order of operations, work out $4 + 3 \times 5 - (2 + 3)$
3 List all of the factors of 12
4 List the first four multiples of 7
5 Write 30% as a fraction and as a decimal.
6 Classify the following angles as acute, obtuse or right:

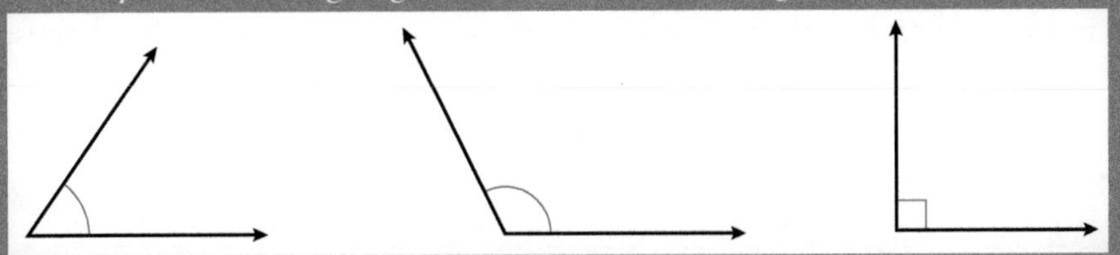

7 Classify the following triangles as equilateral, isosceles or scalene:

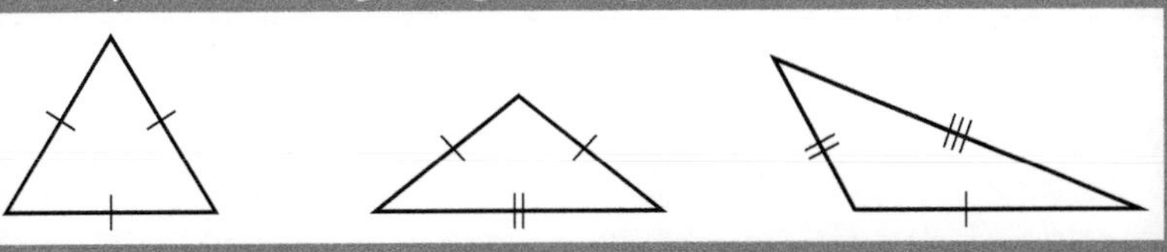

8 If $a = 3$ and $b = 5$, work out the value of $4a - 2b$
9 Find the perimeter and area of this 2D shape.

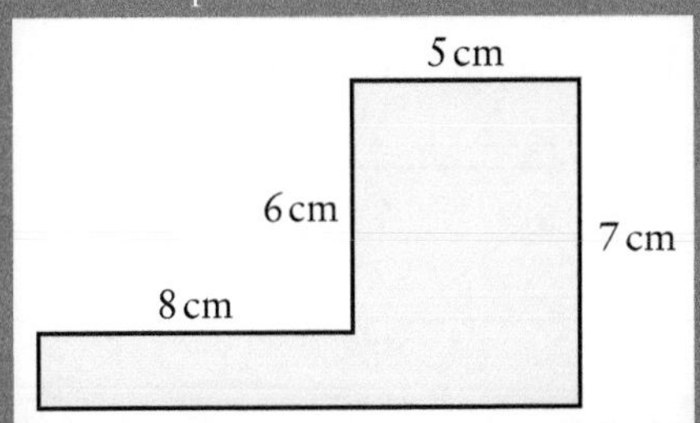

10 Find the volume of this prism.

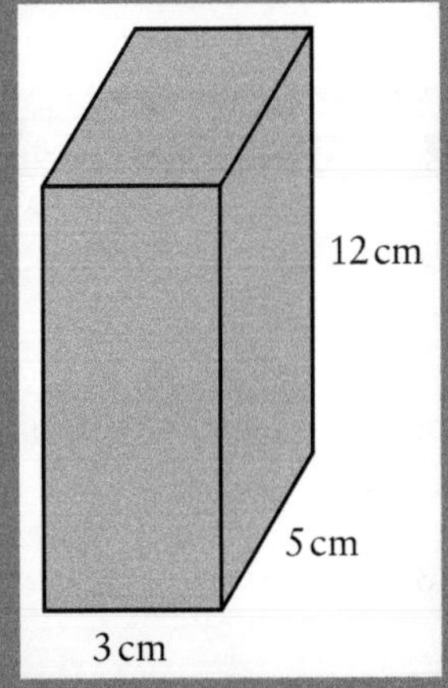

11 Classify the following data as qualitative or quantitative: number of marks scored in a test, time taken to run 100 m, eye colour, favourite food, height.
12 What problem-solving strategies do you know?

1.1 Number basics

We write our numbers using the 10 Hindu–Arabic symbols:

1, 2, 3, 4, 5, 6, 7, 8, 9 and 0

These numerals were invented by Hindus in India around the 7th century and were introduced to Europe around the 12th century. They represented a profound break with previous methods of counting, such as the abacus, and paved the way for the development of algebra.

Explore 1.1

What can you say about $24 \div 6 + 2 \times 5$?

Reflect

What do you recall about the order of operations?

In order to prevent confusion, mathematicians have agreed on an order of operations. This set of rules gives the order in which operations are performed.

Order of operations:

1. brackets
2. $\times$ and $\div$ (left to right)
3. $+$ and $-$ (left to right).

Worked example 1.1

Work out: $5 + 8 \times ((15 - 3) \div 4)$

Solution

Start with the innermost brackets, then work outwards with the remaining brackets. Remember that multiplication comes before addition.

$$5 + 8 \times ((15 - 3) \div 4) = 5 + 8 \times (12 \div 4) = 5 + 8 \times 3$$
$$= 5 + 24$$
$$= 29$$

Practice questions 1.1

1 Write the following as basic numerals:

a 30 000 + 8000 + 400 + 50 + 2

b 500 000 + 8000 + 400 + 5

c 6 000 000 + 400 000 + 7000 + 40

2 Write the following numbers in expanded form:

a 92 673 b 458 062 c 50 432 106

3 State the next integer:

a greater than 345 b greater than −4521

c less than 6897 d less than −84

4 What natural number is 23 less than 97?

5 What integer is 17 less than 4?

6 Copy and complete the table by placing ticks (✓) to show which of the number sets ℕ, ℤ or ℝ these numbers belong to.

Reminder Q6

ℕ: Natural numbers
ℤ:Integers
ℝ:Real numbers

	N	Z	R
−11		✓	✓
$\frac{3}{7}$			
23			
−6.25			
$-\frac{8}{2}$			
3.6			
0			
7821			

7 Write down a real number between:

a 3 and 4 b 12.6 and 12.7

c −62 and −61 d 0.06 and 0.062

e $\frac{4}{9}$ and $\frac{5}{9}$

8 Write the basic numeral for:

a eight thousand and sixty-three

b two hundred and twenty-seven thousand

c fifty three thousand and four

d five million and nine.

9 Write down each numeral in words:

a 4812

b 13 901

c 500780

d 7 082 807

10 Write down the value of 5 in each of the following numbers:

a 53

b 6001.5

c 45732

d 607 588

e 43.765

11 Write each of these expanded numbers as numerals:

a 10 000 + 3000 + 200 + 80 + 6

b 100 000 + 30 000 + 2000 + 60 + 3

c 7 000 000 + 50 000 + 700 + 5

12 Write each group of numbers in ascending order:

a 62 341, 63 812, 63 976, 62 213, 61 093

b 78 302, 78 514, 78 012, 78 365, 78 546

c 302739, 302 645, 302 972, 302 709, 302 963

13 Calculate:

a $8 \times 3 - 6 \times 4$

b $28 - 5 \times 3$

c $3 + 16 \div (4 \times 2)$

d $(16 - 7 + 2) \times 9$

e $(33 + 9) \div (17 - 11)$

f $77 - (56 \div 8) \times 2$

g $100 - (13 - (3 + 8))$

h $\dfrac{63 - 15}{18 \div 3}$

Challenge Q14

14 Calculate $12 \times 5 - 2 \times (18 \div 2) \div 2 \div \frac{1}{2} + 8$

1.2 Divisibility tests

Explore 1.2

Can you list all the digits from 1 to 10 which 3450 is divisible by?
Do not forget to give a reason for each decision.

Divisibility tests are useful because they help us to quickly determine if one number is divisible by another number without doing the division.

The table shows some divisibility tests.

Divisor	Divisibility test	Example
2	The number is even.	538 is divisible by 2 because it is even.
3	The sum of the digits is divisible by 3	412 032 is divisible by 3 because $4 + 1 + 2 + 0 + 3 + 2 = 12$ and 12 is divisible by 3
4	The number formed by the last two digits is divisible by 4	9724 is divisible by 4 because 24 is divisible by 4
5	The last digit is 5 or 0	4 563 195 is divisible by 5 because it ends with a 5
6	The number is divisible by 2 and 3	9006 is divisible by 6 because it is even and the sum of the digits is divisible by 3
8	The number formed by the last three digits is divisible by 8	5112 is divisible by 8 because 112 is divisible by 8
9	The sum of the digits is divisible by 9	412 731 is divisible by 9 because $4 + 1 + 2 + 7 + 3 + 1 = 18$ and 18 is divisible by 9
10	The last digit is 0	9 284 870 is divisible by 10 because it ends with a 0

Worked example 1.2

Is 24 816 divisible by:

a 2 b 3 c 4 d 5

e 6 f 8 g 9 h 10

Solution

a 24816 is an even number. Therefore, 24816 is divisible by 2

b $2 + 4 + 8 + 1 + 6 = 21$ and 21 is divisible by 3. Therefore, 24816 is divisible by 3

c 16 is divisible by 4. Therefore, 24816 is divisible by 4

d The last digit is not 5 or 0. Therefore, 24816 is not divisible by 5

e 24816 is divisible by both 2 and 3. Therefore, 24816 is divisible by 6

f 816 is divisible by 8. Therefore, 24816 is divisible by 8

g $2 + 4 + 8 + 1 + 6 = 21$ and 21 is not divisible by 9. Therefore, 24816 is not divisible by 9

h The last digit is not 0. Therefore, 24816 is not divisible by 10

Reflect

Why is it useful to know divisibility rules?

Why do you think 1 is not included in the table?

Is there a divisibility rule for 7?

The divisibility test for 6 is that the number is divisible by both 2 and 3. What divisibility rule could you use for 12?

Practice questions 1.2

1 Calculate:

a 403×1 b 57×0

c $97 + 0$ d 1×81

e $13 \times 4 \times 0$ f $0 + 42$

2 Write down a numeral in the square to make each number sentence true:

a $9 \times \square = 0$ b $\square \times 1 = 653$

c $\square + 0 = 217$ d $\square \times 0 = 0$

e $7 + 13 = 13 + \square$ f $9 \times 5 = \square \times 9$

g $1.3 \times \square = 0$ h $0.4 + \square = 0.4$

3 Complete the table below by placing a ✓ in the appropriate box for each number that is divisible by 2, 3, 4, 5, 6, 8, 9, 10

Number	Divisible by 2	Divisible by 3	Divisible by 4	Divisible by 5	Divisible by 6	Divisible by 8	Divisible by 9	Divisible by 10
1004	✓							
1485		✓						
1020					✓			
13 476								
91 800							✓	

4 List all of the factors of:

a 24   b 90   c 116   d 105

5 Calculate the highest common factor (HCF) of:

a 45 and 60   b 102 and 153

c 44, 66 and 110   d 27 and 64

6 List the first four multiples of:

a 5   b 9   c 12   d 30

7 Calculate the lowest common multiple (LCM) of:

a 8 and 10   b 9 and 15

c 24 and 30   d 20 and 14

8 List:

a the prime numbers that are less than 30

b the composite numbers between 30 and 40

9 Write each of these numbers as a product of their prime factors:

a 24   b 144   c 225   d 252

10 Write each of the following in expanded form and as a basic numeral:

a 4^2 b 10^3 c 9^2 d 2^4

11 Mary is making identical party bags for each of her friends. She has 32 stickers, 24 pencils and 16 lollipops. She wants each bag to have the same number of each item. What is the greatest number of party bags she can assemble if every item is used?

Reminder Q3

Factors divide natural numbers exactly.

Reminder Q5

HCF is also known as GCD – the greatest common divisor.

Reminder Q8

A prime number has exactly two factors: itself and 1

Hint Q9

We can write a number as a product of its prime factors by continuing to divide by prime factors (starting from the smallest prime factor) until we get a result of 1.

 Challenge Q12

12 Three churches in a town have bells that ring every 3, 5 and 8 minutes respectively. If all of the bells ring together at 08:00 at what time will they ring together again?

 Challenge Q13

13 Write 400 and 1080 as a product of prime factors. Hence, calculate the

a HCF of 400 and 1080

b LCM of 400 and 1080

1.3 Integers

Explore 1.3

What can you say about each of the following? Give your reasons.

a $20 + (-4) - (-9)$

b $3 \times -4 - 5 \times -2$

c $18 \div (-3) - 21 \div (-7)$

Reflect

What are the rules for operating with integers?

Addition and subtraction with integers:

- Adding a negative number is the same as subtracting a positive number, for example:

  $4 + (-1) = 4 - 1 = 3$

- Subtracting a negative number is the same as adding a positive number, for example:

  $4 - (-1) = 4 + 1 = 5$

Multiplication and division with integers:

- Multiplication and division of numbers with the same sign gives a positive answer, for example:

  $3 \times 4 = 12$, $-3 \times -4 = 12$ and $30 \div 6 = 5$, $-30 \div -6 = 5$

- Multiplication and division of numbers with different signs gives a negative answer, for example:

  $3 \times -4 = -12$, $-3 \times 4 = -12$ and $30 \div -6 = -5$, $-30 \div 6 = -5$

Worked example 1.3

Work out $10 + 24 \div -8$

Solution

$10 + 24 \div -8$ — Carry out the division first.

$= 10 + -3$ — Adding a negative number is the same as subtracting a positive number.

$= 10 - 3$

$= 7$

Reflect

We encounter applications of integers regularly in our everyday lives. Common examples are temperature, money and altitude. Discuss some applications of integers you encounter in real life.

Practice questions 1.3

1 Work out:

a $-5 + 2$

b $-4 + 19$

c $4 - 12$

d $-8 - 6$

e $5 - (-14)$

f $14 + (-9)$

g $-2-(-6 + 15)$

h $11 + (-5 - 8)$

i $14 + (-6)$

j $-11 - (-6 - 4)$

k $-13 + (-3 - 7)$

l $0 - (-16)$

2 Work out:

a	$2 \times (-1)$	b	-5×3
c	-6×-7	d	$0 \times (-4)$
e	7×-6	f	8×-8
g	$(-4)^2$	h	$-4 \times -6 \times 2$
i	$-18 \div -3$	j	$-22 \div 2$
k	$28 \div -7$	l	$-42 \div -6$
m	$\frac{-36}{6}$	n	$\frac{-56}{-8}$
o	$\frac{-121}{11}$	p	$\frac{-260}{-10}$

3 Work out:

a $72 \div -12 + 4 \times 5$

b $-5 \times (-4 + 9) + (-3)$

c $5^2 - 18 \div 9 - (-6)$

d $-3 \times (-8 + 2)^2 \div 6 - (-4)$

4 The table below shows the average daily temperature in some European cities in January 2020 in degrees Celsius.

City	Temperature, °C
Amsterdam, Netherlands	3
Belgrade, Serbia	−1
Copenhagen, Denmark	−3
Hamburg, Germany	4
Istanbul, Turkey	6
Kiev, Ukraine	−5

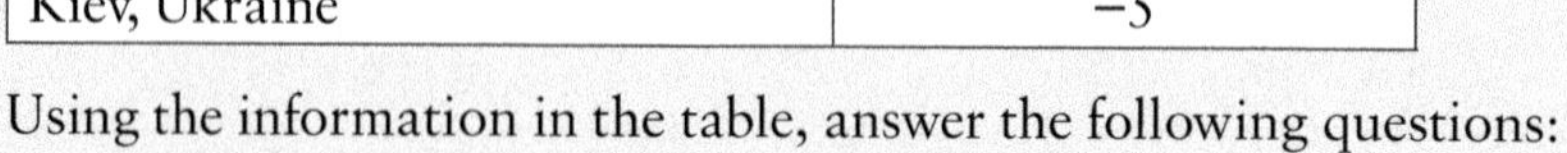

Using the information in the table, answer the following questions:

a On average, which city has the lowest temperature?

b On average, which city has the highest temperature?

c List the cities in order from coldest to hottest.

5 At 20:00, the temperature was −6°C. By 01:00, it had dropped by 5°C. At 08:00, it had increased 7°C from the temperature it was at 01:00. What was the temperature at 08:00?

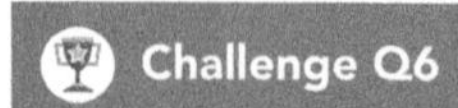

6 Some number x when added to −17 gives a result of 27. Divide x by −11 and then multiply the result by −6. What is the final number?

7 Local time in Kathmandu is 2 hours and 45 minutes ahead of Baghdad and Baghdad is 8 hours behind Sydney, which is 16 hours ahead of New York. What is the time difference between Kathmandu and New York?

8 a Which two numbers have a product of 30 and a sum of -11?

b Which two numbers have a sum of 1 and a product of -42?

1.4 Decimals

Explore 1.4

Sam works out that $31.04 - 23.1 = 8.03$

What do you say? Give your reasons.

When adding and subtracting numbers we align them according to their place values. We can easily align decimal numbers by their place value by aligning the decimal points. An empty space can be filled by a 0

Worked example 1.4

Work out:

a $53.63 + 7.097$　　　b $35.843 - 17$

Solution

a $53.63 + 7.097$

Align the decimal points and fill empty spaces with a 0

Add columns from right to left as normal.

$$\begin{array}{r} 53.630 \\ +\ {}_{1}7.0_{1}97 \\ \hline 60.727 \end{array}$$

Therefore, $53.63 + 7.097 = 60.727$

b $35.843 - 17$

Align the decimal points and fill empty spaces with a 0

Subtract columns from right to left as normal.

$$\begin{array}{r} {}^{2}\not{3}{}^{1}5.843 \\ -\ 17.000 \\ \hline 18.843 \end{array}$$

Therefore, $35.843 - 17 = 18.843$

Reflect

Why can we replace empty spaces with a 0 when adding and subtracting decimals?

Explore 1.5

Lucia works out that $0.14 \times 0.2 = 0.0028$

Rebecca works out that that $0.14 \times 0.2 = 0.028$

What do you say? Give your reasons.

When multiplying decimals, the number of digits after the decimal point in the answer must be the same as the total number of digits that come after the decimal point in the question.

Explore 1.6

Lucia works out that $0.14 \div 0.2 = 0.07$

Rebecca works out that that $0.14 \div 0.2 = 0.7$

What do you say? Give your reasons.

When dividing a decimal by a whole number the point in the answer is aligned with the point in the question. To divide a decimal by a decimal, we change the question so that we are dividing by a whole number.

For example, $0.24 \div 0.8 = 2.4 \div 8 = 0.3$

Worked example 1.5

Work out:

a 5.24×7.1

b $7.816 \div 0.04$

Solution

a 5.24×7.1

Multiply as normal.

There are three digits in total after the decimal in the question so there will be three digits after the decimal point in the answer.

$$
\begin{array}{r}
{}^{1}5^{2}2\,4 \\
\times \quad 7\,1 \\
\hline
5\,2\,4 \\
3\,6_{1}6_{1}8\,0 \\
\hline
37\,2\,0\,4
\end{array}
$$

Therefore, 5.24 × 7.1 = 37.204

b To divide a decimal by a decimal, we need to rewrite the question so that we are dividing by a whole number. In this case we will need to multiply 0.04 by 100 to make it a whole number.
Therefore, we also need to multiply 7.816 by 100

7.816 ÷ 0.04 = 781.6 ÷ 4

$$
\begin{array}{r}
1\,9\,5.\,4 \\
4\overline{)7^{3}8^{2}1.^{1}6}
\end{array}
$$

When dividing a decimal by a whole number the point in the answer is aligned with the point in the question.

Therefore, 7.816 ÷ 0.04 = 195.4

Reflect

When dividing a decimal number by a decimal number why is it possible to change the number we are dividing by to a whole number? Does doing this affect the final answer?

Practice questions 1.4

1 Write each of the following as a fraction:

a 0.6　　b 0.43

c 0.802　　d 4.3

2 Arrange in ascending order:

a 0.61, 0.6, 0.607, 0.619　　b 0.4, 0.44, 0.04, 0.044

c 0.1, 0.09, 0.8, 0.085　　d 0.707, 0.077, 0.77, 0.07

Reminder Q2

Ascending means from smallest to largest.

3 Work out:

a 7.09 + 14.57　　b 12 − 9.06

c 0.6 + 17 + 4.302　　d 1.063 − 0.7

4 Work out:

a 5×0.03 b 54.31×10 c 2.315×4

d 19.2×9 e 3.14×0.7 f 0.6×87.2

5 Work out:

a $62.1 \div 6$ b $0.123 \div 4$ c $7.402 \div 100$

d $349.2 \div 8$ e $3.6 \div 0.9$ f $5.58 \div 0.18$

6 Round to the nearest whole number:

a 4.7 b 56.5 c 33.4 d 14.81

7 Round to 1 decimal place:

a 7.35 b 4.117 c 0.06 d 19.99

8 Round to the nearest hundredth:

a 24.773 b 5.904 c 0.0763 d 3.992

9 Carol brought €87 on a shopping trip. She spent €4.77 on a coffee, €18.50 on a shirt and €12.95 on a book. How much money does she have left?

10 James bought 7.84 litres of diesel at 126.5 cents per litre. How much did he pay?

11 Tom purchased three cinema tickets for $27.81. How much does each ticket cost?

1.5 Fractions

Explore 1.7

What can you say about each of the following? Give your reasons.

a $\frac{3}{2}+\frac{5}{6}-\frac{2}{3}$ b $\frac{5}{6}\times\frac{2}{3}$ c $\frac{5}{6}\div\frac{2}{3}$

Addition and subtraction of fractions:

Before adding or subtracting fractions, we must express each number with the same denominator. This is called a common denominator. We always try to use the lowest common denominator (the LCD).

Multiplication of fractions:

To multiply two fractions, multiply the numerators and multiply the denominators.

Division of fractions:

To divide by a fraction we multiply by the reciprocal.

 **Hint**

We can make the multiplication easier by cancelling common factors in the numerator and the denominator before multiplying.

 **Reminder**

The reciprocal of $\frac{a}{b} = \frac{b}{a}$

Worked example 1.6

Work out:

a $\frac{1}{4} + \frac{2}{5}$ b $3\frac{2}{3} - 2\frac{1}{6}$ c $\frac{4}{9} \times \frac{3}{5}$ d $\frac{4}{5} \div \frac{1}{2}$

Solution

a The LCD of 4 and 5 is 20

$$\frac{1}{4} + \frac{2}{5} = \frac{1 \times 5}{4 \times 5} + \frac{2 \times 4}{5 \times 4}$$

Writing both fractions as equivalent fractions with denominator 20:

$$\frac{1}{4} + \frac{2}{5} = \frac{5}{20} + \frac{8}{20}$$

Now we can add the numerators:

$$\frac{1}{4} + \frac{2}{5} = \frac{13}{20}$$

b We first write the mixed numbers as improper fractions.

$$3\frac{2}{3} - 2\frac{1}{6} = \frac{11}{3} - \frac{13}{6}$$

The LCD of 3 and 6 is 6

Writing $\frac{11}{3}$ as an equivalent fraction with denominator 6:

$$3\frac{2}{3} - 2\frac{1}{6} = \frac{11 \times 2}{3 \times 2} - \frac{13}{6}$$

Now we can subtract the numerators, and simplify:

$$3\frac{2}{3} - 2\frac{1}{6} = \frac{22}{6} - \frac{13}{6}$$

$$= \frac{9}{6} = \frac{3}{2}$$

So, $3\frac{2}{3} - 2\frac{1}{6} = 1\frac{1}{2}$

c We can cancel common factors before multiplying the numerators and the denominators:

$$\frac{4}{9} \times \frac{3}{5} = \frac{4}{\cancel{9}_3} \times \frac{\cancel{3}_1}{5}$$
$$= \frac{4 \times 1}{3 \times 5}$$
$$= \frac{4}{15}$$

So $\frac{4}{9} \times \frac{3}{5} = \frac{4}{15}$

d We need to multiply by the reciprocal of $\frac{1}{2}$

$$\frac{4}{5} \div \frac{1}{2} = \frac{4}{5} \times \frac{2}{1}$$
$$= \frac{4 \times 2}{5 \times 1}$$
$$= \frac{8}{5} = 1\frac{3}{5}$$

So, $\frac{4}{5} \div \frac{1}{2} = 1\frac{3}{5}$

Hint

$\frac{1}{2} \times 8 = 4$ and $4 \div \frac{1}{2} = 8$ are examples of inverse operations.

Reflect

Using the fact that multiplication and division are inverse operations, can you explain why the rule for dividing fractions makes sense?

Reminder Q4

To compare fractions:

1 Find the LCD for the fractions.
2 Convert each fraction into its equivalent with the LCD in the denominator.
3 Compare fractions: If denominators are the same you can compare the numerators. The fraction with the bigger numerator is the larger fraction.

Practice questions 1.5

1 Write each improper fraction as a mixed number:

a $\frac{11}{3}$ b $\frac{8}{5}$ c $\frac{77}{7}$ d $\frac{341}{10}$

2 Write each mixed number as an improper fraction:

a $2\frac{1}{2}$ b $3\frac{3}{5}$ c $10\frac{2}{9}$ d $5\frac{2}{3}$

3 Complete these equivalent fractions:

a $\frac{1}{2} = \frac{\square}{14}$ b $\frac{3}{5} = \frac{\square}{20}$ c $\frac{1}{7} = \frac{\square}{28}$ d $\frac{\square}{18} = \frac{2}{3}$

4 Which of the following <, = or > signs should be placed in each box?

a $\frac{3}{4} \square \frac{4}{5}$ b $\frac{3}{8} \square \frac{9}{24}$ c $\frac{5}{6} \square \frac{7}{9}$ d $\frac{1}{7} \square \frac{14}{100}$

5 Work out:

a $\frac{7}{12} + \frac{1}{12}$ b $\frac{7}{8} - \frac{5}{8}$ c $\frac{2}{3} + \frac{2}{3}$ d $1 + \frac{3}{10}$

6 Work out:

a $\frac{7}{10} + \frac{3}{5}$ b $\frac{2}{3} - \frac{3}{8}$ c $\frac{5}{12} + \frac{3}{8}$ d $\frac{4}{9} - \frac{1}{4}$

7 Work out:

a $8\frac{3}{4} + 2\frac{1}{2}$ b $6\frac{1}{5} - 1\frac{1}{4}$ c $1\frac{7}{8} + 2\frac{3}{5}$ d $7\frac{1}{2} - 2\frac{2}{3}$

8 Work out:

a $\frac{2}{3} \times \frac{9}{16}$ b $\frac{3}{4} \times \frac{9}{10}$ c $1\frac{2}{3} \times \frac{3}{4}$ d $3\frac{1}{5} \times 5\frac{3}{4}$

9 Work out:

a $\frac{5}{8} \div \frac{1}{2}$ b $\frac{7}{10} \div \frac{2}{3}$ c $1 \div \frac{2}{3}$ d $7\frac{1}{4} \div 3\frac{4}{5}$

10 Calculate:

a $\frac{4}{5}$ of 1 m b $\frac{2}{3}$ of 2 hours c $\frac{3}{8}$ of 1 litre d $\frac{9}{10}$ of 3 kg

11 What fraction is:

a 20c of €1 b 60 cm of 3 m

c 5 min of 2 h d 50 g of 3 kg?

12 It takes Ben $\frac{1}{3}$ of an hour to iron a shirt.
How many minutes would it take him to iron eight shirts?

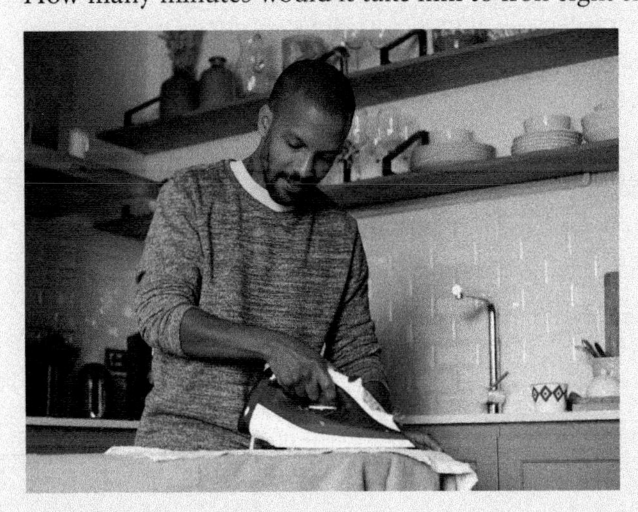

13 Eight students are sharing three pizzas. If the pizzas are shared equally between the students, how much did each student get?

14 Edward spends the same length of time each day doing his homework. After seven days he had spent 9 hours and 20 minutes doing his homework.

How much time did he spend doing his homework each day?

15 In a bag containing purple, red, yellow and orange sweets, there are 33 purple sweets.

$\frac{1}{4}$ of the sweets are red, $\frac{2}{5}$ are yellow and $\frac{1}{6}$ are orange.

How many sweets, in total, are in the bag?

1.6 Percentages

Explore 1.8

Jack saves $5 every month in his savings account at the bank. At the end of the year (12 months later) the bank pays 10% interest on the total amount of money that Jack has saved. Jack works out that he will have $66 in his account at the start of the next year. What can you say?

Reflect

Money can be saved in different ways. For example, in a money box, bank account, savings account. What do you think are the advantages and disadvantages of each?

We encounter percentages daily. Whether we are shopping, saving, working or comparing test scores. Percentages are very useful for making comparisons as they are essentially equivalent fractions with denominator 100.

Worked example 1.7

When downloading a video game Eleanor notices that 32% of the 64 000 kilobytes have been downloaded so far.

How many kilobytes still need to be downloaded?

Solution

We need to work out how many kilobytes still need to be downloaded. In order to do this we first need to work out what per cent of kilobytes still need to be downloaded.

$100\% - 32\% = 68\%$

So we need to calculate 68% of 64 000

Cancel common factors before multiplying 68 by 640:

$$\frac{68}{100} \times 64\,000 = \frac{68}{1\cancel{0}\cancel{0}} \times \frac{640\cancel{0}\cancel{0}}{1}$$

$$= 68 \times 640$$

$$= 43520$$

Therefore, 43520 kilobytes still need to be downloaded.

Reflect

How can we check if the answer is correct?

Practice questions 1.6

1 Write each of the following percentages as a fraction in its simplest form:

a 1% b 25% c 114% d $4\frac{3}{4}\%$

2 Write each of these fractions as a percentage:

a $\frac{17}{100}$ b $\frac{3}{20}$ c $2\frac{5}{8}$ d $\frac{42}{300}$

3 Write each of the following percentages as decimals:

a 14% b 85% c 126% d 9.9%

4 Write each of the following decimals as a percentage:

a 0.42 b 0.3 c 0.09 d 2.05

5 Calculate:

a 10% of €546 b 15% of 52 km

c 2% of 3 hours d 60% of 140 g

6 What percentage is:

a 32 marks of 40 marks b 500 g of 10 kg

c 12 min of 2 hours d 560 cm of 8 m?

Reminder Q1

To write a percentage as a fraction or mixed number, first write it as a fraction with denominator 100, then simplify.

Reminder Q2

To write a fraction as a percentage, first write it as an equivalent fraction with a denominator of 100, then simplify. Alternatively, we can divide the numerator by the denominator and multiply by 100.

Reminder Q6

To write one quantity as a percentage of another:
1 Make sure that the two quantities are in the same units.
2 Write the first number as a fraction of the second.
3 Convert the fraction to a percentage.

7 A jacket costs \$66. In a sale it is reduced by 20%.

a How much money will you save if you buy the jacket in the sale?

b What is the sale price of the jacket?

c The shop decides to discount the reduced sale price of the jacket by a further 10%. What is the new price of the jacket?

8 At a football match there were 20 girls, 16 boys, 18 parents and six teachers.

What percentage of those at the match were:

a teachers b parents c girls

d boys e not boys f boys or girls?

1.7 Probability

Explore 1.9

A coin, flipped four times, landed on heads every time. On the next flip, is it more likely that the result will be tails than heads? Discuss.

Worked example 1.8

Three red and four yellow discs are placed in a bag. One disc is randomly selected.

What is the probability of selecting:

a a red disc b a yellow disc?

Solution

a Three of the seven discs are red so there is a three in seven chance of selecting a red disc.

$P(\text{Red}) = \frac{3}{7}$

b Four of the seven discs are yellow so there is a four in seven chance of selecting a yellow disc.

$P(\text{Yellow}) = \frac{4}{7}$

Practice questions 1.7

1 Describe each of these events as either impossible, unlikely, even chance, likely or certain.

a It will rain where you are today.

b If you throw a dice you will get a number less than 7

c Your maths teacher is less than 40 years old.

d If you roll a dice the result will be a 1, 2 or 3

e You will see a living dinosaur today.

f You will travel in a bus next week.

2 Arrange the parts **a–f** of Question 1 in order from least likely to most likely.

3 There are 20 identical discs in a bag. Four discs are red, eight are blue, two are green and six are yellow. Emma picks one disc at random from the bag.

What is the probability that the result will be:

a red　　b blue

c not yellow　　d green or yellow?

4 A dice is thrown. What is the probability that the result will be:

a a prime number　　b a composite number

c neither prime nor composite　　d a factor of 6

e a multiple of 3?

Hint Q4

1 is neither prime nor composite.

1.8 Geometry

Explore 1.10

What types of angles can you identify in these shapes?

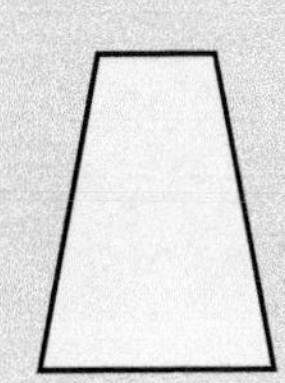

Discuss and compare your answers.

Angles can be classified according to their size, as shown in the table.

Type	acute	right	obtuse	straight	reflex	revolution
Size	less than 90°	90°	between 90° and 180°	180°	between 180° and 360°	360°
Diagram						

Worked example 1.9

Work out the value of the unknown angle, a, in the diagram.

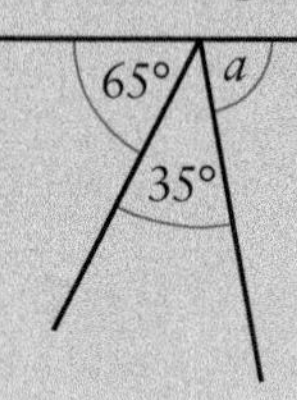

Solution

The sum of the three angles is 180°, a straight line

Therefore, $a = 180° - 65° - 35° = 80°$

Fact

Architects use geometry to define the spatial form of buildings. This can be seen in structures throughout history.

Explore 1.11

Discuss what 3D shapes you can identify in the following pictures:

Worked example 1.10

Draw the net of this rectangular prism on squared paper.

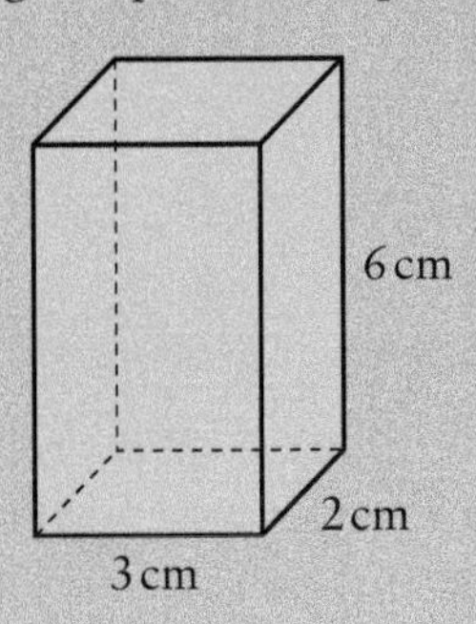

Solution

The solid has six rectangular faces.

Step 1: Draw the bottom. This is 3 cm long and 2 cm wide.

Step 2: Add the two sides that join the bottom.

Step 3: Add the front and the back that join the bottom.

Step 4: Finally add the top. The top can be added to the front, back or either side.

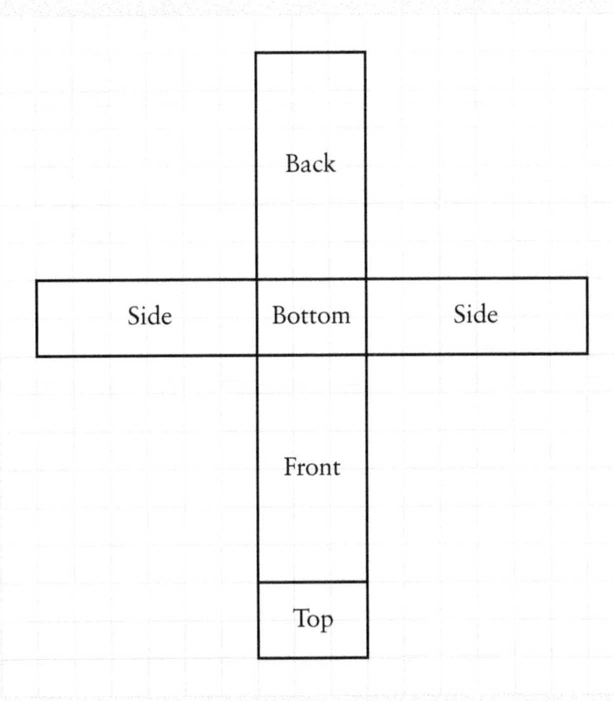

Practice questions 1.8

1 Classify each of these angles as acute, right, obtuse, straight or reflex:

a

b

c 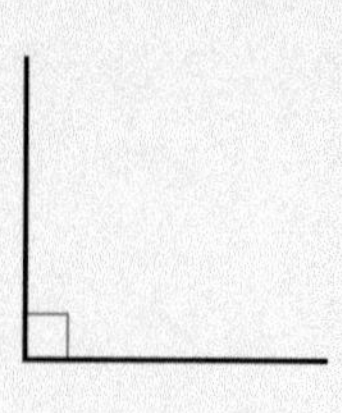

d 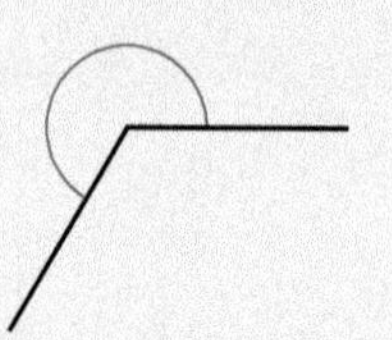

e 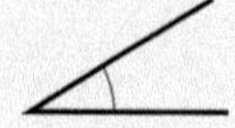

2 Classify each of these angles as vertically opposite, corresponding, supplementary, co-interior, complementary or alternate:

a 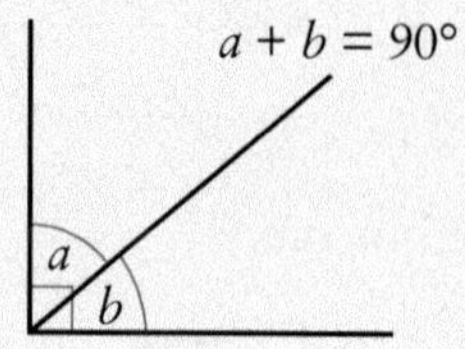

b 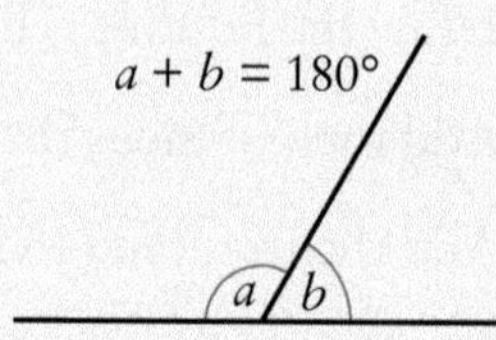

c 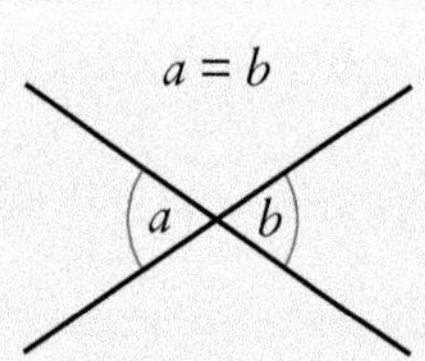

d 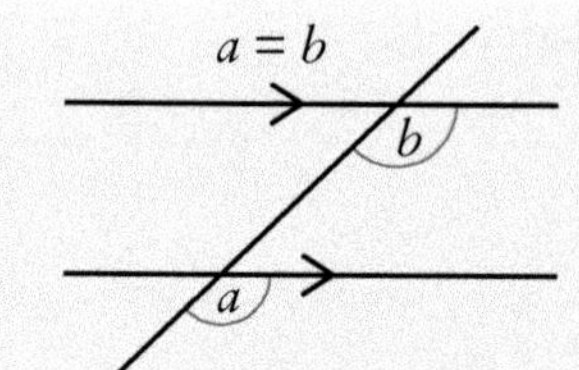

e 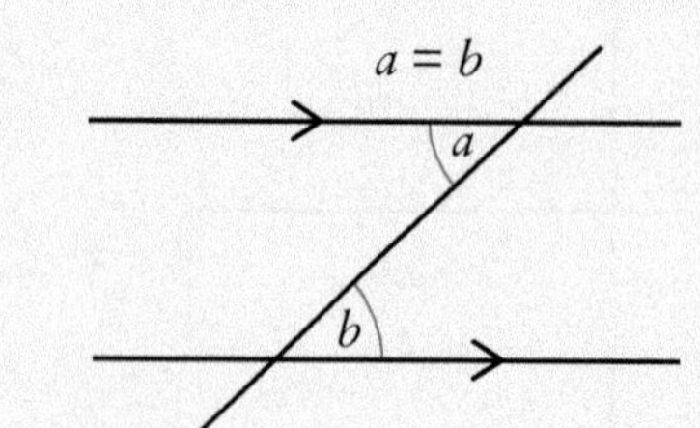

f 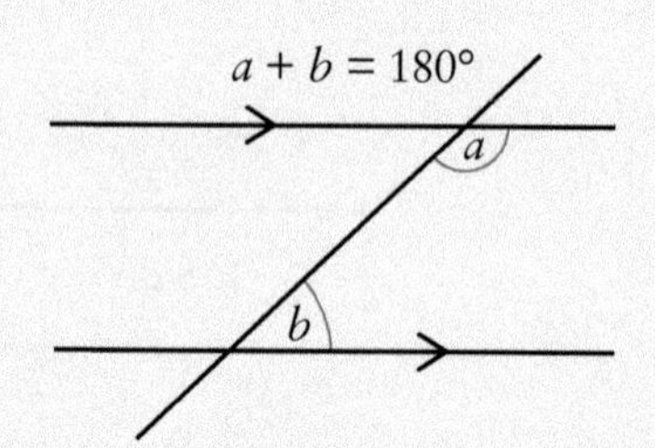

3 Work out the value of the variable in each of the following:

a 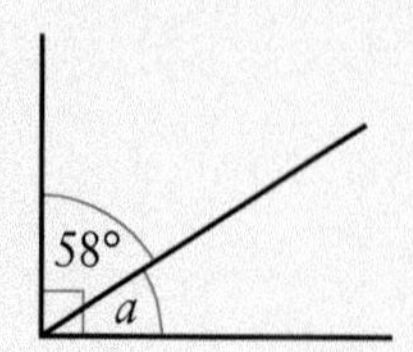

b 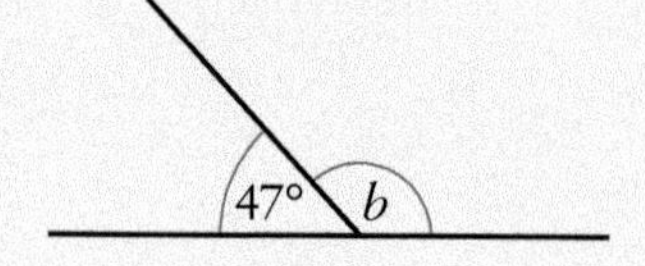

c

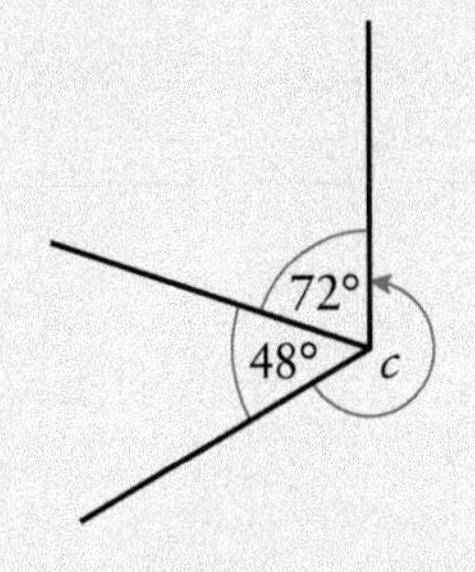

d

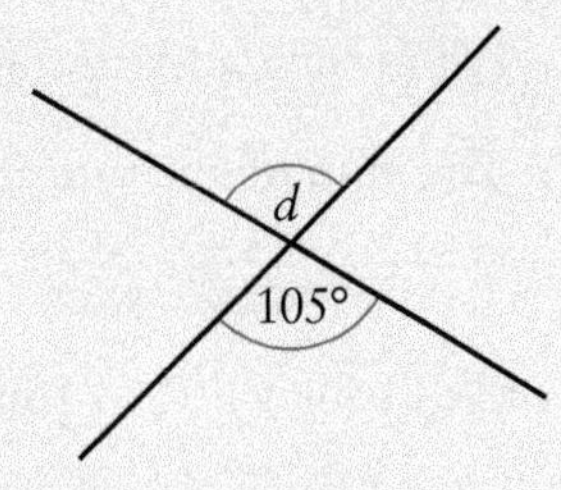

e

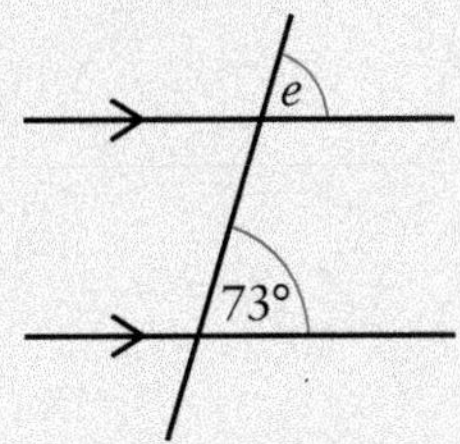

f

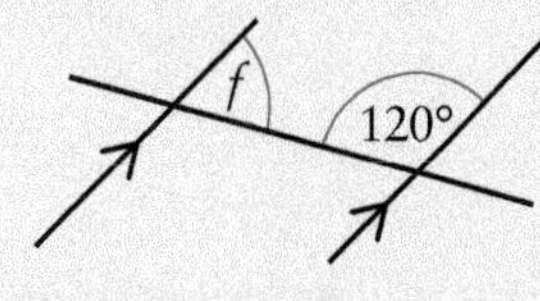

g

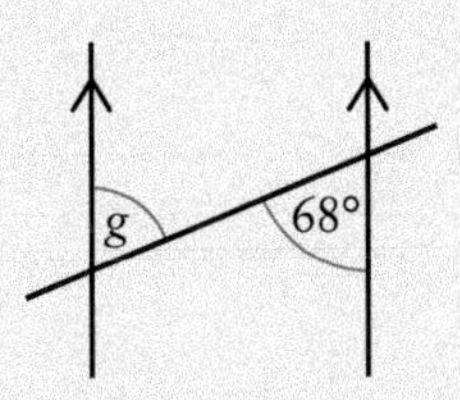

h

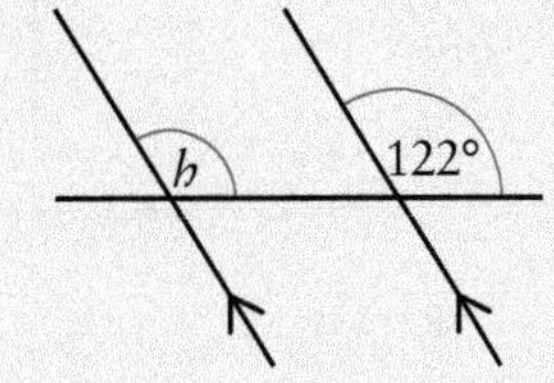

4 Name these shapes.

a

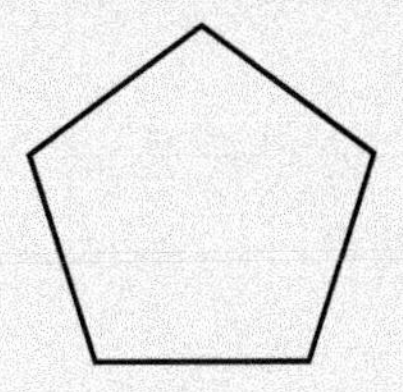

b

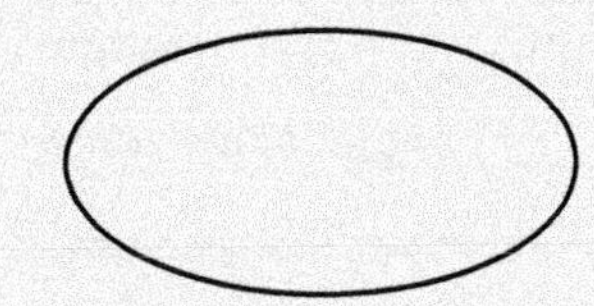

c

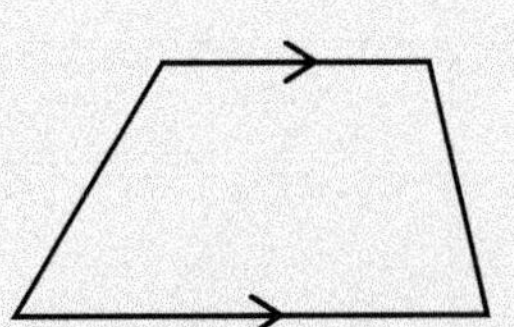

d

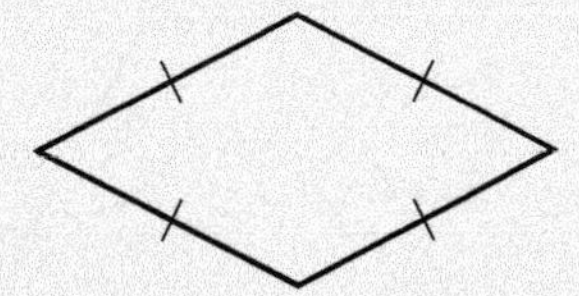

5 Copy and complete this table.

a

Polygon	Number of sides	Sum of the interior angles
Triangle		
Quadrilateral		
Pentagon		
Hexagon		
Heptagon		
Octagon		

b Fill in the blank in this statement:

The sum of the interior angles of any n-sided polygon is ____ × 180°.

6 Work out the value of the missing angle in each of these shapes.

a

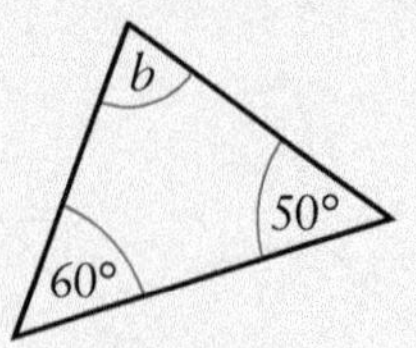

b

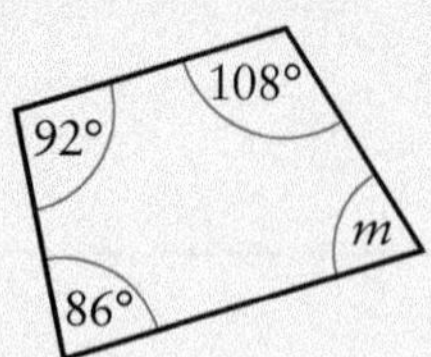

c

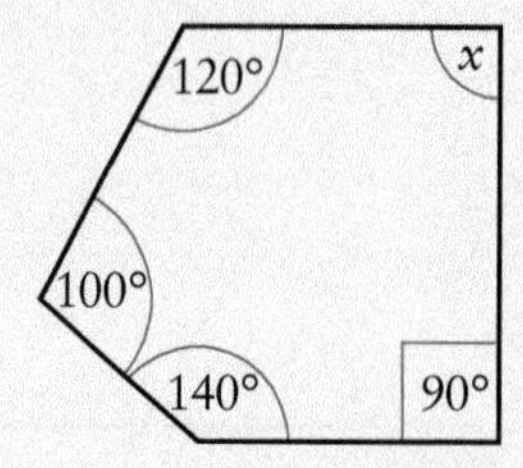

d

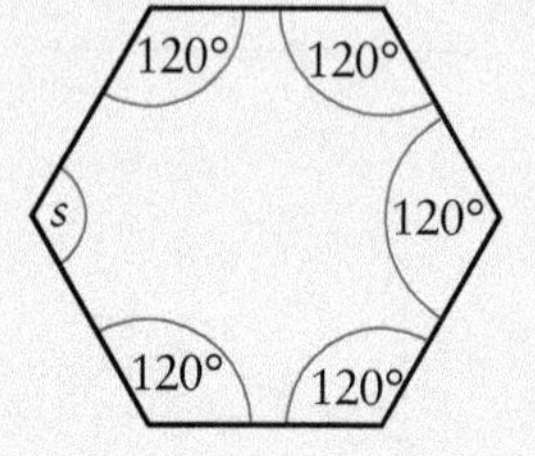

7 For each of the triangles below:

a Work out the value of the missing angle

b Classify each triangle by:

i angle: acute-angled, right-angled, obtuse-angled

ii side length: scalene, isosceles, equilateral.

a

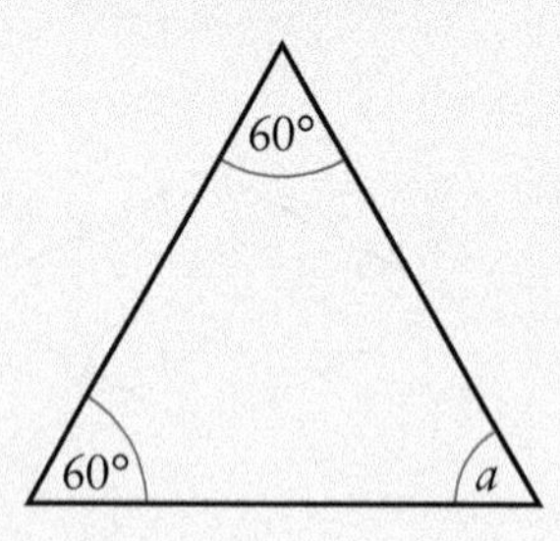

b

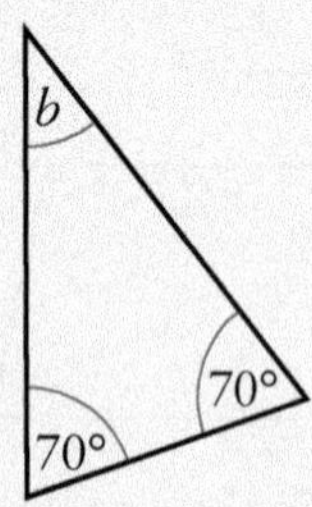

c

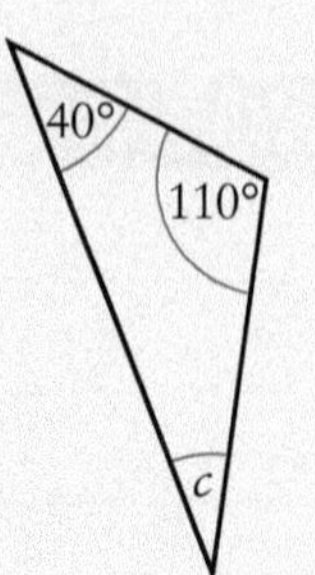

d

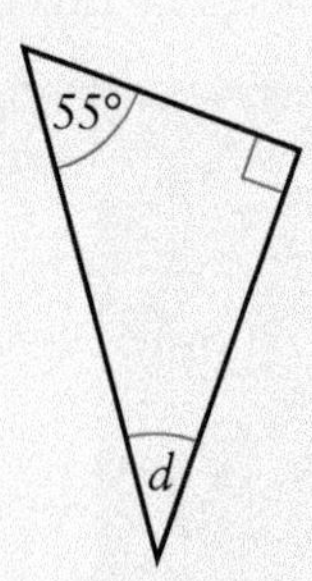

e

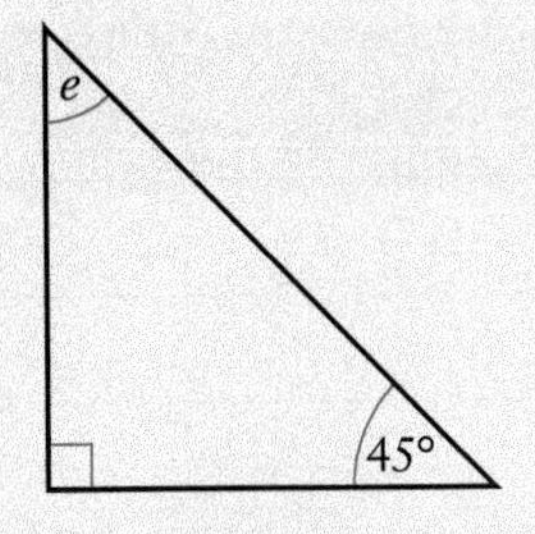

f

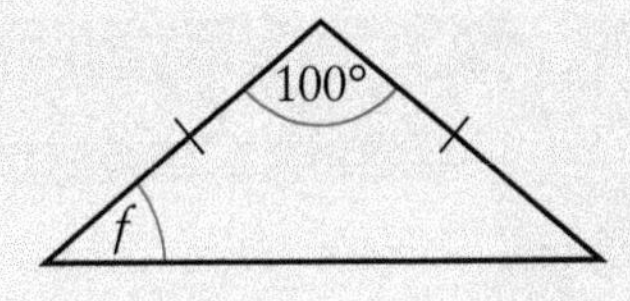

g

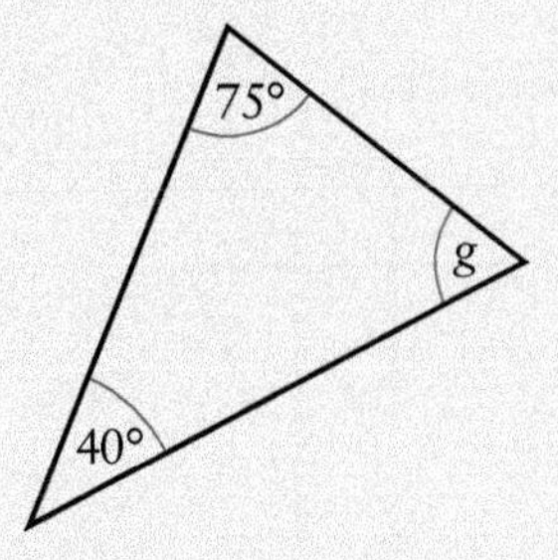

h

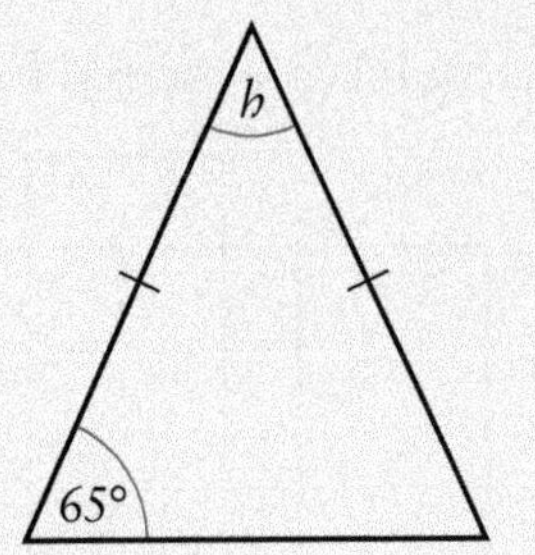

8 Complete the table for each of the solids shown.

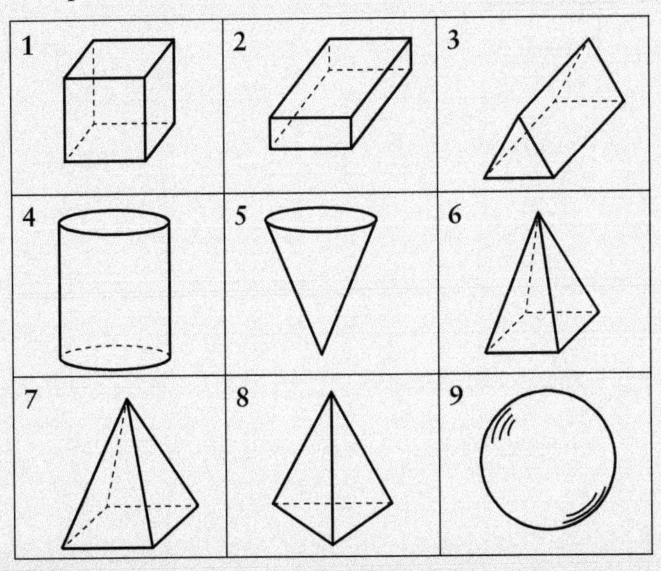

Shape number	Name	Prism	Pyramid	Curved surface	Number of faces	Number of vertices	Number of edges
1	Cube	Yes	No	No	6	8	12
2							
3							
4							
5							
6							

Shape number	Name	Prism	Pyramid	Curved surface	Number of faces	Number of vertices	Number of edges
7							
8							
9							

9 What solids are formed from the following nets?

a

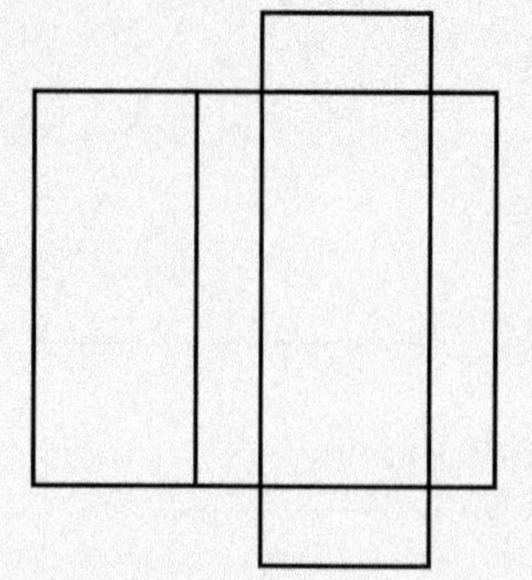

b

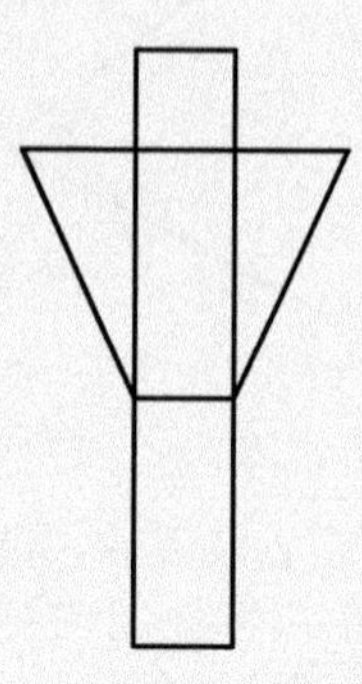

c

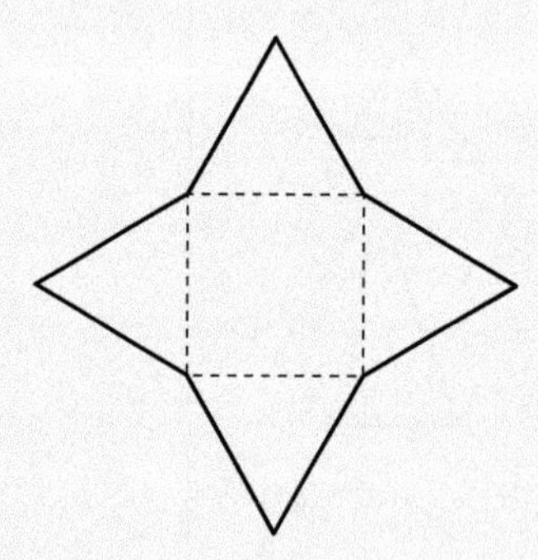

1.9 Measure

Explore 1.12

Can you work out the area and perimeter of each of these two shapes?

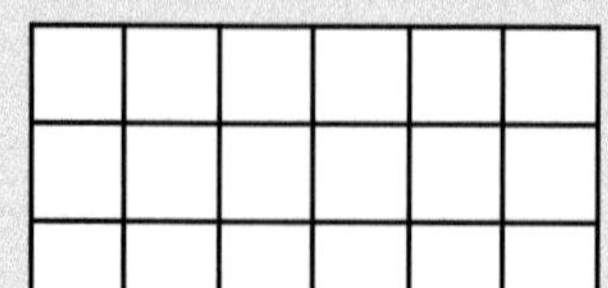

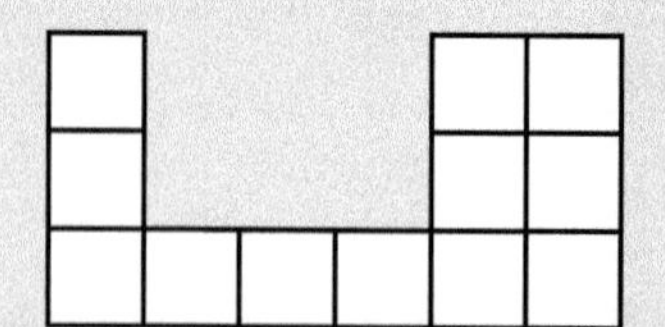

Can you draw a shape where the area is numerically equal to the perimeter?

Can you draw a shape in which the perimeter is numerically twice the area?

Can you draw a shape in which the area is numerically twice the perimeter?

Can you draw some shapes with the same area but different perimeters?

Can you draw some shapes with the same perimeter but different areas?

Fact

The metre is the basic unit of length in the International System of Units (SI).

During the French Revolution, the metre was defined as being one ten-millionth $\left(\frac{1}{10\,000\,000}\right)$ of the distance between the Earth's equator and the North Pole along the Paris meridian.

The metre is now defined as the distance light travels in a vacuum in $\frac{1}{299\,792\,458}$ of a second.

Area is the amount of space inside a region.

Perimeter is a measurement of the distance around the boundary of a figure.

Worked example 1.11

Work out the perimeter and area of this shape.

4 cm

9 cm

6 cm

12 cm

Solution

Work out the lengths of the unknown sides first:

The height of the shape is 9 cm.

Therefore, the unknown height must be 9 cm − 6 cm = 3 cm.

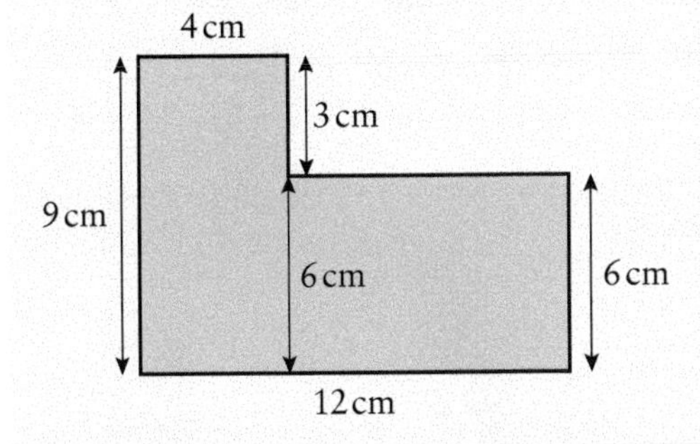

The length of the shape is 12 cm.

Therefore, the unknown length must be 12 cm − 4 cm = 8 cm.

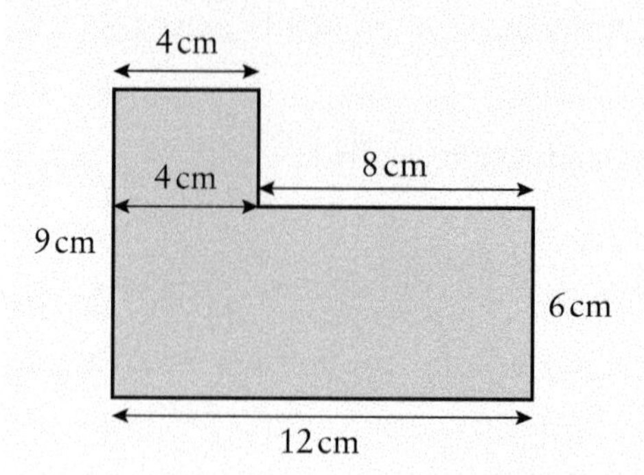

Now that we know all of the side lengths, the perimeter, P, is given by:

$P = 9 + 4 + 3 + 8 + 6 + 12 = 42$ cm.

To find the area we can split the shape into two rectangles and sum the areas of each of the individual rectangles:

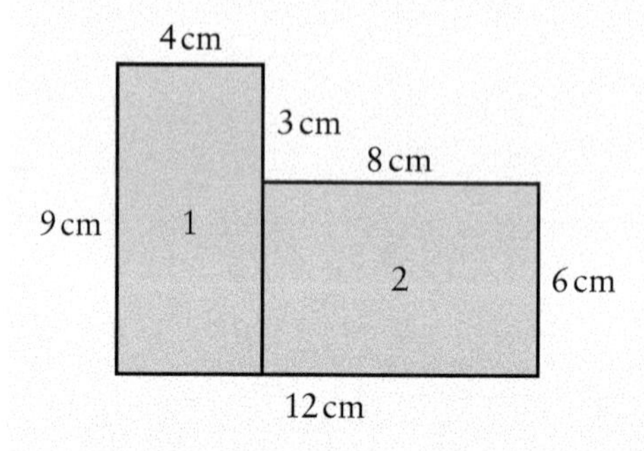

Area of rectangle 1 = $9 \times 4 = 36\,\text{cm}^2$

Area of rectangle 2 = $8 \times 6 = 48\,\text{cm}^2$

Total area of the shape = $36 + 48 = 84\,\text{cm}^2$

Reflect

How else could you have divided the shape to find the total area?

Practice questions 1.9

1 Write each of the following as a 24-hour time.

a 15 minutes past 8 (before noon).

b 40 minutes past 7 (after noon).

c 8 minutes before 4 (before noon).

d 23 minutes before 11 (after noon).

2 I spent 1 h 35 min walking on Monday, 58 min on Tuesday and 46 min on Wednesday. How long did I spend walking altogether?

3 A slow cooker takes $7\frac{1}{2}$ hours to cook dinner. It has been on for 2 h and 20 min. How long until dinner is ready?

4 My exam started at 9:45 and finished at 11:20. How long was the exam?

5 Complete each of the following:

a 800 mm = ☐ cm

b 6345 cm = ☐ m

c 43 m = ☐ mm

d 3.62 km = ☐ cm

6 Simplify:

a 4.3 m + 90 cm + 2 m

b 63 cm + 13 mm

c 3.2 km + 7632 m

d 3 × (2.3 m + 80 cm)

7 Calculate the perimeter of these shapes.

a

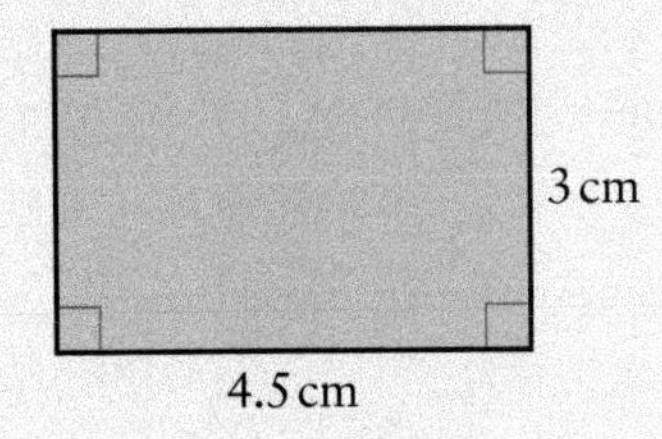

b

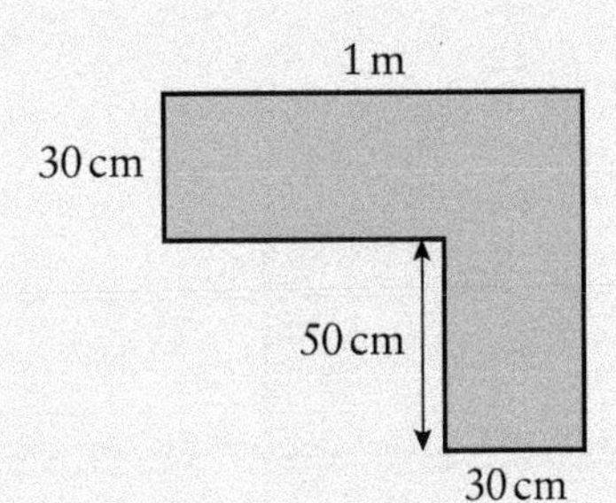

c

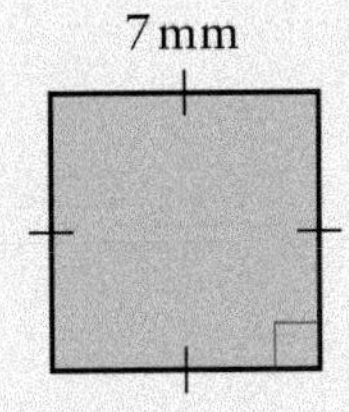

8 Calculate the area of the shapes in Question 7.

9 Calculate the area of each of these triangles.

a

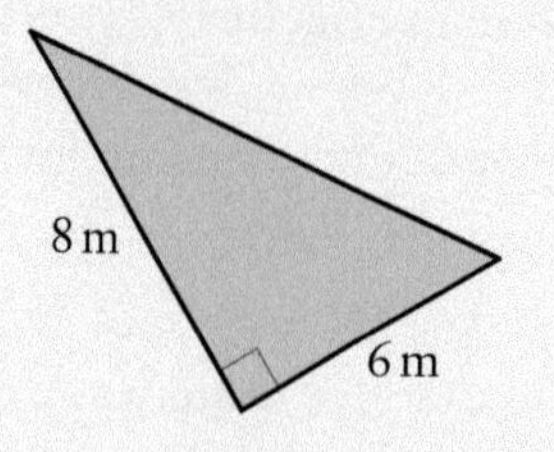

b

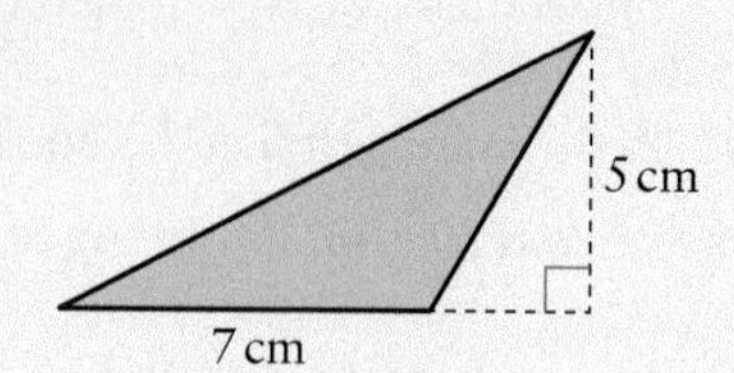

c

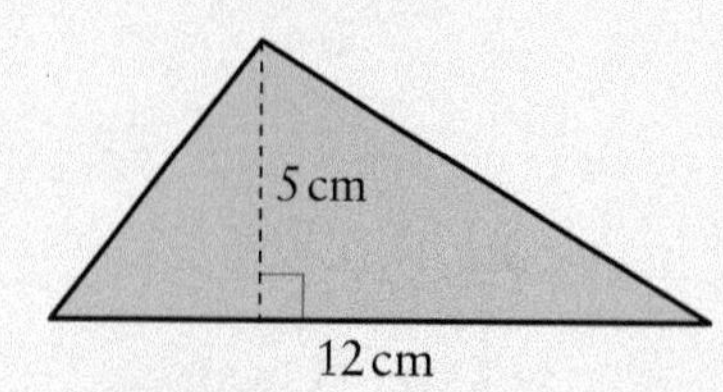

10 Calculate, in m^3, the volume of each of these prisms.

a

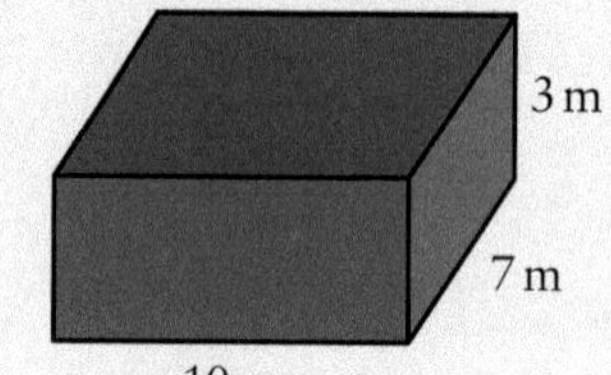

b

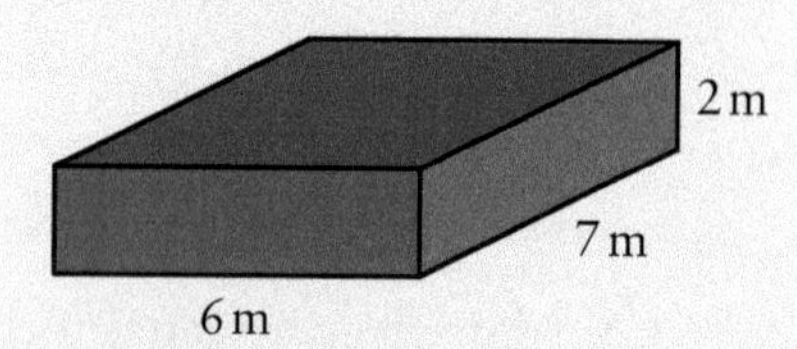

c

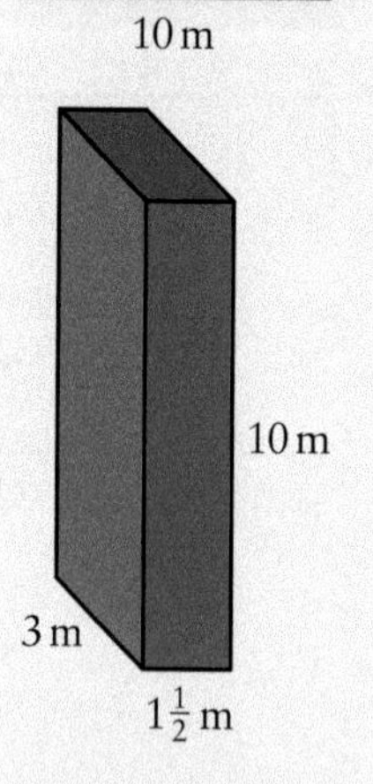

d

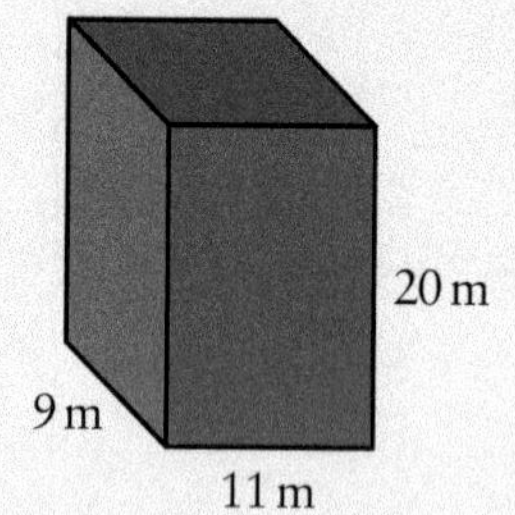

e

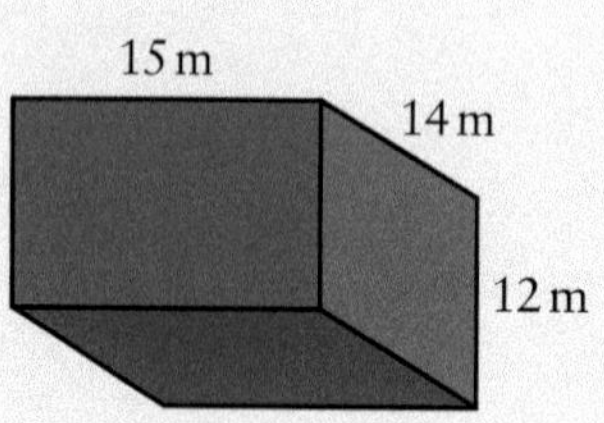

1.10 Patterns and algebra

Explore 1.13

Write down all the words that you can recall in relation to algebra, for example: equation, expression, terms and so on.

Can you write a definition and give an example for each word?

The table shows some key terms used in algebra.

Word	Definition	Example
terms	a single number or variable, or numbers and variables multiplied together, separated by + or – signs	$3x - 2y + 4 - 5x^2$ has four terms: $3x$, $-2y$, 4 and $-5x^2$
like terms	terms with the same variable form	$3xy$ and $7xy$
expression	algebraic expressions are made up of one or more terms joined by + and – signs	$5x + 2y - 1$
equation	a number sentence that contains an '='	$x + 6 = 13$
variable	a symbol, usually a letter, that takes the place of a number	x and y are commonly used variables
coefficient	a number used to multiply a variable	-5 is the coefficient in the term $-5x^2$
constant	a term which does not contain a variable	in the expression $3x - 2y + 4 - 5x^2$, 4 is the constant term.

Worked example 1.12

a If $x = -2$ and $y = 3$ work out the value of $x - 4y$.

b Solve for x: $x - 4 = 11$

Solution

a If $x = -2$ and $y = 3$, work out the value of $x - 4y$.

Substituting in the values for x and y:

$$x - 4y = (-2) - 4(3)$$
$$= -2 - 12$$
$$= -14$$

b Solve for x: $x - 4 = 11$

We need to work out which number minus 4 equals 11.

The answer is 15 as $15 - 4 = 11$

$x = 15$

Alternative method:

Add 4 to both sides of the equation:

$x - 4 = 11$

$x - 4 + 4 = 11 + 4$

$x = 15$

Reflect

Why can we add 4 to both side of the above equation?

Practice questions 1.10

1 Pablo is using matchsticks to make squares.

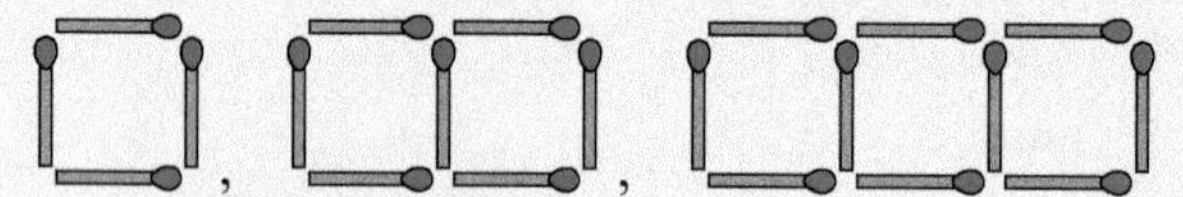

a Complete the table to show how the number of matchsticks is related to the number of squares.

Squares (s)	1	2	3	4	5
Matchsticks (m)					

b Describe the pattern in words.

c Describe the pattern in symbols.

d How many matchsticks would Pablo need to make:

i 20 squares ii 30 squares iii 100 squares?

2 Rewrite each of these expressions without × or ÷ symbols:

a $7 \times k$

b $3 + 4 \times z$

c $6 \div d$

d $x \times y \times z$

e $6 \times h \div z$

f $9 \times (f + 4)$

3 Rewrite each of these expressions showing all × and ÷ symbols where they belong:

a $2b - 7$

b $7r - \frac{s}{3}$

c $8(a + 5)$

d $\frac{p + 4}{5}$

4 If $a = 2$ and $b = 5$, work out the value of:

a ab

b $a + b$

c $3b + 4$

d $4b - 3$

e $-6a + 2b$

f $7a - 2b$

g $4ab$

h $\frac{8b}{5a}$

5 Match each table with the correct rule or formula:

a

x	1	2	3	4
y	−1	2	5	8

b

x	0	2	4	6
y	0	1	2	3

c

x	3	4	5	6
y	12	16	20	24

d

x	5	7	9	11
y	1	3	5	7

e

x	5	6	7	8
y	−4	−6	−8	−10

f

x	−2	−1	0	1
y	−3	−1	1	3

A $y = x - 4$

B $y = 2x + 1$

C $y = 4x$

D $y = 3x - 4$

E $y = \frac{x}{2}$

F $y = -2x + 6$

6 Rewrite these expressions without × symbols:

a $8 \times x \times y$

b $3 \times 5 \times d$

c $a \times 6 \times b$

d $3g \times 9$

e $6 \times -2p$

f $-4 \times 8k$

g $-7m \times -2n$

h $5 \times e - 4 \times f$

7 Rewrite these expressions without ÷ symbols:

a $16c \div 4$

b $42h \div -6$

c $-15t \div 5u$

d $-6a \div -2b$

e $10x \div -2$

f $-24a \div 6$

g $12r \div 8s$

h $9k \div 6l$

8 Simplify by collecting like terms:

a $6v + 5v - 7v$

b $11n - 3n - 6n$

c $14a + 4b - -4a$

d $8k + 3l + 2k + 7l$

e $9f - 4e + 3f - 2e$

f $7p - 2q - 6p + 3q$

g $4t^2 + 2t - t^2 - 6t$

h $3w - 4w^2 - 5w + 7w^2$

9 Expand and simplify where possible:

a $2(3x + 1)$

b $6(3q - 5)$

c $f(3 + f)$

d $4r(r - 8)$

e $7(4a + 3b)$

f $6g(3g - h)$

g $6(2n + 1) + 8$

h $4(t + 4) - 20$

10 Simplify:

a $p \times p \times p$

b $m \times n \times m \times m \times n$

c $5 \times a \times 6 \times a$

d $3 \times g \times 5 \times g \times g$

e $2 \times x \times 10 \times 7 \times y \times x \times y$

f $r \times 5 \times r \times s \times 3 \times r$

11 Solve these equations:

a $h + 6 = 9$

b $5 + d = 12$

c $12 - r = 11$

d $w - 9 = 1$

e $3y = 18$

f $30 = 6z$

g $15 \div s = 5$

h $\frac{q}{5} = 3$

12 Use an equation to solve each of the following:

a Sam is six years younger than Bob. The sum of their ages is 78

How old are Sam and Bob?

b Barney has three times as many books as Tanya. Altogether they have 56 books.

How many books does Tanya have?

c If 8 is subtracted from the product of 3 and a number the result is −14.

What is the number?

d Seven adult tickets and four children's tickets cost €54.
The adult tickets costs €3 more than the children's ticket.

How much does each ticket cost individually?

Explore 1.14

You are given two lists of data about a group of 25 students. One list contains the number of students in the group that are taller than each individual student, and the other list consists of the height of each member of the group. What can you say about the types of data in the lists?

Discrete data can only take on specific values.

Continuous data can take on any value within a range.

Worked example 1.13

State whether each of these data is discrete or continuous.

a The number of students in grade 7

b Your shoe size.

c The weights of students in your class.

d The time taken to run 100 m.

e The number of visitors to a museum.

Solution

a Discrete. The number of students can be counted so therefore can only take on specific values.

b Discrete. Shoe sizes can only take on specific values.

c Continuous. Weight is a measure so can take any value within a range.

d Continuous. The time taken can take any value within a range.

e Discrete. The number of visitors can be counted therefore can only take on specific values.

Practice questions 1.11

1 Classify each of these types of quantitative data as discrete or continuous:

a Number of cars in a car park.

b Weights of students in grade 7

c Number of marks scored in a test.

d Time to run 100 m.

e Volume of water drunk by students in year 2 in one day.

f Number of visitors to a museum in one year.

2 Mr Sweeney surveys his class to find out how many times each student ate in the school canteen last week. The results are shown below.

5, 1, 3, 4, 1, 5, 5, 4, 2, 4, 5, 4, 4, 2, 4, 2, 3, 4, 1, 2, 4, 2, 5, 4, 1, 3, 2, 1, 1

a Copy and complete the frequency table:

Outcome	Tally	Frequency
1	~~IIII~~ I	
2		
3		
4		
5		

b How many students are in Mr Sweeney's class?

c What was the most common number of times a student ate in the canteen?

d Draw a bar chart of the data.

3 The annual sales of a mobile phone company for 10 years are shown in the table below. Draw a line graph to represent the information.

Year	2012	2013	2014	2015	2016	2017	2018	2019
Sales (billions of US $)	203	205	198	209	214	227	275	270

4 The number of each colour sweet is shown in the table below. Draw a pie chart to represent the information.

Colour	Red	Yellow	Green	Purple	Orange
Frequency	10	12	8	24	6

5 Eileen scored the following marks in her last 10 maths quizzes:

6, 8, 9, 4, 5, 8, 9, 10, 7, 3

For this data calculate the:

a range b mode c mean d median.

Investigation 1.1

An infographic is a visual representation of data or information. The infographic below shows the population and migration estimates for the Republic of Ireland in April 2020.

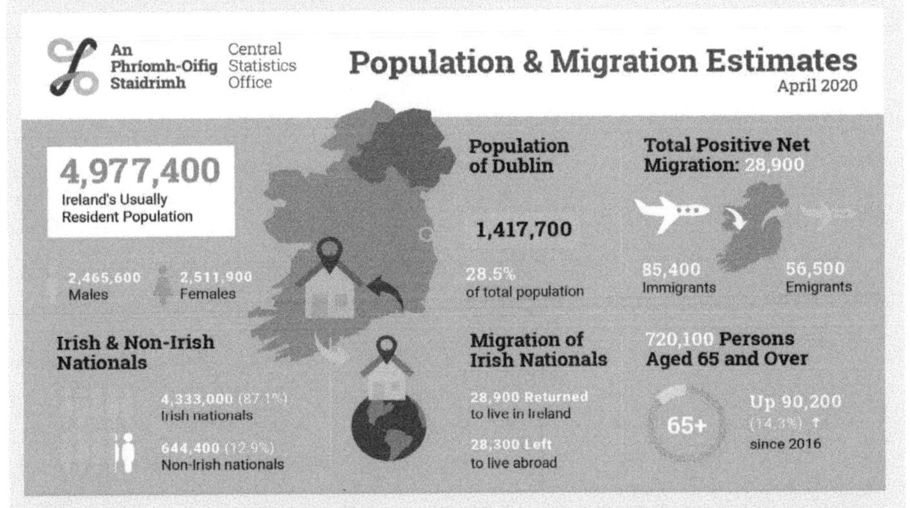

Source: Central Statistics Office of Ireland / An Phríomh-Oifig Staidrimh

Use the above infographic to answer the following questions:

1 What is the main purpose of the infographic?

2 What pieces of information are represented in the infographic? Write down as many as you can.

3 For each piece of information listed in Question 2, explain how the information in the infographic is presented. For example, how is colour used? How are images used? What charts are used?

4 In your opinion, does the infographic do a good job representing the information? Support your answer with examples.

Research skills

5 Create an infographic to communicate information about a country of your choice. Your infographic should have a clear goal and be supported by data from a reliable source. You should use at least three visuals on your infographic and organise your visuals and information clearly.

1.12 Working mathematically

Explore 1.15

A Ferris wheel ride lasts four minutes. The ride starts when cart number 1 is at the bottom (see Picture 1) and rotates clockwise. If after 32 seconds cart number 1 has reached position 5 (see Picture 2), how high will cart number 1 be off the ground at the end of the ride?

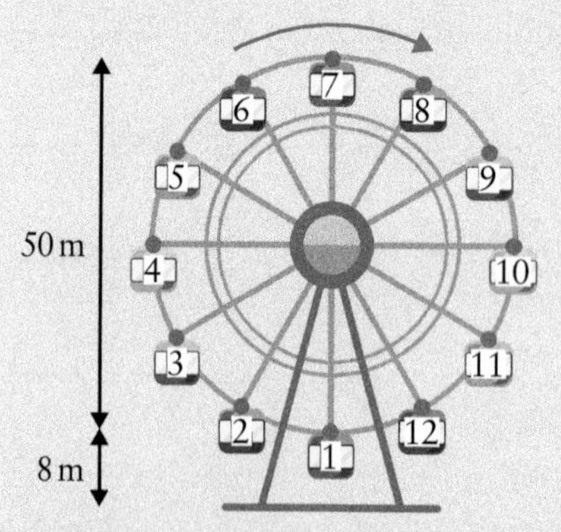

Picture 1: starting position of cart 1

Picture 2: position of cart 1 after 32 seconds

How long will it take for the cart to complete one full revolution?

At what times will the cart be at the highest point?

In what position will the cart be after two minutes?

How long should the ride last so that cart number 1 finishes in the position where it began?

How many full revolutions will the cart do in one hour?

Reflect

What strategy did you use to answer the Explore questions?

Would you use a different strategy if solving a similar problem? Why?

We can use **Pólya's four-step problem-solving process** as a structured approach to unfamiliar problems:

Step 1: Understand the problem

Read and re-read the problem carefully so that you understand it.

Step 2: Make a plan

Choose a strategy to solve the problem.

Here is a list of strategies you learned about last year:

1 Trial and improvement.

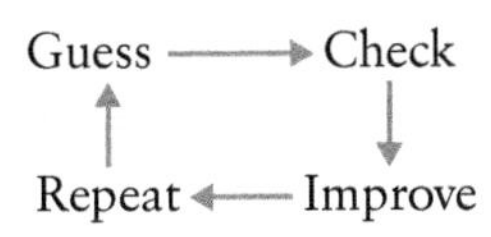

2 Make a list.

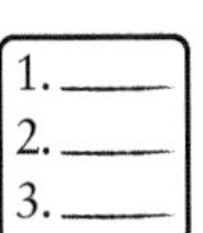

3 Make a table.

	X	Y	Z
Boy			
Girl			

4 Eliminate possibilities.

5 Make a diagram.

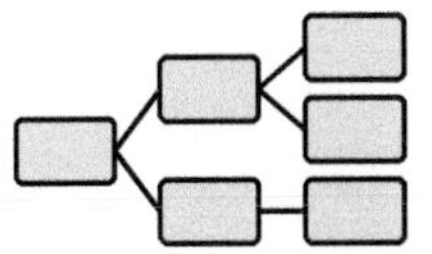

6 Look for a pattern.

7 Work backwards.

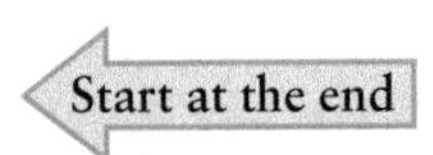

8 Simplify the problem.

The more problems you solve, the easier it will be to choose a strategy.

You will learn more strategies as you work though this course. For example, 'Write an equation' is a useful strategy that you will learn when you study algebra.

Step 3: Carry out the plan

Carry out your plan to solve the problem. Make sure you check your work as you go. Lay out your work clearly so that others can understand and investigate the problem you have attempted.

Step 4: Look back

When you have finished working out the problem you should look back on the work you have done.

- Did you answer the question that was asked?
- Does your answer make sense within the context of the question?
- Is your answer reasonable?

Worked example 1.14

A zookeeper has kangaroos and giraffes. She counts 78 legs and 24 heads in total.

How many kangaroos are there?

Solution

Understand the problem

Identify the important information:

There are 24 kangaroos and giraffes altogether.

A kangaroo has two legs. A giraffe has four legs. There are 78 legs in total.

What are we asked to find?

We need to work out how many kangaroos there are.

Make a plan

We can use the 'trial and improvement' strategy to solve this problem.

We can start by guessing how many giraffes there are. By subtracting our guess from 24 we can work out the corresponding number of kangaroos.

We can work out how many legs there would be in total for our guess.
If the first guess doesn't give us the correct solution, we can use the answer to revise our guess, up or down, and repeat the process.

We can continue to repeat the process until we work out a solution.

Carry out the plan

Guess 1: 16 giraffes

Check 1: If there are 16 giraffes, then:

Number of kangaroos = 24 – 16 = 8

Working out the total number of legs: $(16 \times 4) + (8 \times 2) = 80$

This is too many legs!

Guess 2: 15 giraffes

Check 2: If there are 15 giraffes, then:

Number of kangaroos = 24 – 15 = 9

Working out the total number of legs: $(15 \times 4) + (9 \times 2) = 78$

This matches the information given in the question so our guess must be correct.

Therefore, the answers is: There are nine kangaroos.

Look back

If we have 15 giraffes and 9 kangaroos, we have 24 animals altogether.

We can see from our working above that 15 giraffes and 9 kangaroos have a total of 78 legs.

Does the answer work? Yes!

Practice questions 1.12

1 **Trial and improvement.** John has 87 legs to build 26 stools. Each stool must have either three or four legs and all of the legs must be used. How many stools of each type should he build?

Guess → Check → Improve → Repeat

2 **Make a list.** Three girls, Maria, Julia and Adriana are being considered for girl's captain and vice-captain, and two boys Charlie and Pedro, are being considered for boy's captain and vice-captain. In how many different ways could the captains and vice-captains be chosen?

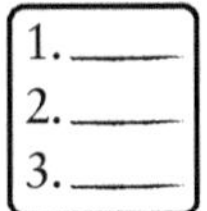

	X	Y	Z
Boy			
Girl			

3 **Make a table.** There are 96 people in grades 1, 2 and 3

We have the following information:

- 37 children cannot swim
- 11 children in grade 1 cannot swim
- 21 children in grade 2 can swim
- there are 30 children in grade 3
- 18 children in grade 3 cannot swim.

How many children in grade 1 can swim?

4 **Eliminate possibilities.** Sarah, Becky and Karen live in the same village but attend different schools. Their last names are Taylor, Lee and Foster, but not necessarily in that order. Sarah attends Averly International, Karen does not attend Rice High School, Foster attends Oaks Hill School and Taylor attends Rice High School. What are the full names of each of the girls?

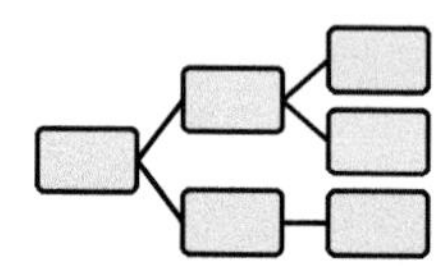

5 **Make a diagram.** A firefighter stands on the middle rung of a ladder, spraying water on a burning house. He then climbs up seven rungs before the heat of the flames causes him to come down 12 rungs. After some minutes he was able to climb 16 rungs to the very top of the ladder. How many rungs does the ladder have?

6 **Look for a pattern.** Sandra is making the following pattern using matchsticks:

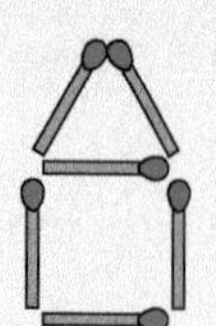

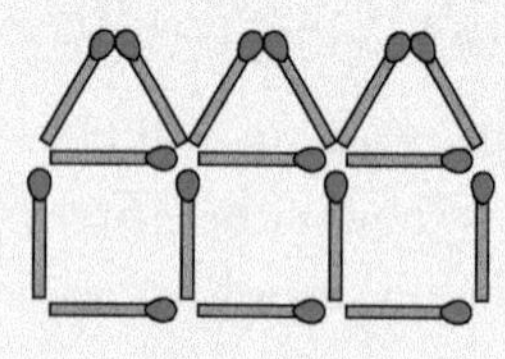

Stage 1 Stage 2 Stage 3

How many matchsticks will they need for:

i Stage 6

ii Stage 10

iii Stage 30?

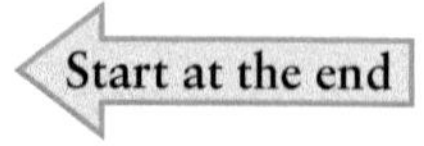

7 **Work backwards.** Matthew, Andrew, Li and Jai are at a barbeque. Matthew is 11 years younger than Andrew, Li is half of Andrew's age, and Jai is 6 years older than Li. If Jai is 35, what age is Matthew?

8 **Simplify the problem.** If it takes 12 people 8 hours to make 50 cookies, how long will it take four people to make 25 cookies?

9 Use any strategy you like to solve the following problems:

a Alexei, Ross, Jamie and Sam have a race. In how many different ways could they finish the race, assuming no dead heats are possible?

b Alice and Miguel get married and have seven children. Five of their seven children get married and they each have four children. Their other two children remain single. Assuming no one has died, how many people are now in the extended family altogether?

c Which two numbers have a sum of 13 and a product of 36?

d Five people are playing in a chess tournament. If each person must play every other person twice, how many games will there be in total?

e Florian is 11 years older than Jan. The sum of their ages is 35. How old is Florian?

f A farmer has chickens and cows. If there are 16 heads and 40 legs altogether, how many chickens are there?

g What is the last digit of 9^{10}?

h Ben has four shirts, three ties and two pairs of trousers. How many unique outfits can he make?

i Mr French, Mr English, Mr Kraft and Mr Musik are four teachers who teach French, English, Craft and Music, however, none of them teaches a subject that sounds like his name. Mr French does not teach Music. When Mr French and Mr Kraft beat the other two at tennis, the English teacher is a bad loser but the music teacher is a gracious winner. What subject does each teacher teach?

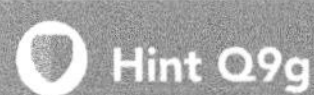

Find out what the last digits of 9^1, 9^2, 9^3, ... are and see if you can find a pattern.

1.13 Problem-solving using Venn diagrams

Sets of things are frequently referred to in problem-solving. The most popular method of representing the relationship between sets is to use Venn diagrams. Each set is represented by a region; usually a circle.

Explore 1.16

Place the numbers from 1 to 20 in the correct position on the Venn diagram.

A = {factors of 20}

B = {factors of 36}

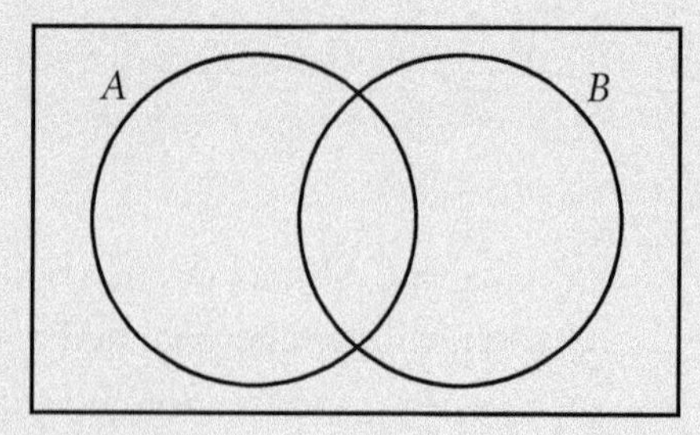

Can you describe each part of the diagram?

Venn diagrams provide a way to represent sets in a simple way. This helps us to solve problems more easily.

Worked example 1.15

In a class of 30 students everyone eats pizza or lasagne or both for lunch.

Given that 17 students eat pizza and 20 eat lasagne, calculate:

a How many students ate both pizza and lasagne?

b How many students ate only pizza?

c How many students ate only lasagne?

Solution

We need to work out how many students ate both pizza and lasagne. Using this, we can work out how many ate only pizza and how many ate only lasagne.

We will let P represent the set of students who ate pizza and L represent the set of students who ate lasagne. If we work out how many lunches were eaten in total, we can subtract 30 from this value to find out how many students ate both pizza and lasagne. Once we know how many ate both we can subtract to work out how many ate only pizza and how many ate only lasagne.

We can represent the information on a Venn diagram:

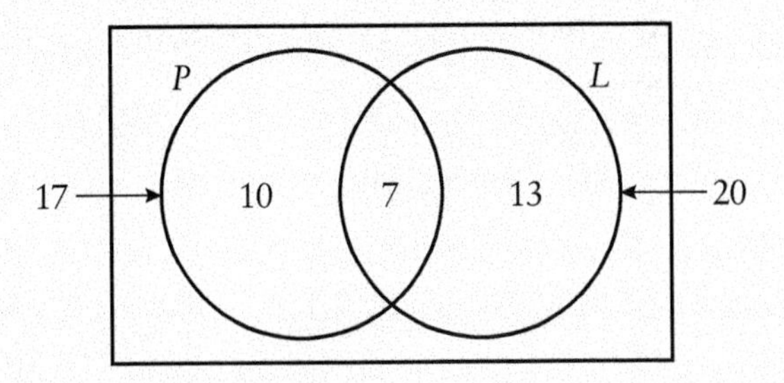

a Total number of lunches eaten = 17 + 20 = 37

If each student only ate one lunch, then 37 − 30 = 7 lunches would remain.

So, seven students ate both pizza and lasagne.

b 17 students ate pizza but seven of these also ate lasagne.

So, 17 − 7 = 10 people ate only pizza.

c 20 students ate lasagne but seven of these also ate pizza.

So, 20 − 7 = 13 people ate only lasagne.

Reflect

In the previous example, how can you show that the answer works?

Practice questions 1.13

1 The Venn diagram shows the number of children in a club who play tennis or football.

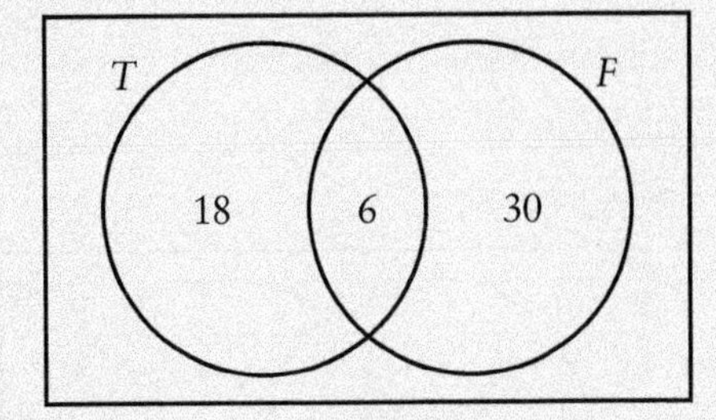

T = {number of children who play tennis}

F = {number of children who play football}

a How many children play both tennis and football?

b How many children play tennis but not football?

c How many children play tennis?

2 In a class of 19 students everyone studies French or Spanish or both. If 12 students study French and 12 study Spanish. How many study both French and Spanish?

3 35 copies of the Leader and 88 copies of the Post were delivered to 100 houses.

a How many houses received two papers?

b How many houses received only the Leader?

c How many houses received only the Post?

4 Leonie is doing a survey to find out how many people passed maths and drama in Year 2. There are 72 people in Year 2 and Leonie surveys all of them. She finds that 67 students passed maths and 53 passed drama. How many students passed both maths and drama?

5 The Venn diagram shows the number of children in a club who play different instruments.

P = {number of children who play piano}

V = {number of children who play violin}

F = {number of children who play flute}

There are 32 children in a class and every child plays at least one of these instruments.

We have the following information:

- Three children play all three instruments
- Five play piano and violin but not flute
- Seven play violin and flute but not piano
- One plays piano and flute but not violin
- a total of 17 play violin
- a total of 13 play piano.

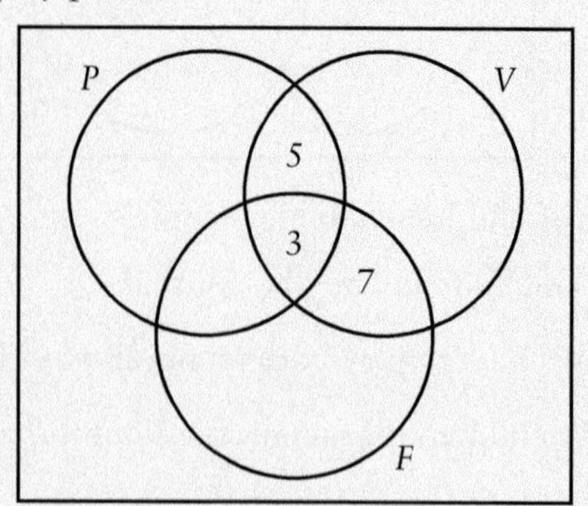

Complete the Venn diagram and calculate how many children play:

a only piano
b only violin
c flute
d violin or flute
e piano and violin.

6 One hundred people were asked what genre of book they like out of fiction (F), non-fiction (N) and poetry (P). 64 like fiction, 80 like non-fiction, two like only poetry, 23 like non-fiction and poetry, 51 like fiction and non-fiction, 34 like only fiction and non-fiction, and 10 like only fiction and poetry.

How many of these people like:

a all three genres?
b only non-fiction?
c fiction and poetry?
d only non-fiction and poetry?
e only fiction?
f none of these genres?

How many of these people do not like:

g fiction?
h non-fiction?
i poetry?

Self-assessment

- I know the different number sets and the notation used to represent each set.
- I know the order in which operations should be carried out.
- I know what a factor is and how to find factors of numbers.
- I know what a multiple is and how to find multiples of numbers.
- I can add, subtract, multiply and divide decimal numbers.
- I can add, subtract, multiply and divide fractions.
- I can convert between fractions, decimals and percentages.
- I can work out the probability of simple events.
- I can classify angles.
- I can classify triangles.
- I can evaluate expressions using algebraic substitution.
- I can find the perimeter of a 2D shape.
- I can find the area of a 2D shape.
- I can find the volume of a 3D shape.
- I know the difference between discrete and continuous data.
- I can use the following problem-solving strategies:
  - Trial and improvement.
  - Make a list.
  - Make a table.
  - Eliminate possibilities.
  - Make a diagram.
  - Look for a pattern.
  - Simplify the problem.

? Check your knowledge questions

1 Copy and complete the following table by placing ticks (✓) to show which of the number sets $\mathbb{N}, \mathbb{Z}, \mathbb{R}$ these numbers belong to.

	$\mathbb{N}$	$\mathbb{Z}$	$\mathbb{R}$
-9			
15			
7.5			
$-\frac{3}{2}$			

2 Work out:

a $(5 + 7) \times (23 - 11)$

b $(7 \times 3) + (4 \times 2)$

c $20 - \{15 + (40 \div 8)\}$

d $3 + 8^2$

e $\dfrac{8 + 20}{14 \div 2}$

f $5 \times 8 - (16 - 4 \div 2)$

3 Work out the highest common factor (HCF) of 64 and 144.

4 Write 675 as a product of its prime factors.

5 Work out the lowest common multiple (LCM) of 24 and 36

6 Work out:

a $-17 + 2$

b $-9 + -4$

c $-5 - -6$

d -5×6

e 7×-4

f -9×-5

g $-40 \div 4$

h $\dfrac{-18}{-3}$

7 What two numbers have a product of -48 and a sum of -2?

8 Work out:

a $3.462 + 19.89$

b $63.43 - 14.5$

c 1.7×41.8

d $8.304 \div 8$

9 Round to the nearest tenth:

a 3.042

b 142.072

c 53.362

d 63.955

10 Work out:

a $3\frac{1}{2} + 2\frac{2}{5}$

b $6 - 2\frac{4}{7}$

c $\frac{3}{5} \times 3\frac{1}{4}$

d $1\frac{1}{2} \div 3\frac{3}{4}$

11 What is $\frac{3}{8}$ of 1 km?

12 What fraction of 1 hour is 5 minutes?

13 Work out:

a 30% of \$50

b 5% of 4 hours

c 18% of 1 litre.

14 What percentage is 72 marks out of 96?

15 Of 60 tickets sold in a raffle, Joseph bought 10, Parsa bought five and Alana bought one. What is the probability that the prize was won by:

a Joseph

b Parsa

c Alana

d none of these three people?

16 Work out the value of the variable in each of the following:

a

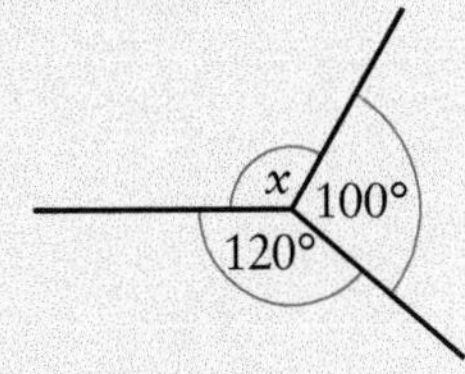

b

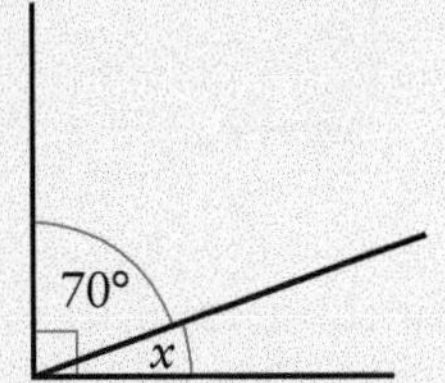

c

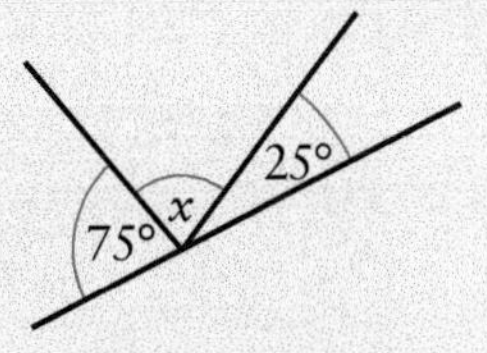

d

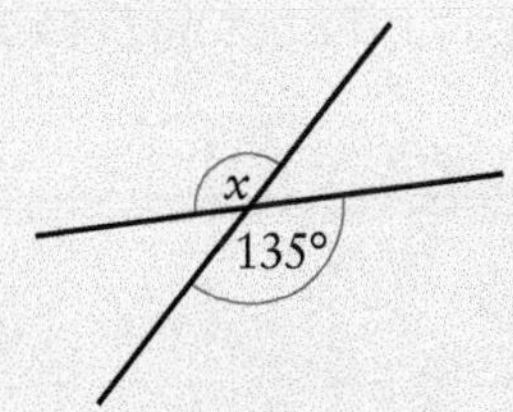

e

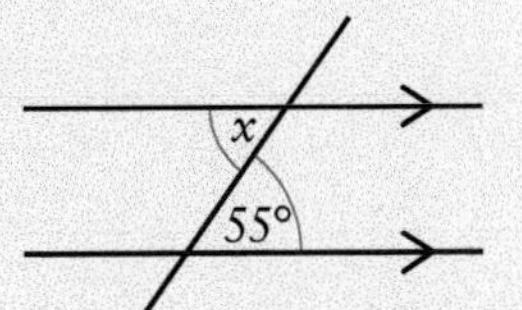

f

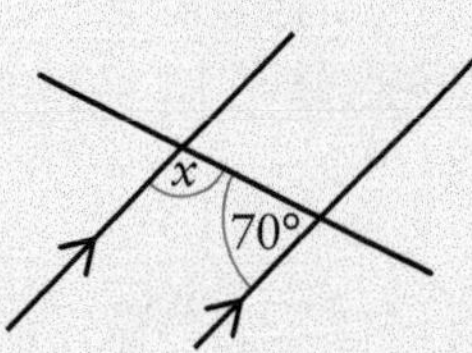

g

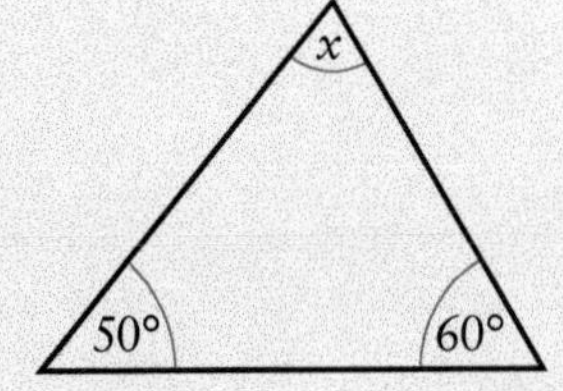

h

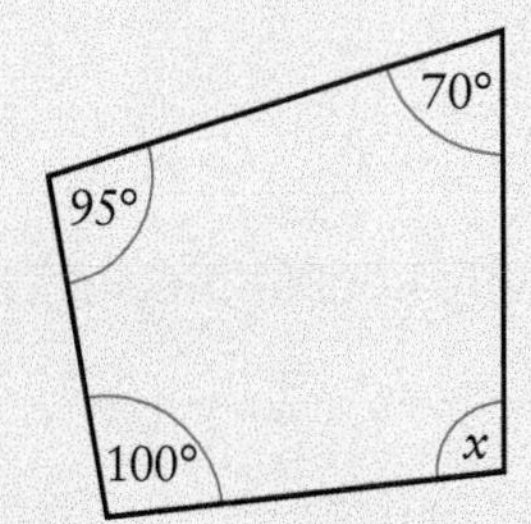

17 Write in 24-hour time:

a 23 minutes past 6 (before noon)

b 7 minutes past 11 (after noon)

c 19 minutes before 10 (before noon)

d 25 minutes before 5 (after noon).

Challenge Q18

18 If my birthday in August 2020 fell on a Wednesday, on what day of the week will my birthday fall in

a 2028 b 2031 c 2033?

19 Work out the perimeter of these shapes.

a

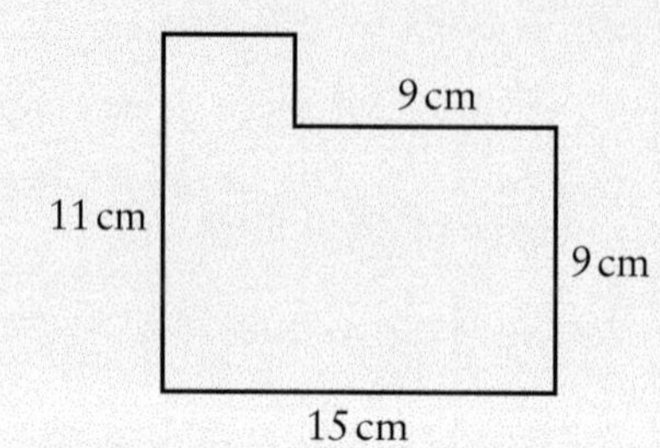

b

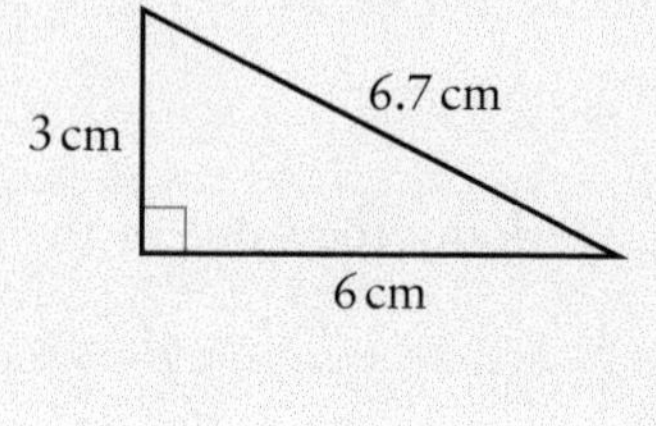

20 Work out the area of the shapes in Question 19.

21 Work out the volume of each of the following prisms:

a

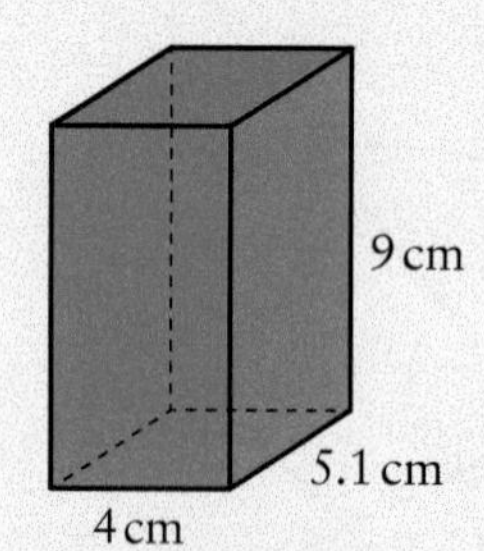

b

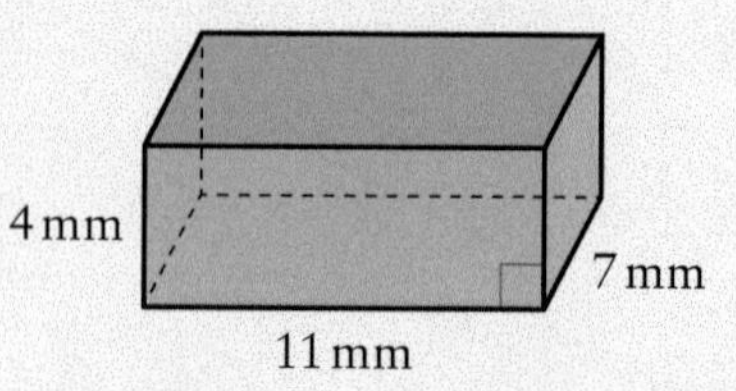

22 If $x = -2$, $y = 3$ and $z = -4$, evaluate:

a $x + y$ b xyz

c $x - y - z$ d $\frac{x + z}{y}$

23 Simplify by collecting like terms:

a $6a + 4b - 2a - b$ b $2x + 3y - 7x - 9y$

c $3r^2 + 2r - r^2 - 5r$ d $k + l - 4k + 3l$

24 Expand:

a $5(x - 3)$

b $(2 - 4d)3$

c $y(2y - 4)$

d $5c(2c + 3b)$

25 Solve:

a $m + 3 = 11$

b $18 - x = 5$

c $w \div 2 = 6$

d $48 = 6m$

Challenge Q26

26 Solve:

a $2x + 1 = 7$

b $5m - 3 = 12$

c $9 - 2a = 3$

d $19 = 10 + 3p$

27 Lilli collects data on the hair colour of the students in her class. The results are shown in the table.

Colour	Blonde	Red	Brown	Black
Frequency	7	2	13	8

a Draw a bar chart to represent the information.

b Draw a pie chart to represent the information.

28 For this set of data: 2,5,2,4,7,6,3,2,3,5,2,7 find the:

a mode

b median

c mean

d range.

29 Use any strategy you like to solve the following problems.

a Entrance to a museum costs \$12 for adults and \$9 for children. A tour group of 27 adults and children pays a total of \$279 for their tickets. How many adults are in the group?

b There are seven people at a meeting. Each person shakes hands with every other person once. How many handshakes are there in total?

c Martin has a bag of sweets. On the first day he eats $\frac{1}{6}$ of the sweets. On the second day he eats $\frac{1}{5}$ of the remaining sweets. On the third day he eats $\frac{1}{4}$ of the remaining sweets. He eats $\frac{1}{3}$ of the remaining sweets on the fourth day. On the fifth day he eats $\frac{1}{2}$ of the remaining sweets. On the sixth day he eats the last four sweets. How many sweets were in the packet originally?

30 Ms Dubois asked her grade 7 class how many of them have visited Sydney and how many have visited Melbourne. When she asked how many had visited Sydney, 22 students raised their hands. When she asked how many had visited Melbourne, 19 raised their hands. When she asked how many had visited both Sydney and Melbourne, eight raised their hands. When she asked how many had visited neither, three students raised their hands. Draw a Venn diagram to represent the information and answer the following questions:

a How many students had visited only one of the cities?

b How many students are in the class?

Challenge Q31

31 Ezra decided to row 10 km up his local river without leaving the river. He rowed 2 km every hour then rested for 10 minutes. Each time he rested the current dragged him back 1 km. How long did it take Ezra to complete the 10 km journey?

2

Percentages, ratio and rates

2 Percentages, ratio and rates

KEY CONCEPT

Logic

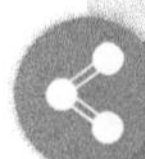

RELATED CONCEPTS

Equivalence, Quantity, Simplification

GLOBAL CONTEXT

Globalisation and sustainability

Statement of inquiry

Using logical processes to simplify quantities and to determine equivalence can help us understand how to manage finite resources.

Factual

- What is the difference between a ratio and a rate?

Conceptual

- How do we establish equivalence between percentages, ratios and rates?

Debateable

- Can percentages and ratios be used interchangeably?

Do you recall?

1 Convert $\frac{2}{5}$ into a decimal.

2 Convert 0.35 into a fraction in its simplest form.

3 Convert 23% into a decimal.

4 Work out 30% of $150

2.1 Review of percentages

With almost eight billion people sharing the resources on our planet, the sustainable use of our resources becomes more important. Percentages, ratios and rates enable us to evaluate how resources are used and shared on a local and on a global scale.

Percentages are useful in all areas of life. Examples of percentages that you might be familiar with include reductions of prices during sales, test results, tax, interest rates, percentage profit or loss and commission rates.

Fact

Per cent comes from Latin 'per centum'. The Latin word 'centum' means 100

Fact

In principle, a 'percentage' is the result obtained by multiplying a quantity by a per cent. So 50 **per cent** of 10 books is five books: the five books is the **percentage**.

However, in practice, both words are used interchangeably.

2.1.1 Converting fractions and decimals to percentages

When we say 'per cent', we are saying 'per 100'.

Explore 2.1

Can you explain why $\frac{2}{5} = 40\%$?

Can you explain why $0.32 = 32\%$?

With that in mind can you express the following as percentages:

a $\frac{3}{25}$ **b** 0.61 **c** $\frac{17}{20}$ **d** 0.04 **e** $\frac{7}{10}$ **f** 0.9?

A per cent is a number expressed as a fraction with a denominator of 100, for example:

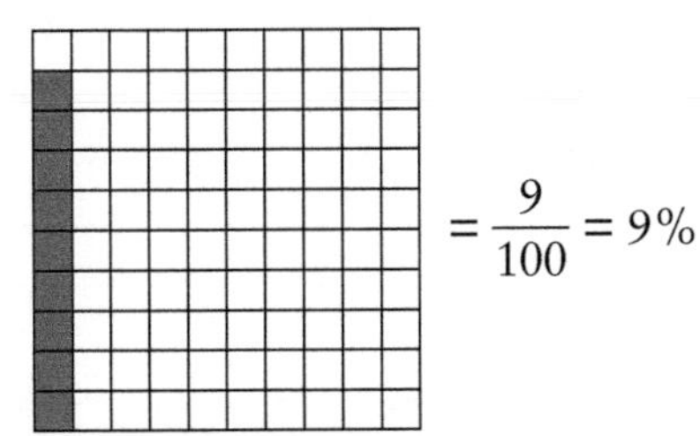

$= \frac{9}{100} = 9\%$

Worked example 2.1

a Write 0.125 as a percentage.

b Write $\frac{11}{20}$ as a percentage.

Solution

a To convert a decimal to a percentage, we multiply by 100%:

$0.125 \times 100\% = 12.5\%$

So, $0.125 = 12.5\%$

b To convert a fraction to a percentage, we multiply by 100% and simplify:

$\frac{11}{\cancel{20}_1} \times \frac{\cancel{100}^5}{1}\% = 55\%$

So, $\frac{11}{20} = 55\%$

Reflect

What is a rule we could use to convert fractions and decimals to percentages?

Practice questions 2.1.1

Challenge Q1g,h

1 Convert these fractions to percentages.

a $\frac{1}{100}$ b $\frac{27}{100}$ c $\frac{3}{10}$ d $\frac{11}{20}$

e $\frac{5}{8}$ f $2\frac{1}{12}$ g $\frac{2}{7}$ h $1\frac{5}{6}$

2 Convert these decimals to percentages.

a 0.05 b 0.66 c 0.1

d 1.45 e 0.015 f 2.375

Reminder Q3

Ascending means to arrange from smallest to largest.

3 Write each set of numbers in ascending order.

a $\frac{1}{5}$, 75%, 0.5 b $\frac{3}{4}$, 72%, 0.63

c 1.8, 18%, $1\frac{1}{8}$ d $\frac{2}{3}$, 65%, 0.68

2.1.2 Converting percentages to fractions and decimals

We also need to be able to convert from percentages to fractions and decimals.

Explore 2.2

Can you explain why the following statement is true?

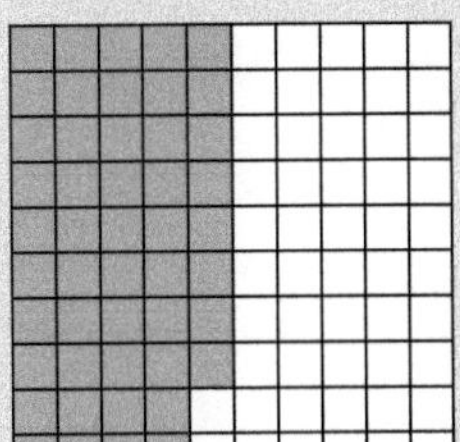

$= 48\% = \frac{12}{25} = 0.48$

You can use fractions, decimals and percentages to represent the same value in different ways, for example:

Percentage	Fraction	Decimal
30%	$\frac{30}{100} = \frac{3}{10}$	$30 \div 100 = 0.3$

Percentages can easily be written as a fraction with denominator 100. If possible, the fraction then can be simplified.

Percentages can be converted to decimals by dividing by 100

Worked example 2.2

a Write 80% as a fraction.

b Write 28% as a decimal.

Solution

a To convert a percentage to a fraction we first write it as a fraction, with denominator 100, and then simplify the fraction.

$80\% = \frac{80}{100} = \frac{8}{10} = \frac{4}{5}$

So, $80\% = \frac{4}{5}$

b To convert a percentage to a decimal we divide by 100:

$28\% = \frac{28}{100} = 0.28$

So, $28\% = 0.28$

Reflect

How can you use a diagram to show that your answers are correct?

Practice questions 2.1.2

1 Convert each percentage to a fraction.

a 9% b 49% c 60%

d 32% e 115% f 325%

2 Convert each percentage to a decimal.

a 14% b 3% c 40%

Challenge Q2e,f

d 500% e 12.24% f $15\frac{1}{2}\%$

2.1.3 Finding a percentage of a quantity

It is useful to be able to calculate a percentage of a quantity. For example, in a sale we might want to work out how much a reduction of 20% is so we know how much we have to pay for an item.

Explore 2.3

Can you work out:

a 5% of €600 b 43% of €600?

There are certain percentages of quantities that we can work out easily, for example, 50%, 10% and 1%. We can then use these amounts to calculate any other percentage.

Another method used to find a percentage of a quantity is to write the percentage as a decimal or a fraction and multiply by the quantity.

These methods are shown in the Worked example below.

Worked example 2.3

Work out 8% of 300

Solution

To find a percentage of a quantity, we can write the percentage as a fraction or as a decimal and multiply.

Method 1: using decimals

8% of 300 = 0.08×300

$= 24$

Method 2: using fractions

$$8\% \text{ of } 300 = \frac{8}{\cancel{100}_1} \times \frac{\cancel{300}^3}{1}$$
$$= 24$$

Both methods give the same answer: 8% of 300 is 24

Reflect

Why can we use both decimals and fractions to find a percentage of a quantity?

Practice questions 2.1.3

1 Write the percentage as a fraction to work out:

a 5% of 1600 g   b 83% of 40 m   c 40% of $300
d 24% of €36.50   e 0.08% of 8000 g   f 0.6% of 350 ml

Challenge Q1e,f

2 Write the percentage as a decimal to work out:

a 15% of 8 kg   b 2% of 3000 mm   c 105% of 800 litres
d 115% of €2000   e 6% of 2425 cm   f $12\frac{1}{2}$% of 320 g

Challenge Q2e,f

3 Which of the following has the greatest value?

a 50% of 10   b 40% of 20   c 30% of 30
d 20% of 40   e 10% of 50

4 A family decides to donate 5% of their earnings in December to famine relief. If they earn £8240 in December, how much money will they donate?

5 Europe produces 20% of global plastic. If global plastic production is 350 million tons annually, how much plastic does Europe produce?

Fact

Recycling plastic consumes 88% less energy than making plastic from raw materials.

6 Europe recycles 30% of the plastic it produces. Using your answer to Question 5, work out how much of the plastic produced in Europe is recycled.

7 What is 50% of 2020 plus 2020% of 50?

Challenge Q7

8 In the first week of February, an investor lost 10% on an investment of €10 000. In the second week of February he lost 15% of what remained. How much money had he left at the start of the third week of February?

Challenge Q8

2.1.4 One quantity as a percentage of another

Percentages are a universal way to compare quantities.

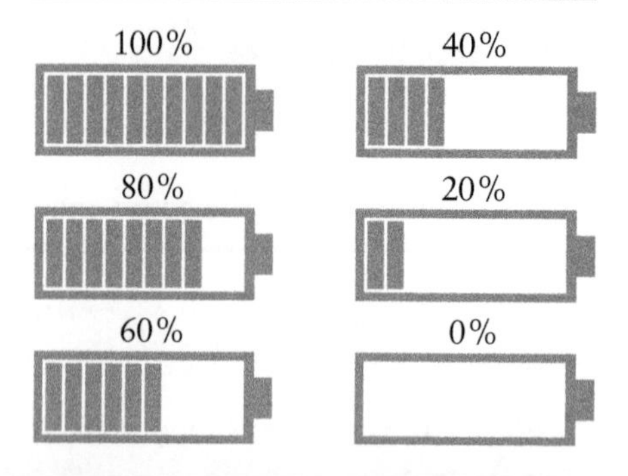

Explore 2.4

Amir and Barbara are practicing basketball. Amir scores 30 shots out of 40 and Barbara scores 35 out of 50.
How can you compare the two scores?

To express one quantity as a percentage of another.

1. Write the two quantities in the same units.
2. Write the first quantity as a fraction of the second.
3. Convert the fraction to a percentage by multiplying by 100%.

Worked example 2.4

Express 15 cm as a percentage of 3 m.

Solution

Understand the problem

This question asks us to express one quantity as a percentage of another.

Make a plan

In order to solve this problem, we need to express 3 m in centimetres, then work out what percentage 15 cm is of 3 m.

Carry out the plan

$3\,\text{m} = 300\,\text{cm}$

$\frac{15}{300} \times 100 = 5\%$

15 cm as a percentage of 3 m is 5%.

Look back

5% of $300\,\text{cm} = 0.05 \times 300 = 15\,\text{cm}$.

Practice questions 2.1.4

1 What percentage of each figure is shaded?

a 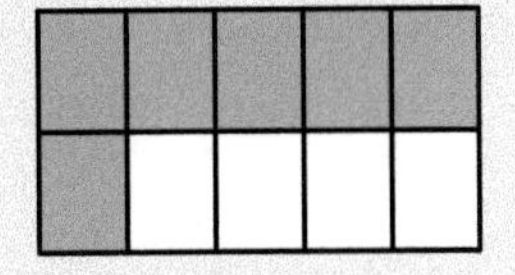

b 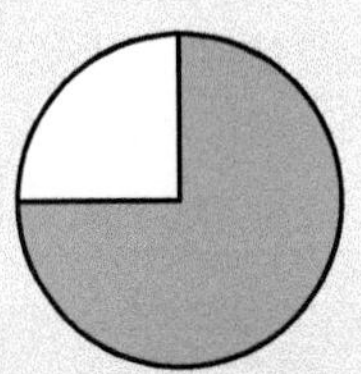

c 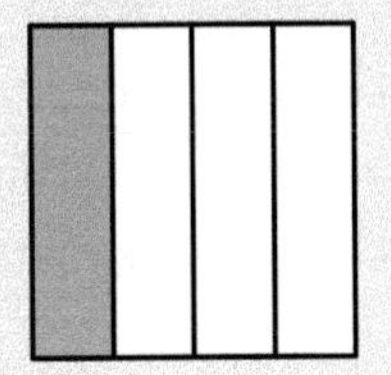

d 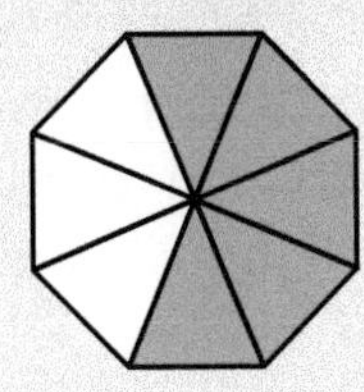

2 For each pair, express the first quantity as a percentage of the second.

a 3 g, 30 g

b 15 ml, 90 ml

c 8 litres, 2 litres

d 70 min, 105 min

e 42 s, 1 min

f 2 m, 60 cm

g €2.20, €8.00

h 40 g, 32 kg

3 Lauren scored 15 out of 25 in a test and Phillip scored 36 out of 48. Who scored the highest percentage?

4 Approximately 820 million people live below the international extreme poverty threshold of $1.90 a day. Express this as a percentage of the total population of the world. Currently, the world population is estimated to be eight billion.

 **Fact**

In 2019 the world's richest 1% owned 44% of the world's wealth.

5 Approximately 1.8 billion people live below a moderate poverty threshold of $2.50 a day. Express this as a percentage of the total population of the world. Currently, the world population is estimated to be eight billion.

 **Hint**

1 billion = 1000 million

6 What percentage of $\frac{1}{4}$ is $\frac{1}{5}$?

 **Challenge Q6**

7 Chantou scored exactly 10 out of the first 15 questions in her maths quiz. She answers all of the remaining questions correctly and receives a final score of 80%. How many questions were on the quiz?

 **Challenge Q7**

2.2 Percentage change

There are many situations where quantities are increased or decreased by a certain percentage. For example, wage increases, decreases in the prices of items in sales, price increases due to tax.

Explore 2.5

Nico and Ben are shopping. Nico likes a pair of runners that cost €80 before 15% tax is added. Using the calculators on their mobile phones, Nico and Ben both work out the cost of the runners using different methods.

Nico's method:

€80 + €12 = €92

Ben's method:

1.15 × €80 = €92

Later in the day, Ben sees a sweater that he would like to buy. The sweater normally costs €45 before a 20% discount is applied. Again, Nico and Ben both work out the cost of the sweater using their different methods.

They both concluded that the sale price is €36.

Can you show how both worked out the solution?

Which method do you prefer? Why?

Worked example 2.5

A pair of jeans costs €60 and the sales tax is 12%. What is the selling price?

Solution

We can do this in two ways.

Method 1: add or subtract the amount of tax

Calculate the amount of the increase

12% of €60 = 0.12 × 60

= €7.20

Add the increase to the cost.

60 + 7.2 = 67.2

That is, the selling price is €67.20.

Method 2: use a multiplier

Work out the multiplier

100% + 12% = 112% = 1.12, that is, the selling price is 112% of the cost.

Multiply the cost amount by the multiplier

112% of €60 = 1.12 × 60

= 67.2

That is, the selling price is €67.20.

There are two methods for increasing and decreasing amounts by decimals.

Method 1: add or subtract the amount

1. Calculate the amount of the increase or decrease.
2. Add or subtract the increase or decrease to or from the original amount.

Method 2: Use a multiplier

1. Work out the multiplier.

   To increase an amount by a percentage, say 30%:

   100% + 30% = 130%

   As a decimal, 130% = 1.3, which is known as the **multiplier.**

   To decrease an amount by a percentage, say 30%:

   100% − 30% = 70%

   As a decimal, 70% = 0.7, which is known as the **multiplier.**

2 Multiply the original amount by the multiplier,

Both methods are shown in the worked examples.

Worked example 2.6

A chef discovers that the number of guests coming to her restaurant that evening is 15% less than she originally estimated. She needs to reduce the 800 g of one of her ingredients accordingly. What is the new amount?

Solution

Method 1: add or subtract the amount

Calculate the amount of the decrease:

15% of 800 g = 0.15 × 800

= 120

Subtract the decrease from the original amount.

800 – 120 = 680

The new amount is 680 g.

Method 2: use a multiplier

Work out the multiplier

100% – 15% = 85% = 0.85, that is, the new amount is 85% of the old amount.

Multiply the original amount by the multiplier

85% of 800 g = 0.85 × 800

= 680

The new amount is 680 g.

Reflect

Which method above do you think is better? Explain your reasons.

Practice questions 2.2

1 Match the percentage increase/decrease to its corresponding multiplier.

a	10% increase	A	0.7
b	15% decrease	B	0.85
c	25% increase	C	1.1
d	15% increase	D	0.99
e	12.5% increase	E	2
f	30% decrease	F	1.25
g	100% increase	G	1.125
h	1% decrease	H	1.15

2 Increase

a €800 by 10%

b 4 litres by 12%

c 60 g by 85%

d 72 min by 100%

3 Decrease

a $80 by 20%

b 3.25 kg by 40%

c 600 mm by 4%

d 588 litres by 90%

4 Projections show that the global population will increase by 40% between 2020 and 2100. If the global population was eight billion people in 2020 what is the predicted population in 2100?

5 A couch is reduced by 30% in a sale. If the couch initially cost €460, work out the sale price of the couch.

6 The value of my laptop depreciated by 18% this year. When I bought it last year, it was $900. What is its value now?

7 Harry's salary increases 3% due to inflation. If he earned €42 000 before the increase, how much will he earn after?

8 Abe bought a car for $9000 but sold it to Sue at a loss of 78% after it was involved in an accident. Sue repaired the car and sold it to Joe for 250% more than she paid for it. What price did Joe pay for the car?

2.3 Finding a quantity if a percentage is known

Sometimes we will know the amount that is equivalent to a certain percentage. For example, if Eve spent 20% of her money on lunch, which cost €10, we can use this information to work out how much money Eve had before she bought lunch.

Explore 2.6

220 girls were born in Sunwra last year. If we know that 55% of all of the babies born were girls, can you work out how many babies were born in total?

If a percentage of a quantity is known, we can find any other percentage of the quantity by first finding 1% and then multiplying by whatever percentage we want to work out.

Worked example 2.7

Ayaan spent 16% of his money on books that cost $24.

How much money did he have before he bought the books?

Solution

Understand the problem

This question asks us to work out 100% of Ayaan's money, that is, how much money Ayaan had before he bought the books.

Make a plan

In order to solve this problem, we need to first work out 1% and then multiply by 100 to work out 100%.

Carry out the plan

We know that

16% = \$24

First, we need to find 1% by dividing by 16

1% = \$24 ÷ 16

= \$1.50

Then we need to find 100% by multiplying by 100

100% = \$1.50 × 100

= \$150

Ayaan had \$150 before he bought the books.

Look back

16% of \$150 = 0.16 × 150 = \$24

Reflect

How can we check that the answer is correct?

Practice questions 2.3

1 Determine the size of the following quantities

a 5% is 20 cm

b 60% is 420 t

c 8% is €108

d 120% is 840 litres

2 I spent 12% of my money on a tablet which cost \$468. How much money did I have?

3 China produces 28% of the world's carbon dioxide. If China produced 11 200 million tons of carbon dioxide in 2018, what was the total amount of carbon dioxide emitted globally?

4 Klara raised €129 for charity at a school fundraiser. If this represented 15% of all of the money raised at the fundraiser, how much was raised at the fundraiser in total?

2.4 Applications of percentages

Percentages are used in many different situations. In this section you will learn about some of their applications.

Explore 2.7

Sasha bought his apartment for \$150 000 and sold it for \$225 000

Can you work out the percentage profit?

Worked example 2.8

Aria buys a bike for €300

a If she sells the bike for €360, what is her percentage profit?

b If she sells the bike for €270, what is her percentage loss?

Solution

Understand the problem

This question is asking us to work out how much profit or loss Aria makes when selling the bike and then to express this profit or loss as a percentage of the cost price.

Make a plan

To solve this problem, we need first to work out the amount of profit or loss Aria makes. We then need to express this profit or loss as a percentage of €300; the cost price of the bike.

Carry out the plan

a We first need to work out how much profit she makes:

profit = selling price – cost price

= €360 – €300

= €60

Now we can write the profit as a fraction of the cost price:

$$\text{profit as a fraction of cost price} = \frac{\text{profit}}{\text{cost price}} = \frac{60}{300} = \frac{1}{5}$$

To convert the fraction to a percentage we need to multiply by 100:

$$\text{percentage profit} = \frac{1}{5} \times \frac{100}{1} = 20\% \text{ profit}$$

Aria will make 20% profit if she sells the bike for €360

b We first need to work out how much loss she makes:

loss = cost price − selling price

= €300 − €270

= €30

Now we can write the loss as a fraction of the cost price.

loss as a fraction of cost price $= \frac{\text{loss}}{\text{cost price}} = \frac{30}{300} = \frac{1}{10}$

To convert the fraction to a percentage we need to multiply by 100

percentage loss $= \frac{1}{10} \times \frac{100}{1} = 10\%$ loss

Aria will make 10% loss if she sells the bike for €270

Look back

20% of €300 = 0.2 × 300 = €60, so the selling price would be:
€300 + €60 = €360

10% of €300 = 0.10 × 300 = €30, so the selling price would be:
€300 − €30 = €270

$$\text{percentage profit} = \frac{\text{profit}}{\text{cost price}} \times 100$$

$$\text{percentage loss} = \frac{\text{loss}}{\text{cost price}} \times 100$$

Reflect

Why do we divide by the cost price to work out percentage profit and percentage loss?

Practice questions 2.4

1 A store advertises skateboards at a discount of 20% and another store advertises skateboards at a 15% discount. Do we know which shop has the cheaper skateboards? Why or why not?

2 A store increases the price of their T-shirts from €13 to €20. The owner then decides to include the T-shirts in a 30% off sale. What is the new selling price of the T-shirts?

3 Lamps are advertised at \$48. If you purchase three or more lamps you will receive a 30% discount. Christopher needed one lamp but couldn't resist the bargain, so he bought three lamps. How much did he pay for the three lamps?

4 A basketball is advertised at €15.80 with a 10% discount in Sports Life and as €20.40 with a 35% discount in Sportland. Which basketball is cheaper, and by how much?

5 A discount of \$84 was given on a mobile which initially cost \$280. Work out:

a the price paid

b the discount as a percentage of the initial price.

6 Liam owns a second-hand antique shop. If he

a buys a mirror for \$176 and sells it for \$184.80, what percentage profit does he make

b buys a table for \$460 and sells it for \$368, what percentage loss does he make?

2.5 Ratio

A ratio is a comparison of two or more numbers. For example, the ratio of apples to oranges is 3 : 2

Explore 2.8

Denny is adding water to a juice concentrate to make up a fruit drink. The instructions say: 'Mix one part juice with three parts water'.

Can you work out how much juice concentrate and water he will he need to make 100 ml of the fruit drink?

To work out the ratio in the Explore question, we could write:

ratio of juice to water = 1 : 3 or ratio of water to juice = 3 : 1

Is the order of the ratio important? If yes, why?

A ratio is a comparison of numbers in a definite order. The quantities are expressed in the same units. The ratio can be written in the form $a : b$ or a to b.

Worked example 2.9

Every day, Sylvia spends 30 minutes practising the piano and two hours doing homework. Express the time spent per day practising piano to the time spent doing homework as a ratio.

Solution

This question is asking us to express one quantity as a ratio of another.

In order to solve this problem, we need to express both quantities in the same units, then work out the ratio of the time spent practising piano to the time spent doing homework.

First, we need to write both quantities in the same units.

2 hours = 120 minutes

So, time spent practising piano to time spent doing homework can be written as:

30 minutes to 120 minutes.

This gives a ratio of $30 : 120$

30 minutes as a ratio of 2 hours is $30 : 120$

Looking back, as Sylvia spends four times longer doing her homework than practising piano, $30 : 120$ is a reasonable answer.

Reflect

Is $30 : 120$ is equivalent to $120 : 30$. Give reasons for your answers.

Worked example 2.10

For the diagram shown, work out:

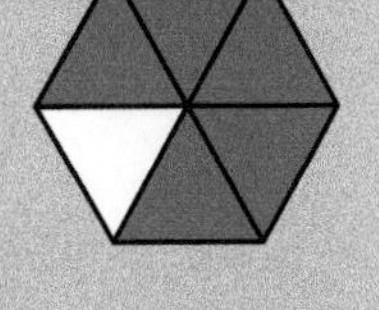

a the ratio of the shaded area to the unshaded area

b the ratio of the shaded area to the total area

c the fraction of the total area which is shaded

d the ratio of the unshaded area to the total area

e the fraction of the total area which is unshaded

f the ratio of the shaded area to the unshaded area to the total area.

Solution

a five triangles are shaded and one is unshaded, so
shaded : unshaded = 5 : 1

b five out of the six available triangles are shaded, so
shaded : total = 5 : 6

c the fraction of the total area which is shaded = $\frac{5}{6}$

d one out of the six available triangles is unshaded, so
unshaded : total = 1 : 6

e the fraction of the total area which is unshaded = $\frac{1}{6}$

f shaded : unshaded : total = 5 : 1 : 6

Reflect

When can you represent ratios using fractions?

Practice questions 2.5

1 Jill buys a book for €5 and sells it for €6. Work out the ratio of her:

a selling price to cost price

b selling price to profit

c cost price to profit

d cost price to profit to selling price.

2 Michael can lift twice as much as Ashar. What is the ratio of:

a the amount Michael can lift to the amount Ashar can lift

b the amount Ashar can lift to the amount Michael can lift?

c Do we know the actual amounts that Michael and Ashar can lift?

3 The ratio of cement to sand in a mortar mix is 2 : 5

a What fraction of the mortar is cement?

b What fraction of the mortar is sand?

4 A class contains 12 girls and 15 boys.

a What is the ratio of boys to girls?

b What is the ratio of girls to boys?

c What is the ratio of boys to the total number of students in the class?

d What fraction of the class is boys?

e What is the ratio of girls to the total number of students in the class?

f What fraction of the class is girls?

5 At a car dealership, cars are sold in the ratio black : red : other colours = 7 : 1 : 4

a What colour is the most common?

b What fraction of the cars sold are black?

c What is the ratio of red cars sold to the total number of cars sold?

6 The recommended ratio of orange fruit juice to water for an orange drink is 1 : 3

a If the drink is mixed in the ratio 1 : 4, is it sweeter than if it was mixed as recommended or not? Justify your answer.

b If the drink is mixed in the ratio 1 : 2, is it sweeter than if it was mixed as recommended or not? Justify your answer.

2.6 Equivalent ratios

Equivalent ratios are two ratios that express the same relationship between numbers.

Explore 2.9

Can you match the ratios with their equivalent forms? One has already been completed.

8 : 16	5 : 4
15 : 12	4 : 5
50 : 25	1 : 3
6 : 18	3 : 5
21 : 35	3 : 1
10 : 6	5 : 3
9 : 3	2 : 1
16 : 20	1 : 2

If we multiply or divide the parts of a ratio by the same (non-zero) number, an equivalent ratio is formed. We can use this to simplify ratios.

Worked example 2.11

Simplify:

a $12:10$ b $2.5:1.5$ c $2\frac{1}{4}:3$ d $1\,\text{kg}:200\,\text{g}$

Solution

a 12 and 10 are both divisible by 2, so:

$$12:10 = (12 \div 2):(10 \div 2)$$
$$= 6:5$$

b Multiplying both sides by 10, to change the decimals to whole numbers, then dividing both sides by 5, gives:

$$2.5:1.5 = (2.5 \times 10):(1.5 \times 10)$$
$$= 25:15$$
$$= (25 \div 5):(15 \div 5)$$
$$= 5:3$$

c Write $2\frac{1}{4}$ as an improper fraction, then multiply both sides by 4 to change the fractions to whole numbers. Finally, divide by 3:

$$2\frac{1}{4}:3 = \frac{9}{4}:3$$

$$= \left(\frac{9}{4} \times 4\right) : (3 \times 4)$$

$$= 9 : 12$$

$$= 3 : 4$$

d Write both quantities in the same units:

$$1\,\text{kg} : 200\,\text{g} = 1000\,\text{g} : 200\,\text{g}$$

$$= (1000 \div 200) : (200 \div 200)$$

$$= 5 : 1$$

Reflect

Explain how you simplify ratios containing fractions and decimals.

Two ratios are equivalent if one of them is a multiple of the other.

Worked example 2.12

Are the ratios $8:3$ and $12:5$ equivalent?

Solution

The lowest common multiple of 8 and 12 is 24 so we can write both ratios as equivalent ratios with the left-hand side equalling 24:

$8:3 = (8 \times 3) : (3 \times 3) = 24:9$

$12:5 = (12 \times 2) : (5 \times 2) = 24:10$

Because $24:9 \neq 24:10$ we know that $8:3 \neq 12:5$

Reflect

Could we have solved the above problem in another way?

Practice questions 2.6

1 Complete the following:

a $2:1 = 8:\square$

b $5:9 = \square:45$

c $20:6 = 10:\square$

d $48:100 = \square:25$

2 Simplify:

a $15:10$ b $72:12$ c $27:18$ d $3\text{ cm}:1\text{ mm}$

e 3.5 litres : 6 litres   f $400\text{ kg}:1$ tonne

g 2 days : 6 hours   h €4 : 00 : 50c : €4.50

3 Simplify:

a $1.4:1$ b $4.8:1.2$ c $\frac{1}{10}:\frac{1}{2}$ d $85:60:25$

4 There are 8 girls and 18 boys in a class. What is the ratio of:

a boys to girls

b girls to boys

c boys to the total number of students in the class

d boys to girls to the total number of students in the class?

5 On a scale drawing 1 mm represents 1 m. Write this as a ratio.

6 An alloy is made by combining 3.5 kg of metal A with 2.5 kg of metal B and 3 kg of metal C. What is the ratio of weights of:

a metal A to metal B

b metal B to metal C

c metal A to metal C

d metal A to metal B to metal C

e metal A to the alloy?

2.7 Dividing a quantity into a given ratio

Dividing quantities into a specified ratio helps to divide quantities fairly.

Explore 2.10

You have a 50 cm piece of wire. Can you bend it into the shape of a rectangle with a width to length ratio of 2 : 3?

Worked example 2.13

In 2018, the ratio of the population of Delhi to Bangalore was approximately 7 : 3. If the population of the two cities were 40 million in total, work out the population of each city.

Solution

This question asks us to work out the populations of Delhi and Bangalore respectively.

To solve this problem, we first need to work out the total number of population parts for Dehli and Bangolore together. By dividing 40 million by the total number of parts we can work out how the population represented by one part and hence multiply to work out the population of each city.

There were 40 million people in total.

Ratio of Delhi : Bangalore = 7 : 3

The total number of parts is 7 + 3 = 10

The size of one part $= \frac{40 \text{ million}}{10} = 4$ million.

The population of Delhi = 4 × 7 = 28 million.

The population of Bangalore = 4 × 3 = 12 million.

Looking back at the results, 28 million + 12 million = 40 million, therefore, the answer is consistent with the information given in the question.

Reflect

How can you check that your answer is correct using fractions?

Practice questions 2.7

1 Divide:

a 24 in the ratio 5 : 3

b 2730 in the ratio 3 : 2

c 585 in the ratio 1 : 3 : 5.

2 A school consists of boys and girls in the ratio of 5 : 7. If there are 660 students in the school, work out how many students are girls.

3 Brigid is dividing €2169 between her two sons in the ratio 4 : 5. How much is the smaller amount?

4 Pewter is made by mixing tin and lead in the ratio 5 : 1. How much of each metal would there be in 30 kg of pewter?

5 In a village the ratio of adults to children is 3 : 5. If the population of the village is 768 people, how many children are there?

6 A dry concrete mix is made by mixing gravel, sand and cement in the ratio 2 : 3 : 4. Work out the mass of each that is required to make 72 kg of the dry concrete mix.

7 The lengths of the sides of a triangle are in the ratio 1 : 2 : 3. Find the lengths of the sides if the perimeter of the triangle is 25.2 cm.

8 a Which point divides the line AK in the ratio:

i 9 : 1 ii 7 : 3 iii 3 : 7 iv 4 : 1 v 1 : 4

A B C D E F G H I J K

b Draw a line 8 cm long and divide it in the ratio 11 : 5

9 The ratio of the populations of town A and town B is 3 : 8, while the ratio of the populations of towns B and C is 5 : 2. If the total population of the three towns is 85 342, work out the population of each town.

2.8 Rates

Rates are used to compare quantities that have different units.

Explore 2.11

I can make four pancakes every five minutes. This is a rate of:

- 4 pancakes per 5 minutes
- 0.8 pancakes per minute
- 48 pancakes per hour
- an hourly rate of 48

etc.

Can you think of two other examples and express them in many ways?

A rate is the ratio between two related quantities in different units. In describing the units of a rate, the word 'per' is used to separate the units of the two measurements used to calculate the rate (for example a heart rate is expressed in 'beats per minute').

Worked example 2.14

Express each of the following as a rate in its simplest form:

a A car travels 300 km in 5 hours.

b A hose fills a 1200-litre tank in 8 minutes.

Solution

a rate $= \frac{300\text{ km}}{5\text{ h}} = \frac{300\text{ km} \div 5}{5\text{ h} \div 5}$

$= \frac{60\text{ km}}{1\text{ h}}$

The car is travelling at a rate of 60 km in 1 hour or 60 km/h.

b rate $= \frac{1200\text{ litres}}{8\text{ minutes}}$

$= \frac{1200\text{ litres} \div 8}{8\text{ minutes} \div 8}$

$= \frac{150\text{ litres}}{1\text{ minute}}$

The hose fills the tank at a rate of 150 litres in 1 minute or 150 litres/minute.

Reflect

What other examples of rates can you think of?

How are rates similar to ratios? How are they different?

How is simplifying rates similar to simplifying fractions and ratios?

Practice questions 2.8

1 Write each of the following rates in simplest form.

a 6 km in 3 h

b 20 kg for \$5

c 50 c for 10 g

d 112 students to 4 teachers

e 20 kg over 8 m^2

f 120 litres in $2\frac{1}{2}$ h

2 Complete the equivalent rates:

a 8 litres/hour = ______ litres/day

b 5 kg/min = ______ kg/h

c $3/g = ______ $/kg

d 2 km/min = ______ km/h

e 4 ml/min = ______ ml/h

f 2 litres/h = ______ ml/h

3 A car travels 260 km in 5 hours. What is its average speed?

Reminder Q3

$$\text{speed} = \frac{\text{distance}}{\text{time}}$$

2.9 Applications of rates

Applications of rates are found in everyday life. For example, we use rates to convert measurements and currencies, work out costs and to solve other problems.

Explore 2.12

A baker can bake 630 cakes in seven weeks (at a constant rate).

Sharon works out that the baker can bake 90 cakes in one week. What do you think?

If you know how many cakes the baker can bake in one week, can you use this to work out how many cakes he can bake in 11 weeks? Explain your method.

Worked example 2.15

A cyclist travels 45 kilometres in three hours. Work out:

a The cyclist's average speed in km/h.

b The cyclist's average speed in m/s. Round your answer to two decimal places.

Solution

a 45 and 3 are divisible by 3, so:

$$\text{average speed} = \frac{45}{3} = \frac{45 \div 3}{3 \div 3}$$
$$= \frac{15}{1}$$

Therefore, the cyclist is travelling at an average speed of 15 km/h.

b We need to convert kilometres to metres and hours to seconds

$$\text{average speed} = \frac{15\,\text{km}}{1\,\text{h}} = \frac{15 \times 1000 \text{ metres}}{1 \times 60 \times 60 \text{ seconds}} = \frac{15\,000\,\text{m}}{3600\,\text{s}}$$
$$= \frac{15\,000 \div 3600\,\text{m}}{3600 \div 3600\,\text{s}}$$

Divide the numerator and denominator by 3600 to work out how many metres the cyclist travels in one second

$$= \frac{4.166\,66\ldots\,\text{m}}{1\,\text{s}}$$

The cyclist is travelling at an average speed of 4.17 m/s.

Reflect

Why do we say 'average speed' rather than just 'speed' in the above example?

Worked example 2.16

A cold tap can fill a bath in five minutes and a hot tap can fill a bath in 10 minutes. How long will it take to fill the bath if both taps are working together?

Solution

This question asks how long it will take to fill the bath with both taps working together.

In order to solve this problem, we need to work out the rate at which the bath will be filled with both taps working together. Using this rate, we can work out how long it will take to fill the bath.

rate of cold tap = $\frac{1}{5}$ bath/minute

rate of hot tap = $\frac{1}{10}$ bath/minute

rate of both taps together = $\left(\frac{1}{5} + \frac{1}{10}\right)$ bath/minute = $\frac{3}{10}$ bath/minute

$\frac{3}{10}$ bath = 1 minute

Converting 1 minute to 60 seconds:

$\frac{3}{10}$ bath = 60 seconds

Divide by 3 to work out the time to fill $\frac{1}{10}$ of the bath:

$\frac{1}{10}$ bath = 20 seconds

Multiply by 10 to work out the time to fill $\frac{10}{10}$ of the bath, that is, the full bath:

$\frac{10}{10}$ bath = 10 × 20 seconds = 200 seconds = 3 min 20 s

Therefore, it will take 3 min 20 s to fill the full bath if both taps are working together.

Looking back at the solution, 3 minutes 20 seconds is less than the time for the cold tap to fill the bath on its own so is a reasonable answer.

Reflect

How can you check your answer is correct?

Practice questions 2.9

1 Complete these equivalent rates.

a 4 km/min = ______ km/h

b 6 litres/h = ______ litres/day

c 40 ml/min = ______ litres/min

d 5 m/s = ______ km/h

2 Which item from each pair is better value?

a 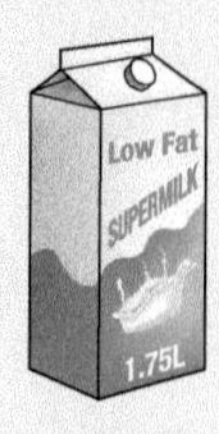

Low Fat Supermilk 1.75 litres €1.99

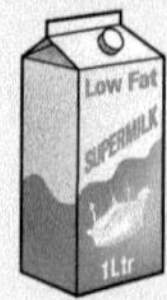

Low Fat Supermilk 1 litre €1.49

b 

450 g €2.99

500 g €3.15

3 A plane travels at 900 km/h. How far will it travel in 35 minutes?

4 Car A consumes 28 litres of fuel when it travels 400 km and car B consumes 20 litres of fuel when it travels 250 km. Which car is more economical? Justify your answer.

5 How long will it take to travel 300 km at 80 km/h?

6 A shearer can shear 18 sheep/hour. How many sheep could she shear in five hours?

7 The exchange rate for converting US dollars to Japanese yen is $1 = ¥106. What is the value of $300 in yen?

8 Matt can make 12 cups of coffee in nine minutes. How long does it take him to make three cups of coffee?

9 Pablo earns $170 in five hours. How much will they earn in 22 hours?

10 June can make 40 bracelets in 25 minutes. How many bracelets can she make in three hours?

11 Light travels at 300 000 km/s.

a How far does light travel in one minute?

b How long would it take light to travel from the sun which is 148 800 000 km away?

12 The exchange rate for converting euros to Australian dollars is €1 = A$1.63. Kian wants to convert €50 to Australian dollars, can you work out how much in Australian dollars he should receive?

Challenge Q13

13 The exchange rate for converting euros to Australian dollars is €1 = A$1.63. The exchange rate for converting Australian dollars to South African rand is A$1 = R12.08. What is the exchange rate for converting euros to South African rand? Round your answer to 2 decimal places.

Challenge Q14

14 Pump A can fill a tank in four minutes and pump B can fill a tank in six minutes. How long will it take to fill the tank if both pumps are working together?

Investigation 2.1

Population density is a measurement of population per unit area. For example, the population density of Dhaka, Bangladesh is approximately 44 000 people/km^2 and the population density of Melbourne, Australia is approximately 450 people/km^2.

Connections

You will encounter population density in geography. It is an important factor to be considering for urban and regional planning.

Research skills

1 Research and explain:

a how population density is calculated

b why is population density a useful measure.

2 Calculate the population densities for the places shown in the table. What observations can you make from your results?

Place	Location	Population	Area (km^2)
Albania	Albania	2 862 000	28 748

Cambodia		16 250 000	181 035
Kenya		47 564 000	580 367
New Zealand		4 917 000	268 021
Colombia		50 340 000	1 141 748

3 Research the population and area to calculate the population densities for the following places: Hong Kong, New York City, Mongolia, Amsterdam, Greenland, Mumbai, Mexico City, Canberra and Johannesburg. Explain the accuracy of your solutions.

4 The following graphics illustrate population densities. Write down at least three observations that you find interesting from each graphic.

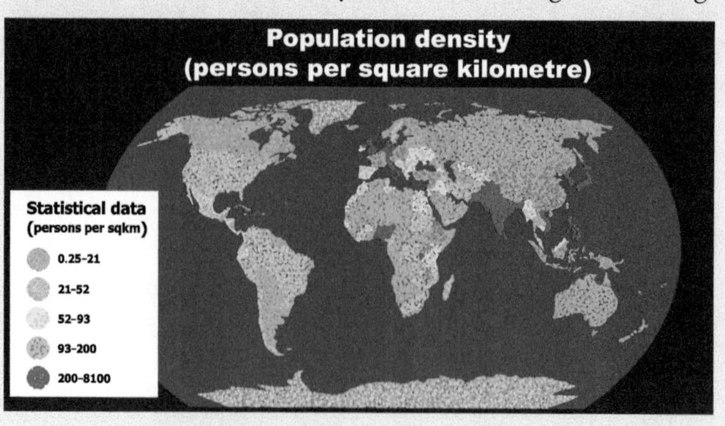

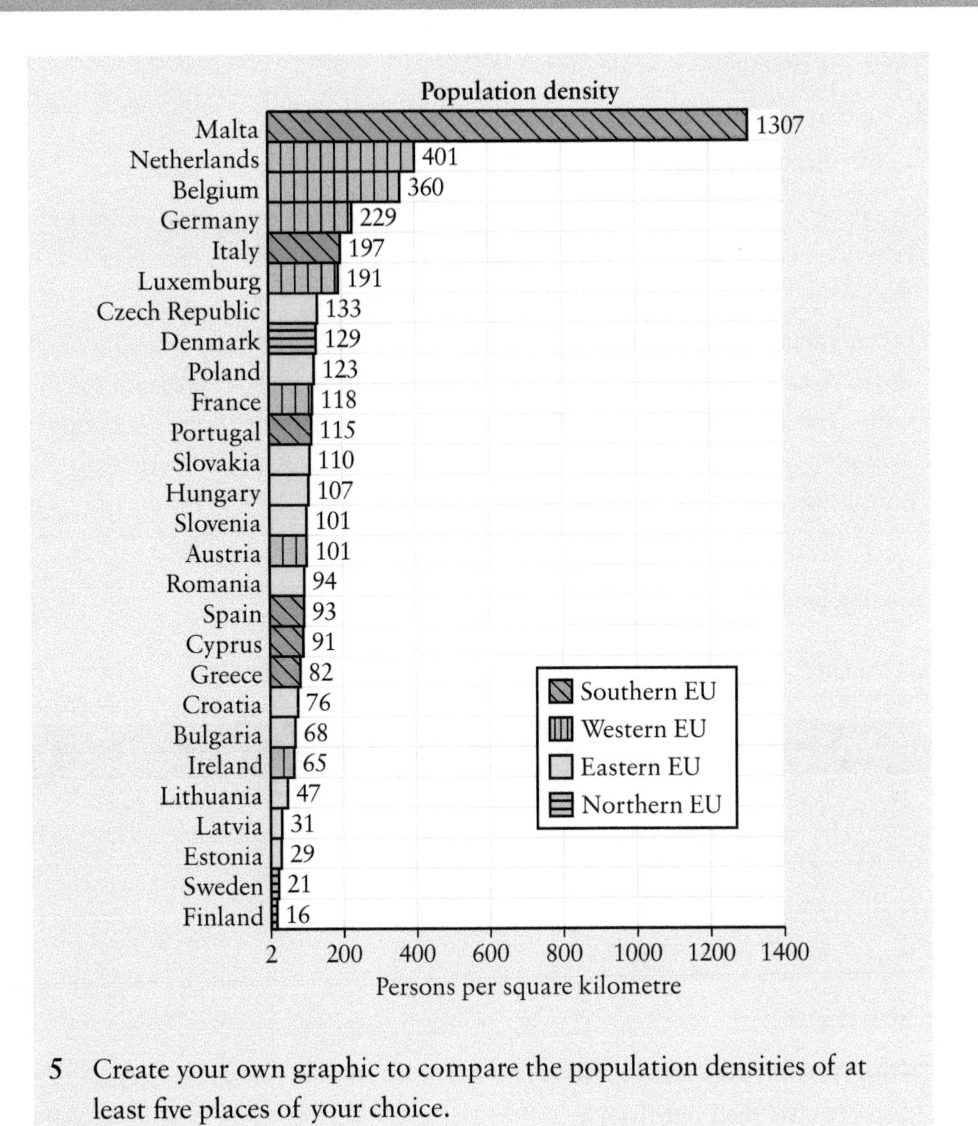

5 Create your own graphic to compare the population densities of at least five places of your choice.

2.10 Scale drawing

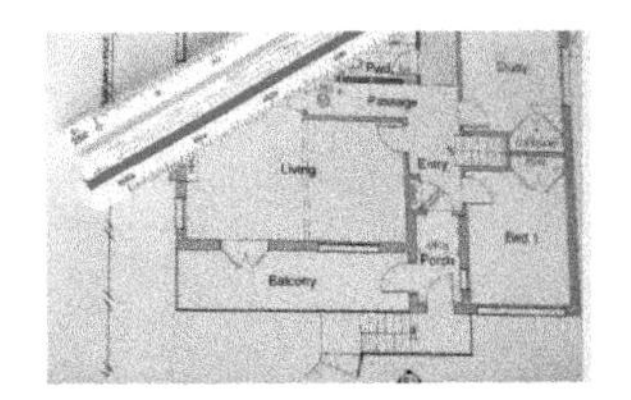

Scale drawings are used to represent the shape of objects with accurate sizes that are reduced or enlarged by a certain amount called a scale. For example, architects and engineers make scale drawings of buildings and bridges.

scale = length on drawing : real length

A scale can be written in two ways, as 1 cm : 1 m or 1 : 100

A scale of 1 : 100 means that the real length is 100 times the length on the drawing.

There are two types of scale drawings:

1. making a scale drawing of an object
2. calculating real sizes of objects from the drawing.

Explore 2.13

An architect is making a scale drawing of a city park. The park is rectangular in shape, it is 150 m by 90 m. He is making a scale drawing of the park and chooses a scale of 1 : 500. On the drawing he represents 150 m with a line of length 30 cm and 90 m with a line of length 18 cm. Can you explain how to convert between the lengths on the drawing correspond and the actual lengths in real life?

$$\text{drawing length} = \frac{\text{actual length}}{\text{scale}}$$

actual length = scale length × scale

Worked example 2.17

The scale on the map shown is 1 : 20 000.

Work out:

a the actual distance from A to B

b the actual distance from C to D.

Solution

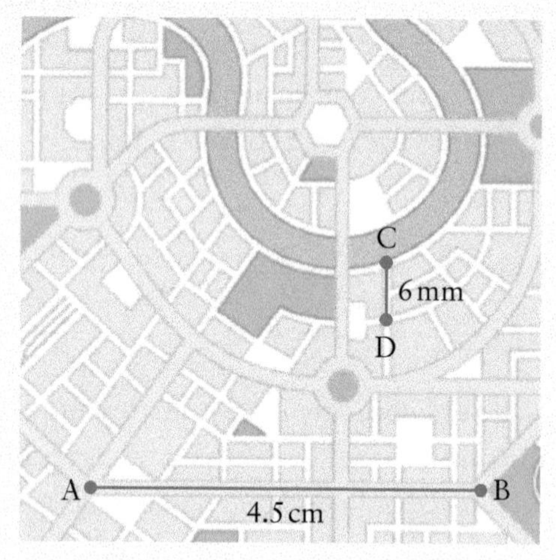

This question asks for real-life distances, worked out from a map that is drawn to scale.

To solve this problem, we need to accurately measure the distances on themap using a ruler. We then need to multiply this value by the scale given to work out the actual distance between the given points.

a Use a ruler to measure the distance between A and B on the drawing.

length on the drawing = 4.5 cm

So, the real length = 4.5 × 20 000 = 90 000 cm

= 900 m.

b Use a ruler to measure the distance between C and D on the drawing.

length on the drawing = 6 mm

So, the real length = 6 × 20 000 = 120 000 mm

= 120 m.

Looking back at our solutions to check whether they make sense:

$$\text{drawing length} = \frac{\text{actual length}}{\text{scale}}$$

a $\frac{90\,000}{20\,000} = 4.5$ b $\frac{120\,000}{20\,000} = 6$

Both answers coincide with the original drawing lengths.

Reflect

What limitations are there to reading measurements from scale drawings?

Practice questions 2.10

1 Complete the following.

a 1 : 500 = 1 cm : _______ cm b 1 : 10 000 = 1 cm : _______ cm

c 1 : 20 000 = 1 m : _______ m

2 Write each of the following scales in ratio form.

a 1 mm to 1 m b 1 cm to 15 m

c 1 mm to 20 m d 5cm to 1 km

3 If the scale is 1 : 500, work out the real distance between two points if they are the following distances apart on the drawing.

a 3 cm b 5.2 cm c 2 mm d 35 mm

4 If you had to make a scale drawing using a scale of 1 : 100, what lengths would be represented by the following actual distances.

a 50 cm b 350 cm c 72 m d 7.5 m

5 Choose an appropriate scale and make scale drawings of the following:

a a square with sides 400 m

b a triangle with sides 9 m, 12 m and 15 m

c a rectangle of width 3 km and length 7 km.

Hint Q5

Consider the dimensions of the page that you are drawing on, and the maximum length of the object to be drawn.

6 The plan of a house is drawn to a scale of 1 : 100. If a room measures 39 mm by 56 mm on the plan, how big is the room in real life?

7 A scale drawing is to be made of an aeroplane with wingspan 64 m and length 76 m. The drawing is to be on a page with dimensions 21 cm by 30 cm, work out an appropriate scale for the drawing.

8 Use the scale on the map to work out the distance from:

a Riga to Cēsis
b Valmeira to Alūksne
c Stende to Valga
d Mērsrags to Ogre

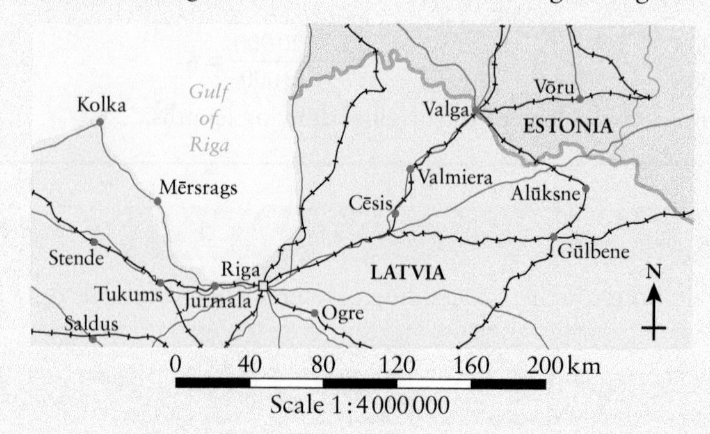

Self-assessment

- I can convert fractions and decimals to percentages.
- I can convert percentages to fractions and decimals.
- I can find a percentage of a given quantity.
- I can express one quantity as a percentage of another.
- I can calculate percentage change.
- I can find a quantity if a percentage is known.
- I can calculate percentage profit and loss.
- I can calculate percentage discount.
- I understand what a ratio is.
- I can simplify ratios.
- I can calculate equivalent ratios.
- I can divide a quantity in a given ratio.
- I understand what a rate is.
- I can use rates to convert between currencies.
- I can use rates to convert measurements.
- I can use rates to problem-solve.
- I can calculate the scale drawing length of an actual length.
- I can calculate the actual sizes of objects from a scale drawing.
- I can choose an appropriate scale to make a scale drawing of an object.

? Check your knowledge questions

1 Convert these fractions to percentages.

a $\frac{33}{100}$ b $\frac{3}{25}$ c $\frac{5}{12}$ d $1\frac{7}{9}$ e $2\frac{1}{6}$ f $\frac{7}{15}$

2 Convert these decimals to percentages.

a 0.09 b 0.72 c 0.3 d 2.67 e 0.0452 f 4.315

3 Convert each percentage to a fraction.

a 10%   b 120%   c 64%   d $2\frac{1}{4}\%$

4 Convert each percentage to a decimal.

a 23%   b 5%   c 60%   d 175%

5 Work out:

a 45% of 7000 g   b 15.5% of 220 cm   c 24% of 1 tonne

d 72% of 2 hours   e 0.8% of 480 ml   f 0.09% of 2 kg

Challenge Q5e,f

6 A farmer produces 150 kg of Fairtrade coffee. There are 60 kg of average quality, the rest of the coffee is good quality. If good quality coffee sells for $0.55 per kg and average quality coffee sells for $0.48 per kg, find how much the farmer earns.

7 For each of the following, express the first quantity as a percentage of the second.

a 50 g, 3 kg   b 45 s, 3 min

c 8 months, 2 years   d 64 marks, 80 marks

8 Increase $260 by 12%

9 Decrease 560 cm by 20%

10 A bicycle is reduced by 35% in a sale. If the bike initially cost €640, work out the sale price of the bicycle.

11 There are 1200 trees in a forest. If 20% of the trees are cut down at the beginning of 2018 and then the number of trees increase by 35% in the next 10 years, how many trees will there be in 2028?

12 Determine the size of a quantity if:

a 4% is 32 g   b 30% is 57 mm

c 28% is $29.12   d 120% is 960 ml

13 Electric cars account for 58% of all car sales in Norway. If 145 000 electric cars were sold last year, how many cars were sold in total?

14 Michael bought a table for €350

a If he sold it for €420, work out the percentage profit.

b If he sold it for €315, work out the percentage loss.

15 Express the following as ratios. Simplify where possible.

a 2 h to 4 min

b 50 c to €10

c 20 kg to 200 g

d 36 seconds to $2\frac{1}{2}$ min

16 An international rapid relief team consists of doctors and general volunteers in the ratio of 2 : 7. If there are 63 general volunteers how many doctors are there?

17 A school has boys and girls in the ratio 5 : 3. If there are 864 students in the school, how many girls are there?

18 Mira wants to divide $521 145 between her three children Albert, Bob and Charles in the ratio 3 : 5 : 1. How much money will each son receive?

 Challenge Q19

19 A chemical solution is made by mixing an acid with water in the ratio 1 : 24. This solution is then mixed with water in the ratio 1 : 3. How much acid would there be in 400 ml of the final solution?

20 Which is better value: 250 g of chocolate for €4.50 or 430 g of chocolate for €8.60?

21 A car can travel 72 km on six litres of diesel.

a How far could the car travel with 35 litres of diesel?

b How many litres of diesel would the car need to travel 624 km?

22 Convert 15 m/s to km/h

23 Convert 60 km/h to m/s

24 The exchange rate for converting Hong Kong dollars to euros is $1 = €0.11. What is the value of $145 in euros?

25 Kristina earns £83.20 in 8 hours. How much will she earn in 13 hours?

 Challenge Q26

26 In making one revolution, a car wheel travels 200 cm. How fast are the wheels spinning, in revolutions per minute, when the car is travelling at 60 km/h?

27 A drawing has a scale 1 : 100. Convert the following scaled distances to real distances.

a 4 cm

b 6.2 cm

c 3 mm

d 37 mm

28 A scale drawing has a scale of 1 : 15 000. Convert the following real distances to scaled distances.

a 15 km

b 28.5 km

c 455 m

d 563.45 m

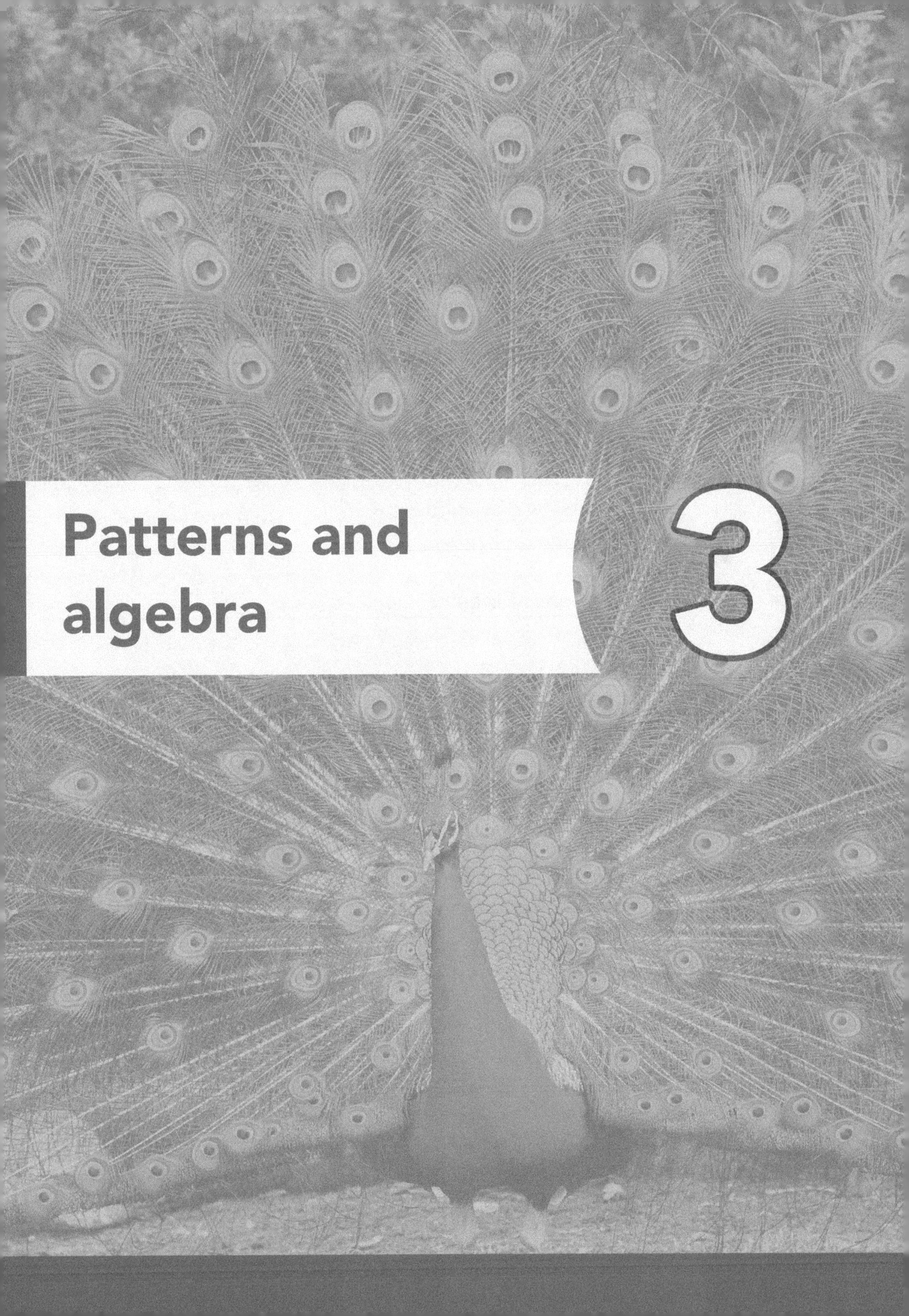

3 Patterns and algebra

Patterns and algebra

KEY CONCEPT

Form

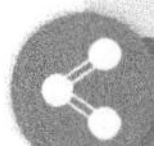

RELATED CONCEPTS

Generalisation, Models, Patterns, Simplification

GLOBAL CONTEXT

Identities and relationships

Statement of inquiry

Representing patterns in different forms allows us to generalise and model relationships in real life.

Factual

- What strategies can you use to simplify expressions?

Conceptual

- How can expressions be used to model real-life situations?

Debateable

- Can you always predict patterns correctly?

Do you recall?

1 Can you add and subtract ordered numbers?

a $4 - 7$ b $8 + -3$ c $-9 - 4$

d $2 + -14$ e $-1 + 5$

2 Can you multiply and divide ordered numbers?

a 3×-9 b -8×-5

c -7×4 d $-2 \times -4 \times -3$

3 How do you write numbers in index notation?

a 5×5 b $2 \times 2 \times 2 \times 2$ c $-4 \times -4 \times -4$

4 What is a variable?

5 What is an expression?

6 What is a term?

7 What are coefficients?

8 What are 'like terms'?

3 Patterns and algebra

3.1 Patterns and puzzles

3.1.1 Number patterns

Ian Stewart, the British mathematician, famously said, 'Mathematics is the science of patterns, and nature exploits just about every pattern there is'. As a result, understanding the patterns of nature can help us to better understand the world around us and to make predictions.

Fact

To exploit something can mean to use it for your benefit.

Understandably, mathematicians have been fascinated by patterns since the start of written mathematics. You will begin by looking at number patterns, which are also called **sequences**.

Explore 3.1

All the number patterns, or sequences, below have special names.

For each sequence, can you

a write down a rule in words to describe the pattern?

b predict the next three numbers in the pattern?

c identify the name of the pattern? What type of numbers does it involve? (If you have never heard of the pattern, think of a name you could call it, making sure you give a reason for your thinking.)

i 1, 3, 5, 7, 9, …

ii 1, 3, 6, 10, 15, …

iii 1, 2, 3, 4, 5, …

iv 1, 4, 9, 16, 25, …

v 1, 1, 2, 3, 5, 8, …

vi 1, 2, 4, 8, 16, …

Hint Qii

It is useful to know the position of a term in a sequence, also called the term number. This tells you where the term comes in the pattern. Knowing the term number can sometimes help uncover the sequence pattern.

Mathematicians call the individual numbers in a number pattern **terms**. Recall that term is also the word used in algebra for the individual elements of an expression. This is an example of words in mathematics having slightly different meanings in different contexts.

Reflect

How are the two meanings of **term** related? How are they different?

Research skills

There are many other number patterns with specific names. Can you find two more?

Most number patterns, however, do not have individual names. Instead, we sort them based on what kind of pattern they follow. In the Explore activity, you saw several types of pattern including those that required addition and multiplication. Mathematicians give these patterns precise names.

The simplest type of addition pattern is one where you add the same number repeatedly, for example, 5, 7, 9, 11, 13, … . Mathematicians call patterns like this **arithmetic sequences**. Each term is found from the previous term by adding a **common difference**. In the example above, the common difference is + 2 because 2 has been added to each term to get to the next. Another example is 2, 5, 8, 11, … which has a common difference of + 3 because 3 has been added to each term to get the next.

There is a similar type of pattern for multiplication. **Geometric sequences** are number patterns where the next term can be found by multiplying the previous term by a specific number. This number stays the same for the whole pattern and is called the **common ratio**. An example of a geometric sequence is 4, 8, 16, 32, 64, … where each term is found by multiplying the previous term by 2.

Worked example 3.1

For each of these number patterns:

a describe the pattern

b write down the next three terms

c identify whether the sequence is geometric or arithmetic.

i 3, 8, 13… **ii** 32, 16, 8 …

iii 1, 3, 9, … **iv** 8, 6, 4…

Solution

i	**a** Add 5	**b** 18, 23, 28	**c** Arithmetic
ii	**a** Multiply by $\frac{1}{2}$	**b** 4, 2, 1	**c** Geometric
iii	**a** Multiply by 3	**b** 27, 81, 243	**c** Geometric
iv	**a** Subtract 2	**b** 2, 0, −2	**c** Arithmetic

Reminder Qii

Dividing by 2 is the same as multiplying by $\frac{1}{2}$

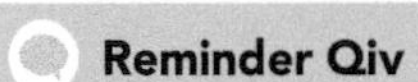

Reminder Qiv

Subtraction is the same as adding a negative number. This means that subtraction patterns are arithmetic sequences.

Practice questions 3.1.1

For Questions 1–9:

a Write down a rule to describe the pattern, using appropriate notation.

b Predict the next three numbers in the pattern.

c State whether the pattern is arithmetic or geometric.

1 3, 6, 12, …

2 1, 5, 9, …

3 10, 20, 30, …

4 81, 27, 9, …

5 8, −1, −10, …

6 $\frac{1}{3}$, 1, 3, …

7 4, 16, 28, …

8 −1, 2, −4, …

9 −5, −12, −19, …

For Questions 10–16:

a Write down a rule in words to describe the pattern.

b Predict the next two terms in the pattern.

c Justify your answers to **a** and **b** with one piece of evidence.

10 1, 12, 123, …

11 M, T, W, T, F, …

12 1, $\sqrt{2}$, $\sqrt{3}$, 2, $\sqrt{5}$, …

13 0, −1, −3, −6, −10, …

14 O, T, T, F, F, …

15 1, 2, 6, 24, …

16 Love, 15, 30, …

Reflect

How are Questions 10–16 different from Questions 1–9?

How did your knowledge of mathematics and patterns help you answer Questions 10–16?

For which questions was mathematical knowledge unnecessary? Give a reason why.

What conclusion could you come to about predicting patterns based on your experience of solving these questions?

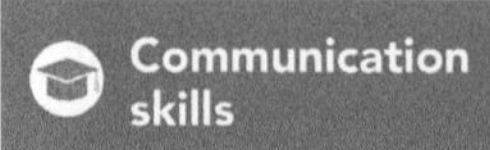

Now answer this debatable question: Can you always predict patterns correctly? Give at least one piece of evidence for your answer.

3.1.2 Patterns with algebra

Now that you have seen some number patterns, consider how you can describe these patterns more clearly. Mathematicians often use algebraic **formulae** to describe patterns. You will learn more about formulae in Chapter 4, but for now, you will see how to use them to describe number patterns.

Worked example 3.2

Complete the table for the formula $y = 2x - 1$ and state the general rule for continuing the sequence.

x	1	2	3	4	5
y					

Solution

We need to substitute each of the values of x listed in the table into the formula so that we can evaluate for the corresponding values of y. Substituting means that a numerical value will take the place of the variable.

Substitute each value of x from the table into the formula, $y = 2x - 1$

It is a good idea to put substituted values into brackets:

$x = 1 \Rightarrow y = 2(1) - 1$

We can use multiple lines of working if necessary:

so, $y = 2 - 1$

therefore, $y = 1$

Now we can add this value to our table.

x	1	2	3	4	5
y	1				

Repeat this process for the other values of x in the table:

$x = 2 \Rightarrow y = 2(2) - 1$

$y = 4 - 1$

$y = 3$

Reminder

When a number is written in front of a variable, it represents multiplication, so $2x$ means $2 \times x$

$x = 3 \Rightarrow y = 2(3) - 1$

$y = 6 - 1$

$y = 5$

x	1	2	3	4	5
y	1	3	5		

Do you recognise the pattern formed by the values of y in the table? Can you predict what the next two numbers will be?

You have already seen this pattern in the Explore section; these are odd numbers (1, 3, 5, 7, 9 ....). Once you have spotted the pattern, you can complete the table without using the formula.

x	1	2	3	4	5
y	1	3	5	7	9

Self-management skills

Hint

You can use your GDC to do all calculations and for the graph below. It is recommended that you learn how to use your GDC effectively.

However, if you do make a prediction, it is always worth checking it with the formula to make sure that it is correct.

Check the values for $x = 4$ and $x = 5$ using the formula.

$x = 4 \Rightarrow y = 2(4) - 1$

$= 8 - 1$

$= 7$ ✓

$x = 5 \Rightarrow y = 2(5) - 1$

$= 10 - 1$

$= 9$ ✓

Note that the pattern of y values in Worked example 3.2 is an arithmetic sequence.

You can also describe the pattern as **linear**, because if you plot the pairs of x and y values from the table as coordinate pairs they form a straight line, as shown on the graph.

You will learn more about these relationships in the next chapter.

Now look at some other applications of **substitution**.

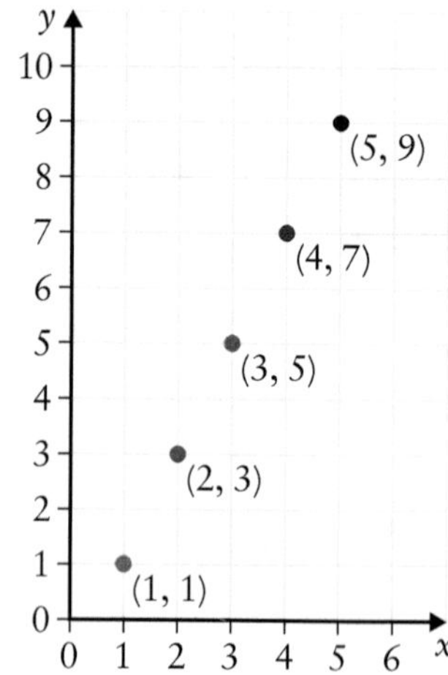

Worked example 3.3

Evaluate these expressions when $a = -3$

a $2a + 5$ b $1 - a^2$ c $4(8 - a)$

Solution

We use brackets around the number when we substitute it.

This will help us to keep track of the order of operations and the negatives.

Use brackets around negative numbers:

a $2(-3) + 5 = -6 + 5$

$= -1$

Squaring a negative number gives a positive result:

b $1 - (-3)^2 = 1 - 9$

$= -8$

Subtracting a negative number becomes an addition:

c $4(8 - (-3)) = 4(8 + 3)$

$= 4(11)$

$= 44$

Reminder

Don't forget to use the correct order of operations.

Practice questions 3.1.2

For Questions 1–5:

a Copy and complete a table, as shown, for each formula.

b Write down a rule for the pattern formed by the y-values.

c Write down which part of the formula helps to predict the rule.

1 $y = 3x - 1$ 2 $y = 4 - 2x$

3 $y = 5x + 3$ 4 $y = -1 - x$

5 $y = -10 + x$

x	1	2	3	4	5
y					

6 Use the table to complete this question.

a Complete a table of values for:

i $b = 1 - a^2$ ii $b = a^2 + a$

a	1	2	3	4
b				

b Explain what makes the two patterns for b different from the patterns for y in Questions 1–5. Give at least one piece of evidence to support your answer.

 Hint Q7

Consecutive means numbers immediately following one another, for example, 1, 2, 3 or 100, 101, 102

 **Reminder 8c**

Don't forget to apply the correct order of operations.

7 Predict the rule in each pattern for g, then complete a table using at least three consecutive values of f to check your predictions.

a $g = 5f + 3$ b $g = 6 - 2f$ c $g = -8 + f$

8 Substitute the following values into the expression $6j - 2k + 5k^2$

a $j = 1, k = 3$ b $j = 3, k = 2$ c $j = -1, k = 0$

d $j = -2, k = 3$ e $j = 5, k = -1$

9 Evaluate the following expressions when $m = -2$ and $n = 4$

a $5n + 3m$ b $3 - 2n$ c $7m - n$

d $2mn$ e $n^2 - m^2$ f $m + 5n + 2$

g $4mn - 6$ h $7m^2 - 2n$ i $-m + (n^2 - 3n)$

3.1.3 Puzzles and problem solving

You have seen how algebra can help you to predict simple number patterns. Now you will see how you can use mathematical principles to solve more complicated problems.

Explore 3.2

There are four teachers in your school whose last names begin with 'S': Mrs Singh, Mr Sánchez, Ms Solarin and Mr Sun. Each teacher teaches one of four subjects: Maths, Art, English and Physical Education. The school building has four floors, and each teacher has their own office in the building. The offices in the school have the following door colours: pink, blue, green and yellow. Use this information and the hints below to work out what Mrs Singh teaches, and on which floor.

- Each teacher's classroom is on a different floor.
- Each floor has a different door colour.
- Floor 1 has pink doors.
- The English teacher's office has a yellow door.
- Mr Sun's office is between the floors with Mr Singh's and Ms Solarin's offices.

 Hint

Follow the four steps of Pólya's method to help you solve the puzzle:

1. Understand the problem.
2. Make a plan.
3. Carry out your plan.
4. Look back (check your work).

- Mr Sánchez teaches Art.
- The maths offices are on the 2nd floor.
- Mrs Singh's office has a green door.
- Ms Solarin teaches on the top floor.

What strategies helped you to solve the Explore problem? Did Pólya's method help you? Often, we might not need each step, but it is useful to think about how these can help you to tackle problems more easily.

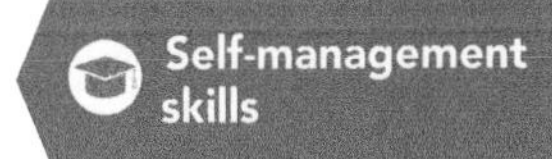

Regardless of the strategy you choose, if you use a system to organise the information step by step you are more likely to be successful. In mathematics, we call this a systematic approach. See if you can use this type of thinking in a different way.

Investigation 3.1

Look at the diagram. Each figure has more black squares than the previous figure.

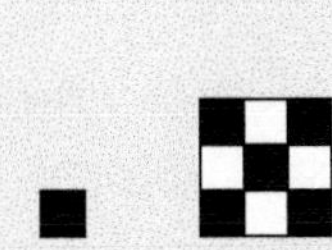

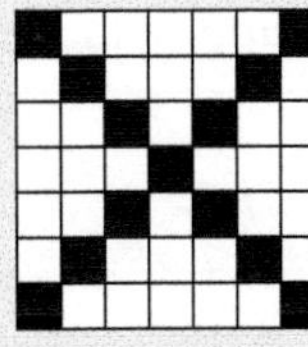
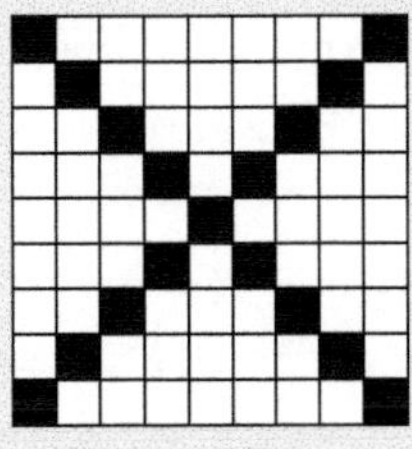

Figure 1 Figure 2 Figure 3 Figure 4 Figure 5

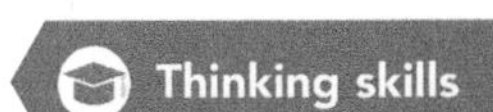

1 Without counting the individual squares, predict what the pattern might be. Give one reason for your prediction.

You can turn this information into a table to help spot the pattern.

2 Copy and complete this table:

Figure number, x	1	2	3	4	5
Number of black squares, y					

3 What kind of pattern or sequence do these shapes form?

One of these formulae linking x and y can enable you to make predictions about the number of black squares in other Figures:

$y = 1 + 4x$ $\quad y = 1 - 4x$ $\quad y = 4 + x$ $\quad y = 4x - 1$ $\quad y = 4 - x$

4 State which formula is correct, giving supporting calculations and at least one reason for your choice.

5 Use your formula to predict the number of black squares in Figure 10 of the pattern. Justify your answer.

Here is another visual pattern, this time with blocks.

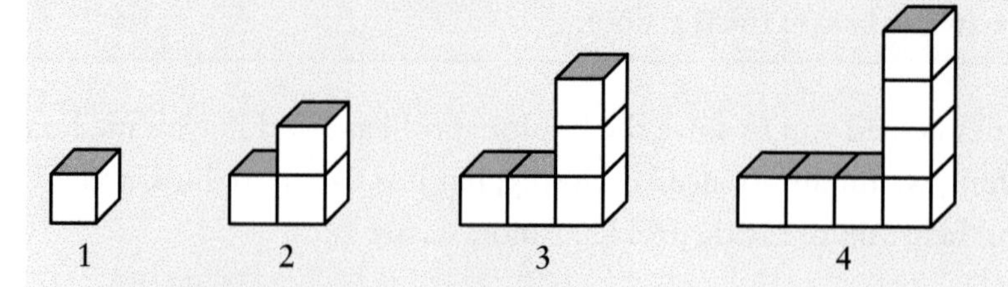

6 Copy and complete the table.

Figure number, x	1	2	3	4	5
Number of blocks, y					

7 Predict the number of blocks in Figure 5 of the pattern. Give a reason and evidence for your prediction.

8 Using your observations for Question 4, write a formula that will help you to predict the number of blocks in other figures of the pattern. Make sure you show supporting calculations.

Reflect

Think back on the Explore and Investigation tasks in this section.

What was the most interesting thing you learnt or noticed by completing these tasks?

How did your understanding of patterns and algebra help you to work through these tasks successfully?

What was the most successful strategy you used to make sure that your work was correct?

Practice questions 3.1.3

1 Use the diagram to answer these questions.

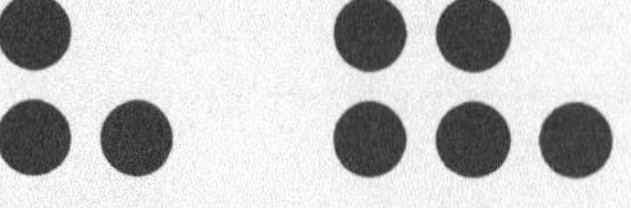

Figure 1 Figure 2 Figure 3

a Predict how the next figure will look.

b What is the pattern for the number of circles used?

c Find a formula to predict the number of circles in the 6th figure.

d Draw the 5th and 6th figures to check your predictions.

2 You can create this pattern using toothpicks.

Figure 1

Figure 2

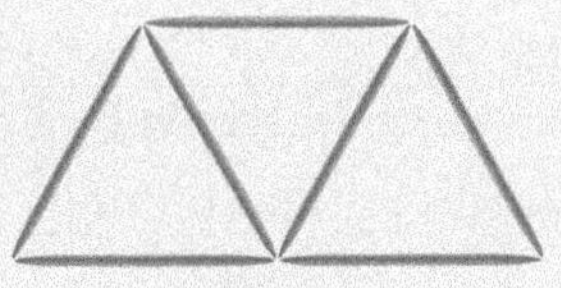

Figure 3

a How many more toothpicks will you need to construct the 4th figure?

b Write a formula that will help you to predict the number of toothpicks needed for the 10th figure.

c How could you check your predictions?

d What similarities are there between this question and Question 1?

3 Lee stacks cans at a shop. These are the first three versions of the stacks of cans:

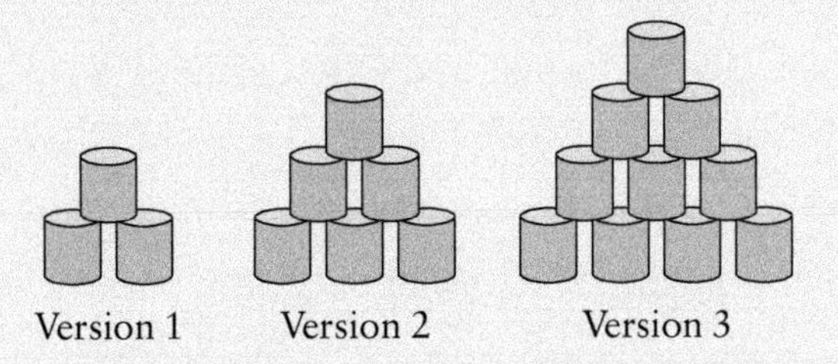

Version 1 Version 2 Version 3

a How many cans will Lee need to make Version 4?

b How many more cans will he need for Version 5?

c Describe, in words, the pattern in the number of cans added to the stack each time a bigger version is formed.

d How is this pattern different from the patterns in Questions 1 and 2?

4 Make your own visual pattern.

Challenge Q4

a Draw the first three figures in your pattern.

b Create a formula that predicts how your pattern continues.

c Check your work by predicting and then drawing the 4th figure in the pattern.

3.2 Operations with variables

3.2.1 Addition and subtraction with variables

Explore 3.3

You are given this addition:

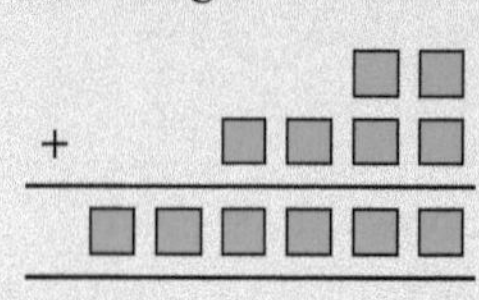

Calling each small blue square b, how do you represent this addition using the letter b?

Now, calling each small yellow square y, how do you represent the operation below?

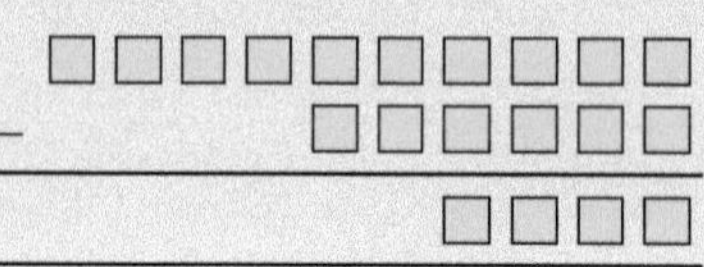

We now have an equation involving the blue and yellow squares from earlier.

What does this represent? Simplify it as far as you can.

When adding and subtracting **variables**, remember that you can only combine **like terms**.

- When you add like terms, you perform the operation using their **coefficients**:

  for example, $2x + 3x = 5x$

  Minus signs belong to the **term** written to their right:

  for example, $6a - 3a = 3a$

- Variables without coefficients have an implied '1' as their coefficient:

  for example, $3b + \underline{b} = 3b + \underline{1b} = 4b$

- Terms that are not like terms cannot be further combined by addition or subtraction:

  for example, $2x + 2y - x = x + 2y$

Reminder

A variable is a letter or symbol used in mathematics to represent a number, and like terms are terms that have the same variable, for example $4a$ and $6a$

Reminder

Coefficients are the numbers in front of variables, for example, 3 is the coefficient of $3a$

Fact

A variable can also be called a pronumeral.

Worked example 3.4

Simplify:

a $4mn - 2mn + mn$ b $5x + 4y + 2x - y$ c $2a - 4a^2 - 5a$

Solution

a Move through longer questions from left to right.

Assume a coefficient of 1 for any variables without a given coefficient:

$$4mn - 2mn + mn = 2mn + 1mn$$
$$= 3mn$$

b Start by identifying like terms. Plus signs belong to the term to their right:

$$5x + 4y + 2x - y = \underline{5x} + \underline{\underline{4y}} + \underline{2x} - \underline{\underline{y}}$$
$$= 7x + 3y$$

c Remember a and a^2 are not like terms:

$$2a - 4a^2 - 5a = \underline{2a} - 4a^2 \underline{-5a}$$
$$= -3a - 4a^2$$

Connections

Often simplifying algebraic equations will use the same skills as operations with ordered numbers.

Look at Worked example 3.4 to identify some useful steps to help 'Make a plan' when you are simplifying expressions.

- First, identify like terms.
- Simplify expressions from left to right.

If there is no coefficient in front of a variable it implies a coefficient of 1 (for example, x is the same as $1x$).

Use this plan when answering the questions in the next Practice questions.

Practice questions 3.2.1

Fact Q1d

The commutative property states that multiplication can happen in any order. In algebra this means that ab is the same as ba.

1 Identify the pairs of like terms in each of these lists:

a $3a, 3b, 2a$ b $2x^2$, $3x$ and x^2

c $2f, 3fg, -fg$ d $5b, -2ab, ba$

2 Simplify by collecting like terms:

a $4q - 2 + 3q$ b $2 + 3p - 2p$

c $6 - 2d + 1$ d $c - 5 + 6c$

e $u + 4 + 6u$ f $1 - a + 3$

3 Simplify by collecting like terms:

a $1 + b + 3 + 2b$

b $5z + 3y + 4z - y$

c $2c - 4d - c + d$

d $a + 2b + 2a$

e $8t - 6 + 2t + 5$

f $4x - 2x + 6 - 5x - 2h + 7h$

g $9k - 6j + 6 + j + 2k - 2$

4 Simplify:

a $1 - a^2 + 10 + a$

b $6g^2 + 2g - 3 + g^2$

c $10r + r^4 + 4r + 2r^4$

d $9k - 4k^3 + 3k + k^3$

e $2z^2 + 3z - 5z^2 + z$

f $7 + u^3 + 3u - 5 + 4u^3$

g $5 - 9y + 8y^2 + 2y - y^2$

5 Simplify:

a $7pqr + 4p^2 - 3pqr + 2pq$

b $5h + 5 - 6h + 4 - h^2$

c $6mn + 4m - 4mn + 2n - 2mn$

d $3a - 7b + ab - 2a + 4ab + b$

e $3z^3 + 2z^2 + z - 3z^2 - 2z - z^3$

f $9xy + 2x^2 - 4xy + 6x^2 - 5xy + y$

g $16f + 8g - g^2 - 7g - 3f + 2g^2 - g$

6 Simplify:

a $mn + 4m^2 - nm + 6m - 3m^2$

b $6wx + 6w - 2x - 2w - 2xw - 4w + wx$

c $2a + 8b + ba - a + 3ab - ab$

d $-4e^2 + 2e - 5ef + fe + 7e - f + 5e^2 - 9e$

3.2.2 Multiplication of variables

At the start of this chapter, you were reminded that in an algebraic term, the coefficient and variable are multiplied together, even if there is no multiplication sign written between them. So, for example, $3y$ means $3 \times y$ So, what happens if you multiply two different terms together?

Explore 3.4

Can you write these expressions in a simpler way?

a $2a^2 \times 3$

b $4m \times 5n \times 2$

Worked example 3.5

Simplify these expressions:

a $2p \times -4q \times 3r$

b $-5x \times -3y$

Solution

a Don't forget to consider the minus signs:

$$2p \times -4q \times 3r = -8pq \times 3r$$
$$= -24pqr$$

b Remember that the product of two negative numbers is positive:

$$-5x \times -3y = 15xy$$

Thinking skills

What do you notice about the order of the combined variables in Worked example 3.5?

They are in alphabetical order. When multiplying variables, listing the results in alphabetical order will help to avoid confusion and to identify like terms. This is an example of a mathematical convention, i.e. a way in which (most) mathematicians have agreed to do things.

Practice questions 3.2.2

1 Simplify:

a $2x \times 3$ b $3h^3 \times 6$ c $a \times 5c$

d $5 \times 6p$ e $5 \times 2u^4$ f $8m \times 9n$

g $7b \times 8a$ h $12fg \times 3h$ i $6s^2 \times 0.8t$

Reminder Q1

Variables should be listed in alphabetical order.

2 Simplify:

a $-a \times 3$ b $2d \times -8$ c $-6m \times -6$

d $-4w \times 7$ e $7f \times -5g$ f $5 \times -11x$

g $-3j \times -9k$ h $-6n \times -5p^2$ i $-9v^2 \times 4w$

3 Simplify:

a $2p \times q \times 4$
b $a \times b \times c$
c $3x \times 2 \times y$
d $4h \times -j \times 3$
e $-4 \times 5z \times 2$
f $-7 \times 2a \times -3b$
g $-6f \times -g \times -2$
h $3w^2 \times -5y \times -5z$
i $-0.1k \times -7m^3 \times -8n^2$

4 Perform the necessary operations then simplify:

Reminder Q4

In the order of operations, multiplication is carried out before addition.

a $2 \times 3x - x$
b $6m + 5 \times 2m$
c $7u - 3 \times u$
d $8b \times -2 + 20b$
e $4st + 8s \times 2t$
f $9r - 3 \times 4r$
g $-2h \times 3 + 7 \times 7h$
h $-3q \times -6 + 4 \times -2q$
i $6u \times -5v + 9 \times 3uv$

5 Evaluate the following for $x = 3$ and $y = -2$

a $x \times 3y$
b $x + 4 \times 2x$
c $5x + 3y \times - x$

3.2.3 Index notation

Fact

While writing a^3 is referred to as index notation, writing out $a \times a \times a$ is called expanded form.

Another type of variable multiplication involves multiplication of like terms or repeated multiplication of the same variable. This type of expression can be simplified using index notation.

Remember that index notation keeps track of repeated multiplication, for example, $2^5 = 2 \times 2 \times 2 \times 2 \times 2$. You can use this concept for variables as well. For example, $a \times a \times a$ can be rewritten as a^3

Worked example 3.6

Write these in **expanded form:**

a h^4
b $3d^2$

Solution

a The index, the 4 in h^4, indicates how many times to multiply:

$h^4 = h \times h \times h \times h$

b Note that the index does not apply to the coefficient:

$3d^2 = 3 \times d \times d$

Worked example 3.7

Simplify:

a $b \times b \times b \times b \times b$ b $2f \times 6f$ c $3m \times 5m^3$

d $(2r)^2$ e $-3xy \times 4y \times y$

Solution

a We can find the index by counting the b

$b \times b \times b \times b \times b = b^5$

b Remember to multiply the numbers

$2f \times 6f = 12f^2$

c Solve harder problems by expanding first

$3m \times 5m^3 = 3m \times 5 \times m \times m \times m$

$= 15m^4$

d Remember: this means $2r$ multiplied by itself

$(2r)^2 = 2r \times 2r$

$= 4r^2$

e $-3xy \times 4y \times y = -12xy^3$

Challenge Qe

Practice questions 3.2.3

1 Write the following using index notation:

a $a \times a$ b $d \times d \times d \times d$

c $r \times r \times r \times r \times r$ d $m \times m \times m \times m \times m \times m \times m$

e $-e \times -e \times -e$ f $5 \times j \times j \times j$

2 Write the following in expanded form:

a z^3 b r^5 c t^2 d $(-p)^5$

3 Simplify:

a $2a^3 \times 5$ b $2w \times 7w$ c $4f \times 3f^2$ d $4k \times 3k \times k$

e $h^3 \times 9h$ f $4r \times 3r^4$ g $7m^2 \times 8m^4$

4 Simplify:

a $mn \times 3$ b $3h \times 2hk$ c $3c \times 8b$

d $4t \times 7st^2$ e $5n^2 \times mn^3$ f $2gh^2 \times 7gh$

5 Simplify:

a $-4k^2 \times 8k^2$ b $-6d \times -7d^5$ c $9b^2 \times -6b^3$

d $-v^2 \times -5v$ e $-4p^2 \times p^3 \times -2$

6 Simplify:

a $(5h)^2$ b $(-3a)^2$ c $(6j^2)^2$

d $(4k^3)^2$ e $(-2n^2)^2$

7 Simplify:

a $8jk \times -6k$ b $2m^2n \times -mn^2$ c $10pq^3 \times 3p^2$

d $-7ab^2 \times 8ab$ e $-9xy^3 \times -7xy$

3.2.4 Division of variables

Division of variables is similar to numerical division and operations with fractions. Remember how to reduce fractions to their simplest form.

For example, $\frac{15}{6}$ can be simplified by dividing the numerator and denominator of the fraction by their highest common factor 3, which gives $\frac{15 \div 3}{6 \div 3} = \frac{5}{2}$

We can do the same with $\frac{6x}{2x}$ by dividing the numerator and denominator of the fraction by the common factor 2, which gives $\frac{6x \div 2}{2x \div 2} = \frac{3x}{x}$

We can further simplify this by dividing both sides by x: $\frac{3x \div x}{x \div x} = \frac{3}{1} = 3$

Explore 3.5

Can you write these expressions in a simpler way?

a $4m \div 8$ **b** $6x \div 10x$

Worked example 3.8

Simplify:

a $18a^3 \div 12a$ **b** $d^2e^2 \div de^3$

Solution

a We can simplify the expression step by step:

$$18a^3 \div 12a = \frac{18a^3 \div 6}{12a \div 6}$$

$$= \frac{3a^3}{2a}$$

If we are not sure about the index, we expand it

$$= \frac{3 \times a \times a \times a}{2 \times a}$$

Divide the numerator and denominator by a

$$= \frac{3 \times \cancel{a} \times a \times a}{2 \times \cancel{a}}$$

$$= \frac{3 \times a \times a}{2}$$

Now write the answer in index form

$$= \frac{3a^2}{2}$$

b First write the expression as a fraction:

$$d^2e^2 \div de^3 = \frac{d^2e^2}{de^3}$$

Write in expanded form

$$= \frac{d \times d \times e \times e}{d \times e \times e \times e}$$

Divide by the common factors d and e

$$= \frac{\cancel{d} \times d \times \cancel{e} \times \cancel{e}}{\cancel{d} \times \cancel{e} \times \cancel{e} \times e}$$

$$= \frac{d}{e}$$

These steps will help us to make a plan for answering division problems.

- Convert the division into a fraction.
- Divide by the numerator and denominator by their highest common numerical factor.
- Expand indices if necessary.
- Divide the numerator and denominator by their highest common variable factor.

Although this might seem like a long process, once you have practised it and have gained experience, you will begin to combine steps, which will make the process smoother.

Practice questions 3.2.4

1 Simplify:

a $3f \div 6$ b $21 \div 7b$ c $6x \div 8$ d $20 \div 12h$

2 Simplify:

a $8n \div 20n$ b $12x \div 9x$ c $30m \div 18m$ d $15z \div 5z$

3 Simplify:

a $2k^3 \div 10k$ b $14t \div 21t^4$ c $8w^3 \div 14w^2$ d $6z^2 \div 15z^5$

4 Simplify:

a $jk \div k^4$ b $3mn^2 \div 18n$

c $15y^2z^2 \div 20yz^3$ d $42a^2b \div 28ab^4$

3.2.5 The order of operations with variables

The inquiry question for this section was 'What strategies can you use to simplify expressions?'. So far, you have learnt how to:

- combine like terms by addition and subtraction
- multiply terms, writing them in index notation if necessary
- divide terms.

Now it is time to combine everything you have learnt so far in this chapter. Here are some strategies on how to solve problems involving a combination of operations.

- Look at the problem carefully: look for any like terms or negative values.
- Follow the order of operations: remember that you need to multiply and divide before you add or subtract.
- Show all your steps: set out your working clearly so that you can look back and review it; this means you'll be less likely to make a mistake.
- Check your work: substituting values into your expression is one way of checking that your initial and final expressions are equivalent.

Worked example 3.9

Using the expression $\frac{x^2y^2}{xy} + 3xy - 2x \times y$

a Evaluate the expression for $x = 1$ and $y = -2$

b Simplify the expression.

c Evaluate your simplified expression using the same variables as in part **a** to check your work.

Solution

a We use brackets around the values we are substituting:

$$\frac{(1)^2(-2)^2}{(1)(-2)} + 3(1)(-2) - 2(1) \times (-2) = \frac{1 \times 4}{-2} + (-6) - (-4)$$

There are several negative signs to look out for

$$= \frac{4 \div 2}{-2 \div 2} + (-6) + 4$$

Simplify the fraction, taking into account the signs, and working from left to right

$$= -2 - 6 + 4$$

$$= -4$$

b Rewrite indices in expanded form:

$$\frac{x^2y^2}{xy} + 3xy - 2x \times y = \frac{x \times x \times y \times y}{x \times y} + 3 \times x \times y - 2 \times x \times y$$

$$= xy + 3xy - 2xy$$

Combine like terms

$$= 2xy$$

c When $x = 1$ and $y = -2$

$2xy = 2(1)(-2) = -4$

Reflect

Copy this table. Look back at Worked Example 3.9 and use it to complete the table.

	Substituting immediately	Simplifying first, then substituting
Advantage(s)		
Disadvantage(s)		

Which method do you prefer for solving more complicated substitution problems, substituting immediately, or simplifying first, then substituting? What about for simpler problems?

Worked example 3.10

Add these algebraic fractions and then simplify:

a $\frac{y}{3}+\frac{2y}{3}$ b $\frac{j}{2}+\frac{3j}{14}$ c $\frac{4u}{5}+\frac{2}{u}$

Solution

a The denominators are already equal $\frac{y}{3}+\frac{2y}{3}$

The numerators are like terms $=\frac{y+2y}{3}$

We can simplify this $=\frac{3y}{3}$

The common factor is 3 $=\frac{3y \div 3}{3 \div 3}$

A number divided by 1 is itself $=\frac{y}{1}=y$

b Here we need to make the denominators equal $\frac{j}{2}+\frac{3j}{14}=\frac{j \times 7}{2 \times 7}+\frac{3j}{14}$

$=\frac{7j}{14}+\frac{3j}{14}$

The numerator are like terms $=\frac{7j+3j}{14}$

This can be simplified further $=\frac{10j}{14}$

Divide by the common factor 2 $=\frac{10j \div 2}{14 \div 2}$

Leave the answer as a fraction $=\frac{5j}{7}$

c We need to change both denominators $\dfrac{4u}{5} + \dfrac{2}{u} = \dfrac{4u \times u}{5 \times u} + \dfrac{2 \times 5}{u \times 5}$

Now we can add the numerators $= \dfrac{4u^2}{5u} + \dfrac{10}{5u}$

This cannot be simplified further than $= \dfrac{4u^2 + 10}{5u}$

Practice questions 3.2.5

For Questions 1–3:

a Evaluate the expression using $a = -1$, $b = 3$ and $c = 4$

b Simplify the expression.

c Evaluate your simplified expression using the same variables to check your answer.

1 $5a \times 2ab - 3a^2b$

2 $-2a \times 2c + 5ac$

3 $\dfrac{6ac}{2a} + 4a - 3c$

4 Simplify:

a $4y \times 7x - 8xy$

b $\dfrac{4m^2 - m^2}{6m}$

c $9jk - 7k \times 8j$

d $\dfrac{3f}{2} + \dfrac{5f}{4}$

e $\dfrac{3f}{2} - \dfrac{5f}{4}$

5 Simplify:

a $5s \times 3t + 2t \times -4s$

b $\dfrac{8d^3 - 5d^3}{2d \times 3c}$

c $\dfrac{2a}{7} + \dfrac{b}{a}$

d $13uw - 2w \times 5u + 4uw$

e $\dfrac{7h^2}{6} + \dfrac{2h \times -4h}{3}$

Reminder Q4a

Listing variables in alphabetical order will make simplifying like terms easier.

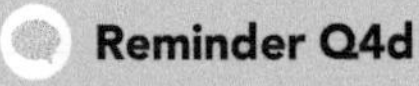

Reminder Q4d

Adding fractions requires writing them over a common denominator.

3.3 Algebraic expressions

3.3.1 Expanding brackets

There are three types of algebraic expression that you will meet.

- Monomials, which are expressions with one term such as $2x$
- Binomials, which are expressions with two terms such as $2x + 1$
- Trinomials, which have three terms such as $3x^2 + 2x + 1$

The shorter the expression, the easier it is to substitute values for variables. So, it is often a good idea to simplify expressions before you substitute as it is easier and, in most cases, faster.

There is one more type of operator that you still need to learn to simplify, namely, brackets. For example, you may have seen equations like this: $y = 2(x - 7)$. How do you simplify this?

The method used is called **distribution.** It removes the brackets by multiplying all the terms within the brackets by the term in front. This means you don't actually need to perform the operation in the brackets.

So, for $y = 2(x - 7)$, you multiply both the x and the 7 by the 2 in front of the bracket.

Hint

The minus sign belongs to the 7

A common way of showing this is $y = 2(x - 7)$ where each line represents multiplication of the terms it links.

This results in $y = 2x - 14$

Reminder

Equations are expressions that contain an equal sign.

You can apply this concept to expressions as well as **equations**.

Worked example 3.11

Expand and simplify:

a $5(1 - g)$ b $2b(3a + 2b)$ c $-4(6y - 3)$

d $x(2x + 4) - 5(3 - x)$ e $3m(2m - n + 4)$

Solution

a It is a good idea to draw the lines first; don't forget the minus sign:

$5(1 - g) = 5 - 5g$

b Remember the rules for multiplying variables.

It is important to list variables alphabetically:

$2b(3a + 2b) = 6ab + 4b^2$

c Minus signs belong to the number written after them. The product of two negative numbers is positive:

$-4(6y - 3) = -24y + 12$

d Expand each bracket separately.

Multiply each term in the second pair of brackets by -5 and combine like terms:

$x(2x + 4) - 5(3 - x) = 2x^2 + 4x - 15 + 5x$

x^2 and x are not like terms

$$= 2x^2 + 9x - 15$$

e We can apply the same principle to trinomials. There is a third term in the expansion:

$3m(2m - n + 4) = 6m^2 - 3mn + 12m$

Practice questions 3.3.1

1 Expand:

a $3(5 + 3x)$ b $5(4f - 2)$ c $h(h + 1)$

d $t(7t + 9)$ e $2b(1 - b)$ f $8g^2(7g + 9)$

2 Expand:

a $-(w + 2)$ b $-6(5 - x)$ c $-3(-2p + q)$

d $-t(8 - t)$ e $-2c(4c - 7d)$ f $-h^2(-4 - 5h)$

3 Expand and simplify:

a $2(x + 4) + 3(2x + 1)$ b $3(6 - 2a) - 4(a + 2)$

c $4j(k + 2) + k(3k - j)$ d $z(5 - z) - 9z^2(z - 7)$

e $8(m - 7) + 3m(9 - m)$ f $r^2(4r - 1) - (2 - r)$

4 Expand:

a $5(p + 2q - 3)$ b $9(u - 7v - 8)$ c $-(6 + pq - 9p)$

d $3z(2x - 2z + 1)$ e $-4(2x + x^2 - 8)$ f $-2j(j - k + 7)$

5 Expand:

 Challenge Q5

a $9f(6f - 2g + 3) + 8g(2f + g - 5)$

b $-7(8k - 4k^2 + 1) - 2k(k - 1)$

3.3.2 Factorising

Have you noticed that most mathematical operations have an inverse? For example, addition and subtraction are inverse operations, as are multiplication and division, or converting between decimals, fractions and percentages. Expanding brackets also has an inverse operation, called **factorising**.

Recall that a factor is a number that divides exactly into another number without leaving a remainder. When you simplified fractions earlier in this chapter, you divided the numerator and denominator by their highest common factor (HCF), for example $\frac{15x}{6} = \frac{15x \div 3}{6 \div 3} = \frac{5x}{2}$, since 3 is the HCF of 15 and 6

You can use a similar method with algebraic expressions. The result looks a lot like the initial problems you solved in the last section.

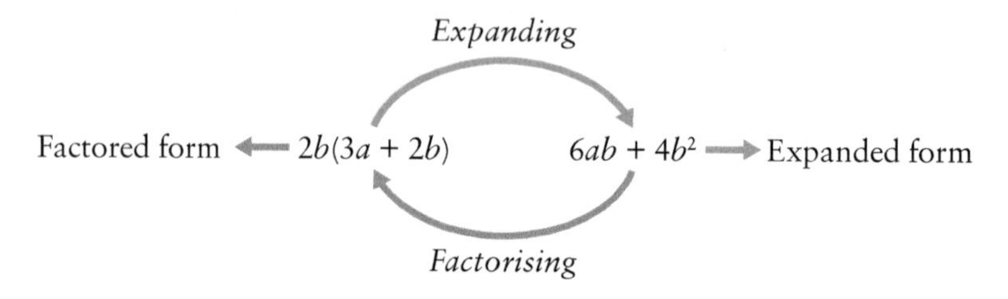

When factorising an expression, it is useful to break the terms into three parts:

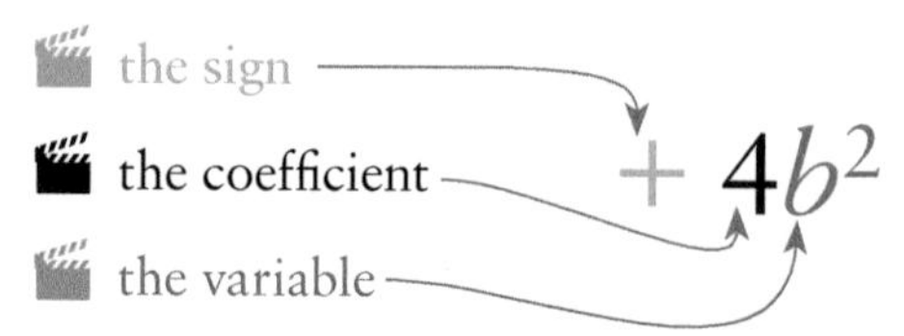

Considering whether all the terms have any of these parts in common will help you to factorise them more easily.

Hint

Factorising completely means there are no more common factors left between terms.

Worked example 3.12

Factorise these expressions completely:

a $3a - 12$

b $2g^2 + 8g$

c $-5xy - 15x$

d $6p + 3p^2 - 15pq$

Solution

a Only the coefficients have common factors. We can rewrite the terms as products of the HCF, which is 3

$3a - 12 = 3 \times a - 3 \times 4$

Finally, group the a and the 4 in brackets

$= 3(a - 4)$

b Both terms share the variable g and the factor 2. We can write this in expanded form to show the HCF:

$2g^2 + 8g = 2g \times g + 2g \times 4$

Write 2g in front of the bracket

$= 2g(g + 4)$

c The terms share a sign, a numerical and a variable factor:

$-5xy - 15x = -5x \times y - 5x \times 3$

The common factor −5 goes in front of the bracket, so the 3 in the bracket is positive

$= -5x(y + 3)$

d The terms have a numerical and a variable common factor:

$6p + 3p^2 - 15pq = 3p \times 2 + 3p \times p - 3p \times 5q$

$= 3p(2 + p - 5q)$

Practice questions 3.3.2

1 Identify which of the following have been factorised completely. For those that have not been, factorise completely.

a $5g + 20 = 5(g + 4)$

b $14t^3 + 13t^2 = t(14t^2 + 13t)$

c $-3x - 9y = -(3x + 9y)$

d $8q - 28pq = 4q(2 - 7p)$

e $12m - 6m^2 = 3m(4 - 2m)$

f $-6jk - 15k = 3k(-2j - 5)$

2 Factorise completely:

a $2w - 4z$

b $5u + 15u^2$

c $-8f - 7g$

d $9x + 3x^2$

e $5ab - 20b$

f $4mn + 8mn^2$

g $14r^3 - 35r^2$

h $3pq^3 - 6q^2$

3 Factorise each expression, then check your answers by expanding:

a $3j + 18jk$ b $-18m + 24mn$

c $42q^2 + 28qr$ d $72h - 16h^3$

e $21wx - 14w$ f $-36st + 28s^2$

4 Factorise:

a $-7j - 2k$ b $-2xy - 6z$

c $-mn - 3n$ d $-7g - g^2$

e $-8pq - 12q$ f $-40s^2t - 24s^3t^2$

5 Factorise these trinomials completely:

a $2j - 3jk + j^2$ b $6h^2 + 21h - 36$

c $-3 - 2r - 5s$ d $9d + 15cd + 3d^2$

e $14m^3 - 28m - 7mn$ f $-5uv - 6u^3 - 7u^2v^2$

3.3.3 Forming expressions

The algebra skills you have been learning in this chapter have many applications. One of the most useful is making generalised solutions of real-world problems. Consider the following dilemma.

Explore 3.6

Marta is ordering her school equipment for the year. She is planning to use her budget to buy as many pens and highlighters as she can. Her favourite brand of highlighters cost €1.29 each and her favourite pens cost €1.09 each.

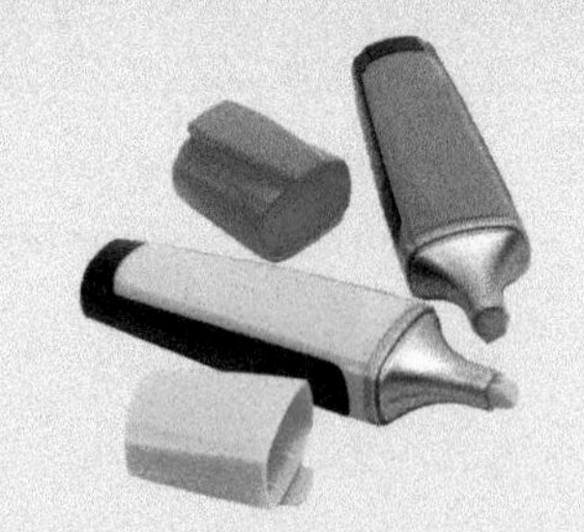

1 Can you find out how much it would cost to buy three highlighters and six pens?

2 What about h highlighters and p pens?

3 Marta has €10 to spend. Can you work out at least three different combinations of pens and highlighters that she could purchase without exceeding her budget?

4 Can you suggest to Marta a combination of pens and highlighters that allows her to buy the maximum possible number of items with her €10?

Hint

Perhaps Pólya's method can help here:

- Understand the problem
- Make a plan
- Carry out your plan
- Look back (check your work)

Thinking skills

Reflect

What other factors might Marta need to consider when making a final decision?

The expression you found in Question 2 is called a generalised expression. This means that it can be applied to many situations. In this case, it represents the number of each stationery item Marta purchases.

You will learn more about how to solve real-life problems like the one in the Explore feature in the next unit. Here are some other uses of the skills you have been practising.

Worked example 3.13

For each of the following shapes, write an expression for:

a the area

b the perimeter.

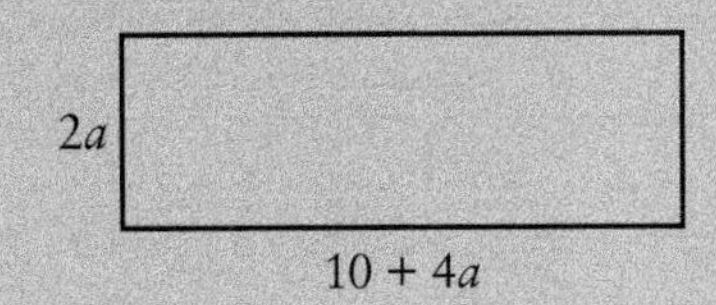

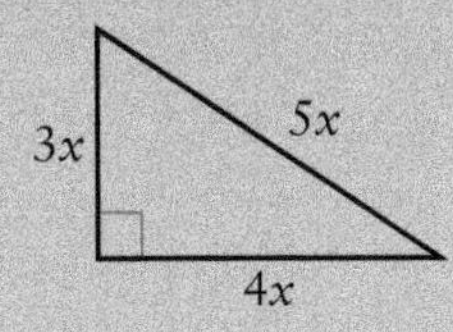

Solution

Understand the problem

Do you understand what you need to calculate and what each side length is?

a area of rectangle $= b \times h$

$= 2a(10 + 4a)$

$= \overset{\frown}{2a(10 + 4a)}$

$= 20a + 8a^2$

$= (20a + 8a^2)$ units^2

area of triangle $= \frac{1}{2} b \times h$

$= \frac{3x}{2} \times 4x$

$= 6x^2$ units^2

Reminder

The perimeter of a shape is the total length of all the sides.

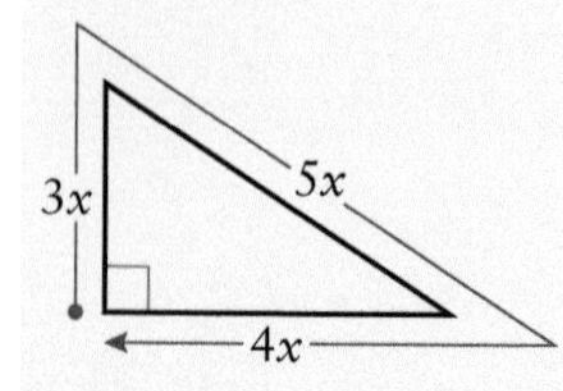

b perimeter of rectangle = sum of side lengths

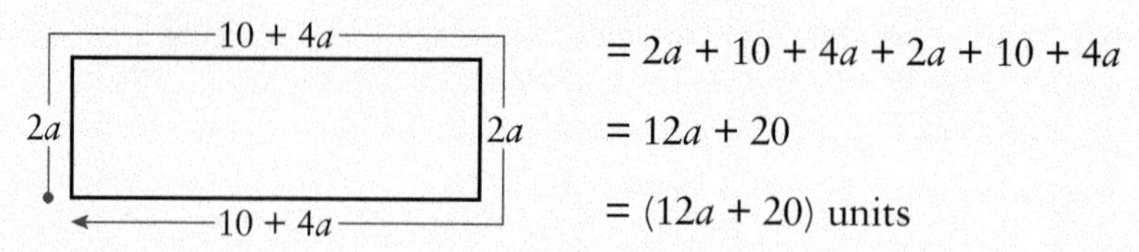

$= 2a + 10 + 4a + 2a + 10 + 4a$

$= 12a + 20$

$= (12a + 20)$ units

perimeter of triangle = sum of side lengths

$= 3x + 5x + 4x$

$= 12x$ units

Recall the conceptual question for this unit: How can expressions be used to model real life? Having studied this unit, you should now have a better idea of how to answer this.

Practice questions 3.3.3

1 Francois is planning his 13th birthday party. He is going to buy one cupcake, costing £1.90, for each of his guests and one for himself. He is also planning to buy some pizzas, which cost £8.50 each.

a Write an expression to describe how much he will spend if he buys c cupcakes and p pizzas.

b Francois wants to invite eight friends. He will order four pizzas. How much will he spend altogether on food?

2 Ms Pandit is buying art supplies for her class. They buy two art notebooks and one watercolour set for each student. There are 21 students in her class.

a Write an expression for the total cost of these art supplies, given that notebooks cost €n each and each watercolour sets costs €w.

b Calculate how much it would cost Ms Pandit if the notebooks cost €2 each and the watercolour sets costs €11 each.

For the shapes in Questions 3–5, write expressions for:

a the area

b the perimeter.

3

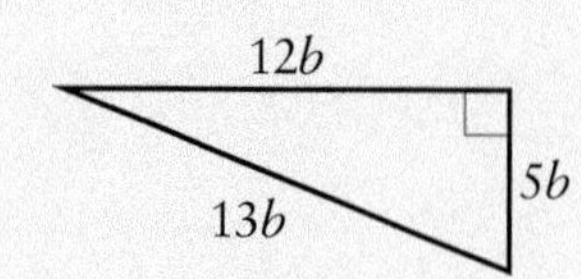

4

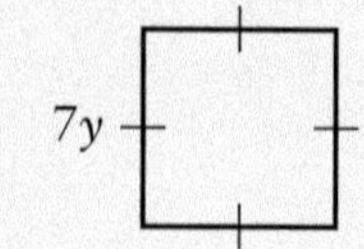

5

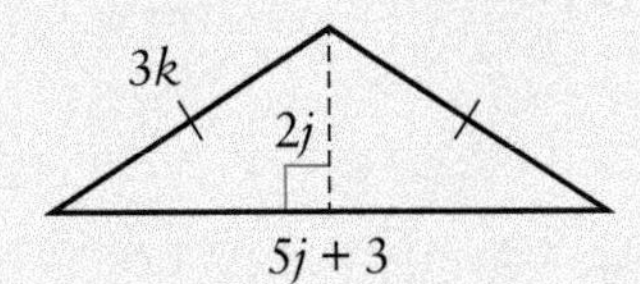

For Questions 6–9:

a Draw and label a diagram of the shape described.

b Find and simplify an expression for the area of the shape.

6 A triangle with height $3x$ cm and base length $7x$ cm.

7 A rectangle with height h metres height and length $4h$ metres.

8 A square with sides of length $10a$ units.

9 A right-angled triangle is stacked on top of a square, without overlapping, to form a trapezium. Both the triangle and square have a height and base length of x units.

Self-assessment

- I can identify number patterns involving addition and multiplication.
- I can classify number patterns as arithmetic or geometric.
- I can use a formula to fill in a table of values.
- I can substitute values into expressions in place of variables.
- I can identify like terms.
- I can add and subtract like terms.
- I can multiply terms together.
- I can square terms.
- I can use index notation with variables.
- I can divide one term by another.
- I can combine algebraic terms and simply expressions using the priority of operations.
- I can expand expressions involving brackets.
- I can factorise expressions using the highest common factor of all the terms.
- I can apply algebra skills to word problems.

Check your knowledge questions

1 Using the numbers 8, 4, 2, …

a describe the pattern in words

b write down the next three terms

c identify whether the pattern is geometric or arithmetic.

2

x	1	2	3	4	5
y					

a Copy and complete the table using the formula $y = 4x + 2$

b Write down a rule for the pattern formed by the y-values.

c Identify whether the pattern is geometric or arithmetic.

d Write down which part of the formula helps to predict the rule.

e Write down the next three terms in the pattern of y-coordinates.

3 Using the formula $n = 2f + 3$, draw the first four figures of this sequence in a visual pattern of dots, where f is the figure number and n is the number of dots in each figure.

4 Evaluate each expression when $a = 3$ and $b = -2$ and $c = 1$

a $2a + 3b$ b $b^2 - 4ac$ c $(4c)^2$

d $3b \times 2c + c \times -2b$ e $4a(2 - 3b + c)$

5 Simplify:

a $2a^2 + 2a - a^2 + a - 2$ b $8pqr + 4qr - 5pqr + 2qr$

c $(7h^2)^2$ d $-8m \times 9m^4$ e $(-5z)^2$

f $5f \times -2 + 13f$ g $5x + 7 \times 2x - 6x$ h $8jk^3 \div 12k$

i $(2w)^2 - w^2$ j $6s^3t \div 15st^4$ k $\frac{3y}{8} - \frac{y}{4}$

Reminder Q5c

Rewrite squaring as a product of terms, for example, $(7h^2)^2 = 7h^2 \times 7h^2$

6 Expand and simplify:

a $3(p + 2)$ b $8x(x - 1)$ c $2z(z^2 - z + 3)$

d $b(5 - b) + 2(b - 3)$ e $3 \times 5a - a(6 - a)$

7 Factorise completely:

a $10b^2 - 14ab$ b $6s^2t + 15st - 3t^3$

c $-30z^3 - 48z - 18z^2$

8 A rectangle has one side with length $3x$ and another with length $5x$

a Draw the rectangle and label the sides.

b Find and simplify an expression for the area of the rectangle.

c Find and simplify an expression for the perimeter of the rectangle.

4

Equations, inequalities and formulae

Equations, inequalities and formulae

KEY CONCEPT

Relationships

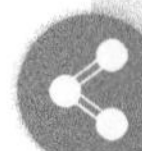

RELATED CONCEPTS

Patterns, Representation, Systems

GLOBAL CONTEXT

Scientific and technical innovation

Statement of inquiry

Representing complex scientific patterns with generalised systems allows us to express the relationships in equivalent forms.

Factual:

- How can we solve linear equations?

Conceptual:

- What makes a formula different from an equation?

Debatable:

- What is the best way of presenting the solution to an inequality?

Do you recall?

1 What is an equation?

2 Find the value of $5a$ when $a = 2$

3 Find the value of $4 - x$ when $x = -5$

4 Find the value of $r^2 + r$ when $r = -3$

5 Expand the index notation for

a 2^3

b $(-1)^5$

c 7^2

6 Use the correct order of operations to show how to find the value of

a $8 + 3^2$

b $5 - 2 \times 6$

c $(3 - 7) \div (-2)$

4.1 Solving equations

4.1.1 Inverse operations

Explore 4.1

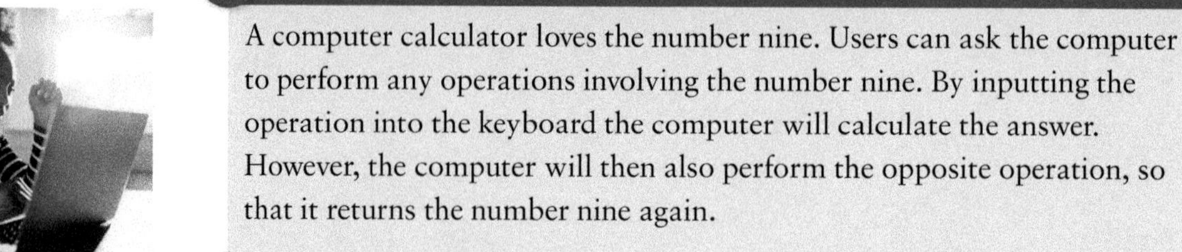

A computer calculator loves the number nine. Users can ask the computer to perform any operations involving the number nine. By inputting the operation into the keyboard the computer will calculate the answer. However, the computer will then also perform the opposite operation, so that it returns the number nine again.

Here are two examples.

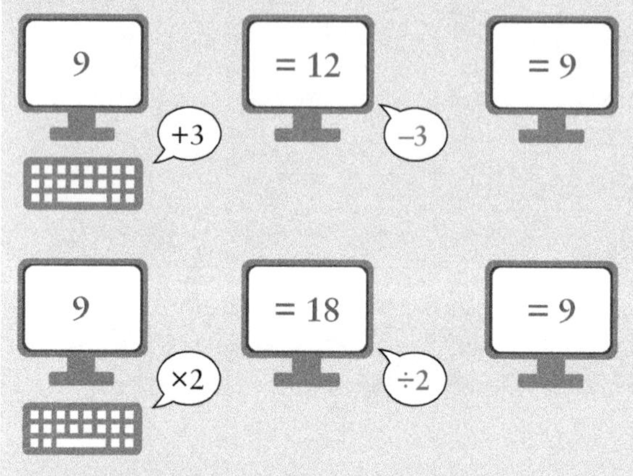

Note that while in the second example, the computer could have also chosen to subtract nine, it always chooses the opposite operation.

a Can you predict what operation the computer will perform given the following two inputs?

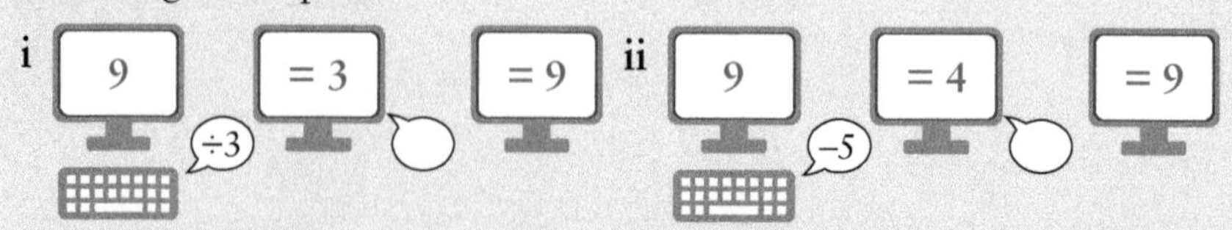

A user can also input multiple operations as in the example shown below:

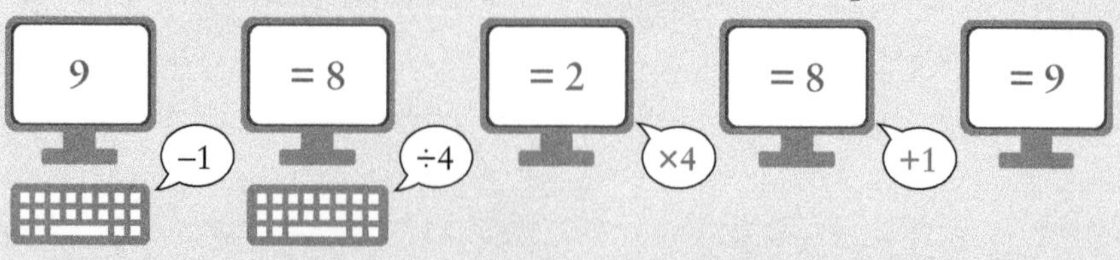

Note that not only are the operations opposite, but the order in which they are performed is opposite.

b Can you predict what operations are missing, and in which order?

i

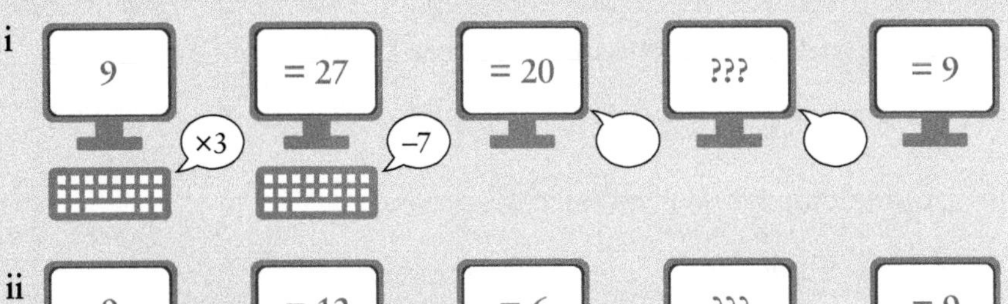

ii

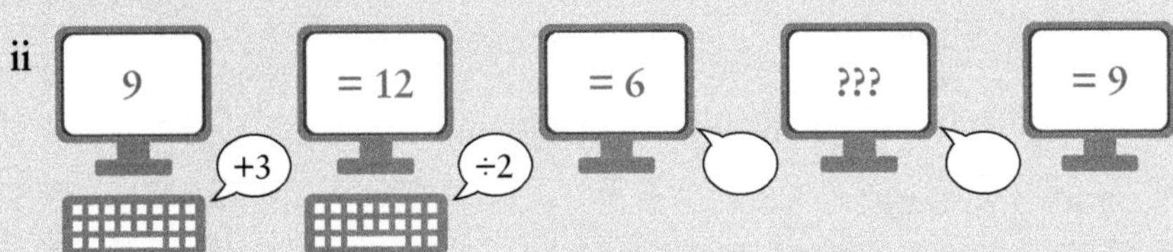

iii What would you predict to appear in place of ??? for parts **a** and **b**?

c What will the computer do in response to a user squaring nine?

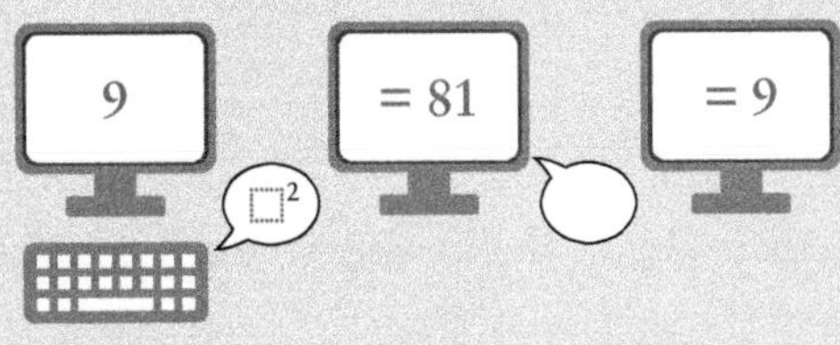

In Chapter 3, when we learnt about factorising expressions, we considered how factorising was the opposite of expanding. Just like factorising and expanding are opposites, each operation that can be performed in mathematics has an 'opposite'. For example, the opposite of addition is subtraction, the opposite of division is multiplication, and squaring is the opposite of taking the square root. In mathematics, we call these pairs **inverse operations**.

As you saw in the Explore activity, when performing pairs of inverse operations in order, the second operation will 'undo' the first operation. Understanding how this works can be very useful in algebra, so let's see how we can apply this.

Worked example 4.1

What operation goes in the blank to give the final answer?

a $1 \_\_\_\_ = 1 + z$

b $b \_\_\_\_ = -3b$

c $x + 7 \_\_\_\_ = x$

d $5a \_\_\_\_ = a$

e $2 - j \_\_\_\_ = 2$

f $\frac{4}{h} \_\_\_\_ = 4$

g $8d \_\_\_\_ = 8$

Hint

There may be other possible solutions than those given here. For example, part b could be "$-4b$" and d "$-4a$".

Solution

a The + z balances both sides of the equals sign:

$1 + z = 1 + z$

b Multiplying by -3 gives our final product:

$b \times -3 = -3b$

c Since we want to remove + 7 to leave only x, we subtract 7:

$7x + 7 - 7 = x$

d $5a$ means multiplication of a by 5. The inverse of multiplication by 5 is division by 5:

$5a \div 5 = a$

e $2 - j + j = 2$

f Since the original fraction divides 4 by h, the inverse is multiplication by h:

$\frac{4}{h} \times h = 4$

g $8d \div d = 8$

Note that in examples **c–g**, the solutions are the inverse operations of whatever needs to be removed to get to the final answer.

Let's see if we can repeat this process with multiple steps.

Worked example 4.2

What operations goes in the sets of blanks to give the final answer?

a i $1 + 2x$ _____ $= 2x$ ii $2x$ _____ $= x$

b i $4k - 5$ _____ $= 4k$ ii $4k$ _____ $= k$

c i $7 - 2y$ _____ $= -2y$ ii $-2y$ _____ $= y$

d i $6 + \frac{p}{3}$ _____ $= \frac{p}{3}$ ii $\frac{p}{3}$ _____ $= p$

e i $2(3 - m)$ _____ $= 3 - m$ ii $3 - m$ _____ $= -m$

iii $-m$ _____ $= m$

Solution

a In the first step, we just want to remove the 1:

i $1 + 2x - 1 = 2x$

$2x$ means multiplication by x, whose inverse is division:

ii $2x \div 2 = x$

b Here the −5, has the inverse +5:

i $4k - 5 + 5 = 4k$

ii $4k \div 4 = k$

c Careful, again the 7 here is positive, so the opposite is subtraction:

i $7 - 2y - 7 = -2y$

Since the term is negative, dividing by a negative gives a positive:

ii $-2y \div -2 = y$

d i $6 + \frac{p}{3} - 6 = \frac{p}{3}$

Here the variable p is divided by 3, so the inverse is multiplication

ii $\frac{p}{3} \times 3 = p$

e In this three-step problem, we start by dividing

i $2(3 - m) \div 2 = 3 - m$

ii $3 - m - 3 = -m$

iii $-m \times -1 = m$

We could also divide by −1, for part **iii**.

Hint

There may be other possible solutions than those given here. For example, part a ii could be "$-x$"; b ii could also be "$-3k$"; d ii "$+2p/3$"; e ii "$+3y$"; or e iii "$+2m$".

Reflect

Look back at Worked example 4.2

What patterns do you notice about the order of the two steps performed?

How is the three-step equation in part **e** different?

List the main ideas that the examples have demonstrated.

List one question you still have.

Hint

Look back at the Explore activity to check for similarities.

Practice questions 4.1.1

1 Write down the inverse operation to:

a subtracting 2

b multiplying by 7

c dividing by 3

d adding 15

e squaring a number.

2 Fill in these blanks with the appropriate operation.

a $2\square = 2 + c$

b $q\square = 7q$

c $x - 3\square = x$

d $-8k\square = k$

e $6 - u\square = 6$

f $\frac{w}{9}\square = w$

g $3f\square = 3$

3 Fill in these pairs of blanks with the appropriate operations.

a i $2 + 3x\square = 3x$ ii $3x\square = x$

b i $7g - 2\square = 7g$ ii $7g\square = g$

c i $-6 + \frac{j}{2}\square = \frac{j}{2}$ ii $\frac{j}{2}\square = j$

d i $8 - m\square = -m$ ii $-m\square = m$

e i $1 + \frac{b}{4}\square = \frac{b}{4}$ ii $\frac{b}{4}\square = b$

f i $2(1 + h)\square = (1 + h)$ ii $1 + h\square = h$

g i $3 - 2f\square = -2f$ ii $-2f\square = f$

h i $\frac{4 + w}{3}\square = 4 + w$ ii $4 + w\square = w$

Hint Q3h

A fraction has implied brackets on the top and bottom.

i i $5 - \frac{q}{6}\square = -\frac{q}{6}$ ii $-\frac{q}{6}\square = q$

j i $-3(c - 5)\square = (c - 5)$ ii $c - 5\square = c$

k i $\frac{z - 8}{7}\square = z - 8$ ii $z - 8\square = z$

For Questions 4–8:

a Using Question 2 as a model, fill in the blanks for the pairs of equations with the appropriate operations.

b Fill in the boxes with the appropriate terms. (Note: boxes in the same line should contain the same term.)

4 $2 + 3f$ ______ $= \square$ and $\square$ ______ $= f$

5 $4(m - 1)$ ______ $= \square$ and $\square$ ______ $= m$

6 $8 - j$ ______ $= \square$ and $\square$ ______ $= j$

7 $1 + \frac{k}{2}$ ______ $= \square$ and $\square$ ______ $= k$

8 $3 - 5n$ ______ $= \square$ and $\square$ ______ $= n$

9 Write down the operation(s) and order required to convert the given expression to the variable:

a $4a$

b $b + 1$

c $2c - 1$

d $\frac{d}{2}$

e $-\frac{e}{3}$

f $-9 - f$

g $7 - 2g$

h $8(h - 1)$

i $\frac{2 - i}{5}$

j $(6 - 2j)^2$

Challenge Q9j

4.1.2 One- and two-step equations

In the last section, we learnt about inverse operations. Inverse operations include:

- addition and subtraction
- multiplication and division
- squaring and taking the square root.

We also started to look at how we might apply these to help transform expressions, and the order in which we might need to perform operations to do this successfully. Let's look at this in more detail.

We know we can transform $4(3 + 2m)$ and reduce it to m through the following three steps:

$4(3 + 2m) \div \mathbf{4} = 3 - 2m$	$3 + 2m - \mathbf{3} = -2m$	$2m \div \mathbf{2} = m$
Step 1: divide by **4**	Step 2: subtract **3**	Step 3: divide by **2**

But how do we decide the order of these steps? To explain that, let us substitute the value $m = 1$ into the original expression:

$4(3 + \mathbf{2} \times (1))$	$= 4(\mathbf{3} + 2)$	$= \mathbf{4} \times (5)$	$= 20$
multiply by **2**	add **3**	multiply by **4**	

Explore 4.2

Patterns in reducing expressions

a Now it's your turn.

i Can you transform the expression $5x - 1$ to x, writing out the operation in each step?

ii Substitute $x = 3$ into $5x - 1$ and write out the operations in each step.

b Compare the steps required to reduce the expression to x and the steps required for substitution.

i What do you notice about the order of the numbers used in the operations?

ii What do you notice about the actual operations performed?

c Do your own calculations in Question a reflect what your observations in Question b were? Explain your reasoning.

d The process to move $5x - 1$ to x is the same as we will use to solve equations in this chapter.

What generalisation can we make about the order we use to solve equations?

As you also noticed in the Explore activity, when we substitute a value into an expression we follow the order of operations. However, when we reduce an expression, we reverse that order! Reversing the order of operations is a key strategy in solving equations.

Worked example 4.3

Find the value of the variable which satisfies the given equation.

a $8f = 56$ b $b - 4 = -17$ c $4a - 1 = 23$

d $\frac{x}{3} + 7 = 15$ e $2 - h = -21$

Hint

To satisfy an equation, the value of the variable must take a value that makes both sides equal.

Solution

a This can be solved by inspection, but let's practice using inverse operations:

$8f = 56$

Dividing both sides by eight will maintain the balance in the equation:

$\frac{8f}{8} = \frac{56}{8}$

$f = 7$

Looking back, we see this result fits with our initial inspection:

$8(7) = 56$ ✓

b Again, we can use the inverse of the −4, performing the operation on both sides keeps them equal:

$b - 4 = -17$

$\quad +4 \quad +4$

$b = -13$

Looking back, let's substitute this into the original equation:

$-13 - 4 = -17$

Both sides are equal, confirming our result, above, of $b = -13$:

$-17 = -17$ ✓

c For this equation we will need two steps. As with the Explore activity, we will start 'farthest' from a:

$4a - 1 = 23$

$\quad +1 \quad +1$

Next, we divide to remove the multiplication with 4:

$$\frac{4a}{4} = \frac{24}{4}$$

$a = 6$

Again, we can check our answer by substitution:

$4(6) - 1 = 23$

$24 - 1 = 23$ ✓

d $\frac{x}{3} + 7 = 15$

$\quad -7 \quad -7$

$\frac{x}{3} = 8$

Multiply by 3 as this is the inverse of division by 3:

$\frac{x}{3} \times 3 = 8 \times 3$

$x = 24$

It is always important to look back:

$\frac{24}{3} + 7 = 15$

$8 + 7 = 15$ ✓

e This problem is harder than it looks. Remember that even after subtracting 2, the h is still negative. Solve this last step by multiplying by −1.

$2 - h = -21$

$-2 \quad -2$

$-h = -23$

$h = 23$

Again, we can look back and check. Checking our answer gives confidence in our skill:

$2 - 23 = -21$ ✓

Practice questions 4.1.2

In Questions 1–5, solve the equations to find the value of the variables, then check you answer by substituting back into the original equation:

1 $2 + p = -9$

2 $7 + 3q = 1$

3 $9 - r = 13$

4 $s - 6 = -10$

5 $15 - 2t = 3$

6 Solve:

a $a + 3 = 4$

b $3 + b = -1$

c $c - 5 = 2$

d $9d = 72$

e $\frac{e}{3} = 7$

f $-2f = -102$

g $g - 8 = -11$

h $\frac{h}{-2} = 6$

7 Solve:

a $2u + 3 = 7$

b $3v + 1 = -5$

c $8 - w = -2$

d $2 - x = 5$

e $\frac{y}{3} + 1 = 4$

f $5 - \frac{z}{3} = 1$

g $\frac{j-1}{2} = 3$

h $4 + 3k = -11$

i $3 - \frac{m}{2} = 10$

j $-9 - 3n = 3$

k $5(9 + s) = -15$

l $(r + 2)^2 = 25$

Challenge Q7l

Reflect

Look back at Practice questions 4.1.2 and check your answers using the solutions provided at the back of the book.

What type of questions were the easiest?

What type of questions were the most challenging? Give one reason why.

What type of questions were the easiest to make mistakes on?

Use the word 'because' in your reflection.

Explain why checking your work might be important.

Give one example to support your thinking in your response.

4.1.3 More types of equations

Explore 4.3

A mathematician thinks of a number n. They multiply the number by 9 then subtracts 6. They get the same result by finding the sum of 19 and the square of n. Can you find n?

Reflect

What vocabulary is necessary to complete the Explore activity?

What are two advantages of breaking down the question into separate parts?

Hopefully, at this point you are on your way to answering: How can we solve linear equations? Let's put the skills we have been learning so far to the test!

Worked example 4.4

Solve for x:

a $3(7 - x) = 18$

b $\frac{2x - 5}{7} = 3$

c $7x + 3 = 3x + 11$

d $4x + 7 = 22 - x$

Solution

a $3(7 - x) = 18$

We have seen equations like this before, and it will require three steps.

Divide first:

$$\frac{3(7 - x)}{3} = \frac{18}{3}$$

$7 - x = 6$

$-7 \quad -7$ At this stage, remember to subtract.

$-x = -1$

$x = 1$

We can look back, and check our work:

$3(7 - 1) = 18$

$3(6) = 18$ ✓

b $\frac{2x - 5}{7} = 3$

Multiply first:

$$\left(\frac{2x - 5}{7}\right) \times 7 = 3 \times 7$$

$2x - 5 = 21$

$+5 \quad +5$

$$\frac{2x}{2} = \frac{26}{2}$$

$x = 13$

c $7x + 3 = 3x + 11$

To solve an equation with variables on both sides, combine them first. Apply the inverse operation, as usual, then solve:

$7x + 3 = 3x + 11$

$-3x \quad -3x$ Subtract $3x$ from both sides and combine like terms.

$4x + 3 = 11$

$-3 \quad -3$

$$\frac{4x}{4} = \frac{8}{4}$$

$x = 2$

Again, try looking back.

$7(2) + 3 = 3(2) + 11$

$14 + 3 = 6 + 11$

$17 = 17$ ✓

d $4x + 7 = 22 - x$

Try making a plan for which variable to move.

$4x + 7 = 22 - x$

$+x \qquad\qquad +x$

$5x + 7 = 22$

$\quad -7 \quad -7$

$\frac{5x}{5} = \frac{15}{5}$

$x = 3$

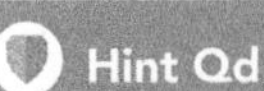

It can help to move the variable which leads to a positive result, as this avoids including more negatives in your solution than necessary.

Practice questions 4.1.3

For Questions 1–5, solve for x, then check your answers by substituting into the original equations.

1 $4(5 + 2x) = 12$

2 $5x - 3 = 1 + 3x$

3 $\frac{3x + 1}{2} = 5$

4 $10 - x = 9 - 2x$

5 $2(7 - 4x) = 10$

6 Solve these equations:

a $4a - 1 = 3a + 2$

b $7e + 1 = 3e - 3$

c $\frac{12 - b}{2} = 4$

d $2(5 - c) = 12$

e $5\left(\frac{d}{2} - 7\right) = -20$

For Questions 7–14, form and solve an equation for each statement.

7 A mathematician thinks of a number x, multiplies it by 3, then subtracts 2. He can get the result by subtracting 6 from the number x. Find x.

Reminder Q8

Draw a diagram to help you understand the problem!

8 An equilateral triangle has sides of length s and a perimeter of 21 mm. Find the length of each side.

9 Tomas has £p in cash and Alice has £13 more than him. Together they have £29. How much cash does Tomas have?

10 A French class has f students while a Spanish class has twice as many. The Spanish class has nine more students than the French class. Find the value of f.

Hint Q11

Even if a question does not suggest a variable, make sure to add one where necessary.

11 A rectangle has perimeter 36 cm. If the height is half the width, find the height of the rectangle.

12 There are two advisory classes in Year 2, 7A and 7B. There are five more students in 7A than in 7B, and there are a total of 39 students in all of Year 2. How many students are in 7B?

13 Sean's sister is half his age plus three years. She is also four years younger than him. How old is Sean?

Challenge Q14

14 Quinn, Sarah, and Juan play DnD and all have their own sets of dice. Quinn owns eight more dice than Juan, and Sarah has twice as many as Juan. In total they have 56 dice. How many dice does Juan own?

4.2 Inequalities

4.2.1 Interpreting and graphing inequalities

So far in this chapter, we have learnt all about how to solve different type of equations. But what about inequalities?

You may already be familiar with inequalities, but let's review the basic symbols:

Meaning	Greater than	Less than	Greater than or equal to	Less than or equal to	Not equal to
Symbol	$>$	$<$	$\geqslant$	$\leqslant$	$\neq$

Hint

A trick to remember the difference is: the larger, open side of the symbol is always closer to the larger quantity, for example $100 > 2$ or $5 \leqslant 24$.

With each of these inequalities, we can reverse the order of the symbol as long as we reverse the quantities too. For example, $100 > 2$ can also be written as $2 < 100$.

We can also begin to involve algebra, so $x > 5$ can also be written as $5 < x$, although convention is usually to write the variable first. Even so, what does $x > 5$ mean?

Reminder

Convention is when mathematicians agree to do something a particular way, usually to make things easier.

Explore 4.4

Consider the inequality: $x > 5$

a Can you write down its meaning in words?

b Can you list all of the numbers that can take the place of x?

You might be able to already see that listing all the numbers might take a long time.

c Think of the smallest number that might replace x. Why might this be a difficult number to identify?

d Can you think of an alternative way of writing the solution?

e How might we differentiate between $x > 5$ and $x \geqslant 5$?

Generally, when graphing an inequality on a number line there are three steps.

1. Draw and label the number line.
2. Identify and label whether the value in the inequality is included or not.
3. Identify and label which side(s) of the point is included.

As an example, let's look at the steps to graph three different inequalities:

Inequality	$x > 7$	$x \geqslant 7$	$x \neq 7$
1. Drawing and labelling the number line.	First, draw number line and label it with your variable and value: 7 x You might notice this is common for all three inequalities, so this step is the same.		
2. Identify and label whether the value in the inequality is included or not.	In this case 7 is not included so we use an open circle to show it is not part of the answer: 7 x	In this case 7 is included, so we use a filled in circle to show it is not part of the answer: 7 x	Again 7 is not included so an open circle is used: 7 x
3. Identify and label which side(s) of the point is included.	Greater than 7 indicate values on the right-hand side of the number line, so we shade this section: 7 x	As this is also greater than, we again shade the right-hand side: 7 x	All numbers that are not 7 are included, so we shade both sides of the number line (with the exception of 7 of course): 7 x

Worked example 4.5

Graph these inequalities on a number line:

a $y \geqslant 4$ **b** $c \neq 0$ **c** $k < -7$

Solution

a $y \geqslant 4$

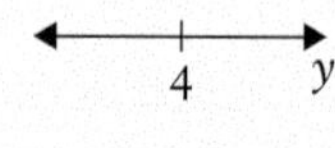

First draw and label the number line.

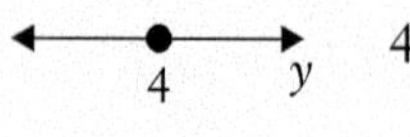

4 is included in this inequality.

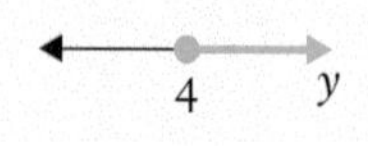

Shade the right-hand side as it is greater than 4

b $c \neq 0$

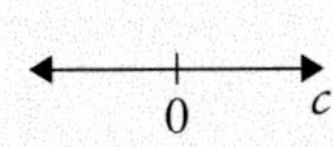

0 c

0 c Remember, everything except 0 is included.

c $k < -7$

-7 k

-7 k

-7 k Shade only the left-hand side as it is strictly less than -7

Practice questions 4.2.1

1 State whether each of these inequalities is true or false:

a $2 > 7$

b $4.5 \leqslant 5$

c $0 \geqslant -1$

d $2 \neq 0.2$

e $2 < -2$

f $0.3 \neq \frac{1}{3}$

g $-5 - 2 > -7$

h $-5 - 2 \geqslant -7$

2 Match these number lines to the correct inequalities:

a 3 x

b 3 x

c 3 x

d 3 x

e 3 x

i $x > 3$ ii $x \geqslant 3$ iii $x \neq 3$

iv $x < 3$ v $x \leqslant 3$

3 Represent these inequalities on a number line:

a $y > 4$ b $b \neq 5$ c $c \geqslant 8$ d $x < -2$

e $z \neq 0$ f $t \leqslant -3$ g $a < 1$

4.2.2 Solving inequalities

Explore 4.5

a Can you show how we find that the solution to $4x + 8 = 20$ is $x = 3$?

b Can you predict which numbers satisfy $4x + 8 < 20$?

So now we know what inequalities represent, but as you can imagine, they can become more complex. As you saw in the Explore activity, inequalities start to look similar to the equations we were solving earlier in this chapter. For example:

$30 - 2a > 24$

It might seem difficult to imagine what this represents; however, we can simplify the inequality, using the same methods that we used to solve equations. There is only one specific instance where a different technique needs to be used. To figure out what this is, let's try solving our example:

$30 - 2a > 24$

$-30 \qquad -30$

$-2a > -6$

At this stage, if this were an equals sign we would divide by −2 and we would have our answer.

But let's check what happens if we do this with an inequality:

$$\frac{-2a}{-2} > \frac{-6}{-2}$$

$$a > 3$$

Choose any number greater than 3 to check.

$a = 5, \quad 30 - 2(5) > 24$

$30 - 10 > 24$ ✗

So why doesn't this work?

Because for inequalities, any time you divide or multiply by a negative value, you must reverse the direction of the inequality symbol:

$-2a > -6$

$$\frac{-2a}{-2} < \frac{-6}{-2}$$

$$a < 3$$

We can also graph this to visualise it:

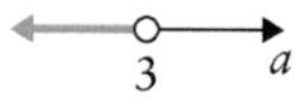

This can also help us choose a number less than 3 to check our work:

$a = 1, \quad 30 - 2(1) > 24$

$30 - 2 > 24$

$28 > 24$ ✓

Worked example 4.6

Solve and graph these inequalities:

a $4 + 5y < 19$ b $7 - x \neq 10$ c $6 - 4z \geqslant -22$

Solution

a We solve as we would a normal equation:

$4 + 5y < 19$

$-4 \qquad -4$

$\frac{5y}{5} < \frac{15}{5}$

$y < 3$

We remember the steps to graph our inequality:

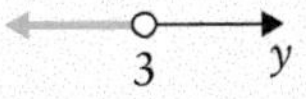

b $7 - x \neq 10$

$-7 \qquad -7$

$-x \neq 3$

Since it is a 'not equal to' symbol, we do not need to worry about reversing the sign, so:

$x \neq -3$

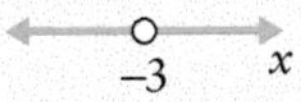

c $6 - 4z \geqslant -22$

$-6 \qquad -6$

$-4z \geqslant -28$

Since we divide by a negative here, reverse the inequality sign:

$\frac{-4z}{-4} \leqslant \frac{-28}{-4}$

$z \leqslant 7$

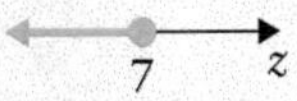

Now that you have some experience with inequalities, try to answer this section's debatable question: What is the best way of presenting the solution to an inequality?

 Thinking skills

Do you think the graph is necessary for understanding? Explain your reasoning.

Practice questions 4.2.2

For Questions 1–5, solve the inequalities, then perform a check to confirm your answer.

1 $x + 4 > 7$ 2 $g - 3 \neq -1$ 3 $7 - b \leqslant 12$

4 $2x + 6 < -8$ 5 $4 - 3a \geqslant 10$

6 Solve these inequalities:

a $2 + k \neq 9$ b $f - 9 > -2$ c $8 + b \leqslant 13$

d $2m - 5 < -7$ e $4 + 3j \geqslant -8$

7 Solve these inequalities:

a $1 - 5n < -9$ b $5 - 3f \geqslant -7$ c $4 - r \neq 6$

d $2 - 2x > 14$ e $\frac{-a}{5} \leqslant 1$

8 Solve these inequalities, then graph their solutions on a number line:

a $2a + 3 > 7$ b $3b + 1 \leqslant -5$ c $\frac{c}{3} + 2 \neq 4$

d $9 - d > -1$ e $7 - e \geqslant 10$ f $3 - \frac{f}{3} < -1$

g $\frac{g - 1}{3} \neq 4$ h $7 - 3h \geqslant -8$ i $4 - \frac{i}{2} > 11$

 Challenge Q8k

j $-6 - 2j \neq 2$ k $3(2 - 4k) > 6$

4.3 Formulae

 **Connections**

Physics is a field in science which studies patterns in the physical world. Biology and chemistry do the same for the patterns in the natural and chemical world, respectively. Perhaps this is the reason Galileo said: 'Mathematics is the key and door to the sciences'.

4.3.1 Using formulae

Consider the equation $E = mc^2$. This is a very famous **formula**, first developed by the legendary scientist Albert Einstein to help describe the relationship between energy, represented by the E, and mass, represented by m. The symbol, c, stands for the speed of light, a constant value at approximately 300 million metres per second.

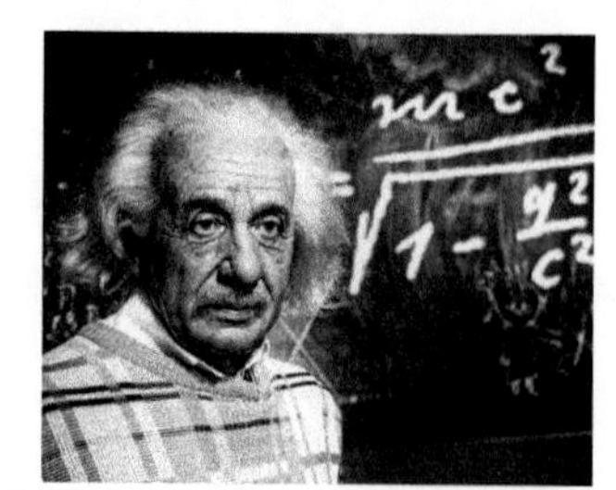

While learning more about algebra, it is important to keep in mind how we will one day apply the knowledge we are learning. The use of formulae, particularly but not limited to those we find in the sciences, is a very useful application of the skills we are practicing in this chapter. But what is a **formula**? As you might be able to tell from the example of $E = mc^2$, a formula is a type of equation that expresses a specific relationship. This means that formulae are specific types of equations.

Explore 4.6

There are two common measures for temperature: Celcius denoted with °C, and Fahrenheight, °F.

a Can you write a formula that helps us find °C when we know °F?

b Try to apply your formula on facts you know about these measures.

c Can you do the reverse, that is find °F when you know °C?

Another example of a formula can be adapted from the last Explore activity of Chapter 3, where you came up with an expression to calculate the cost for p pens and h highlighters for Marta: $1.29h + 1.09p$. We can turn this into a formula by adding C to represent total cost: $C = 1.29h + 1.09p$. This formula tells us the relationship between the number of pens and highlighters purchased, and the total cost. We can use it to calculate different combinations of highlighters and pens.

Worked example 4.7

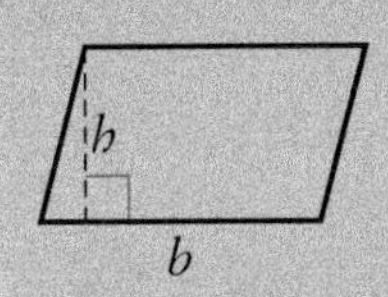

The area of a parallelogram, A, can be calculated using the formula $A = b \times h$, where b is the base length and h is the height.

Find the area of a parallelogram with:

a a base of 10 cm and a height of 5 cm

b a base of 2 m and height of 3 m

c a base of 1.2 cm and height of 0.6 cm.

Solution

a Substitute the given values into the formula:

$A = (10\,\text{cm}) \times (5\,\text{cm})$

$A = 50\,\text{cm}^2$

Note that the units are also squared.

b $A = (2\,\text{m}) \times (3\,\text{m})$

$A = 6\,\text{m}^2$

c We remember our decimal multiplication:

$A = (1.2\,\text{cm}) \times (0.6\,\text{cm})$

$A = 0.72\,\text{cm}^2$

In the example above you can see how useful having a formula to show the relationship between variables can be. But what if, in this example, we were not interested in calculating the area, but rather, the height? This is where being able to rearrange a formula comes in handy. We can use what we have learnt about algebraic equations and inequalities, and apply it to rearrange formulae.

Let's go back to Marta and her stationery purchases.

Say she was interested in knowing how many pens she could buy, given a certain cost and number of highlighters.

She could make her calculation easier by making p the subject of the formula.

Starting with her original equation, she will try to isolate p:

$$C = 1.29h + 1.09p$$

She will subtract the highlighter portion of the formula:

$$C = 1.29h + 1.09p$$

$$-1.29h \quad -1.29h$$

$$C - 1.29h = 1.09p$$

Finally, to eliminate the coefficient of p, she will divide by 1.09:

$$\frac{C - 1.29h}{1.09} = \frac{1.09p}{1.09}$$

$$\frac{C - 1.29h}{1.09} = p$$

This cannot be simplified, so we simply write the formulae with the subject on the left-hand side of the equation:

$$p = \frac{C - 1.29h}{1.09}$$

Now the formula can use the cost and number of highlighters to give Marta the number of pens she can purchase.

Worked example 4.8

a Rearrange the formula $A = b \times h$, to make h the subject.

b Hence find the height of the parallelogram where:

i The area is $56\,\text{cm}^2$ and the base length is $7\,\text{cm}$.

ii The area is $1.6\,\text{m}^2$ and the base length is $0.4\,\text{m}$.

iii The area is $5\,\text{cm}^2$ and the base length is $2\,\text{cm}$.

Solution

a First make height the subject of the formula:

$A = b \times h$

By dividing by b, the b will cancel from the left-hand side:

$$\frac{A}{b} = \frac{b \times h}{b}$$

$$\frac{A}{b} = h$$

For simplicity let us reverse the formula, ready to use for the problems:

$$h = \frac{A}{b}$$

b i The area is $56\,\text{cm}^2$ and the base length is 7cm:
(note that the units also cancel)

$$h = \frac{56\,\text{cm}^2}{7\,\text{cm}}$$

$h = 8\,\text{cm}$

ii The area is $1.6\,\text{m}^2$ and the base length is $0.4\,\text{m}$:

$$h = \frac{1.6\,\text{m}^2}{0.4\,\text{m}}$$

$h = 4\,\text{m}$

iii The area is $5\,\text{cm}^2$ and the base length is $2\,\text{cm}$:

$$h = \frac{5\,\text{cm}^2}{2\,\text{cm}}$$

We can rewrite this as a decimal:

$h = 2.5\,\text{cm}$

Reflect

Look back at Worked example 4.8. Write down one advantage of rearranging the formula before beginning the problems. Write down a possible disadvantage. Discuss your answers in small groups or with your class.

Practice questions 4.3.1

For Questions 1–3 use the fact that the perimeter of a rectangle, P, can be calculated using the formula $P = 2b + 2h$, where b is the base length and h is the height.

1 Find the perimeter of a rectangle with:

a a base of 8 cm and a height of 5 cm

b a base of 3 m and height of 4 m

c a base of 1.1 cm and height of 0.8 cm.

2 Find the base length of a rectangle with:

a a perimeter of 10 cm and height of 2 cm

b a perimeter of 4 km and height of 1 km

c a perimeter of 5.2 m and height of 1.2 m.

3 Find the height of a rectangle with:

a a perimeter of 22 cm and base length of 3 cm

b a perimeter of 6 m and base length of 1 m

c a perimeter of 7.6 m and base length of 2.5 m.

4 Make x the subject of the formula:

a $a = x + 3$ b $b = 2x$ c $c = 3x - 1$

d $d = \frac{x}{4}$ e $e = 9 - x$ f $f = \frac{x - 2}{5}$

g $g = 2x + 2y$ h $h = 3(x - 1)$ i $i = 1 - \frac{x}{2}$

Challenge Q4k

j $j = 3xy - 1$ k $k = x^2$

5 Use the formula $T = 3m - n$, to find the value of m when:

a $T = 1, n = 2$ b $T = 20, n = 1$ c $T = -4, n = -2$

d $T = 7, n = -1$ e $T = 3, n = -15$

6 Use the following formulae to find the value of y, when $a = 2$ and $b = -1$:

a $y = 2a$ b $y = a - b$ c $y = b^2$

d $y = \frac{a}{b}$ e $a = 2y + 2b$ f $a = y - b$

g $b = \frac{y}{3} + 1$ h $b = 4a - y$ i $a = 2(y - 3)$

4.3.2 Constructing formulae

It is not only important to be able to apply and rearrange formulae, it is also fundamental to be able to construct your own formulae from a given real-life situation.

Connections

How is this related to Chapter 3: Patterns and Algebra?

Explore 4.7

A famous mathematical riddle goes like this:

'A mathematician goes into a dog park and notices that between the owners and the dogs there are 16 heads and 50 legs. Can the mathematician calculate how many humans and how many dogs there are without counting anything else?'

Hint

Assume that all humans in this park have two legs and all dogs have four legs.

Worked example 4.9

For the following shapes, write down a formula to help you express the perimeter of:

a a triangle with sides lengths a, b and c

b a square with side length x

c a parallelogram with sides s and b.

Solution

a These problems require visualisation.

The best way to visualise questions is to draw a diagram representing them:

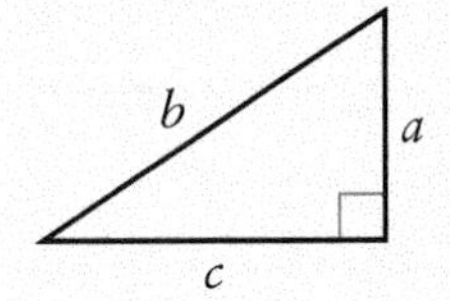

Construct a drawing of the situation:

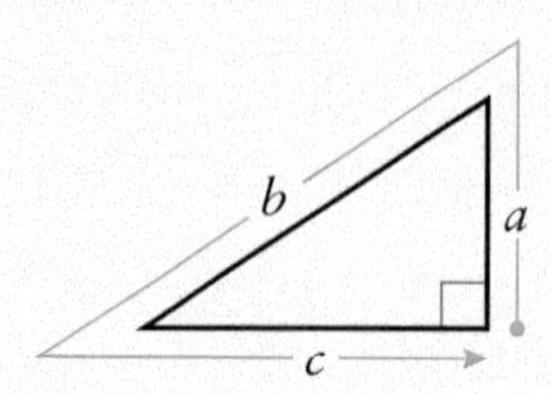

Perimeter measures the distance around the shape:

$P = a + b + c$

Looking back, this agrees with what we know about perimeter. We can also check our work with an example such as

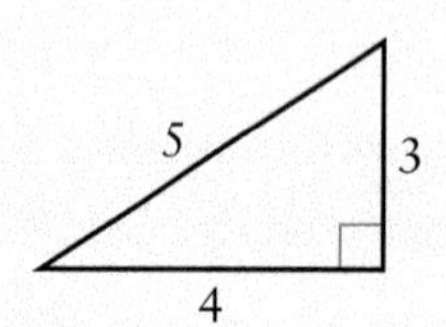

We can see our formula above would give us the correct perimeter, namely

$3 + 5 + 4 = 12$ units

b A square with side length x:

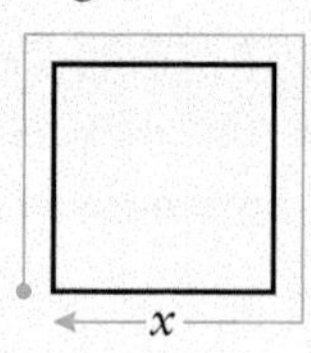

Remember to combine like terms:

$P = x + x + x + x$

$P = 4x$

c A parallelogram with sides s and b:

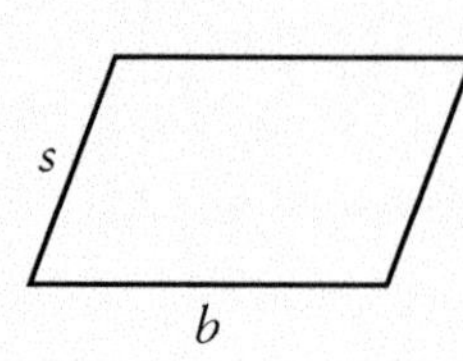

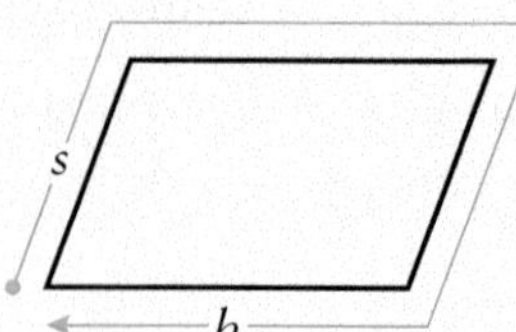

$P = s + b + s + b$

$P = 2s + 2b$

Practice questions 4.3.2

1 Write a formula to express the area of:

a a triangle with height h and base length b

b a square with side length x

c a rectangle with sides length r and t.

Reminder Q1

It is almost always helpful to draw and label a diagram to help understand the problem.

2 An isosceles triangle has two sides of length a and the third with length b.

a Draw and label a diagram of the shape described.

b Write a formula to express the perimeter of the shape described.

c Calculate the perimeter given the two equal sides have length 12 cm and the shortest side has length 7 cm.

3 A regular hexagon (6 equal sides) has sides of length x.

a Sketch and label a diagram of the hexagon.

b Write a formula for the perimeter of the hexagon.

c Calculate the perimeter when each side is 9 cm long.

d Calculate the side length if the perimeter were 6.6 m long.

4 To find the average, x, of two numbers, a and b, you need to divide their sum by two.

a Write a formula to express the average of a and b.

b Find the average of 3 and 11

c Find the average of 23 and 18

d Make a the subject of your formula.

e If the average of a and 18 is 24, find a.

5 Luis and Livia are organising a bake sale. Luis will bake cookies that will sell for €1 each and Livia will bake brownies for €2 each.

a Write a formula to express the total amount raised, T, based on the number of cookies, c, and number of brownies, b, sold.

b Calculate how much money they would raise with 22 cookies and 15 brownies.

c They raise €63 by selling 18 brownies and an unknown number of cookies

i Write down a rearranged formula to help you solve for the number of cookies

ii How many cookies did they sell?

Hint Q5c

Note what the subject of the equation is when you want to solve for a specific variable.

6 Yasmin is bringing apple spritzer to a class party. To make it, she needs apple juice and sparkling water. Each bottle of sparkling water is £0.89 and each box of apple juice is £1.49

a Write a formula to express the cost, C, based on the number of bottles of sparkling water, s, and apple juice boxes, a.

b Using your formula calculate the cost of three bottles of sparkling water and three boxes of apple juice.

c Using your formula calculate the cost of four bottles of sparkling water and six boxes of apple juice.

d Make s the subject of the formula.

e Use the formula in part **c** to calculate the number of bottles of sparkling water she could purchase with £10, if she needs five boxes of apple juice.

f Why might you want to round down your answer in part **e**?

Self-assessment

I can identify opposite operations.

I can solve one- and two-step equations.

I can check my answer to equations.

I can differentiate between $\leqslant$ and $<$.

I can graph inequalities involving $<$, $\leqslant$, $>$, $\geqslant$, or $\neq$ on a number line.

I can solve one- and two-step inequalities algebraically.

I can solve inequalities with negative coefficients.

I can substitute into formulae.

I can change the subject by rearranging formulae.

I can construct formulae from real-life scenarios.

Check your knowledge questions

1 Write down the inverse operation:

a Divide by 4

b Subtract 7

c Multiply 3

d Add 15

2 Write down the operations required to convert the given expression to x.

a $5x$

b $x - 6$

c $\frac{x}{3}$

d $7 - x$

e $1 + 2x$

f $3(x - 5)$

g $\frac{2 + x}{7}$

3 Solve:

a $a - 7 = 9$

b $3b + 1 = 7$

c $6 - c = 8$

d $\frac{d}{2} - 1 = 5$

e $6(2 - e) = 36$

f $\frac{2f - 4}{5} = 4$

g $9g - 8 = 7g - 2$

4 Write and solve an equation to describe these scenarios:

a Bruce finds the product of a number, n, and 3 and finds it is the same as the same number, n, added to 8. What is the number?

b Tyler has six more pens in his pencil case than Janine. Together they have 20 pencils. How many pencils does Tyler have?

c A rectangular garden bed is twice as wide as it is long. If the fence going around it is 12 metres long, how wide is the bed?

5 Represent these inequalities on a number line:

a $s \neq 2$ b $t \geqslant 7$ c $u > -1$ d $v \leqslant 0$ e $w < 5$

6 Solve these inequalities:

a $8 + x > -1$

b $\frac{d + 1}{2} \neq 7$

c $1 - f \geqslant 5$

d $\frac{y}{2} - 4 > 1$

e $4 - \frac{m}{5} < -2$

f $-2 - s \geqslant 7$

7 The formula $F = ma$, describes the relationship between force (F), mass (m) and acceleration (a). Calculate the force exerted by:

a A 50 kg student, running with an acceleration of 1 m/s^2.

b A 180 kg gorilla falling from a branch with an acceleration of 10 m/s^2.

c A bird of 1.1 kg, flying with an acceleration of 4.2 m/s^2.

8 The formula $v = u + at$ is commonly used in a branch of physics called kinematics. Rearrange the formula to make the following variable the subject:

a u

b a

c t

9 A rectangle with height h and base length b.

a Draw a diagram of the shape described.

b Write a formula to express the perimeter of the rectangle.

c Write a formula to express the area of the rectangle.

d For a rectangle with perimeter 5.2 m and height 0.8 m, calculate

i the base length

ii the area.

10 Thomas is ordering monogramed gifts for his tennis team. Scarves cost €8.90, gloves cost €12.80 a pair and visors cost €14.50

a Write down a formula for the total cost of s scarves, g gloves and v visors.

b Using technology find out how much it would cost to buy each member all three items if there are

i 3 members

ii 5 members

iii 10 members.

c The final order was €271.30. Given that there are 10 members, and each wanted a visor and half wanted a pair of gloves, how many scarves were ordered?

The number plane and graphs of linear equations in one and two variables

The number plane and graphs of linear equations in one and two variables

KEY CONCEPT

Logic

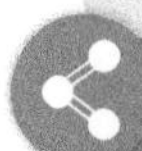

RELATED CONCEPTS

Change, Representation, Systems

GLOBAL CONTEXT

Orientation in space and time

Statement of inquiry

Representing the orientation of objects in space with logical systems allows us to measure the changes in their relationships over time.

Conceptual

- What is meant by an abstract idea or concept or pattern?
- What are the advantages of using a grid of perpendicular lines (at right angles) rather than lines which are not perpendicular for identifying a position in space?

Factual

- What are the differences between the equations of horizontal, vertical and diagonal lines?

Debatable

- Which is the better way of directing someone to a location: Using compass directions and distances or using a grid reference?

Do you recall?

1 Solve the linear equation $3x + 5 = 17$

2 Graph the linear inequality $x \leq 6$ on a number line.

3 Find the value of a in the formula $v = u + at$ when $t = 5$, $u = 30$ and $v = -19$

5 The number plane and graphs of linear equations in one and two variables

5.1 Coordinates

Connections

Map reading is important in Geography (as well as for not getting lost!).

You often need to describe your position to another person or find the location of a particular place. Although you can do this using compass directions and a distance from a starting point, this is not the most efficient method. It is simpler and more precise to describe the location of a person or place in terms of their position on a map with an overlaid **grid**. The location on the grid can be described by a pair of **coordinates**.

5.1.1 Grids

Reminder

Perpendicular means that two lines or line segments meet at a right angle.

A **grid** is a series of parallel lines in two different directions, which are usually **perpendicular** to one another, forming rectangles or squares.

Investigation 5.1

Here is a map of a zoo.

Describe the locations of the different animals on this map.

Where is the Flower garden located?

Both the Penguins and the Children's Zoo are in square D2. Think of two different ways to explain to someone the exact locations of the Penguins and the Children's Zoo.

What are the advantages and disadvantages of this system of grid references?

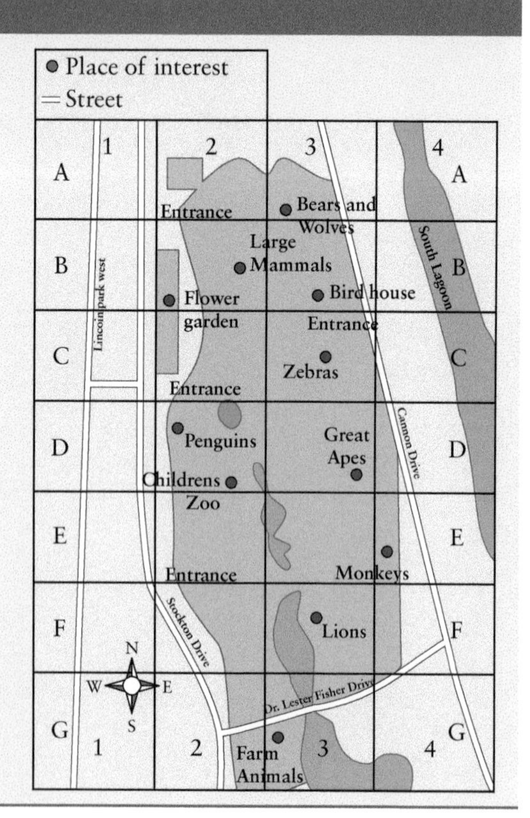

Explore 5.1

what3words is a project that has assigned a unique three-word address to every three-metre by three-metre square in the world! Download the ***what3words*** app, and check it out for yourself.

Start by finding out whose statue is located at ***packing.gulls.aquarium*** and what he is known for; then find the address of your mathematics classroom.

What are the advantages and disadvantages of this system of grid references?

How does it compare with GPS?

Explore 5.2

Here is part of a USNG (US National Grid) map of Washington, DC.

How is this grid different from the grid of the zoo?

The road intersection shown in the circle has grid reference 267019

What could be the grid reference of the hospital (H)? What is the difference between this and your answer for the location of the Flower garden in the zoo?

What are the advantages of this system?

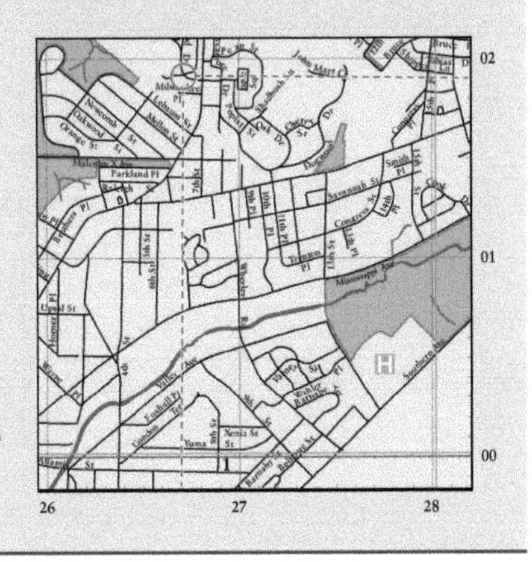

Reflect

You should already be familiar with another system of grid referencing – the grid overlaid on a globe. This consists of lines of latitude and longitude.

How is this system different from the grids of the zoo and the USNG?

The grids in these figures form squares or rectangles. What shapes are formed by the lines of latitude and longitude on a globe?

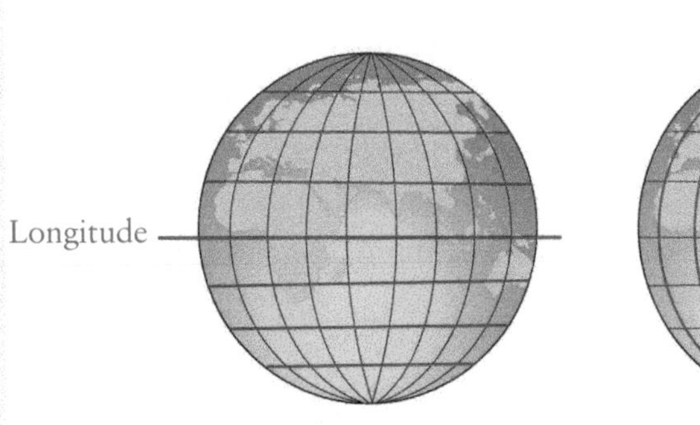

5.1.2 Cartesian plane

In Chapter 4, you used a **number line** to graph inequalities, such as $x > 5$

In this context, the number line is called the x-axis.

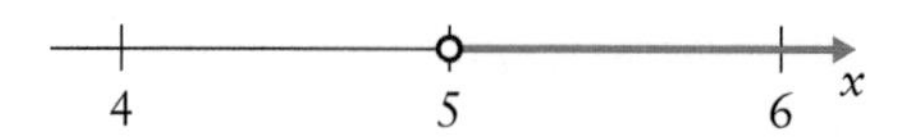

Hint

See the first Explore activity in Section 5.1.1

A **number plane** is obtained when another number line, the y-axis, is placed perpendicular to the x-axis so that the **axes** overlap at the point where both x and y are equal to 0. This point is called the **origin**. A **plane** is a flat, two-dimensional surface that extends infinitely in all directions. The x–y number plane is also known as the **Cartesian plane**. Can you guess who it is named after?

Hint

The x- and y-axis are also called horizontal and vertical axes.

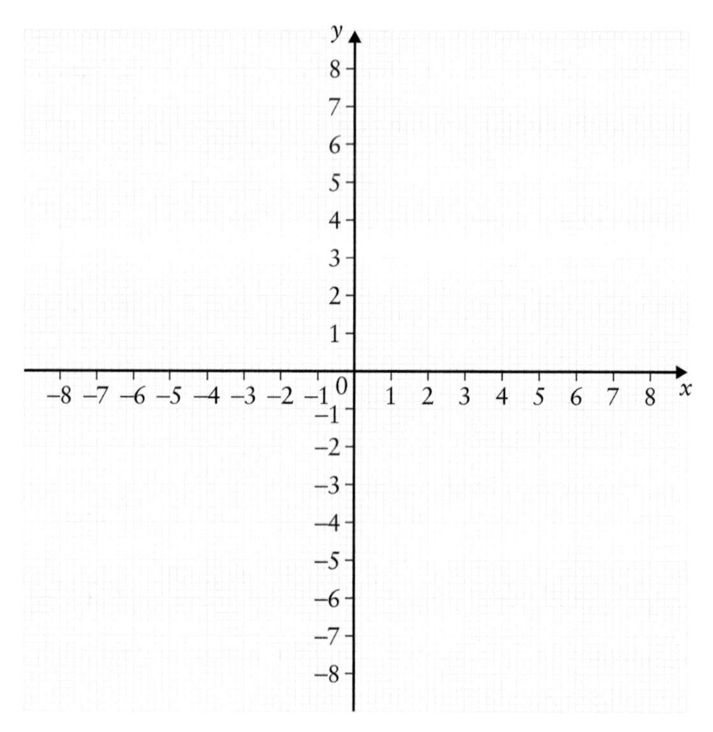

You can identify the location of points on the plane by referring to their **coordinates**. Coordinates are a pair of numbers that represent the grid reference for points on the Cartesian plane. Just as the letter x precedes the letter y in the alphabet, you write the x-coordinate before the y-coordinate when defining a point. The form (x, y) is also called an **ordered pair**.

What ordered pair represents the origin?

Explore 5.3

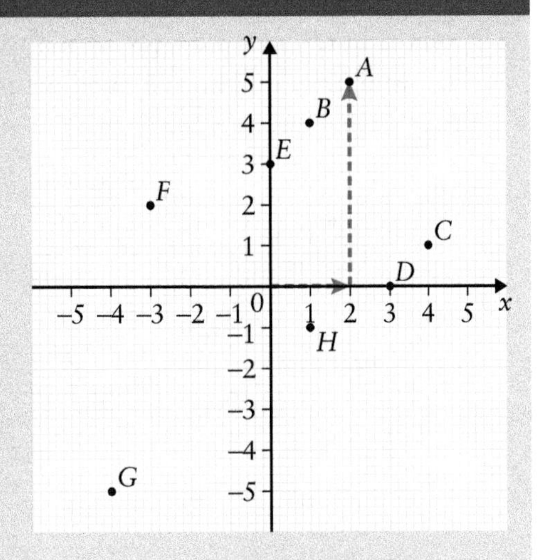

On the graph, point A has coordinates $(2, 5)$. It is above the 2 on the x-axis and to the right of the 5 on the y-axis.

When you want to move to a point given by its coordinates, you always start at the origin. The origin is the name of the point $(0, 0)$, where the x and y-axes cross. If point A has coordinates $(2, 5)$, you need to move 2 units right and 5 units up from the origin to get to A.

Which point on this grid has coordinates $(4, 1)$?

Why is it not enough to say 'E is at 3'? How would you describe point E fully?

Can you describe the positions F, G and H?

If a point has a negative y-coordinate, what does this tell you about its position? What if it has a negative x-coordinate?

Practice questions 5.1

1 This is a plan showing the lettered rows of seats at a cinema.

The seats closest to the screen are A1–A10

What is the seat indicated by the:

a triangle

b rhombus

c circle

d pentagon?

	1	2	3	4	5	6	7	8	9	10
H										
G								⌂		
F										
E		△								
D										
C							◊			
B	○									
A										

Screen

2 On the map below, Smirnoff Hill is exactly on the line marked 35 on the horizontal axis, giving it an x-coordinate of 350. Its vertical coordinate would be 492, making its grid reference 350492

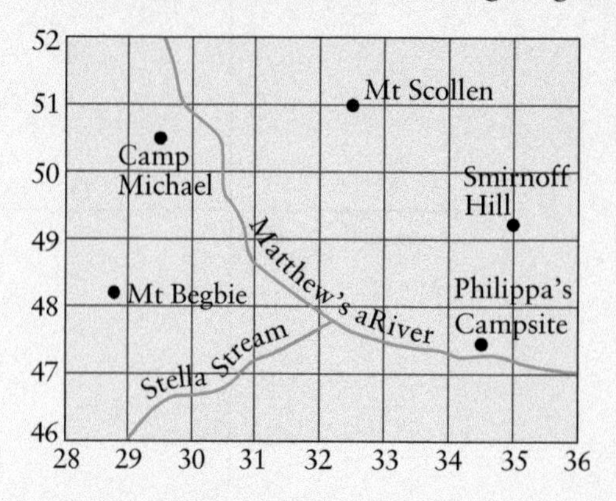

a What is located at grid reference:

i 325 510 ii 322 477?

b What is the grid reference of Camp Michael? Mt Begbie?

c My friend wrote instructions, but I spilled water on them and some numbers have faded. 'Starting at _ _ 5_ _ 5, go to 3 _ 5 _ _ _'.

What are the possible journeys my friend was recommending?

3 On the map of Central America, Panama City appears to be located at 9°N, 80°W, to the nearest degree.

a To the nearest degree, which places are located at

i 14°N 89°W ii 12°N 83°W.

b What are the latitude and longitude of the following places, to the nearest degree:

i Managua, Nicaragua ii San José, Costa Rica?

4 Connect the points on the grid, in the order given.

a Join: (1, −3) to (1, 3); (3, −3) to (3, 3); (4, 0) to (3, 0); (5, 2) to (5, 1); (5, −2) to (6, −3); (−2, 3) to (−4, −3); (0, −3) to (−2, 3); (−3, 0) to (−1, 0); (7, −2) to (7, 0); (0, 3) to (2, 3); (4, −3) to (4, 3); (7, 2) to (6, 3); (5, 1) to (7, 0).

b You will see a word forming. List the coordinates in the order you need to join them to complete the first letter of the word.

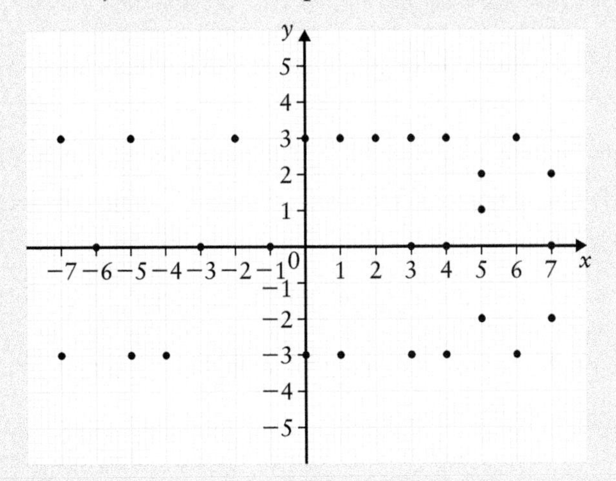

5 Coordinates practice – work with a partner.

a On squared paper, two copies of the axes, as shown; or open two graphs on GeoGebra, or other graphing software, to play a shorter version of the sinking ships game. Label them My ships and Your ships.

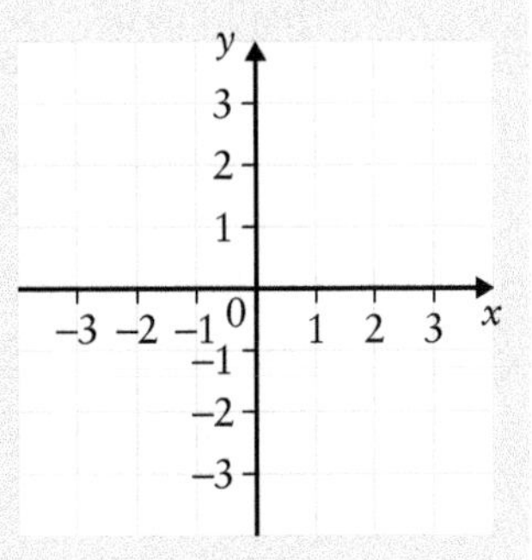

b Choose where your ships will be, and mark the points on your My ships grid. The destroyer covers two points, the submarine covers three points and the battleship covers four points. For example, in the diagram, the destroyer covers points (2, 0) and (2, −1).

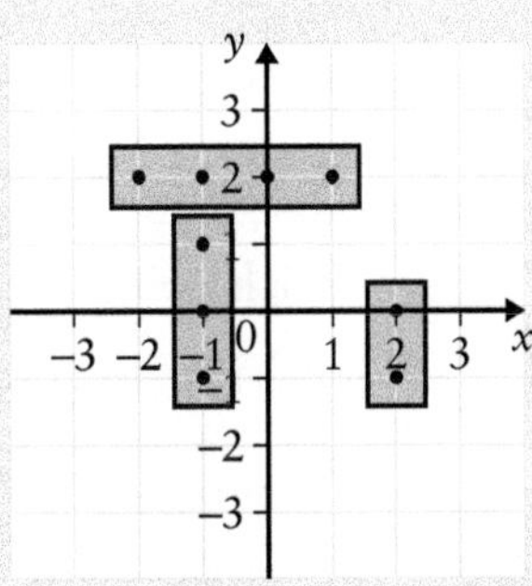

The coordinates of your ships should be integers from −3 to +3, inclusive. Ships cannot overlap. When your opponent hits one of your ships you must say so (but do not tell your opponent which ship they have hit). To sink a ship you must identify every square it covers.

c Start playing.

6 Plot the following points on the same set of axes: A (−2, −1), B (1, −2), C (1, 0), D (−4, 5), E (−1, 6).

a Write the name of the special shape you make when you join the points in each of the orders indicated:

i ABC ii ACD iii $ACED$.

b A, B and C are the vertices of rhombus $BACF$. Write the coordinates of point F.

c A, B and C are the vertices of a parallelogram (which is not a rhombus) whose fourth vertex is G. Give two possible ordered pairs for G.

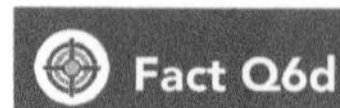

The midpoint, M, of two points P and Q lies on the line through P and Q at an equal distance from both points.

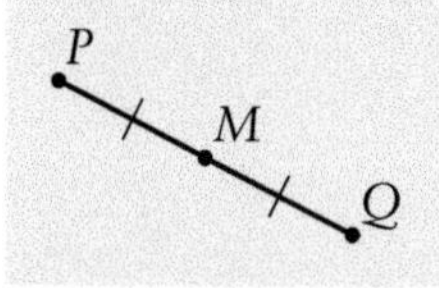

d On which axis does the midpoint of the line segment CE lie?

7 The sections of the coordinate plane are called quadrants. Each quadrant is numbered as such:

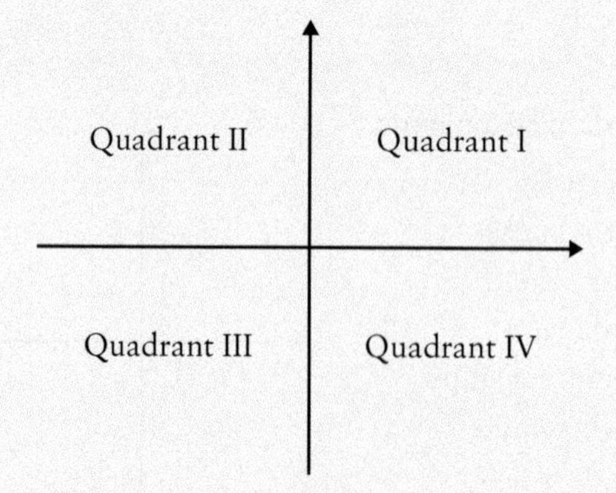

Referring to the points A, B, C, D and E in Question 6 above answer the following:

a Which quadrant is point A in?

b Which points are in Quadrant II?

c Point C lies between which two quadrants?

d The product of the x- and y-coordinates is positive for points in which quadrants?

8 Plot the points $P(-6, -1)$ and $Q(-2, 3)$.

a Point R has coordinates $(a, -1)$ and is such that PQR is an isosceles triangle. Find all possible values of a, where a is an integer.

b Point S has coordinates (b, b), and is such that PQS is an isosceles triangle with base PQ. Find b.

5.2 Lines parallel to the axes

5.2.1 Horizontal and vertical lines

Write the coordinates of points A, B, C and D. What do you notice? What kind of line do A, B, C and D lie on?

Write the coordinates of points D, E, F and G. What to you notice? What kind of line do D, E, F and G lie on?

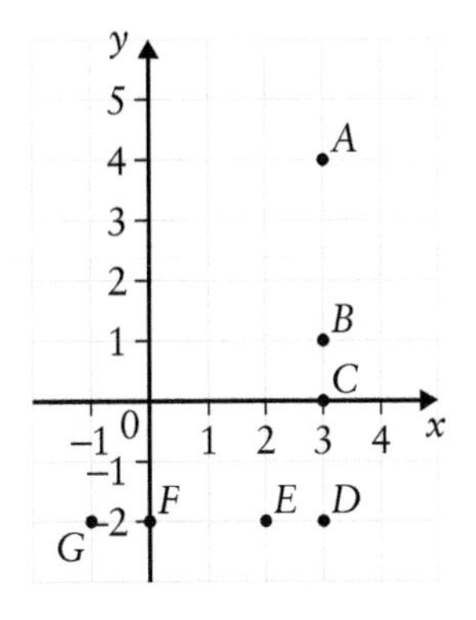

Each set of points forms a line parallel to one of the axes.

Lines parallel to the x-axis are called **horizontal** lines and those parallel to the y-axis are **vertical** lines.

The points A, B, C and D have coordinates $(3, 4)$, $(3, 1)$, $(3, 0)$ and $(3, -2)$, respectively. They all have x-coordinate 3, so the gridline on which they lie is called the line $x = 3$.

 **Fact**

Vertical lines have equations of the form $x = k$, where k is a constant (a fixed number).

The points D, E, F and G have coordinates $(3, -2)$, $(2, -2)$, $(0, -2)$ and $(-1, -2)$, respectively. They all have y-coordinate (-2), so the gridline on which they lie is called the line $y = -2$.

Fact

Horizontal lines have equations of the form $y = c$, where c is a constant.

 **Hint**

Horizontal lines go in the same direction as the x-axis but have equation $y = c$. Vertical lines go in the same direction as the y-axis but have equation $x = k$.

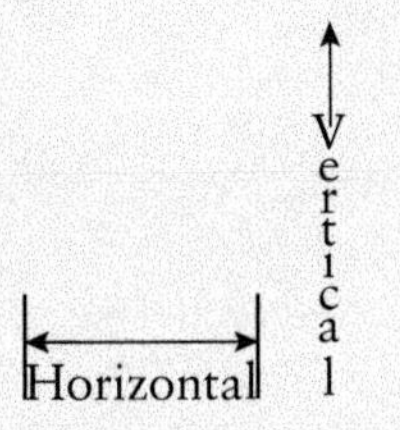

Investigation 5.2

Recall the word you found by connecting the dots in Practice questions 5.1.

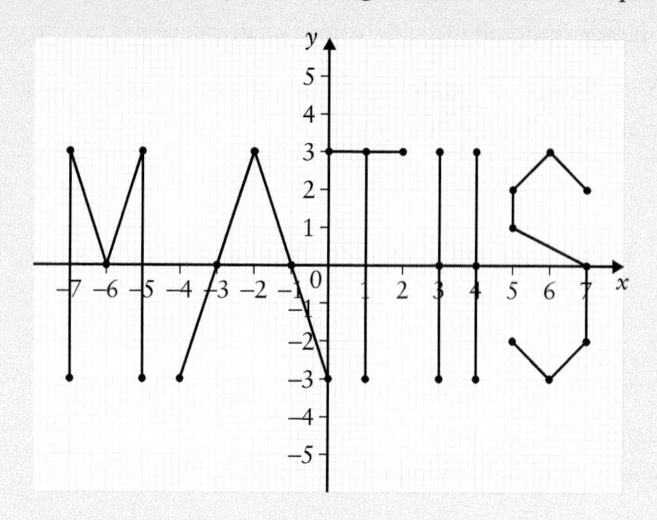

> **Reminder**
>
> In mathematics, a line is infinitely long, which is why we often draw lines with arrows. A part of a line with two endpoints is a line segment. A ray starts at one endpoint and continues infinitely in one direction (like the Sun's rays).

In which letter is there a **line segment** from the line $x = 3$?

What about the line $y = 3$?

What are the equations of the lines forming the letter T and the letter H?

Hence, what is the equation of the x-axis? What is the equation of the y-axis?

Identify a point on the line $y = 3$ between (1, 3) and (2, 3). Identify two more points on this line. How many such points are there?

One of these is a ray, one is a line, and one is a line segment. Which is which?

Practice questions 5.2

1 a Plot the points A (1, 3), B (5, 2) and C (3, −4) on a coordinate grid. Draw a vertical and a horizontal line through each of them.

b Write the equations of the vertical lines through A, B and C.

c Write the equations of the horizontal lines through A, B and C.

2 a Copy this graph:

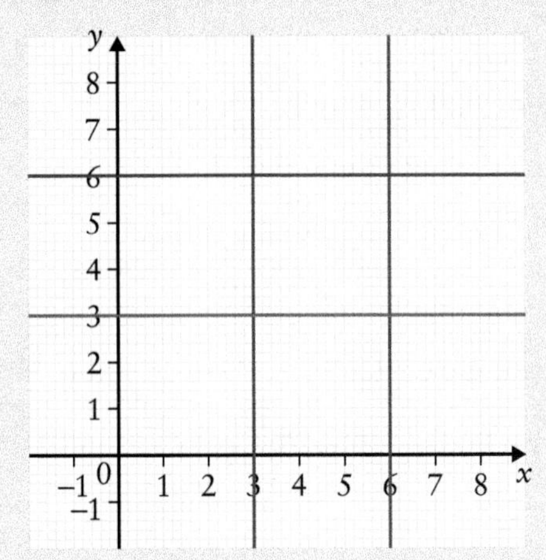

Mark, with an X, the point where the lines $y = 3$ and $x = 6$ meet.

b Mark, with a dot, the point where the lines $x = 3$ and $y = 6$ meet.

3 Write down the coordinates of three points which lie on the line $y = 9$

4 Use the graph below to answer the following questions.

a Does the horizontal line through E pass through A or C?

b Write the equation of the horizontal line through E.

c Write the equation of the vertical line through D.

d Write the equation of the line parallel to the x-axis that passes through B.

e Write the equation of the line parallel to the y-axis that passes through A.

f Write the equation of the vertical line through C.

g Where do the lines you mentioned in parts **d** and **e** meet?

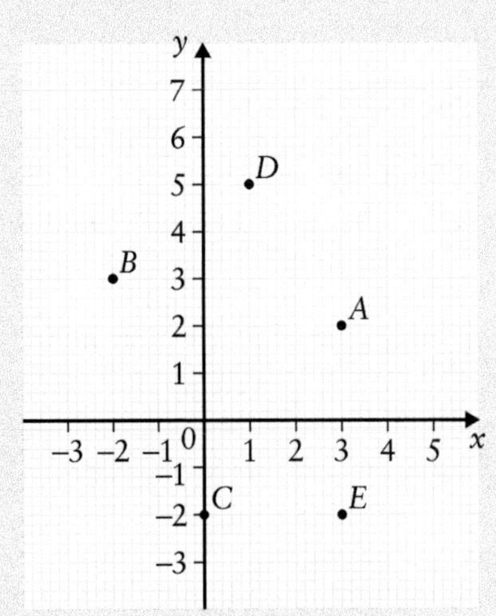

5 Complete the following steps on a single graph. Use a scale from −6 to 6 on both axes.

a Draw the vertical line through the point (−4, 3).

b Draw the horizontal line through the point (2, −5).

c Draw the line $y = 0$

d Draw the line $x = -1$

e Find the area of the rectangle enclosed by the four lines.

f What are the coordinates of the point where the lines you drew in parts **c** and **d** meet?

6 A digital number 6 looks something like this.

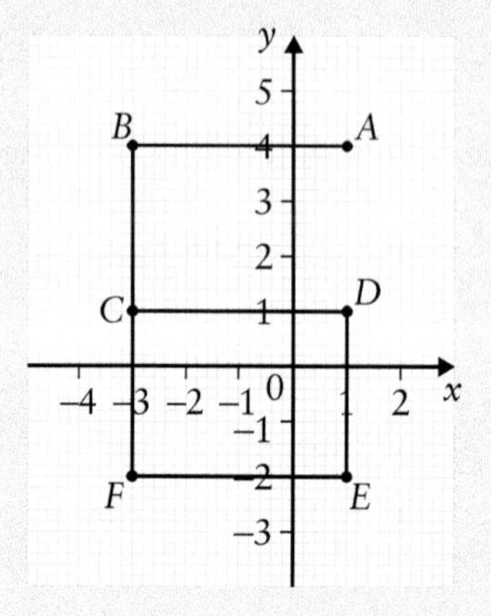

a Which is longer: the total length of the horizontal line segments in the 6 or the total length of the vertical ones?

b Write the equation of the line segment EF.

c Write the equation of the horizontal line passing through B.

d The line $x = 1$ passes through which points?

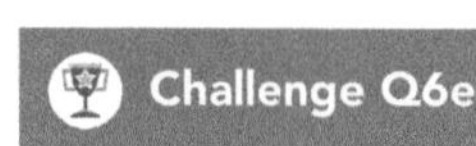
Challenge Q6e

e Write the coordinates of the midpoint of AB, which we will call M. Hence, write the equation of the line passing through M and perpendicular to AB.

Challenge Q6f

f Write the coordinates of the midpoint of BC, which we will call N. Hence, write the equation of the line passing through N and perpendicular to BC.

5.3 Line graphs and travel graphs

Line graphs are made up of joined line segments. Many line graphs represent real-life situations, and it is important to be able to **interpret** these **in context.** Most often, line graphs present how the value of some quantity is changing over time. When time is one of the two quantities, it is always on the horizontal axis.

Investigation 5.3

In the distance–time travel graph below, we see the bicycle journeys of two couples: Bali and Brett in blue, and Reilly and Robin in dotted red. They have decided to meet for a picnic at a popular spot in the countryside.

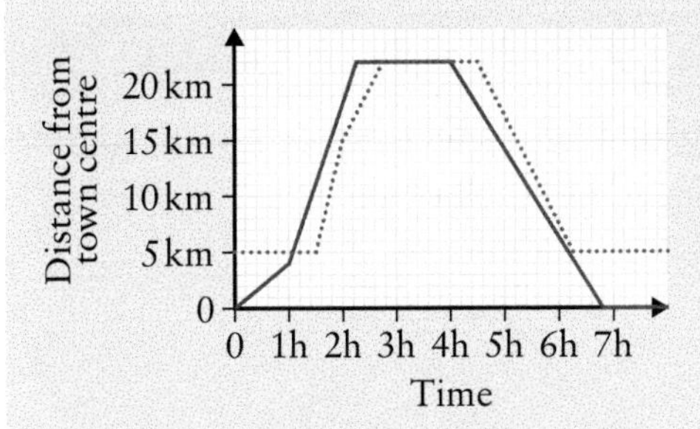

Reilly and Robin (R and R) live outside of town, near the cycle path, and have some errands to run in the morning, so they will set off a bit later and meet Bali and Brett (B and B) there for lunch. Bali and Brett live in town, so they set off earlier (on the same cycle path) at 10:00

1 How far do R and R live from B and B? What time do they set off?

2 How far do B and B travel in the first hour? So, what is their speed? What about the second hour? Why might they have been able to travel faster in the second hour?

3 How far is the picnic spot from B and B's home? How far is it from R and R's home?

4 How long do B and B wait for R and R to arrive at the picnic spot?

5 How long does each couple spend at the picnic spot?

6 What time do R and R get back home? How much shorter was their time away from home than B and B's time away from home?

7 How many times faster do B and B cycle back home than they travelled in the first hour of their trip? How do you know?

8 How fast did R and R cycle home, in km/h? Was this their fastest speed of the day? How do you know?

9 How far did B and B travel altogether? How long did it take them to cycle this distance? Hence, what was their average speed for the journey?

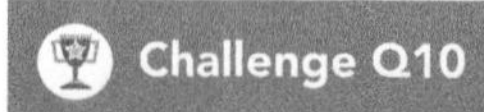

10 At what clock time did B and B cycle past R and R's home in the morning?

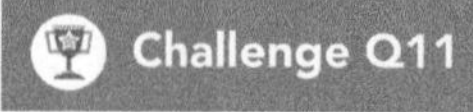

11 R and R almost caught up to B and B on the cycle path during the ride home. How close did they get? How much more time would they have needed to catch them if all had maintained a constant speed?

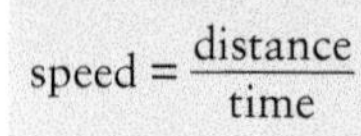

Fact

$$\text{speed} = \frac{\text{distance}}{\text{time}}$$

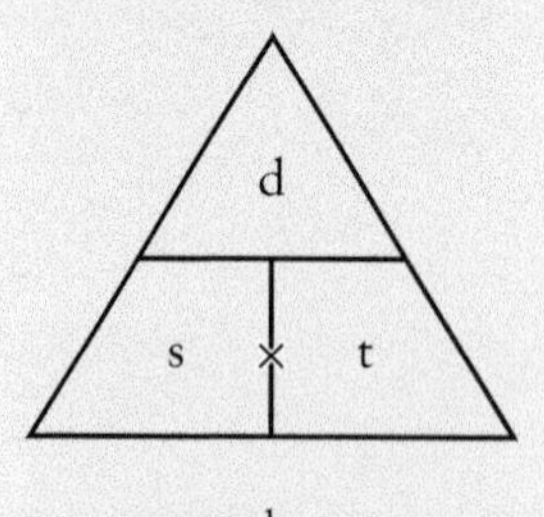

average speed

$$= \frac{\text{total distance}}{\text{total time}}$$

Travel graphs not only have the time on the horizontal axis, but the distance on the vertical axis always represents the distance of the people or objects away from a particular point, not the actual distance travelled.

Explore 5.4

The following graphs show the relationship between temperature and time in different contexts. What situations might each represent?

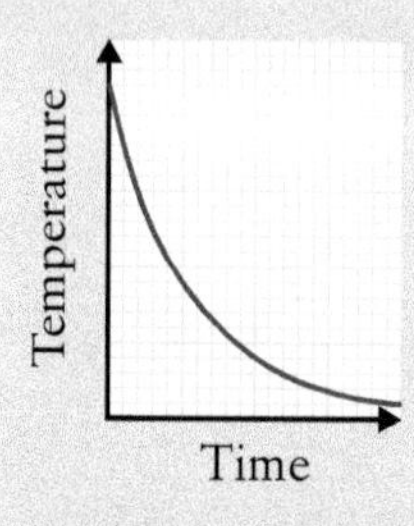

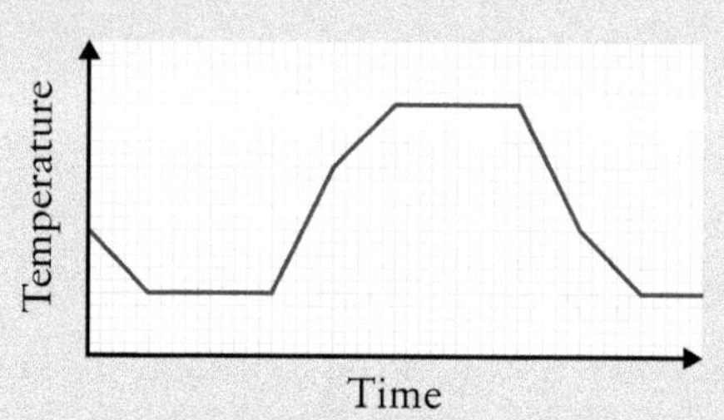

Worked example 5.1

The graph shows how the value of a typical car depreciates over time. For example, a car that is one year old would be worth approximately 80% of its price when new, so it loses 20% of its value in the first year!

After two years, it is worth approximately 70% of its original price, so it loses another 10% of its original value in the second year. The value continues to fall year by year, as shown on the graph.

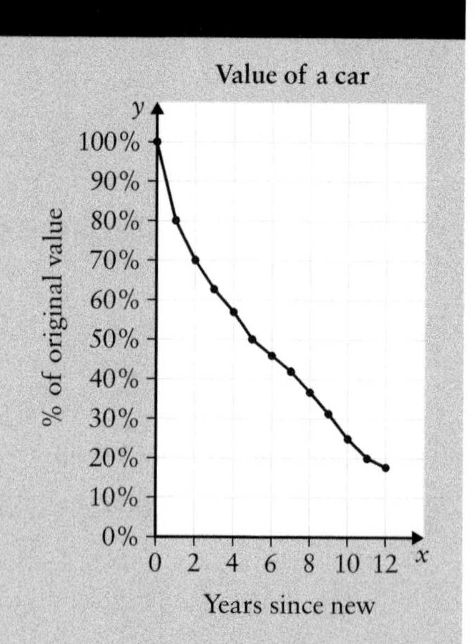

a The graph is steepest between zero and one year. What does this tell you?

b How many years does it take for a typical car to depreciate by:

i half of its original value

ii three-quarters of its original value?

c According to the graph, what percentage of the original value of a car is lost during the first six months? Do you think this is accurate, or do you think this is an over- or underestimate? Justify your reasoning.

d Between five years and six years, and between six years and seven years, the segments joining the points have the same steepness. What does this tell you about the value of the car during these years?

Solution

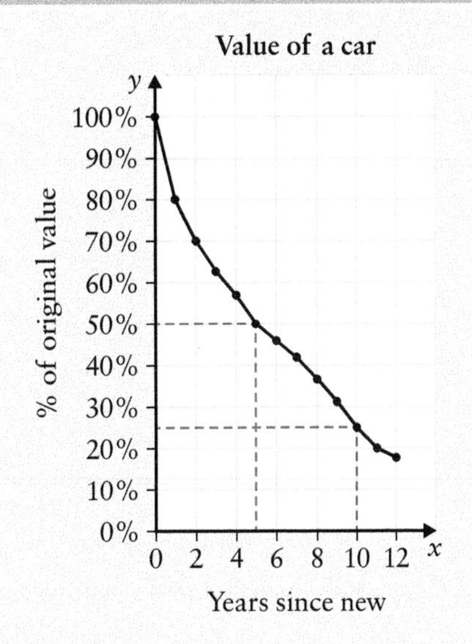

a The value of a car falls by the greatest amount during the first year.

b i The car depreciates by half, or by 50%, of its original value after five years.

ii Losing three quarters of its value means it will be worth one quarter, or 25%, of its original value. This is after 10 years.

c Since 20% is lost during the first year, and the line segment is straight, the graph suggests that 10% is lost during the first six months. In reality, changes are likely to be more gradual, resulting in a smooth curve rather than a jagged line graph. So the value at six months given by the graph is likely to be an overestimate of the true value, as shown here:

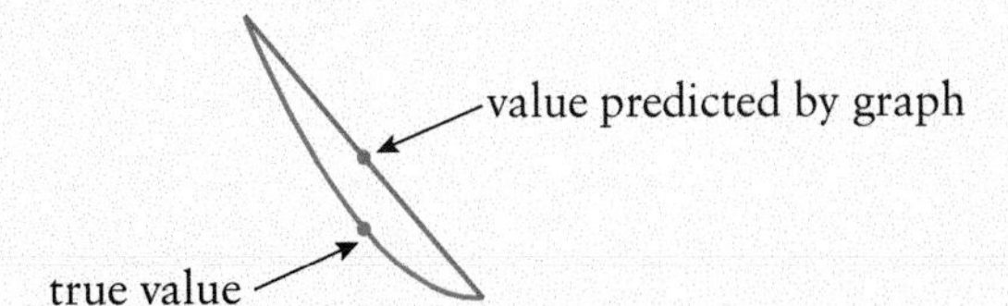

d The value decreases by the same amount during the sixth year as it does during the seventh year.

Practice questions 5.3

Communication skills

Thinking skills

Provide a reason for each of your answers to the following questions.

1 The graphs show three people's trips today.

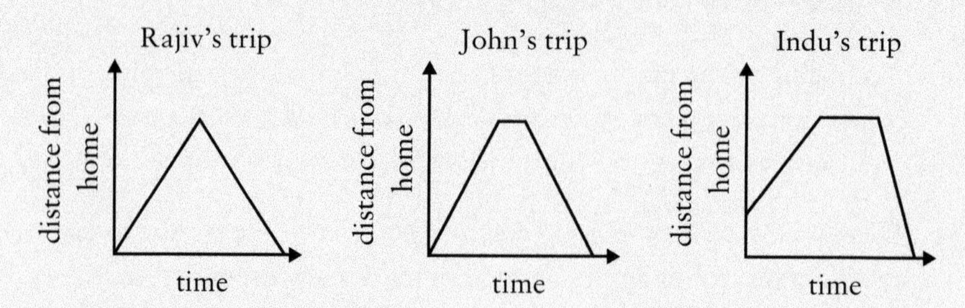

a Who started from home?

b Who did not stop to rest?

c Who travelled faster in one direction than they did in the other?

2 The line graph below shows the marks that three students achieved in four Mathematics tests last year.

a Which student improved the most during the course of the year? Explain how you know. By what percentage did they improve overall?

b Which student was the most consistent? How can you tell this from the graph?

c After the third test, which student had achieved the best results in the class? Explain how you work this out.

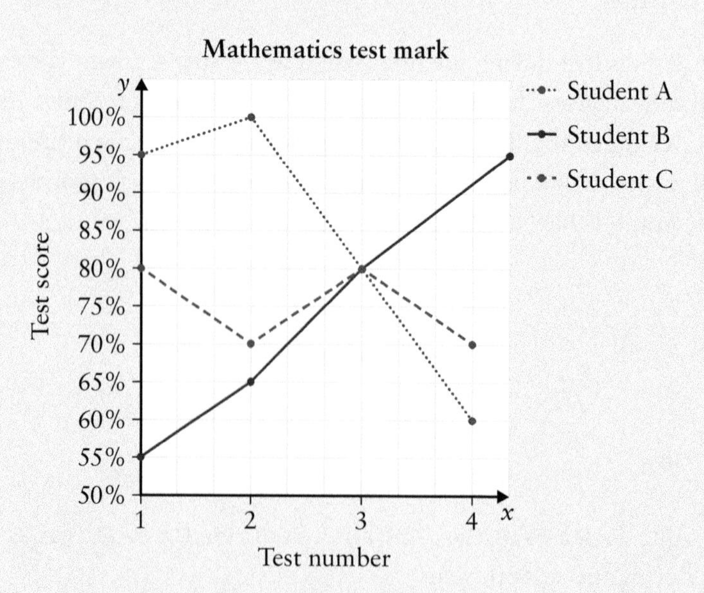

3 The graph shows how much time adults in the US spent watching television and using their mobile devices each day between 2014 and 2021. It was produced in 2019.

a All the following statements about the graph have a mistake. Rewrite them as true statements.

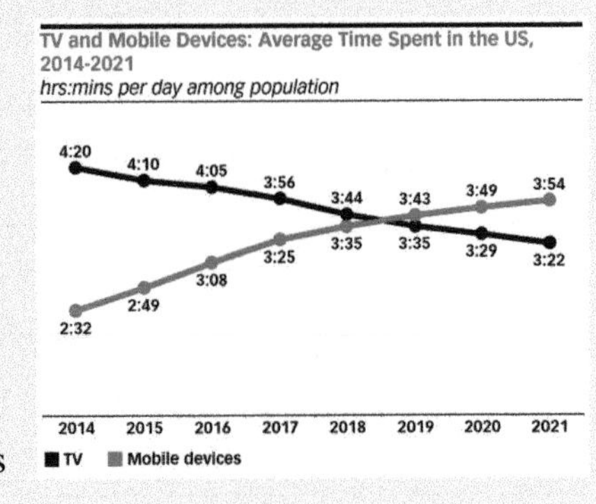

i In 2014, adults in the US watched TV for an average of 4.2 hours each day.

ii In 2019, adults in the US spent an average of 3 hours and 35 minutes a day on their mobile devices.

iii In 2015, adults in the US spent an average of 6 hours and 59 minutes each day looking at a screen.

iv The total screen time per day for adults in the US is predicted to increase significantly from 2014 to 2021

b Explain how the graph can have points for 2019, 2020 and 2021 if it was created in April 2019

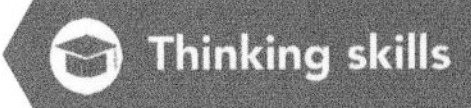

4 Alena and Brianna took part in an orienteering race. They had to use a map and compass to find checkpoints hidden in the forest or countryside and go from one checkpoint to the next to complete the course. Brianna began 5 minutes after Alena. The race started and finished at the same location. The travel graph shows each girl's distance from the start, in kilometres, at different times throughout the race.

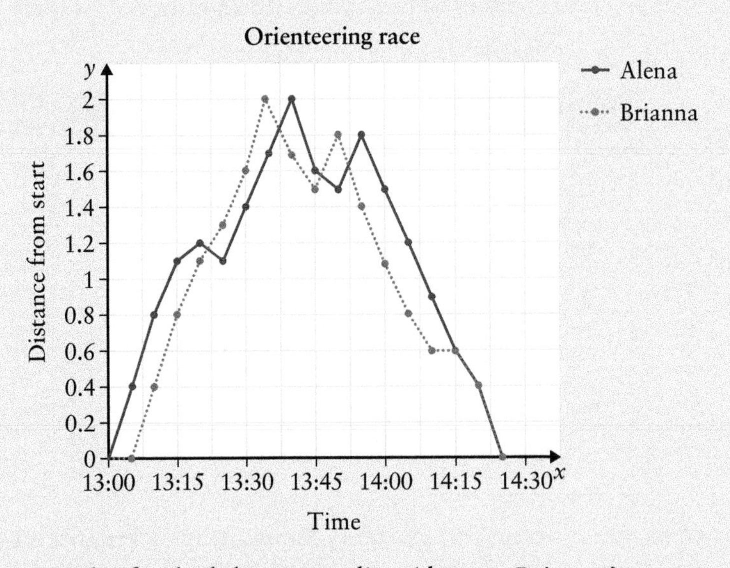

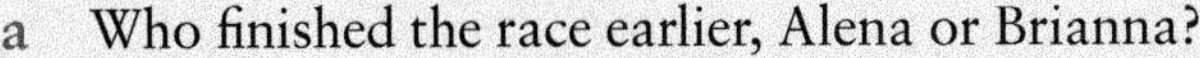

a Who finished the race earlier, Alena or Brianna?

b The first checkpoint was 500 metres from the start. Who arrived there first, Alena or Brianna?

c The furthest checkpoint was 2 km from the start. Brianna arrived there first. How many minutes later did Alena arrive there?

d One of these runners got lost during the race. Who do you think it was, Alena or Brianna? Explain how you know.

e The girls could only see one another during the last 500 metres of the race. Explain why it does not always mean that they could see each other when their graphs cross.

f The length of the race was 6 km. Calculate Brianna's average speed for the race.

5.4 Straight-line graphs as solutions of linear equations in two variables

In Chapter 4, you solved simple equations such as $3x - 5 = 7$. The solution to this equation is $x = 4$. You now know that $x = 4$ is also the equation of a line. So lines can represent solutions to equations.

Investigation 5.4

The solutions to the equation $x + y = 3$ are all the pairs of x- and y-values which make the equation a true statement. You have probably realised that this is true when $x = 1$ and $y = 2$ and when $x = 2$ and $y = 1$. But these are not the only values. What about when x or y are equal to 0? What if they have decimal or fractional values? They could also have negative values.

Some of the possible values are given in this table.

Work out the missing values and complete the table.

x	y
1	2
2	1
0	
$\frac{1}{2}$	
	0.2
−2	
	−1.5

The pairs of solutions can be written as ordered pairs, which represent points on the number plane.

(x, y)
(1, 2)
(2, 1)
(0,)
(,)
(,)
(,)
(,)
(,)

Draw a set of axes, each from −6 to 6, and plot the points given in the table. What do you notice about the points you have plotted?

If you join all the points in the Investigation, they will lie on a straight line. The coordinates of the other points on this line, between and beyond those you plotted, would also make the equation $x + y = 3$ a true statement. So, there are an infinite number of solutions to the equation, and they are represented by the infinite number of points on the line. So this line is called $x + y = 3$

$x + y = 3$ is a **linear equation**, because its graph is a straight line.

Worked example 5.2

Draw axes with these ranges: $-2 \leq x \leq 6$ and $-2 \leq y \leq 12$

Now graph the line $2x + y = 10$

Solution

Understand the problem

We need to draw the graph of a straight line, so we need the coordinates of at least two points on the line. The solutions to the equation of the line correspond to the coordinates of the points on the line. The equation of the line is a formula for calculating these solutions.

Make a plan

Choose three x- or y-values, and then substitute these into the equation and solve for the other variable. This will give the coordinates of three points on the line, which we can plot on the grid. The range $-2 \leq x \leq 6$ means that we need to draw a horizontal number line from −2 to 6 as the x-axis, and the range $-2 \leq y \leq 12$ means that we need to draw a vertical number line from −2 to 12 as the y-axis. Finally, we need to draw a straight line through the points to obtain the graph.

Carry out the plan

For example, choose $x = 4$, $y = 4$ and $x = -1$ (we can choose any values; integers are easier to work with than fractions or decimals). We can show these values in a table:

x	y
4	
	4
−1	

Substitute each chosen value into the equation to find the other coordinate.

When $x = 4$,

$2 \times 4 + y = 10$

$8 + y = 10$

$y = 2$

Do the same for $y = 4$ and $x = -1$:

$$2x + 4 = 10 \qquad 2(-1) + y = 10$$
$$2x = 6 \qquad -2 + y = 10$$
$$x = 3 \qquad y = 12$$

This gives the following table of values and coordinate pairs:

x	y	(x, y)
4	2	(4, 2)
3	4	(3, 4)
−1	12	(−1, 12)

Plotting the points and joining them with a straight line gives this graph.

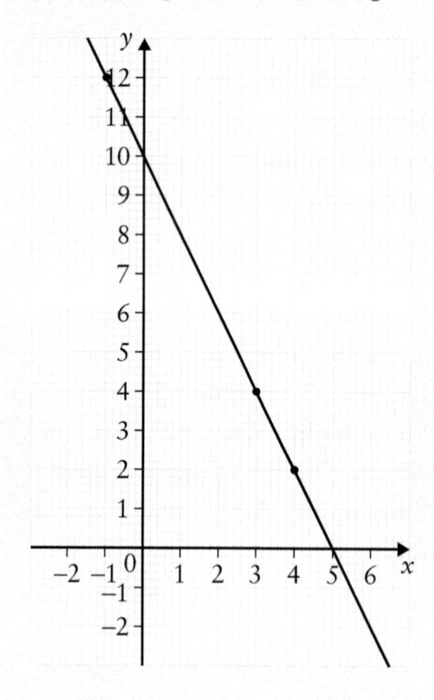

Look back

Can you see any other points with integer coordinates on the graph? Check that the pairs of x and y values satisfy the equation $2x + y = 10$ to verify that these points do lie on the line.

For example, the line seems to pass through the point (6, −2). If so, $x = 6$ and $y = -2$ should make the equation a true statement:

$$\text{LHS} = 2 \times 6 + (-2) = 12 - 2 = 10 = \text{RHS}$$

Since the left-hand side (LHS) is equal to the right-hand side (RHS), it confirms that this ordered pair is a solution of the equation. LHS and RHS are abbreviations that are sometimes used in formal proofs in mathematics.

For example, if we had chosen x to be 3 and y to be 5,
$\text{LHS} = 2 \times 3 + 5 = 6 + 5 = 11 \neq 10 = \text{RHS}$, means that the point (3, 5) is not on the line $2x + y = 10$, since LHS ≠ RHS (because $11 \neq 10$).

Why do we usually choose three points if two are sufficient to draw a line?

Reflect

The selection of x- and y-values used in the previous example was quite random. Can you think of a logical approach to choosing values? Would you choose x- or y-values or a combination of both? Does it depend on the equation?

To draw more lines more accurately, it is a good idea to find two points as far apart as possible, and then connect them with a straight line using a ruler. Try drawing the line $y = x$ by connecting just the points $(-8, -8)$ and $(-7, -7)$.

Does your line also pass through (8, 8)? It should, but it can be difficult to make sure that is does. Plotting $(-8, -8)$ and (8, 8) and then joining them would help make sure that the line is correctly drawn. Why do you think this is?

What do you think is the best approach to choosing values of x or y for the table of values?

Practice questions 5.4

1 Here is the graph of $x - 3y = -6$. Use it to write down three solutions to the equation $x - 3y = -6$ as ordered pairs, (x, y).

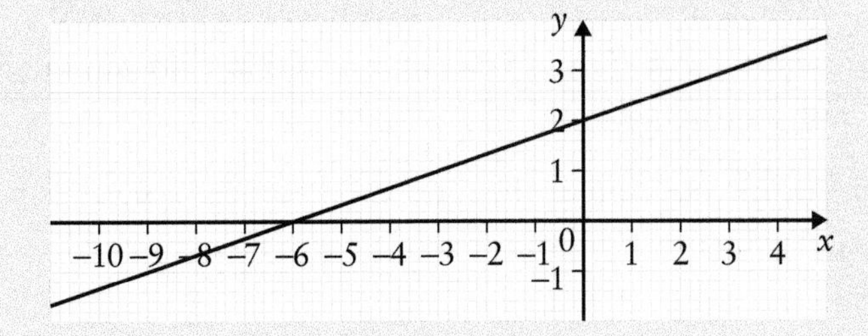

2 Draw axes with these ranges:
$-2 \leq x \leq 6, -5 \leq y \leq 5$,
then complete the table of values to find solutions of the equation $3x - 2y = 10$.
Hence graph the line $3x - 2y = 10$.

x	y
	1
6	
	−2

3 Draw axes with these ranges: $-8 \le x \le 8$, $-8 \le y \le 8$. Use a table of values to find three points on each of the following lines. Then draw their graphs

a $x - y = 6$ b $x + 3y = 8$ c $2x - y = -7$ d $3x + y = 5$

4 This graph shows three lines labelled a, b and c. The equations of the lines, but in a different order, are $2x - 6y = 12$, $x + 2y = -4$ and $x + y = 2$

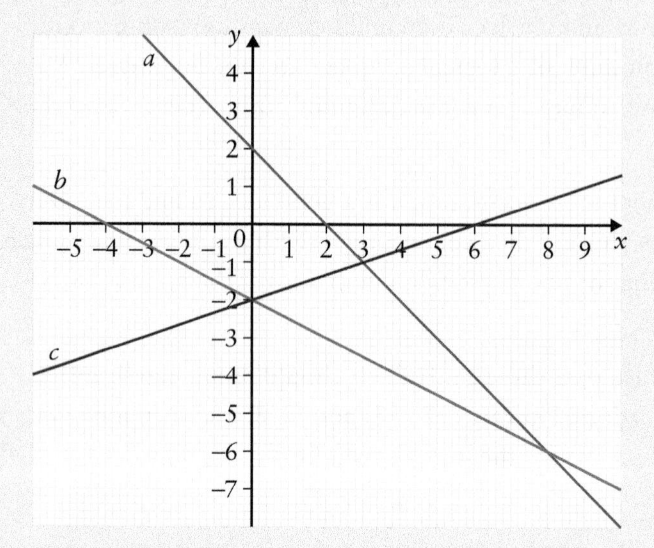

a Which line does the point $(2, -3)$ lie on: a, b or c?

b Which equation is $x = 6$, $y = -4$ a solution to?

c Which ordered pair is a solution to both $2x - 6y = 12$ and $x + y = 2$?

d Line c crosses the x-axis at $(6, 0)$. This is called the x-intercept of the line. What is the y-intercept of the line $x + y = 2$?

e Which equation is $x = 30$, $y = 8$ a solution to? How do you know without doing a calculation?

f The points of intersection of the lines a, b and c are the vertices of a triangular region. Find the area of the triangle formed by the three lines.

5 a Graph the line $y = 2x + 7$

b Your graph in part **a** is the same as one of the graphs you drew in Question 3. Identify which graph this is. Explain why they are the same?

c Rewrite $2x + 3y = 6$ in the form y …

6 Add the graphs of the following lines to those that you plotted in Question 3. Use the same set of axes. What do you notice? Describe how these equations are related to the equations in Question 3. How are the graphs related?

a $x - y = -6$ b $x + 3y = -8$

c $2x - y = 7$ d $3x + y = -5$

5.5 Straight-line graphs of the form $Ax + By = C$

Mathematics involves patterns. A pattern is something that repeats or follows a rule. Noticing similarities between things is how we often learn in life; for example, recognising speech patterns enables you to learn a language. You should look for patterns in mathematics to make your work more efficient.

You can find the coordinates of several points on the graph of an equation by choosing values for x or y and substituting them into the equation to work out the value of the other variable. In this section, you will learn to recognise graphs of lines from their equations, and to write equations of lines by looking at their graphs, without having to use algebra.

Sometimes you can work things out more easily and more efficiently by thinking of them in a different way, especially when the original presentation is **abstract** – something that you have difficulty visualising. It helps if you can make it more **concrete** – something you can touch or see.

Explore 5.5

A mechanical pencil costs €9 and a 50-page notebook costs €4. You need to buy one pencil and a few notebooks. You need to decide how much money you have to have before going to buy the notebooks. Can you set up an equation that enables you to decide on the amount of money you should have depending on the number of notebooks you can buy?

Worked example 5.3

You are given two numbers whose sum is five.

Write down several pairs of numbers that would work here, and hence draw the graph of the line $x + y = 5$

Solution

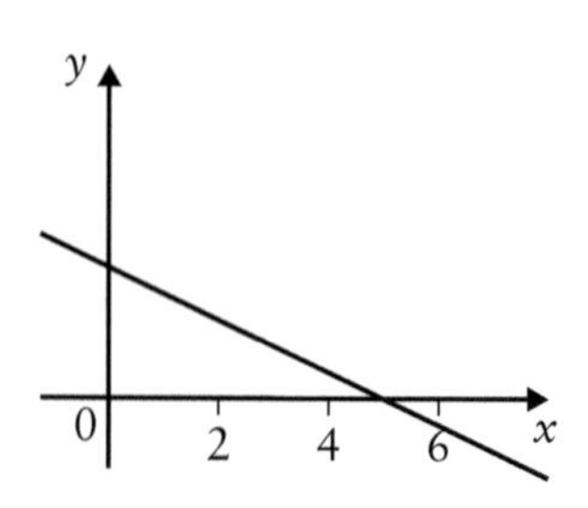

We need two numbers that add up to five. We can start by considering some convenient values for x and find the value of y that corresponds to each. This means that we need to solve the equation $x + y = 5$ for y.

In order to carry out our plan, we can set up a table where we record the values of y for each value of x.

$x + y = 5$ implies that $y = 5 - x$. Here is a table with some values:

x	−2	0	1	2
y	7	5	4	3

Investigation 5.5

For each of the following equations find three ordered pairs that satisfy it:

$x + y = 7$ $\quad$ $x + y = 4$ $\quad$ $x + y = 0$ $\quad$ $x + y = 2$

Using the ordered pairs you found for the four equations above, plot four lines on a single set of axes. Make sure that your axes are long enough to be able plot all the points defined by your ordered pairs: both axes should go from −3 to 8 at least.

What do you notice about the lines? What do they all have in common? How does the constant (number which is not a variable) on the right-hand side of each equation relate to its graph?

Write a sentence describing an easy way to graph any line of the form $x + y = C$, where C is a constant.

Describe three differences between your graphs and the graph of the equation $x - 3y = -6$, shown here:

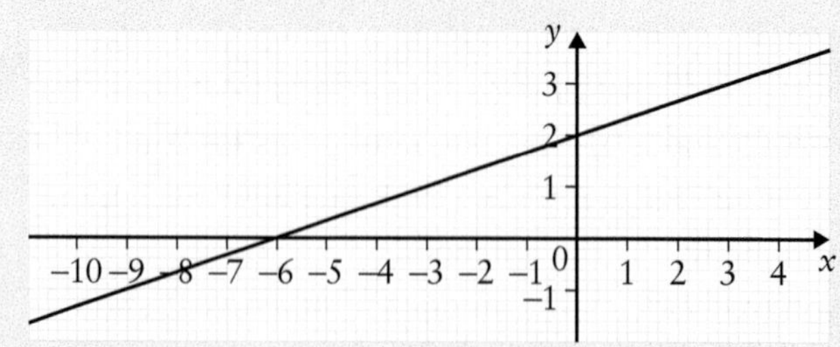

Investigation 5.6

If x is an integer, what kind of number is $2x$?

Hence, write a sentence translating the equation $2x + 2y = 14$ into words.

Can you think of some ordered pairs that would make this equation true?

Plot them on a coordinate grid and join them with a straight line.

Do you recognise this line?

Try the same with these equations:

$2x + 2y = 12$

$2x + 2y = -8$

$2x + 2y = 5$

Write a sentence describing what you have found.

Will this work for the equation $2x + 3y = 6$? Explain why or why not?

Practice questions 5.5

1 What are the x- and y-intercepts of the following lines? Hence, sketch the lines on the same set of axes. Choose appropriate scales.

a $x + y = 12$ b $x + y = 1$ c $x + y = -3$ d $x = 8 - y$

2 Which of the following equations represent the same line?

a $y = x + 4$ b $x + y = 4$ c $y = x - 4$

d $x + y = -4$ e $y = 4 - x$ f $x = 4 - y$

3 Write the equation of each of the lines shown.

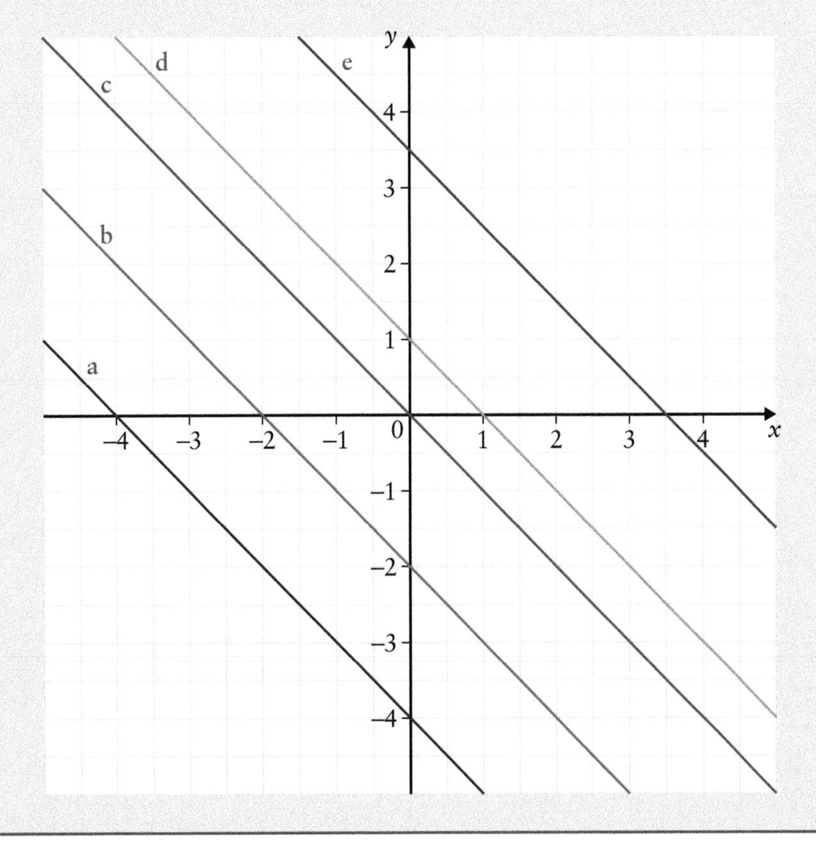

5 The number plane and graphs of linear equations in one and two variables

Self-assessment

- I can identify ordered pairs on the Cartesian plane.
- I can plot points on the Cartesian plane.
- I can describe horizontal and vertical lines using equations.
- I can graph lines of the form $x = k$ and $y = c$.
- I can verify whether or not a point lies on a line.
- I can find points on a line using its equation.
- I can graph a straight line given its equation.
- I can interpret line graphs in context.
- I can create a line graph to represent a real-life situation.

Check your knowledge questions

1 Follow the grid references in the order given and record the letters you find at each location: E1, B3, D1, E4, C1, A2, E3, B4

Use your answer to complete this question:

What is the problem with ______________?

Can you answer the question?

	A	B	C	D	E
1	U	B	M	A	F
2	A	I	R	E	O
3	J	L	W	D	P
4	K	S	C	G	T
5	T	V	H	N	E

2 Use the graph to answer the following questions.

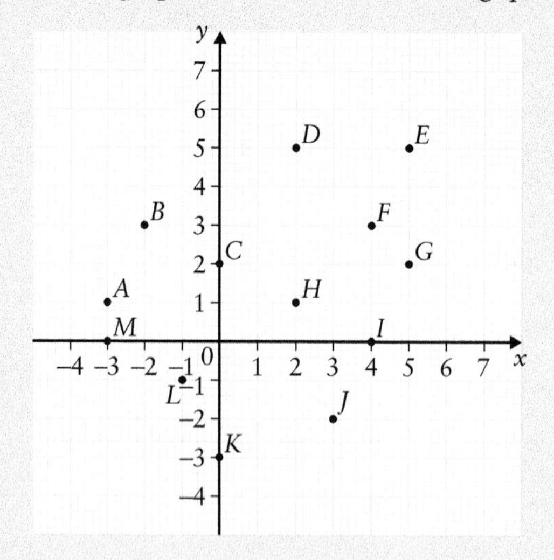

a Which points are on the line $x = 2$?

b Which points are on the line $y = 2$?

c The horizontal line through F also passes through which point? What is the equation of the line?

d The vertical line through E also passes through which other point? What is the equation of the line?

e Which points are on the line $x + y = -2$?

f What is the equation of the line through points D, F and G?

g Which points are on the line $x - y = 1$?

h Point J is on a line of the form $x + y = C$. What other point is on the line, and what is the equation of the line?

i What is the equation of the line through points B, C, H and I?

 Challenge Q2i

 Challenge Q3

3 Some say that the lifespan of a car is around 12 years. Assume that this is true. So, after this point, it is likely to break down and cost more to repair than the value of the car, so it becomes worthless to the owner.

The dotted blue graph in the diagram is the same as that of the one in Worked example 5.1. The orange graph shows a linear fall in the value of the car over its 12-year lifespan. Comparison of the two graphs shows that if you buy a car when it is 8, 9, 10, 11 or 12 years old you would pay a higher percentage of its original value than the percentage of its lifespan remaining.

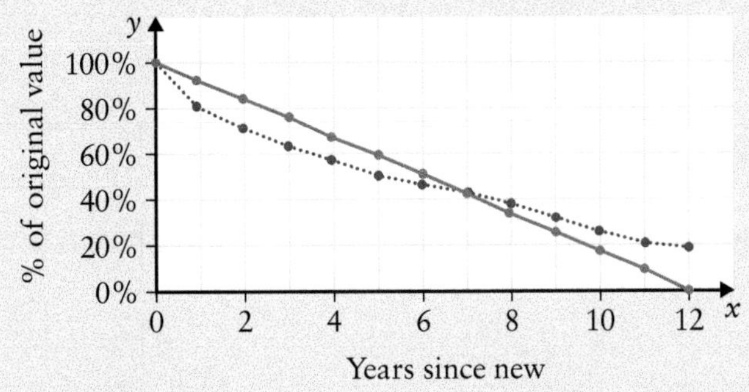

a Tarek is considering buying a seven-year-old car. Would you advise them to buy the car? Explain your answer.

b At what age does a car have the best value? What feature of the graph shows this?

c A typical car loses half of its value after the first five years, and then it loses half of this value after five more years. If this pattern continues, what percentage of the original value will it be worth after

i 15 years ii 20 years?

4 Find the equation of the line parallel to the x-axis which passes through the point (2, 4). Is this a vertical or horizontal line?

5 The values $x = 4$ and $y = 2$ are solutions to both $2x + y = 10$ and $x - 3y = -2$

a What does this tell you about the graphs of these lines?

b Verify this by graphing both lines on the same set of axes (draw axes with these ranges: $-6 \le x \le 10, -2 \le y \le 10$).

c Draw the lines $2x + y = -3$ and $2x + y = 0$. Compare your graphs with the graph of $2x + y = 10$. What do you notice?

6 Graph the lines $2x + y = 0$, $x - y = 0$ and $x + 3y = 0$ on the same set of axes. What point is on every line of the form $Ax + By = 0$, where A and B are constants? Explain why this is.

7 The graph shows three of the lines $x - y = 1$, $y = 1$, $x + y = 3$ and $x + 2y = 2$

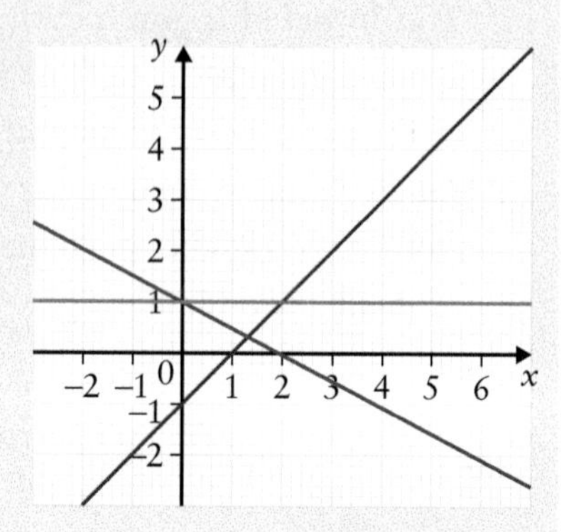

a Copy the graph, and add the line that is missing.

b Think of a reason why three of the four equations form a set but the fourth does not belong to that set.

$x - y = 1$	$y = 1$
$x + y = 3$	$x + 2y = 2$

Hint Q7b

Find a property that three of the equations have in common, but the fourth does not share. Can you make different sets so that each of the equations becomes the odd one out?

For example, in the following square:

1	27
64	256

1 doesn't belong because it is not a composite number (as 27, 64 and 256 are).

27 doesn't belong because it is not a square number (or a power of 2).

64 doesn't belong because the sum of its digits is not odd.

256 doesn't belong because it is not a cube number.

Do you see why saying 'because it is a three-digit number' would not be a good reason? The others must have some characteristic in common that makes them the members of some set. Stating why something is different is not the same as stating why it is not the same.

6 Reasoning in geometry

6 Reasoning in geometry

KEY CONCEPT

Logic

RELATED CONCEPTS

Quantity, Space

GLOBAL CONTEXT

Identities and relationships

Statement of inquiry

By using logical reasoning, the relationships between quantities within shapes can be used to solve problems in space.

Factual

- What is the angle sum of a quadrilateral?

Conceptual

- What are the properties of angles on parallel lines with a transversal cutting the parallel lines?

Debatable

- Can you see mathematics in all structures?

Do you recall?

1 What do the angles in a right angle add up to?

2 What do the angles on a straight line add up to?

3 What do the angles in a full turn add up to?

4 What are complementary angles and supplementary angles?

5 What is a straight line crossing two parallel lines called?

6 What are the properties of an equilateral triangle, an isosceles triangle and a scalene triangle?

6.1 Reasoning with adjacent angles

6.1.1 Adjacent angles

The facts used to calculate angles covered in Chapter 4 from Book 1 will be used in this chapter to develop reasoning.

Some Greek letters are frequently used to represent the size of an angle.

Fact

Greek letters are used to represent the sizes of angles. Find out which letters are used, what the letters look like, how to say them and how to write them.

Explore 6.1.1

Can you work out the value of a in each of these diagrams?

Include your reasoning for each diagram.

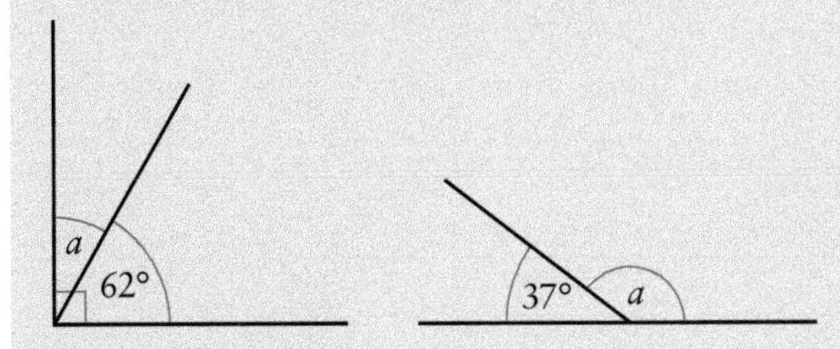

Missing angles often have more than one reason for the method used. Can you use different reasoning to find either value above?

Worked example 6.1

a Find the value of α, giving a reason for your answer.

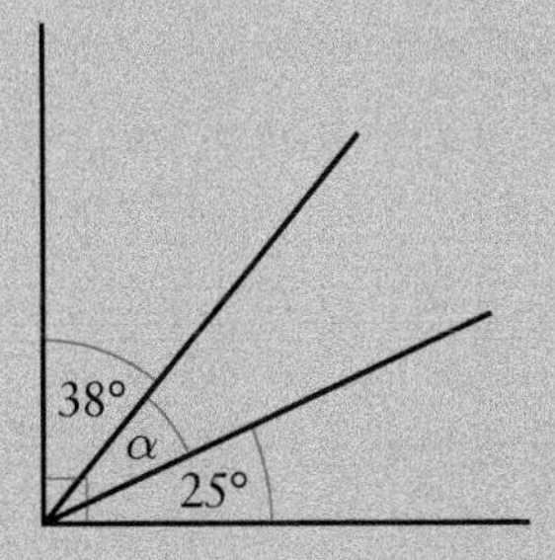

b Find the value of β, giving a reason for your answer.

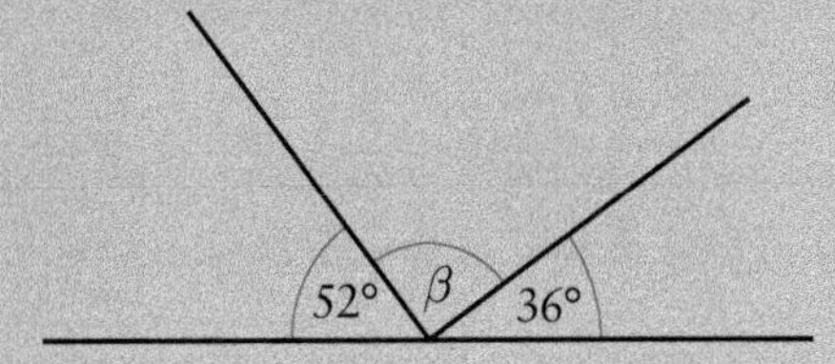

Solution

a Understand the problem

We need to work out the value of α.

Reminder

Angles in a right angle sum to 90°.

Make a plan

We can use the fact that the angles make a right angle so total 90°.

Carry out the plan

$\alpha + 25 + 38 = 90$

$\alpha = 27°$.

Look back

Does the answer work?

$27 + 25 + 38 = 90$

b Understand the problem

We need to work out the value of β.

Reminder

Angles on a straight line sum to 180°.

Make a plan

We can use the fact that the angles make a straight line so total 180°.

Carry out the plan

$\beta + 52 + 36 = 180$

$\beta = 92°$.

Look back

Does the answer work?

$92 + 52 + 36 = 180$

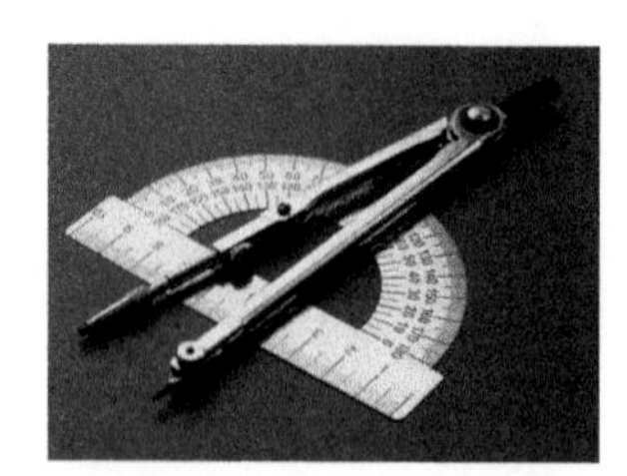

Reflect

Can you explain why using a protractor to measure the sizes of the angles in the Explore activity and the Worked examples does not necessarily give the correct answer?

Can you explain why the angles on a straight line total 180°?

Practice questions 6.1.1

1 In each of these diagrams write down the value of the variable. Explain your reasoning.

a

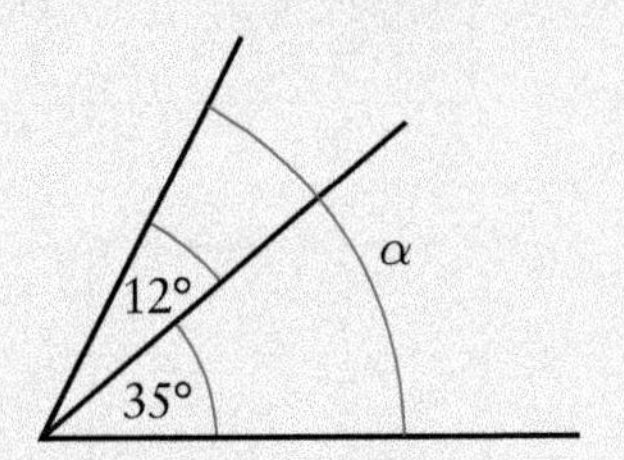

b

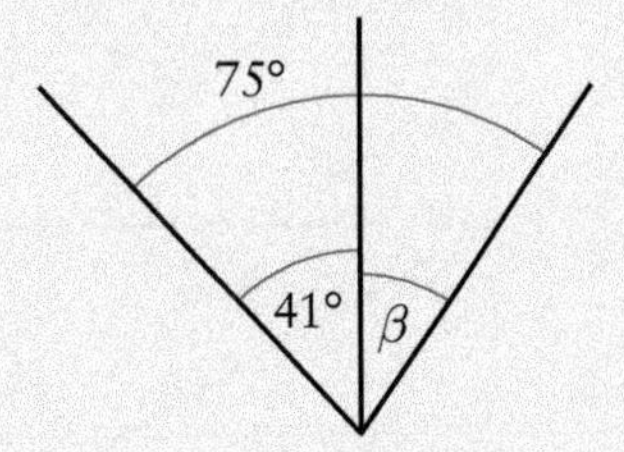

c

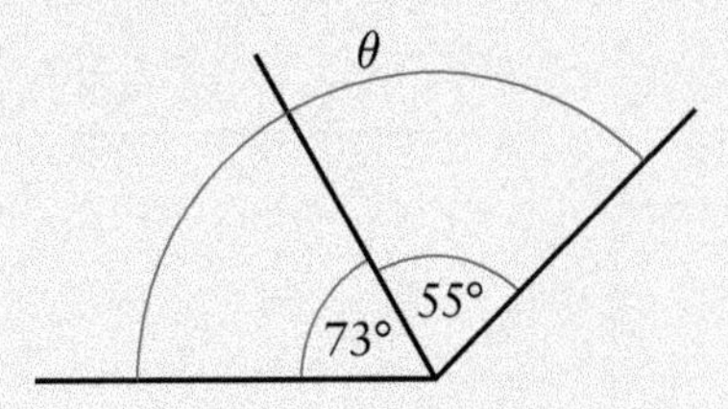

d

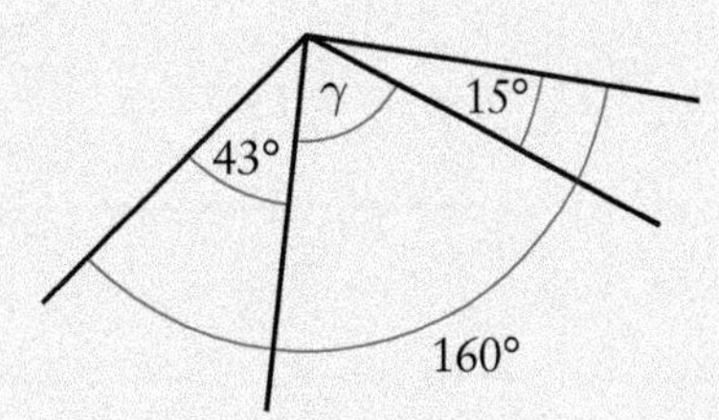

e

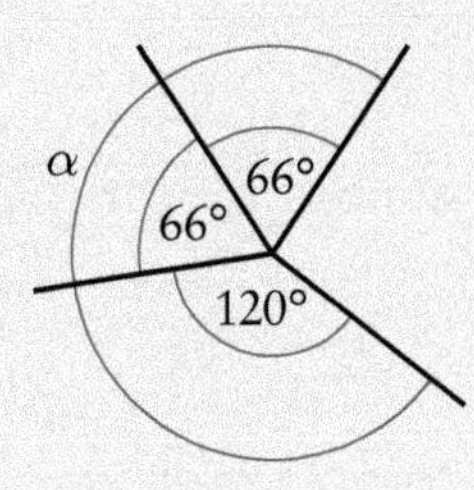

f

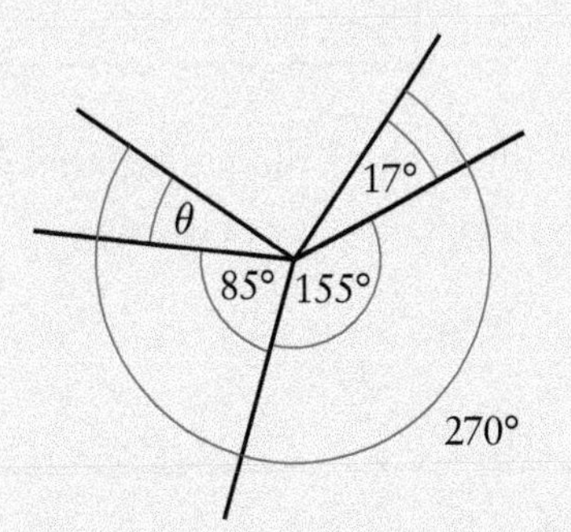

2 In each of these diagrams write down the value of the variable. Explain your reasoning.

a

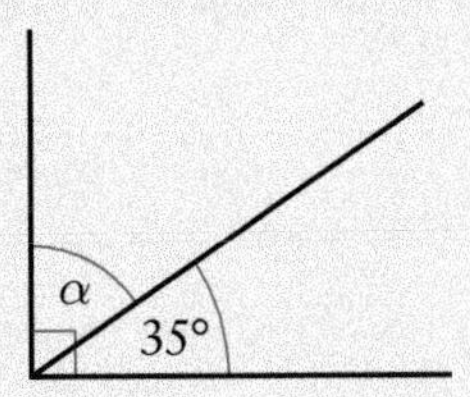

b

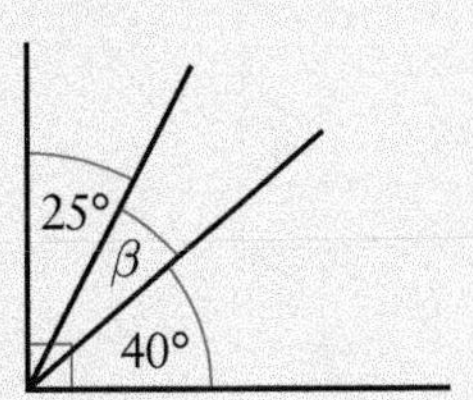

c

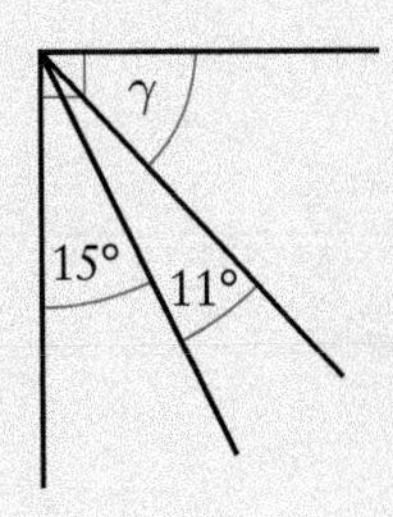

d

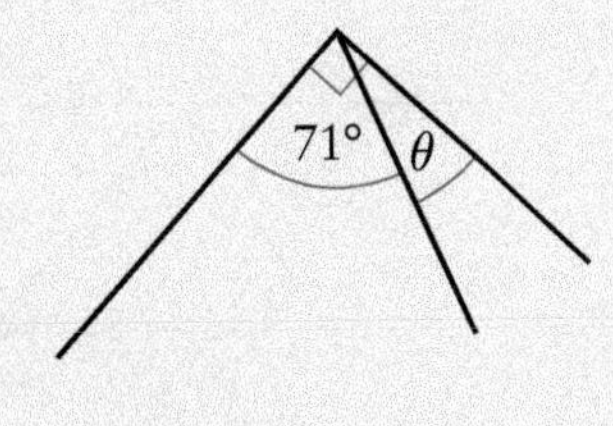

e

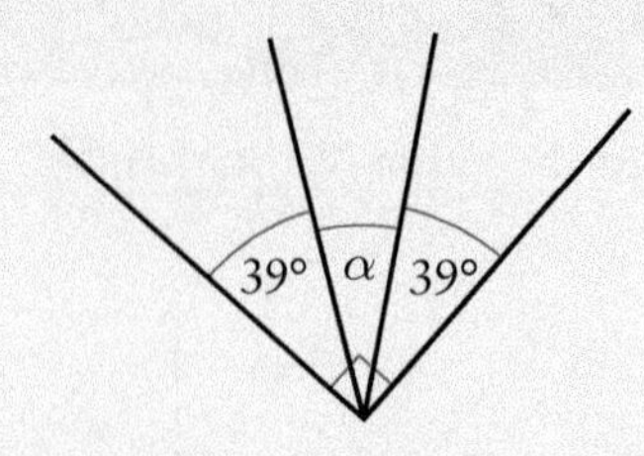

f

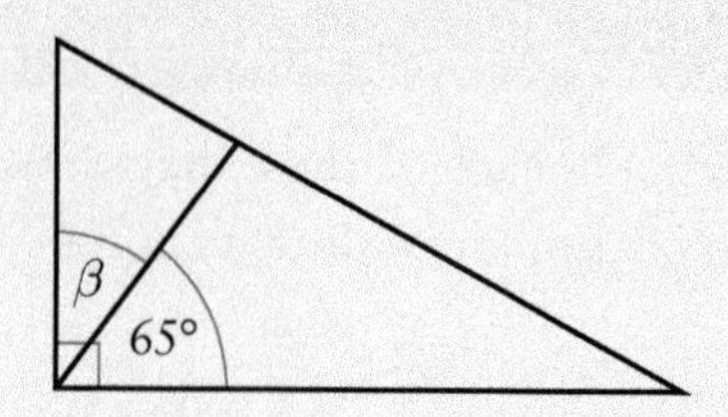

g

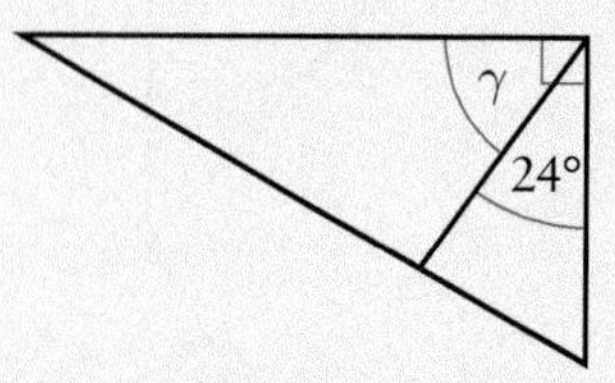

h

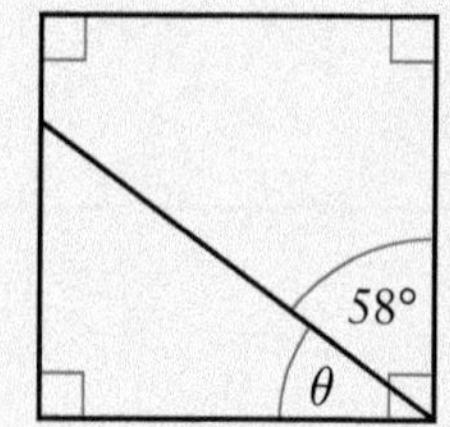

i

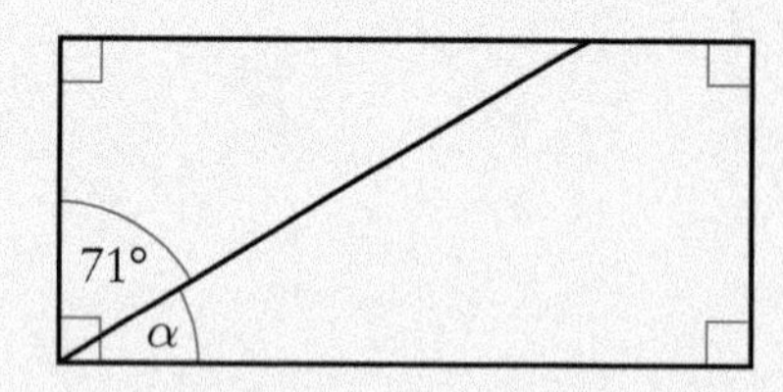

3 In each of these diagrams write down the value of the variable. Explain your reasoning.

a

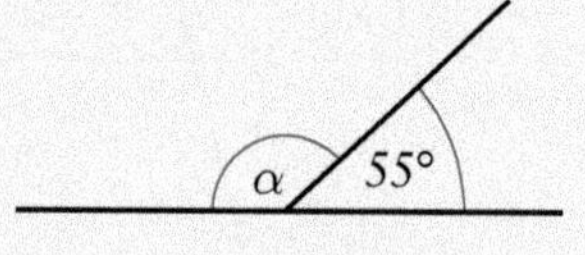

b

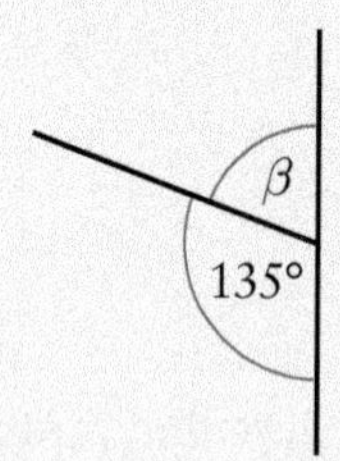

c

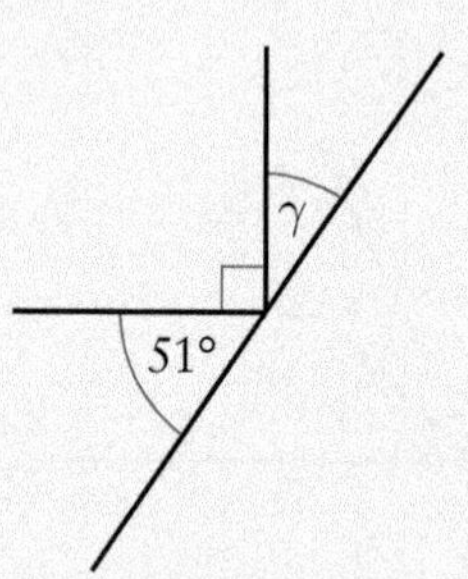

d

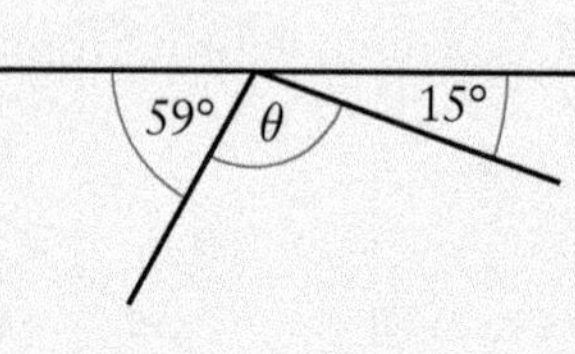

e

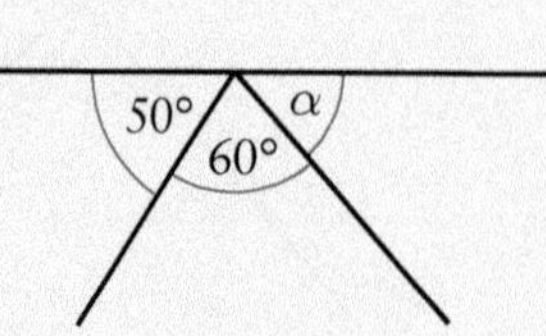

f

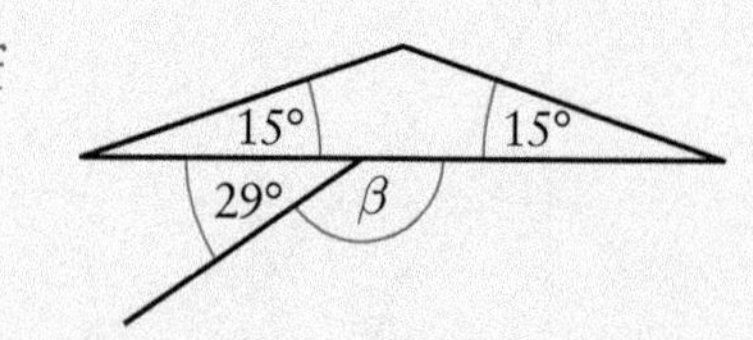

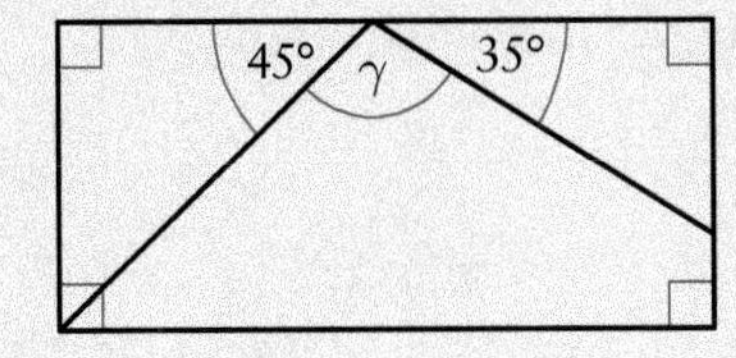

h

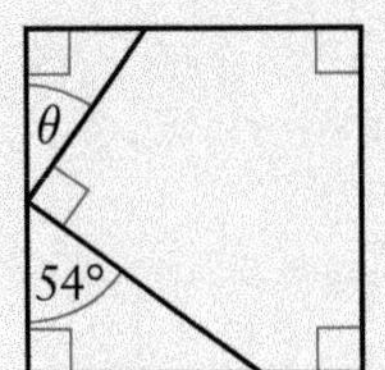

i

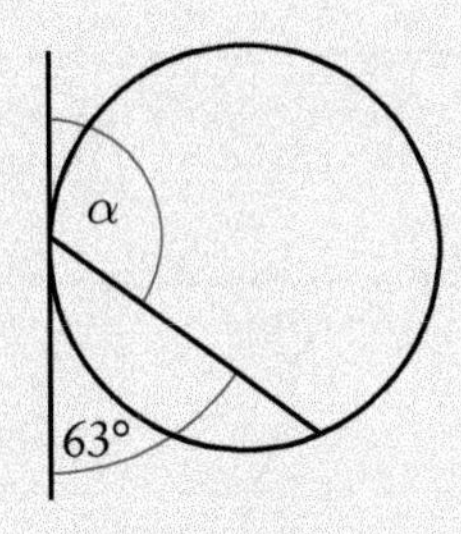

4 In each of these diagrams write down the value of the variable. Explain your reasoning.

a

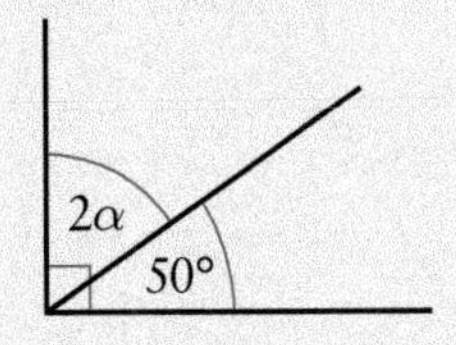

b

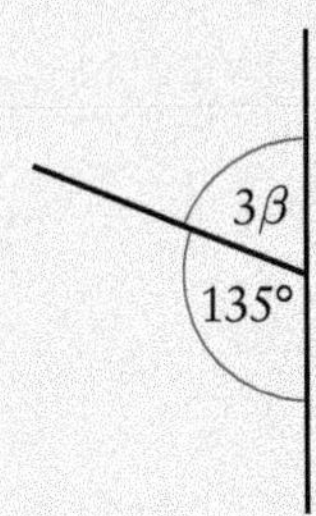

c

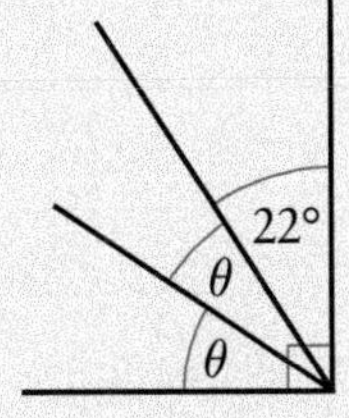

d

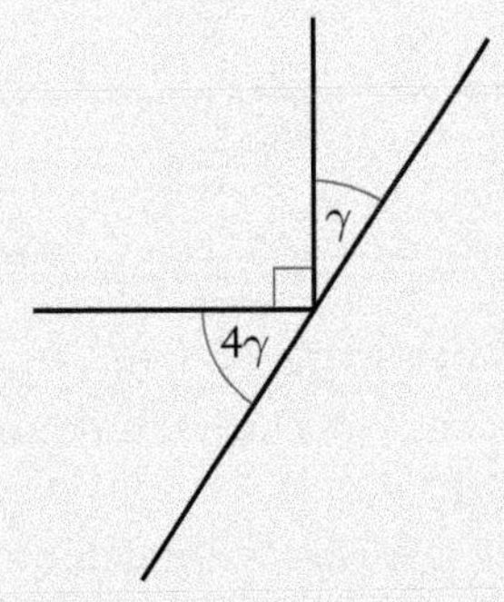

e

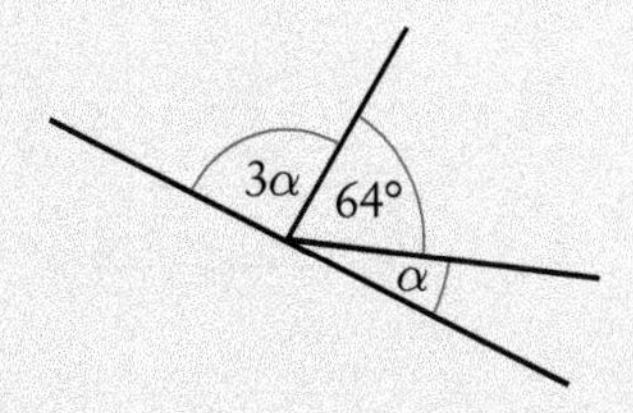

5 Work out the size of each of the angles in each diagram.
Show your working.

a
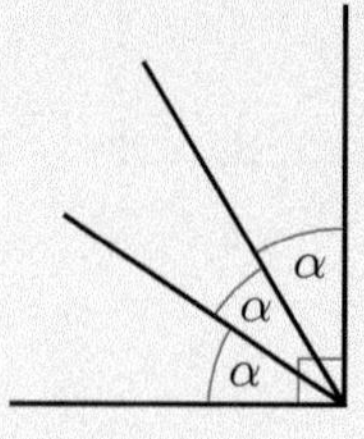

b
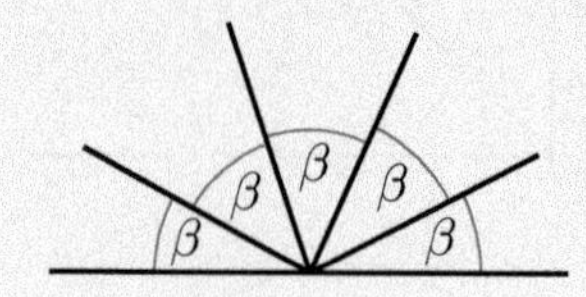

c
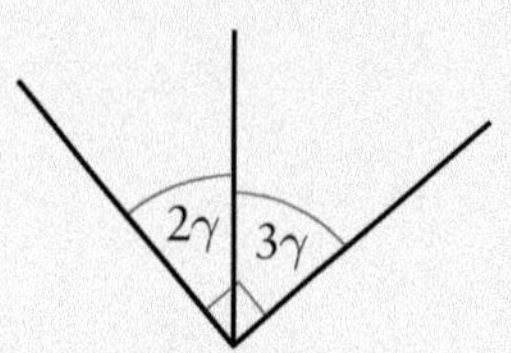

d
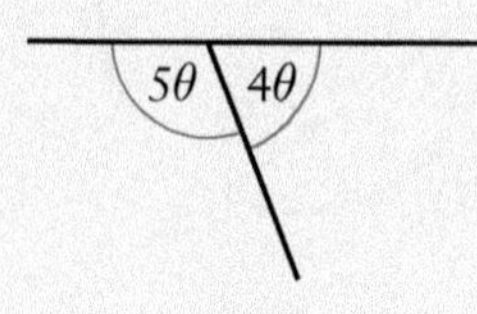

e
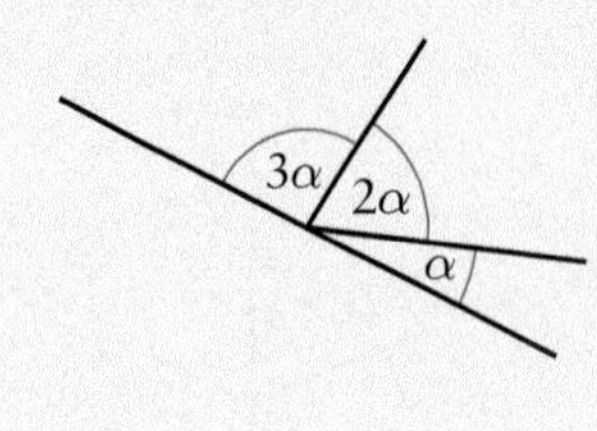

f
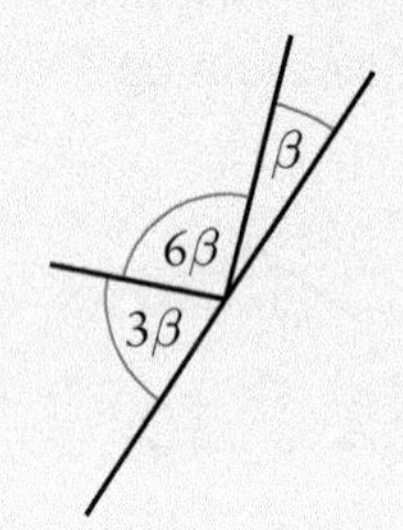

6 Darian is asked to work out the value of α using this diagram.

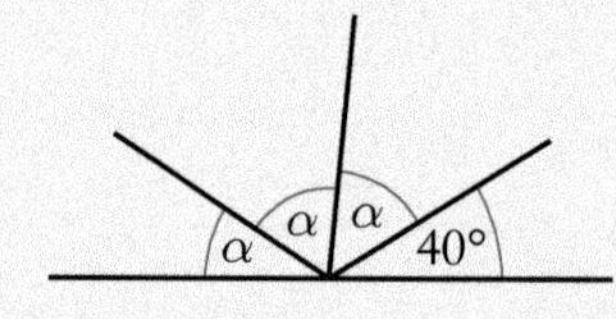

Darian writes, '$\alpha = 40°$, because all the angles are the same'.
Explain why Darian is incorrect.

7 Work out which is greater, α or θ? Explain your answer.

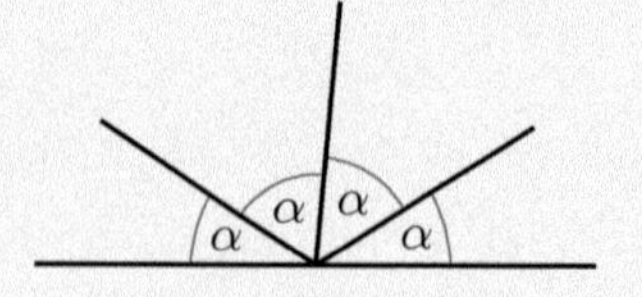

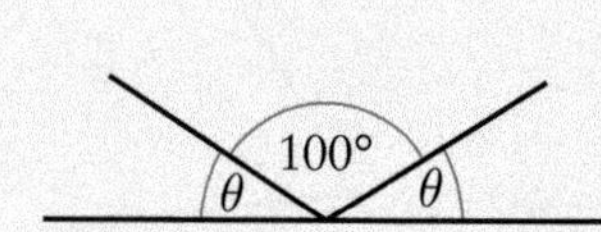

8 Look at this diagram.

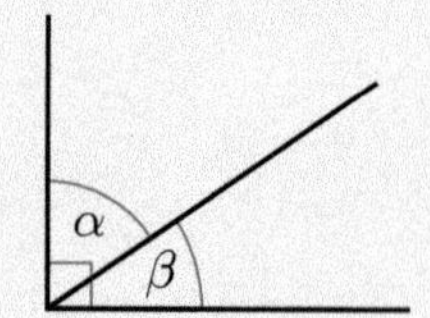

Aisha says, 'α and β must both be the same as the right angle is split into two'.

Give an example to show Aisha is incorrect.

9 The diagram shows a square.
One of the vertices of the square lies on a straight line.
Fahad says, '$\theta = 45°$'.
Is Fahad correct? Explain your answer.

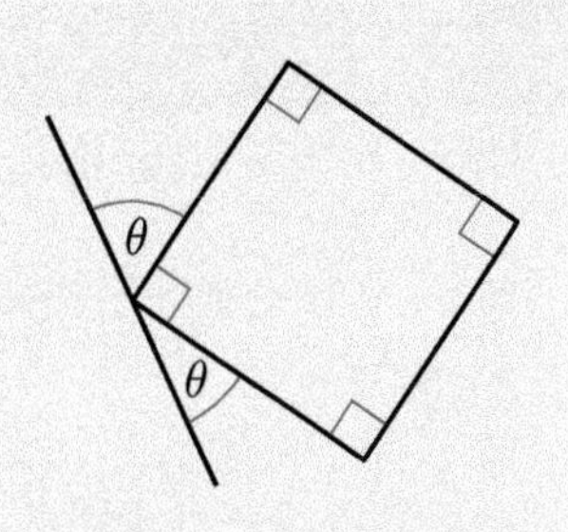

10 Here is a diagram of a right angle.
θ is $\frac{2}{3}$ the size of α. Work out the values of α and θ. Show your working.

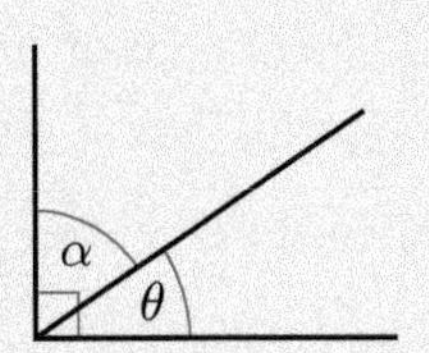

Challenge Q10

11 In the diagram the values of α and β are in the ratio $\alpha : \beta = 7 : 3$. Work out the size of both angles.
You must show your working.

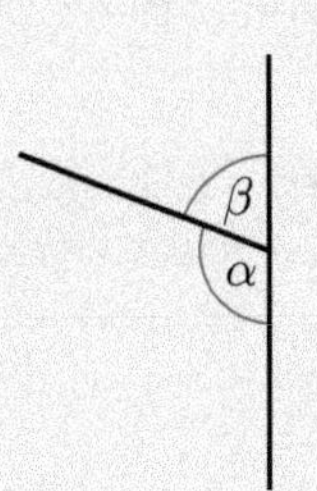

Challenge Q11

12 In the diagram, β is 4% of the value of θ.
Work out the value of α and β.
You must show your working.

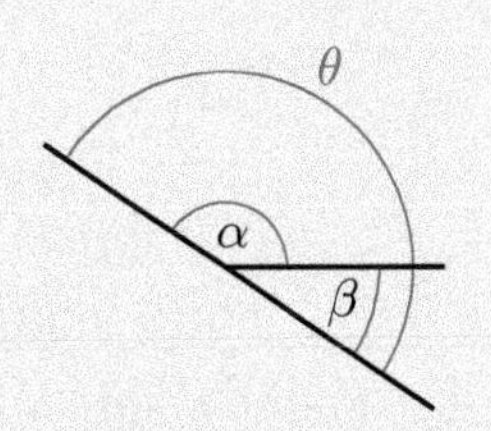

Challenge Q12

6.1.2 Angles at a point

PQ and ST are straight lines that intersect at point Z.

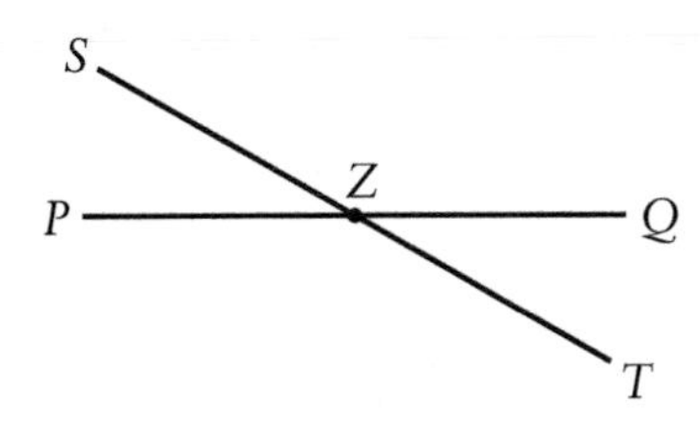

$\angle PZS + \angle SZQ + \angle QZT + \angle PZT = 360°$

$\angle PZS$ is vertically opposite to $\angle QZT$

$\angle SZQ$ is vertically opposite to $\angle PZT$

Explore 6.2

Draw two straight lines that cross like this:

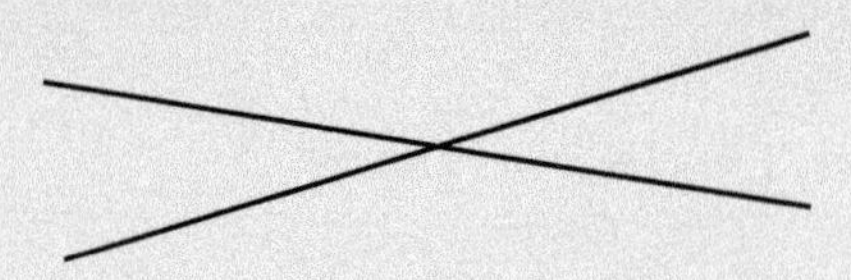

Measure each of the four angles and write your measurements on your diagram.

What do you notice about the angles on your diagram?

Repeat this for lines that cross at different angles. Does the same happen? Can you explain why?

Worked example 6.2

a Work out the value of α, giving a reason for your answer.

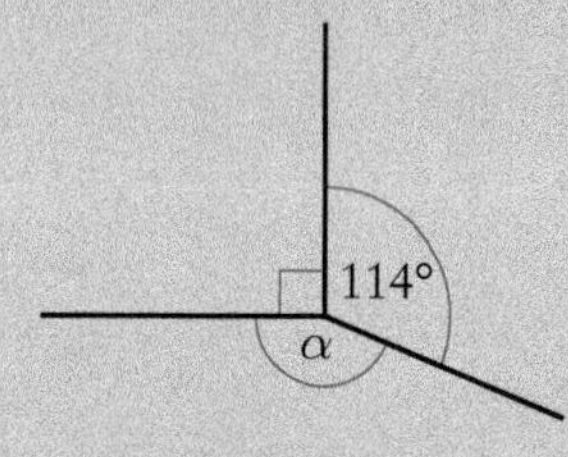

b Find the values of the variables, giving reasons for your answers.

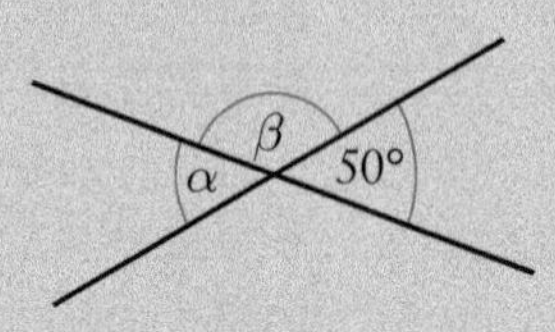

c Work out the value of θ, give a reason for your answer.

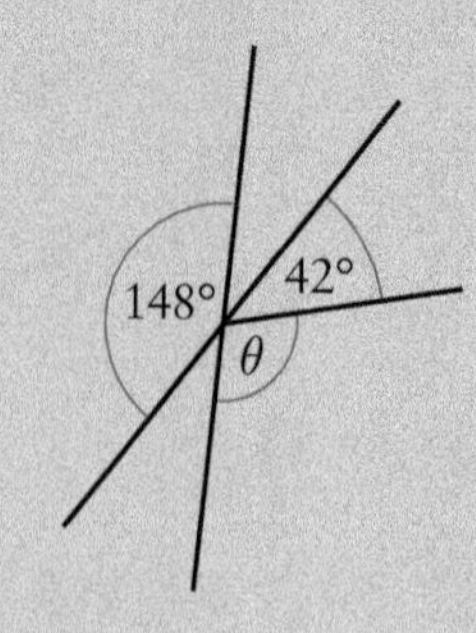

Solution

a Understand the problem

We need to work out the size of the value of α.

Make a plan

We can use the fact that the angles make a full turn so total 360°.

 **Reminder**

Angles at a point sum to 360°.

Carry out the plan

$\alpha + 90 + 114 = 360$

$\alpha = 156°$.

Look back

Does the answer work?

$156 + 90 + 114 = 360$

b Understand the problem

We need to work out the size of the values of α and β.

Make a plan

We can use the fact that the angles on a straight line total 180° to work out the value of β. We can use the fact that vertically opposite angles are equal to write down the value of α.

 **Reminder**

Vertically opposite angles are equal.

Carry out the plan

$\alpha = 50°$ as vertically opposite angles are equal.

The adjacent angles total 180° as the angles make a straight line:

$\beta + 50 = 180$

$\beta = 130°$.

Look back

Does the answer work?

$\alpha + \beta = 50 + 130 = 180$

c Understand the problem

We need to work out the value of θ.

Make a plan

We can use the fact that vertically opposite angles are equal.

Carry out the plan

Since vertically opposite angles are equal:

$\theta + 42 = 148$

$\theta = 106°$.

Look back

Does the answer work?

$106 + 42 = 148$

Reflect

Can you explain why angles at a point have a sum of 360°?

Can you explain why vertically opposite angles are equal?

Can you explain why you cannot work out the value of α in this instance?

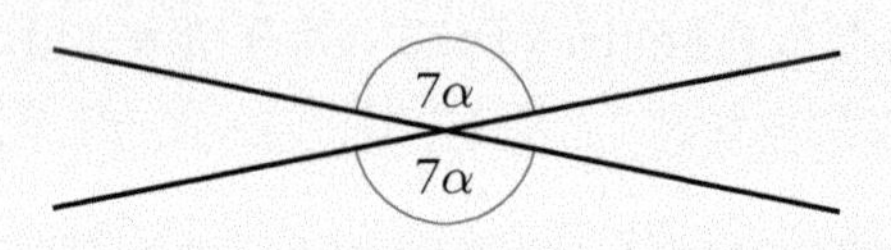

Practice questions 6.1.2

1 For each of the diagrams, work out the value of the variable.
You must show your working.

a

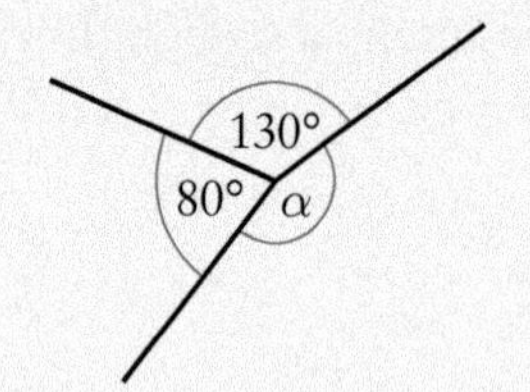

b

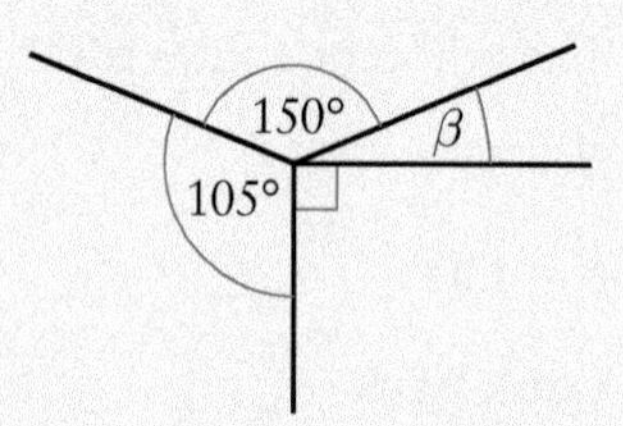

c

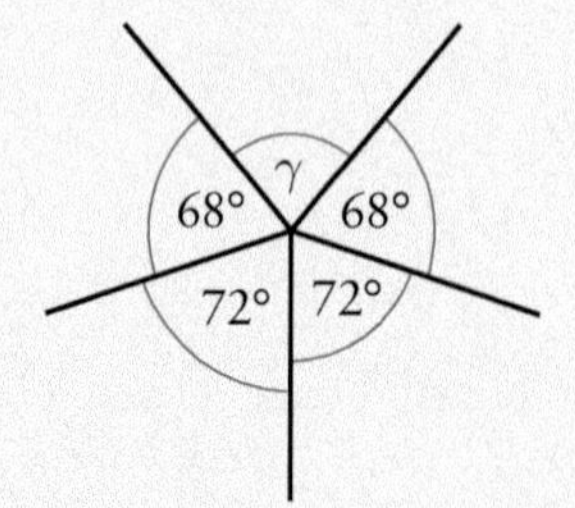

d

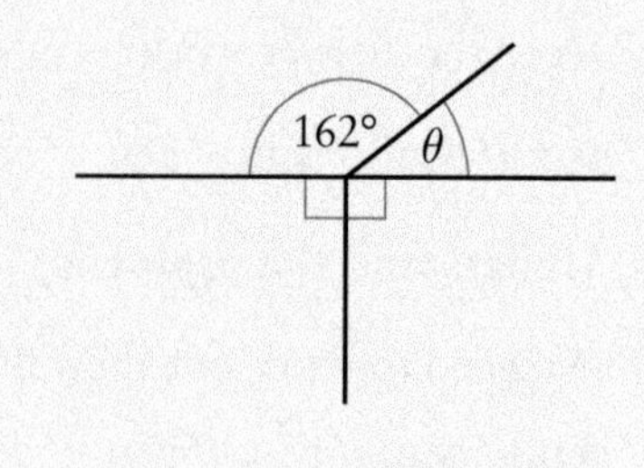

e

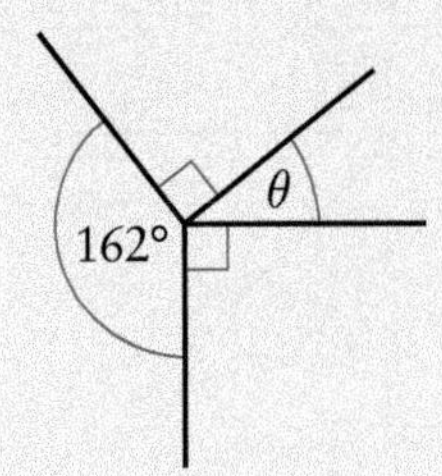

2 For each of the diagrams, work out the value of the variable.
You must show your working.

a

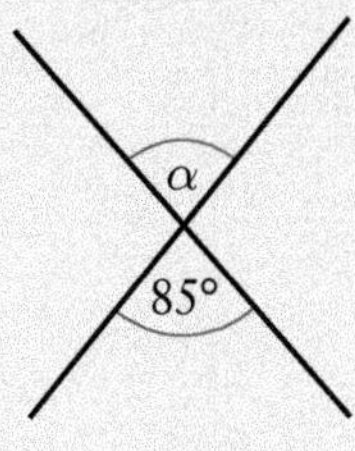

b

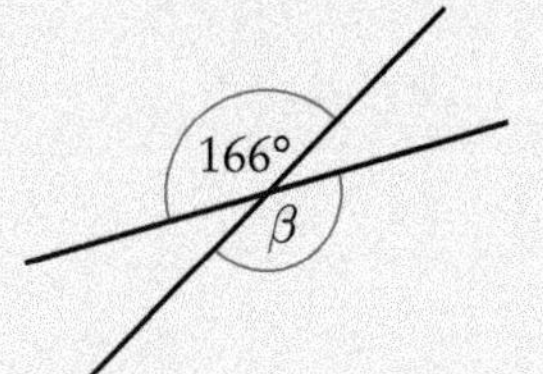

c

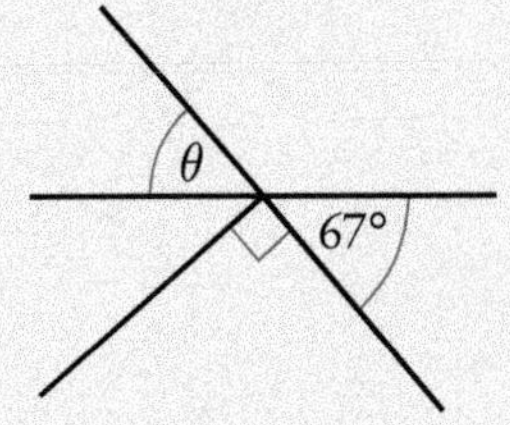

d

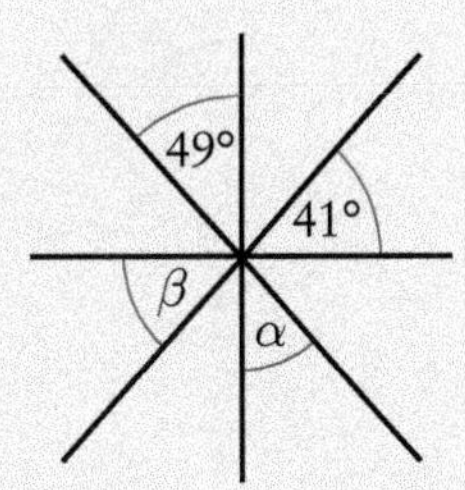

e

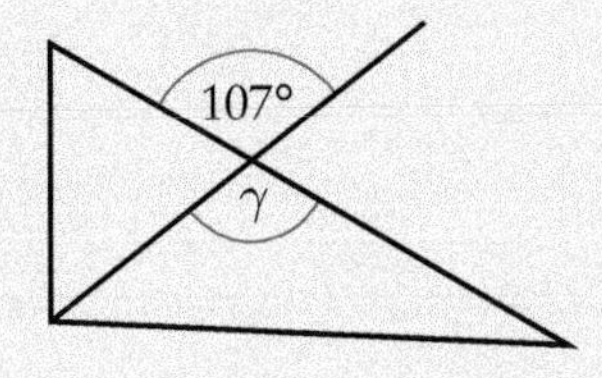

f

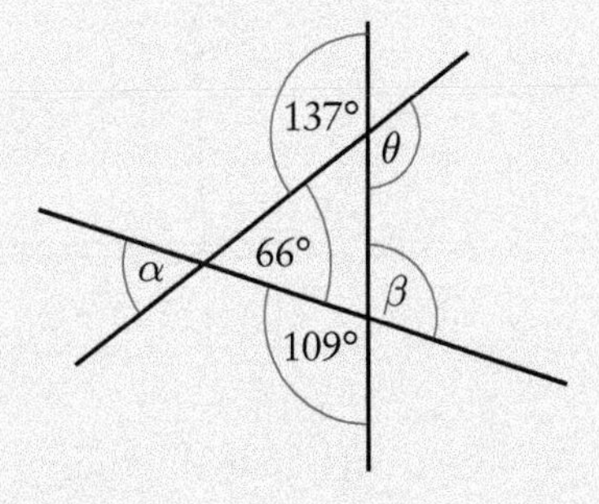

3 For each of the diagrams, work out the value of the variable.
You must show your working.

a

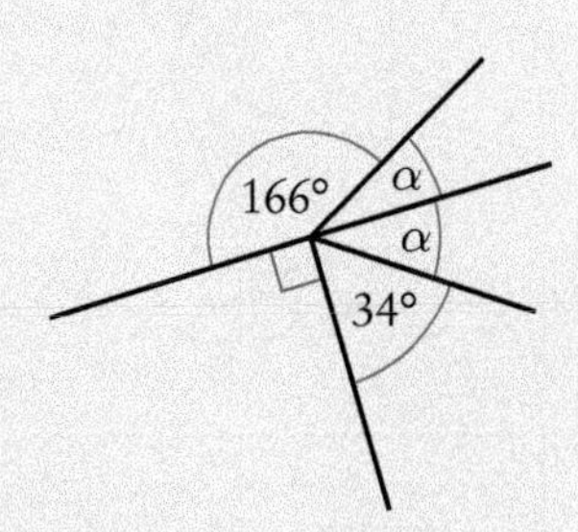

b

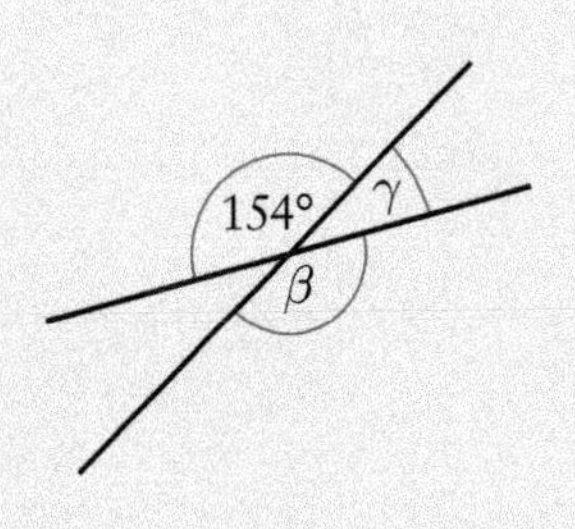

c

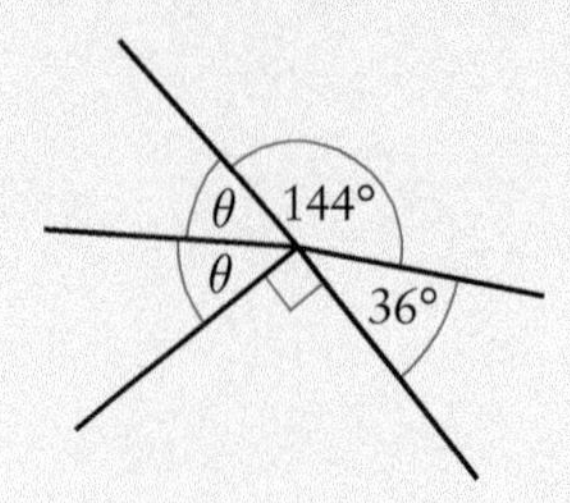

d

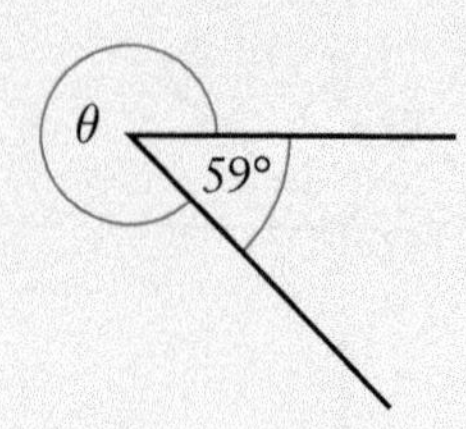

e

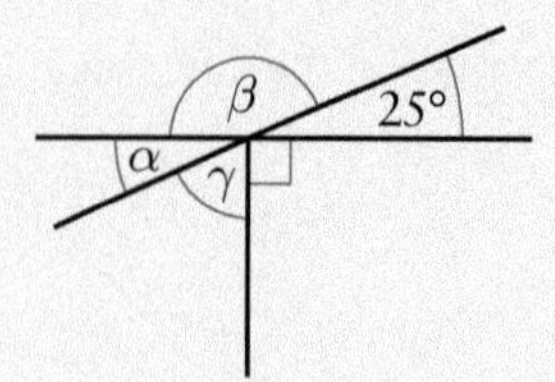

f

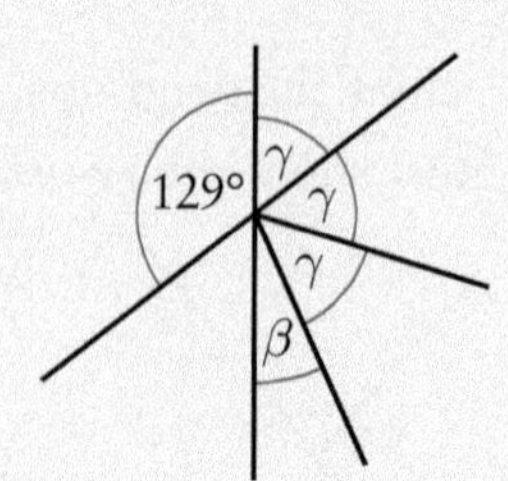

g

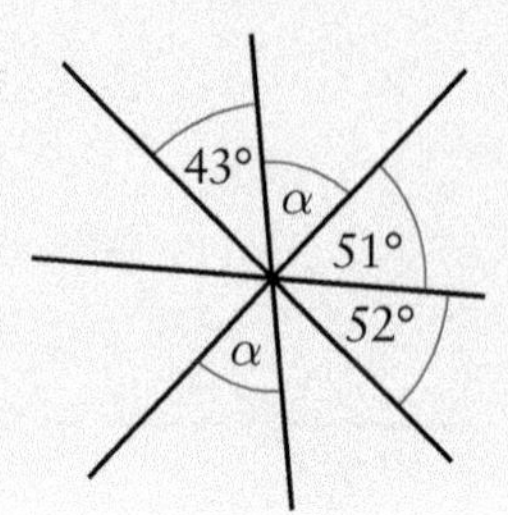

h

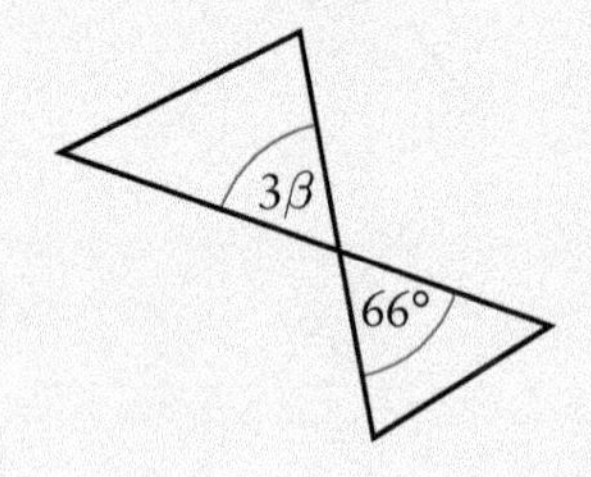

i

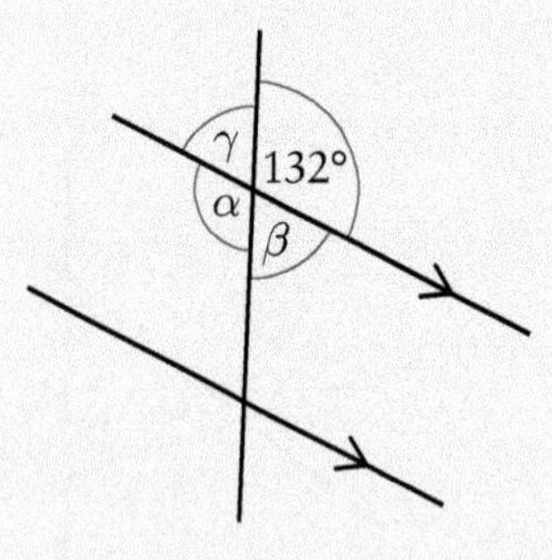

4 For each of the diagrams, work out the value of the variable.
You must show your working.

a

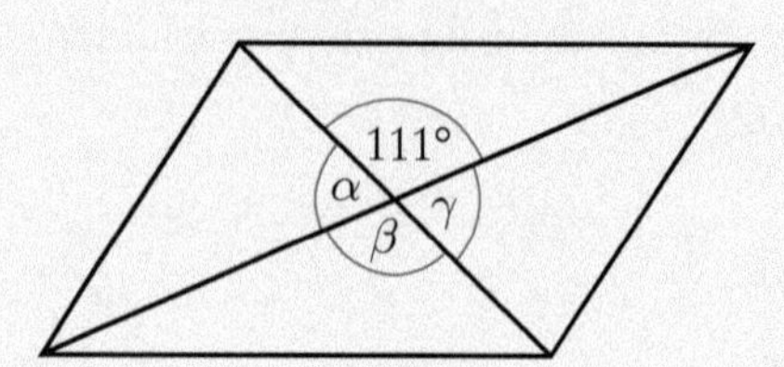

b

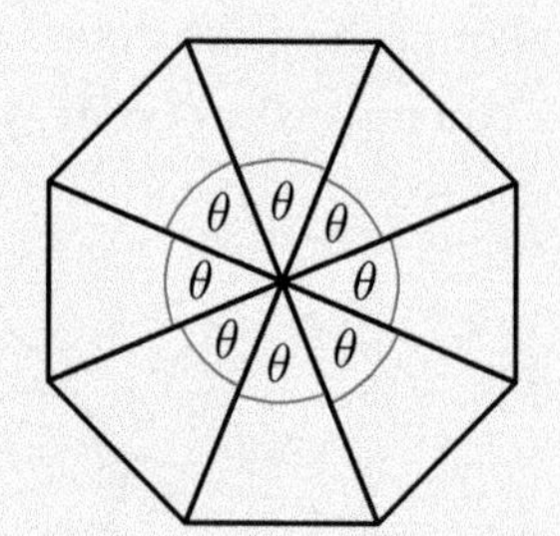

c

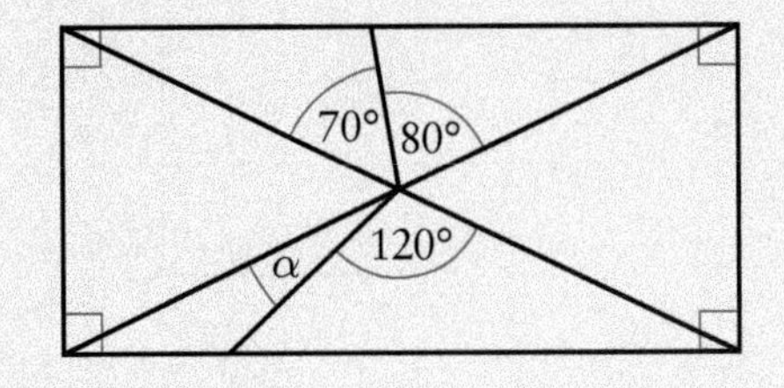

d

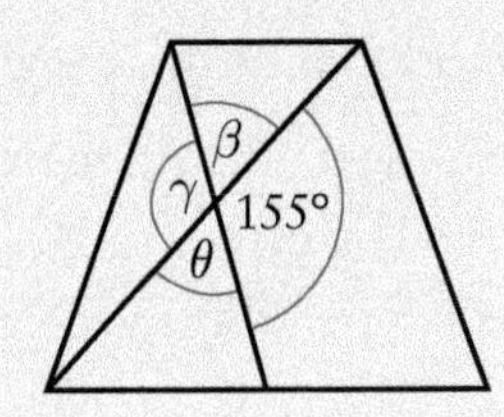

5 For each of the diagrams, work out the value of the variable.
You must show your working.

a

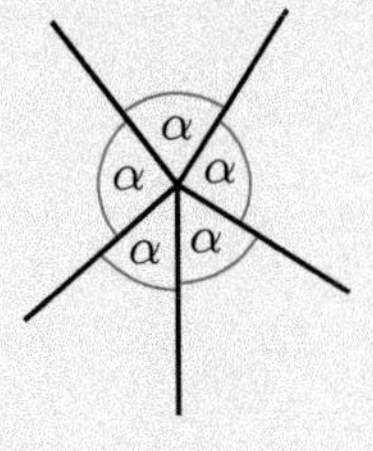

b

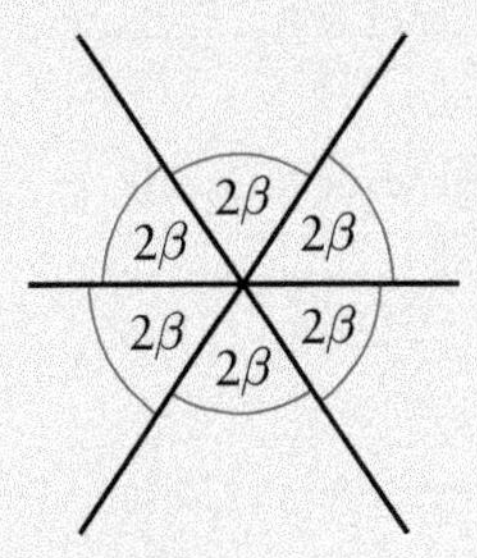

c

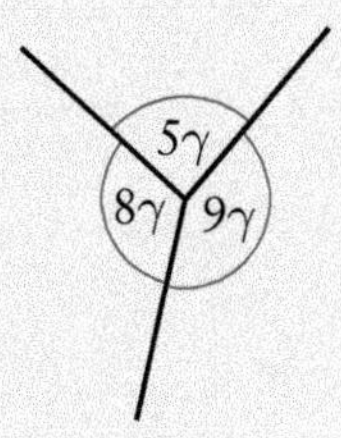

d

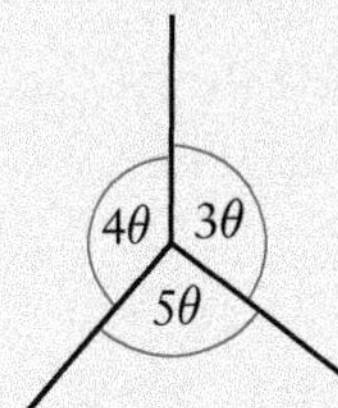

e

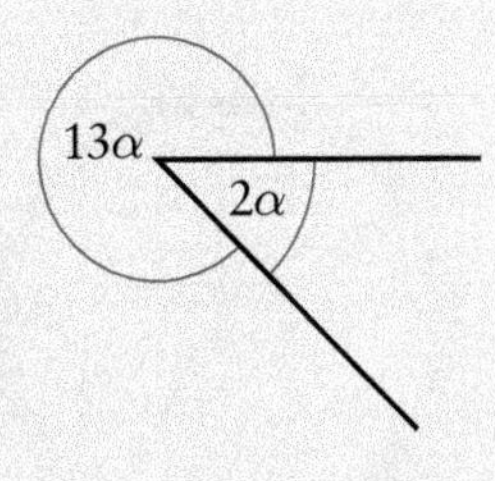

f

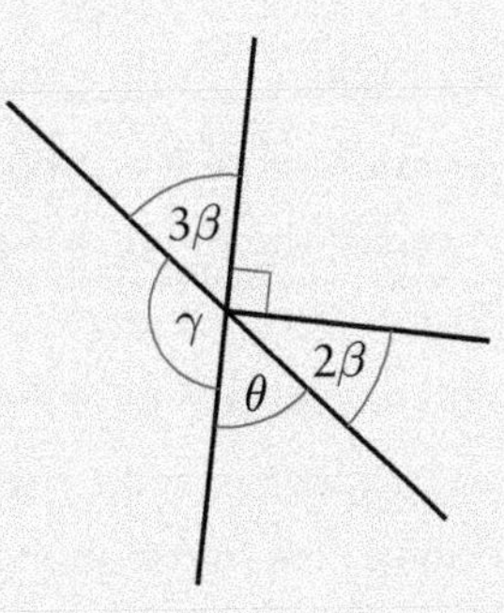

6 In the diagram the values of α and β are in the ratio $\alpha : \beta = 7 : 13$

Challenge Q6

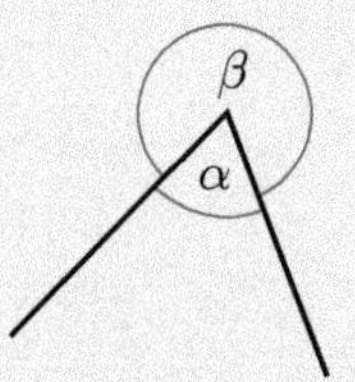

Work out the value of α and β. You must show your working.

7 Anya looks at this diagram.

Anya says, 'the value of θ is 90°'.
Is she correct?
Explain your answer.

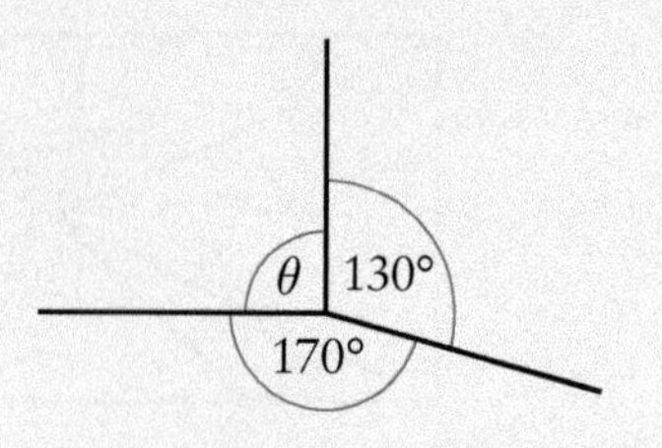

8 Dara is asked to work out the value of θ using this diagram.

Dara says, 'θ = 15° as it is half of 30°'
Is he correct?
Explain your answer.

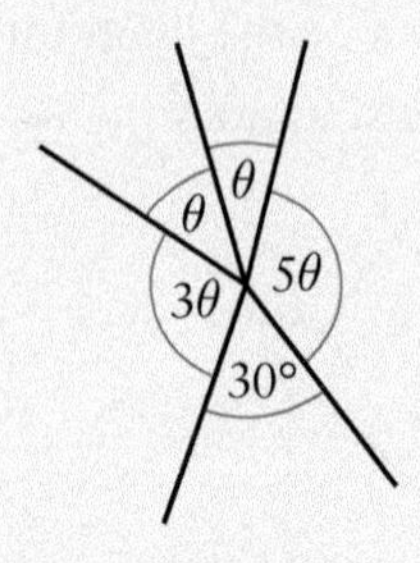

Challenge Q9

9 On the diagram the values of α, β and γ are in the ratio $\alpha:\beta:\gamma = 2:8:7$

Work out the values of α, β and γ.
You must show your working.

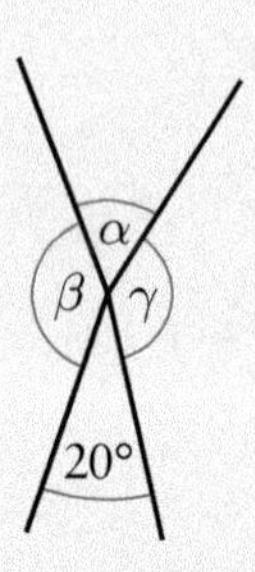

Challenge Q10

10 Abasi looks up at the clock on the wall.

The hands show it is 5 o'clock.
He notices the hands form an obtuse angle and a reflex angle.

Work out the size of the obtuse and reflex angles formed by the hands at 5 o'clock. You must show your method.

Challenge Q11

11 In the diagram the value of β is 10% larger than the value of α.

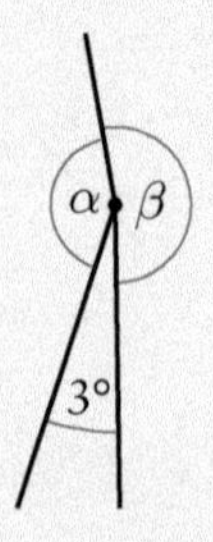

Work out the values of α and β. You must show your working.

12 This structure is made using regular hexagonal glass panels that fit together with no gaps. Work out the size of one of the interior angles of each glass panel.

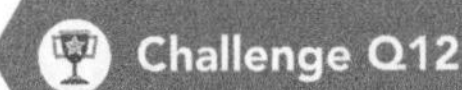

Hint Q12

A regular hexagon is a six-sided shape with six equal sides and six equal angles.

6.2 Angles and parallel lines

Using your knowledge of vertically opposite angles from the previous lesson, you can see that this diagram has four pairs of angles with equal measures:

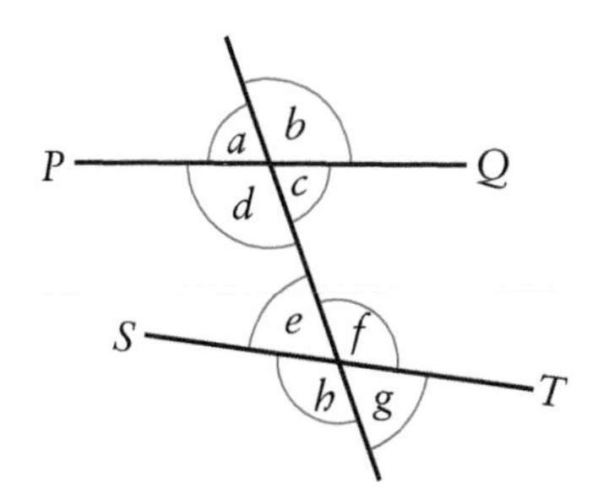

In this lesson we are going to explore what happens if the lines PQ and ST are parallel.

Explore 6.3

Draw a diagram with a pair of parallel lines cut by a transversal (see diagram).

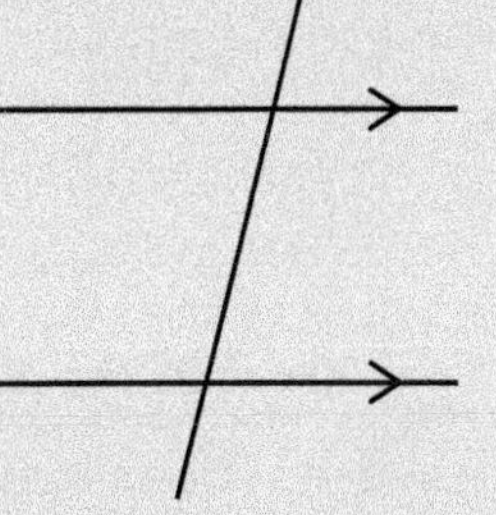

Measure the size of all eight angles and write the measurements on your diagram.

What do you notice?

Does this work for all cases when you have a pair of parallel lines cut by a transversal? Can you explain why?

Fact

A pair of parallel lines cut by a transversal, forms these four pairs of equal corresponding angles:

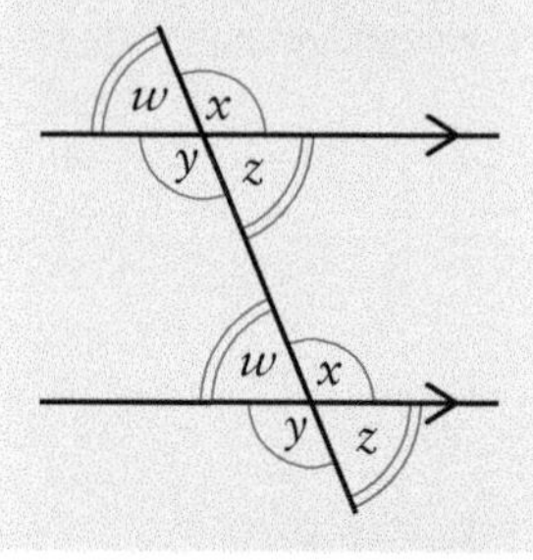

Fact

A pair of parallel lines cut by a transversal, forms these two pairs of equal alternate angles:

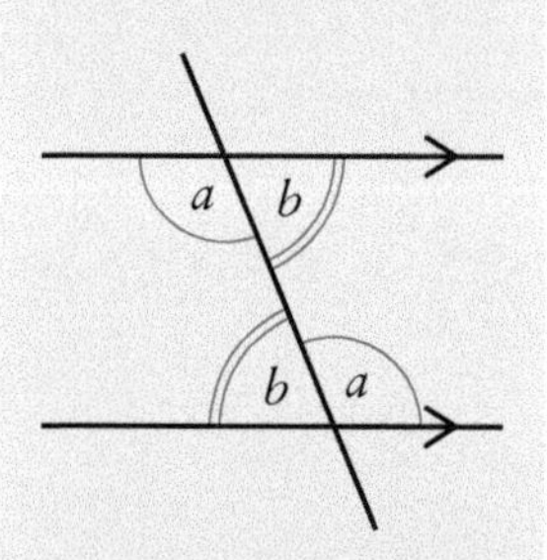

Fact

A pair of parallel lines cut by a transversal, forms these two pairs of supplementary co-interior angles *c* and *d*; and *e* and *f*:

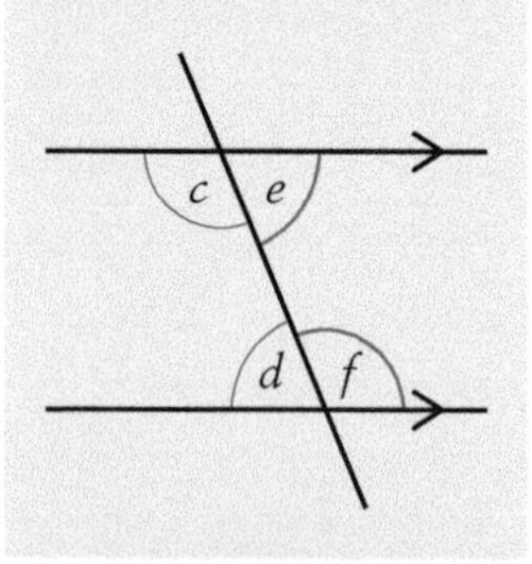

Reminder

Alternate angles have equal measures.

Worked example 6.3

a Find the value of α, giving a reason for your answer.

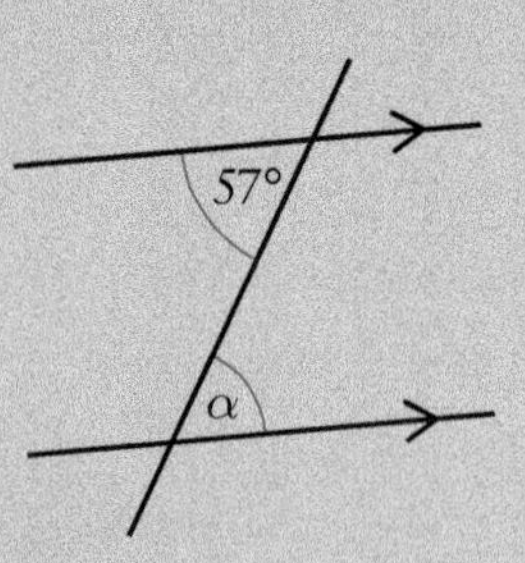

b Find the value of β, giving a reason for your answer.

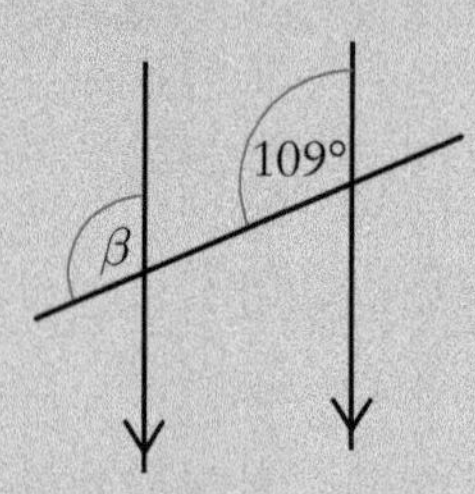

c Find the value of the θ, giving a reason for your answer.

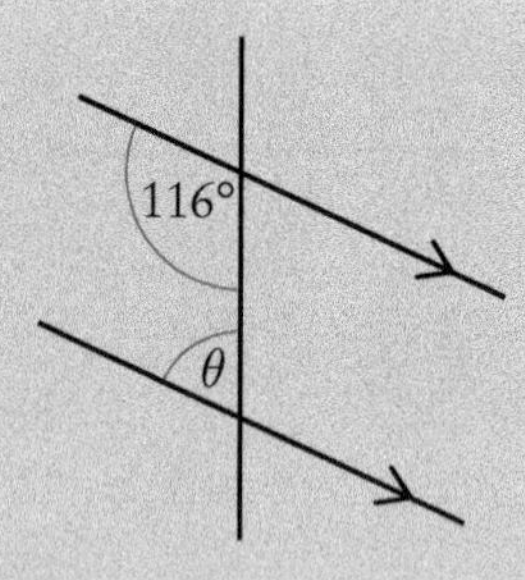

Solution

a **Understand the problem**

We need to work out the size of the value of α.

Make a plan

We can use the fact that alternate angles are equal to write down the value of α.

Carry out the plan

$\alpha = 57°$ as alternate angles are equal.

b **Understand the problem**

We need to work out the value of β.

Make a plan

We can use the fact that corresponding angles are equal to write down the value of β.

Carry out the plan

$\beta = 109°$ as corresponding angles are equal.

c **Understand the problem**

We need to work out the value of θ.

Make a plan

We can use the fact that co-interior angles total 180° to work out the value of θ.

 **Reminder**

Co-interior angles are supplementary and therefore sum to 180°.

Carry out the plan

$\theta + 116 = 180$ as co-interior angles total 180°.

$\theta = 64°$.

Look back

Does the answer work?

$64 + 116 = 180$

Reflect

Can you come up with a way of remembering the three different types of angles involving parallel lines: corresponding angles, alternate angles and co-interior angles? Share your method of remembering with others in your group.

Practice questions 6.2

1 Copy each of these the diagrams. For each diagram write in the sizes of all eight angles on the diagram.

a

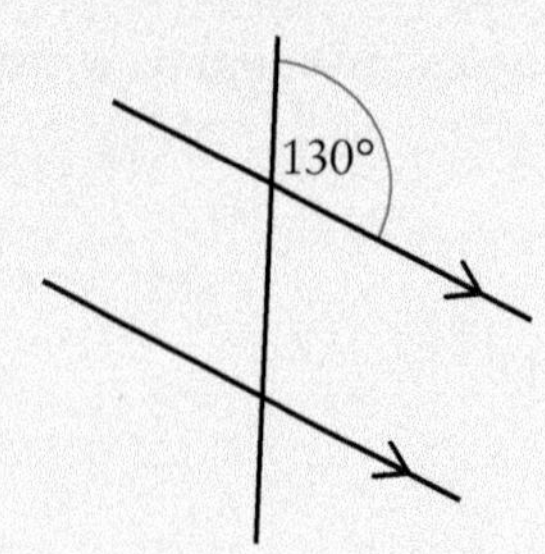

b

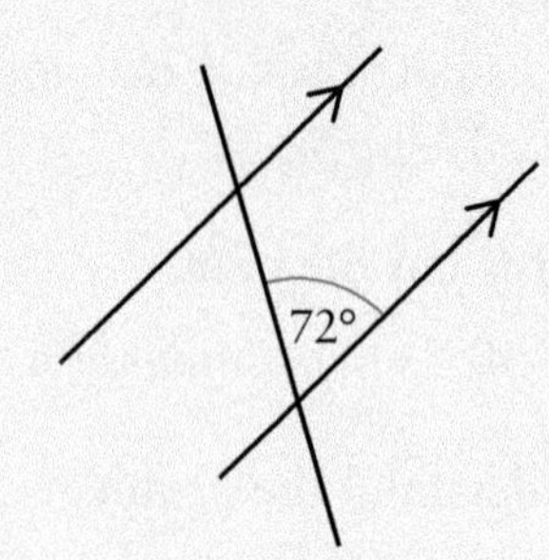

c

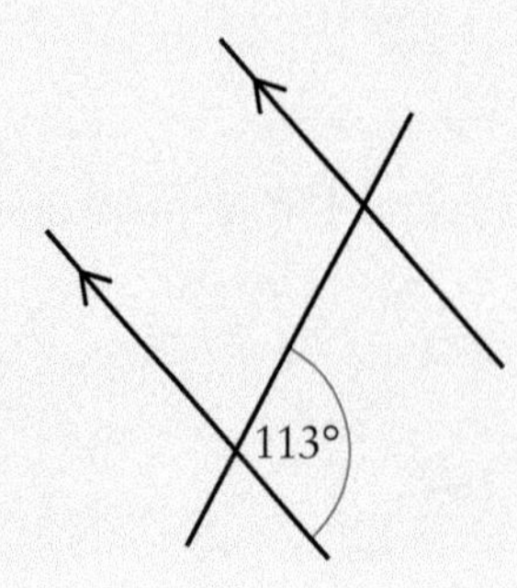

2 Copy each of these the diagrams.

For each diagram, mark the four pairs of corresponding angles.

Use four different colours.

a

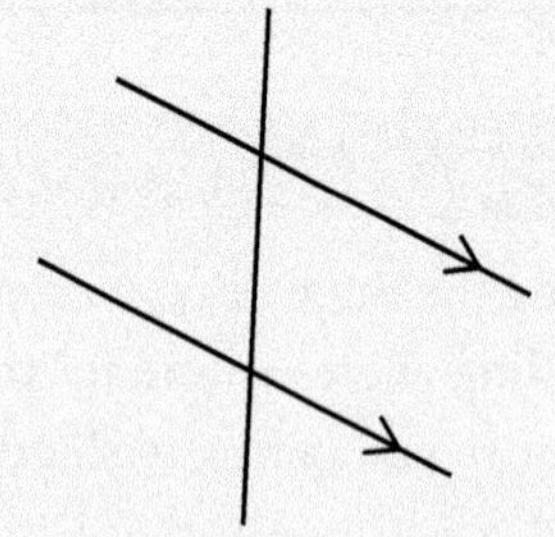

b

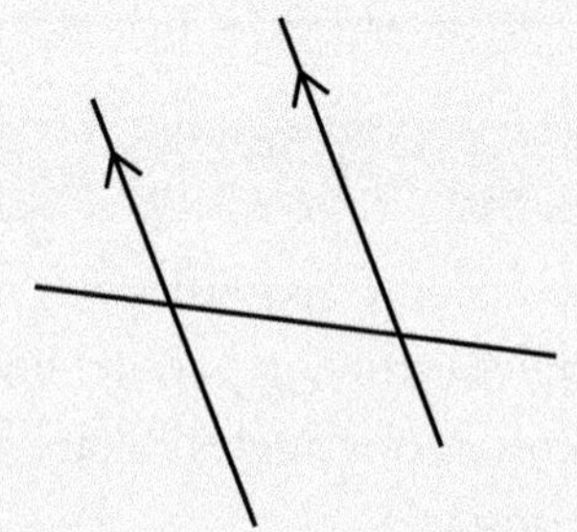

c

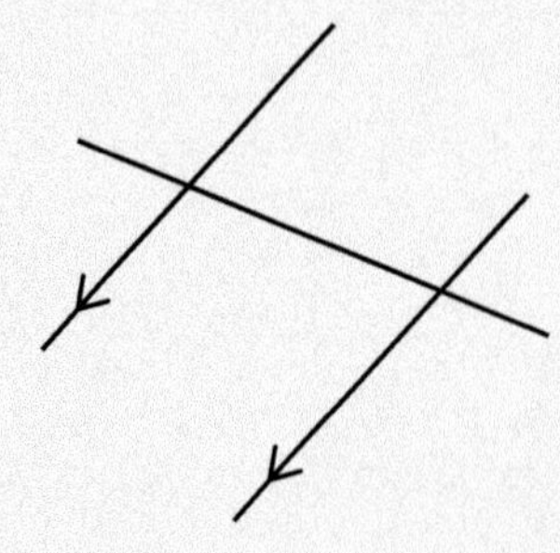

3 For each diagram, find the value of the variable.
You must state your reasons.

a
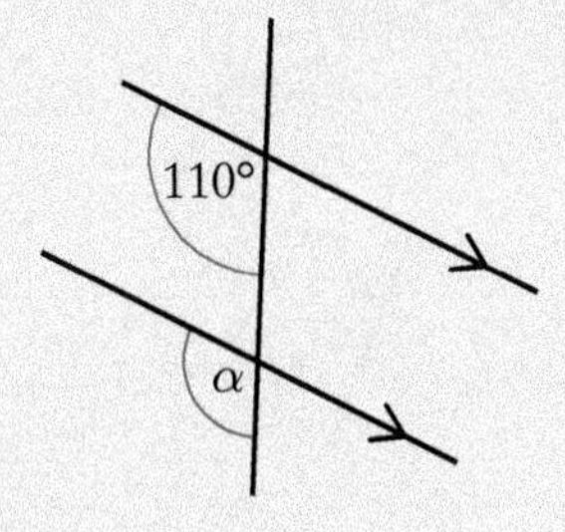

b
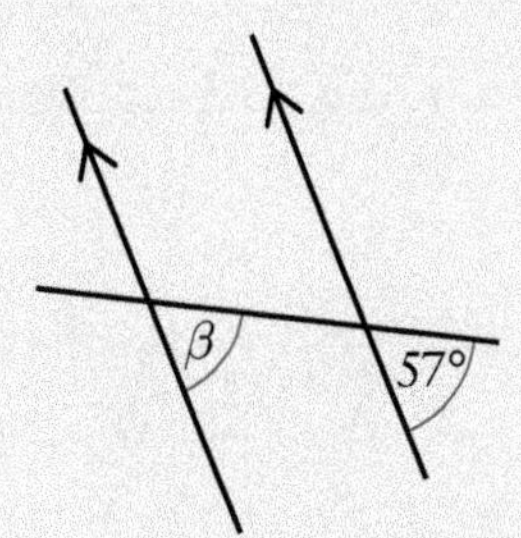

c
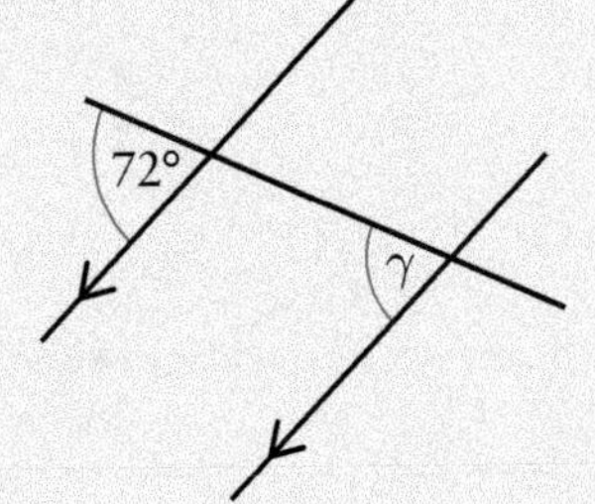

d
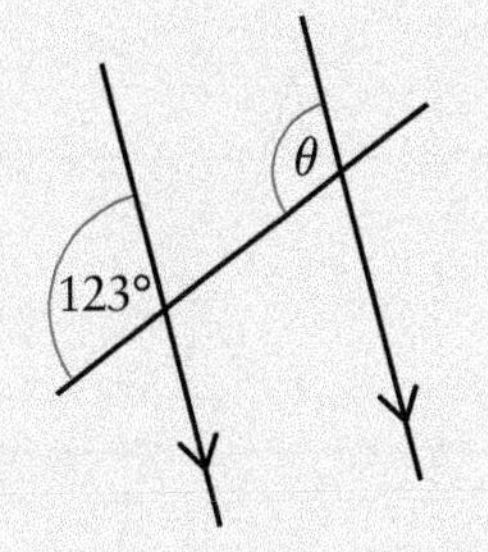

e
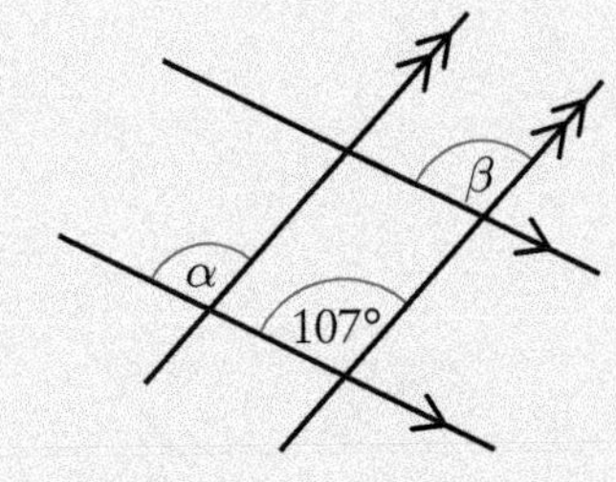

4 Copy each of these diagrams. For each diagram, mark the two pairs of alternate angles. Use two different colours to identify the angles.

a
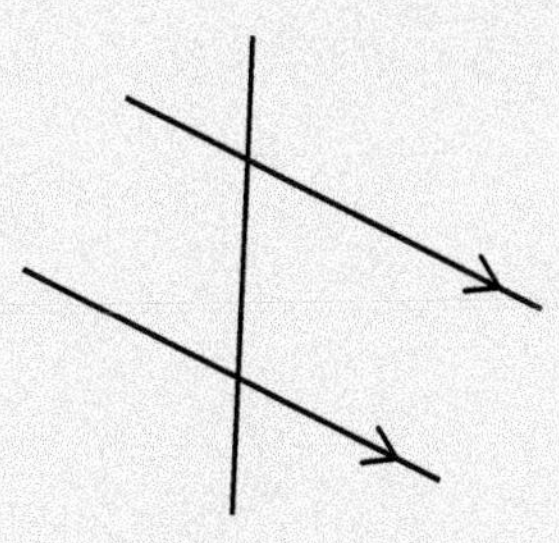

b
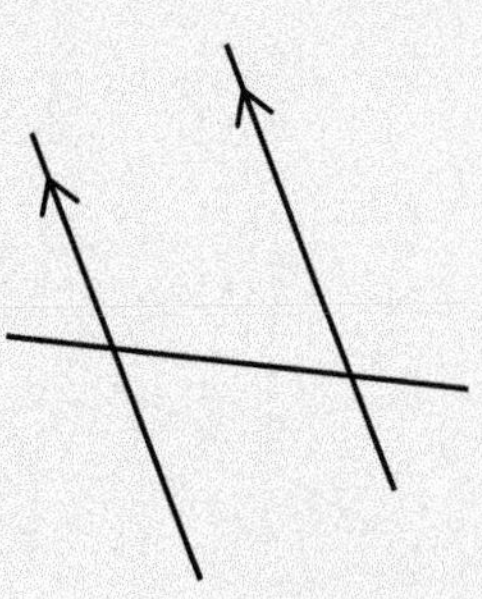

c
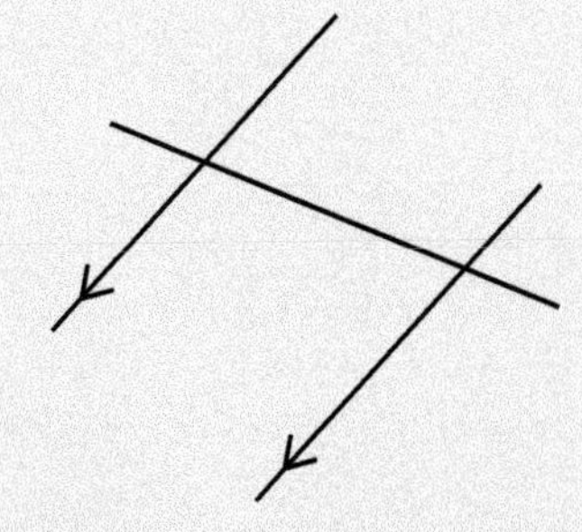

5 For each diagram, find the value of the variable.
You must state your reasons.

a
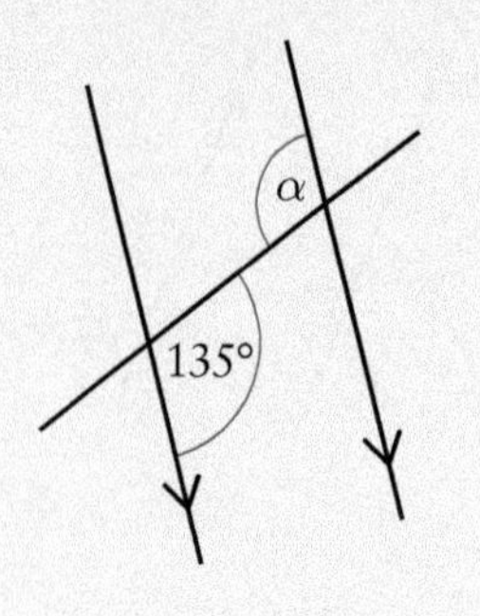

b
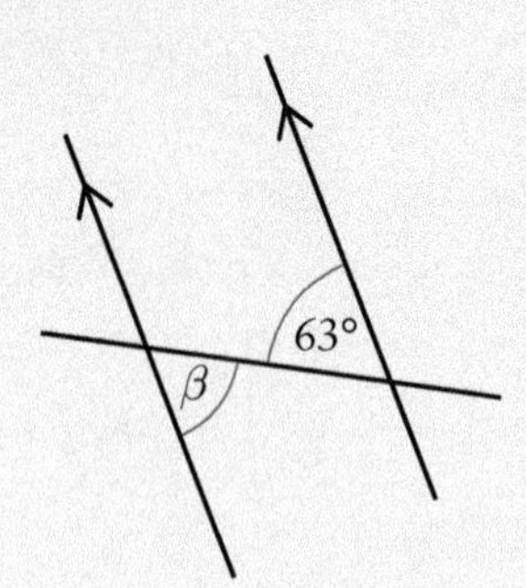

c
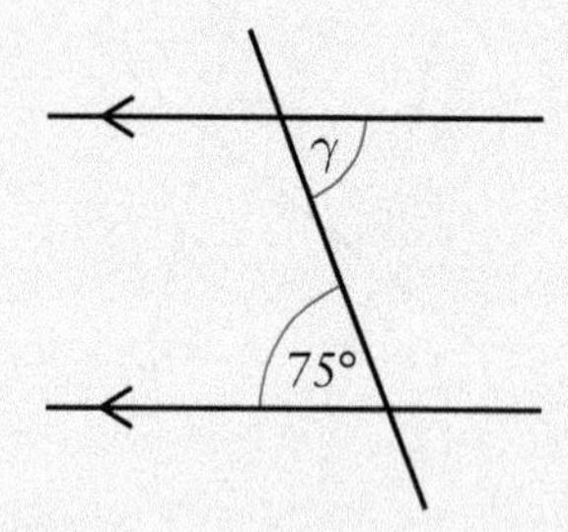

d
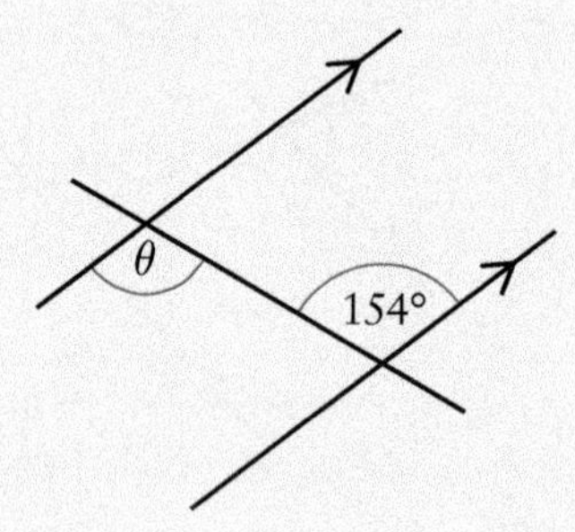

e
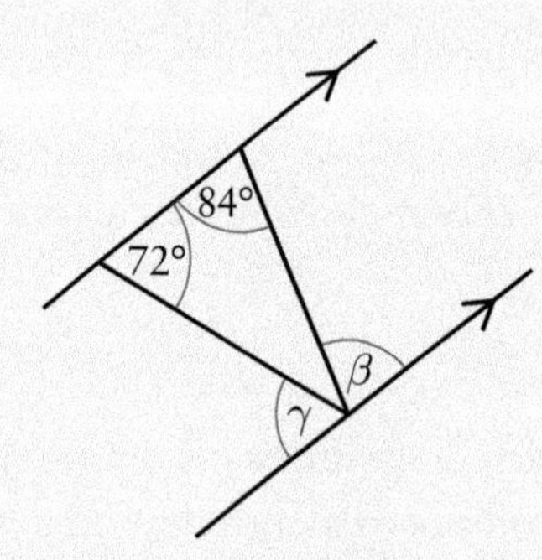

f
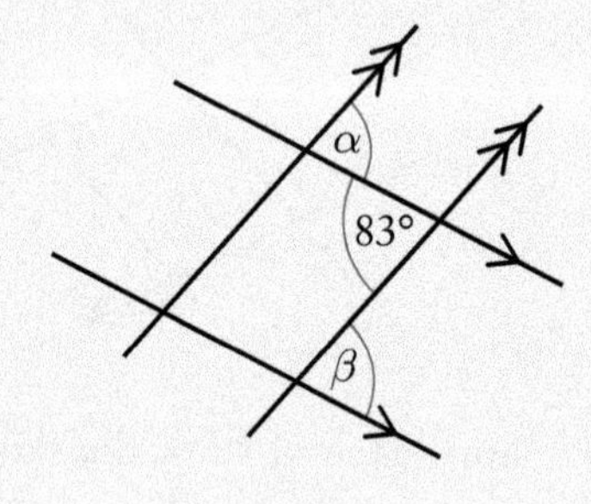

g
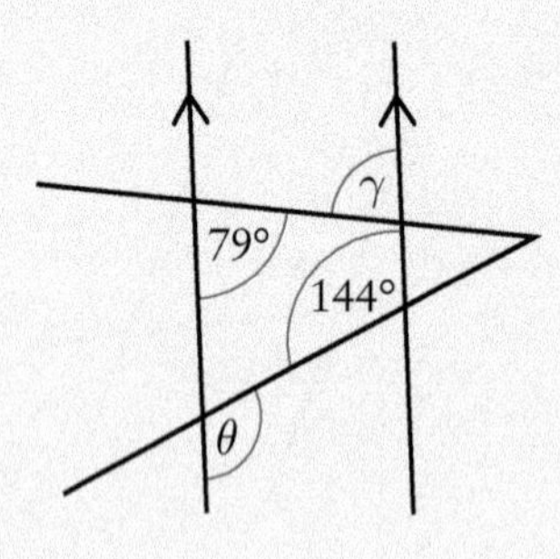

6 Copy the diagrams. For each diagram, mark the two pairs of co-interior angles. Use two different colours to identify the angles.

a

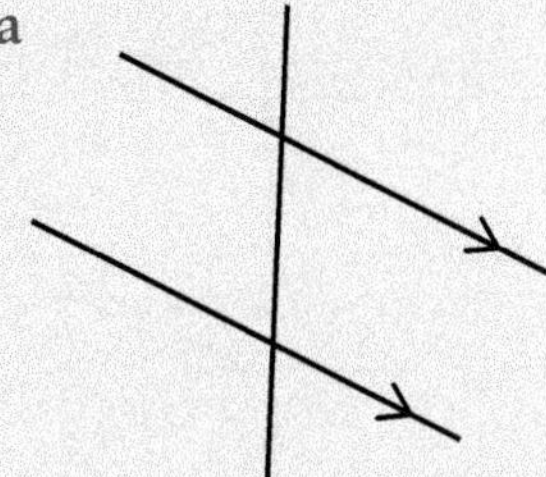

b

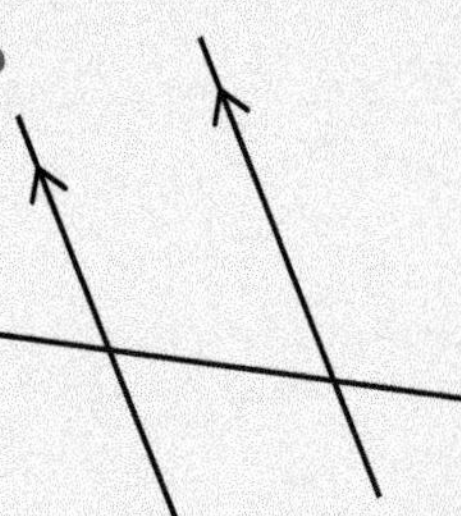

c

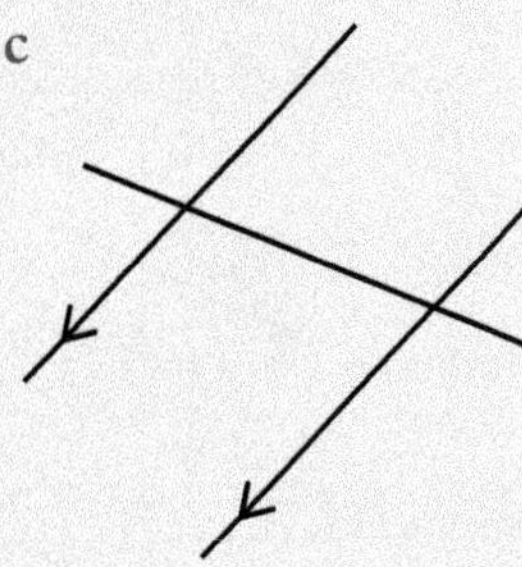

7 For each diagram, find the value of the variable.
You must state your reasons.

a

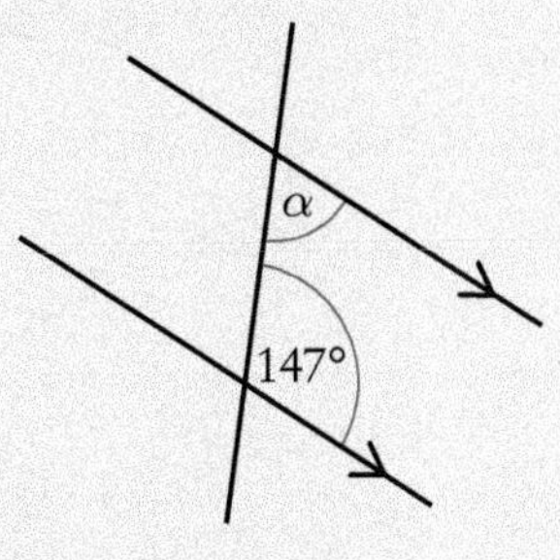

b

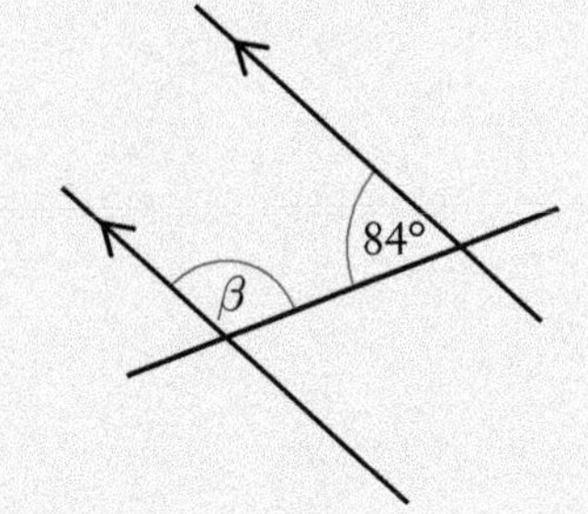

c

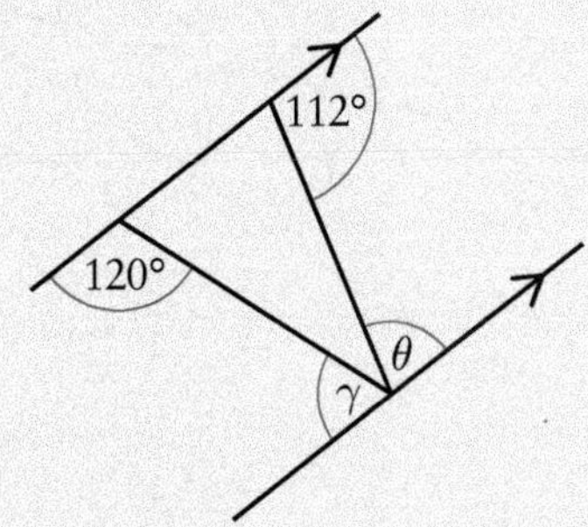

d

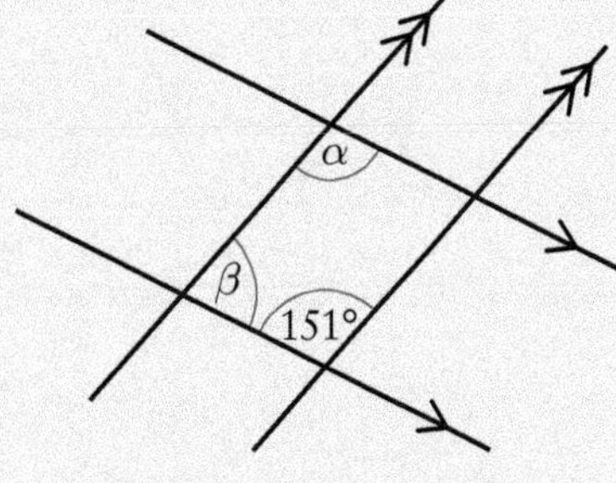

e

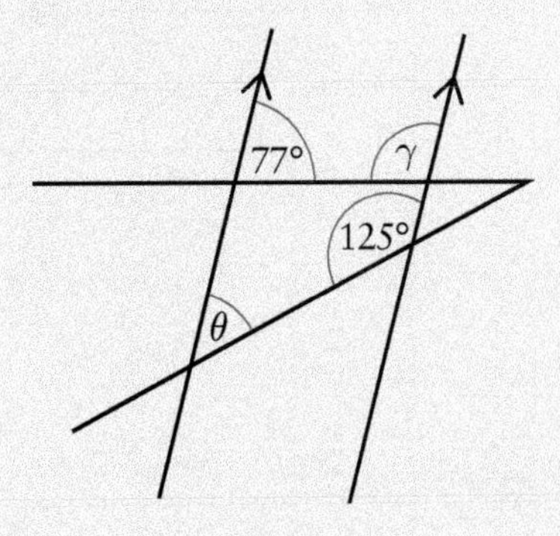

8 For each diagram, find the value of the variable. You must a reason for each answer.

a

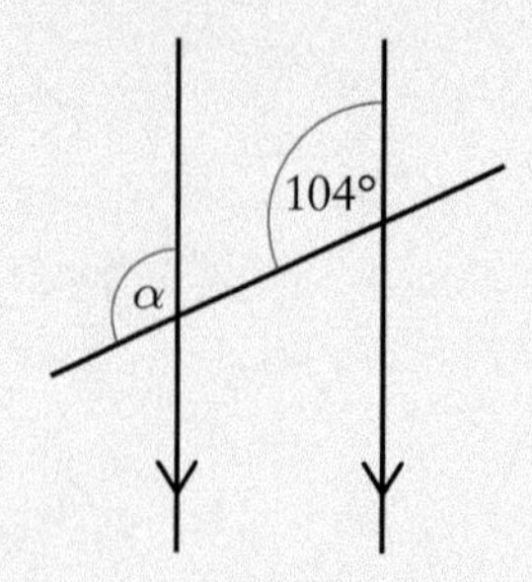

b

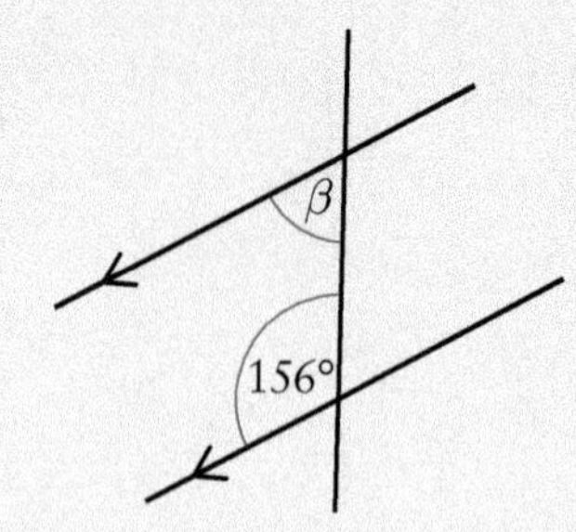

c

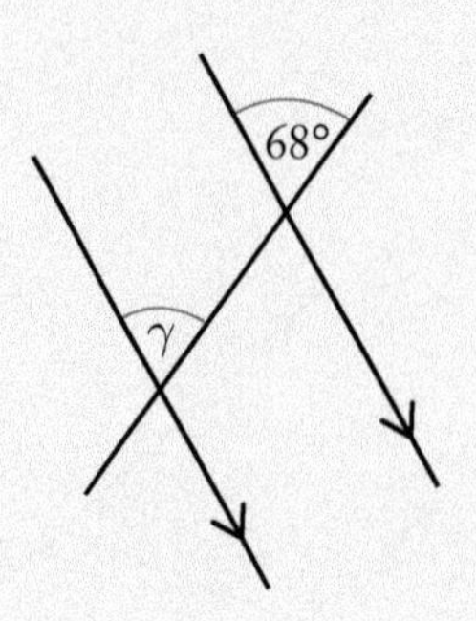

d

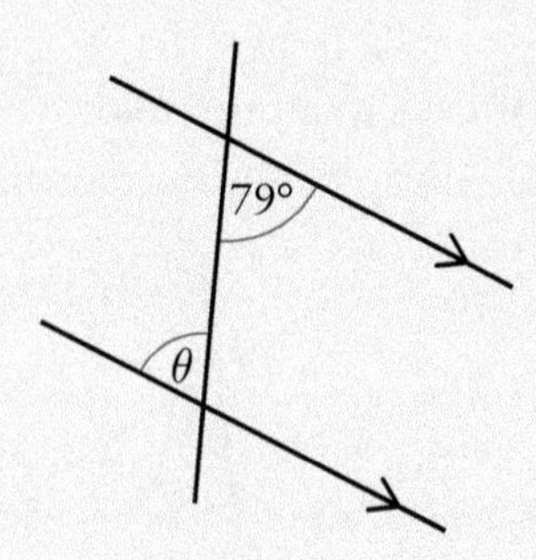

e

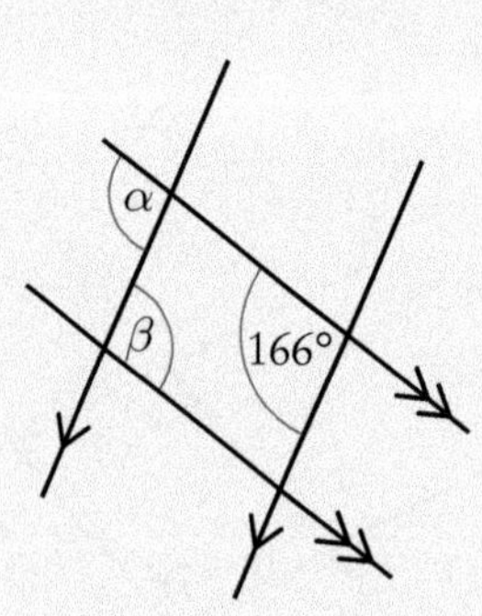

f

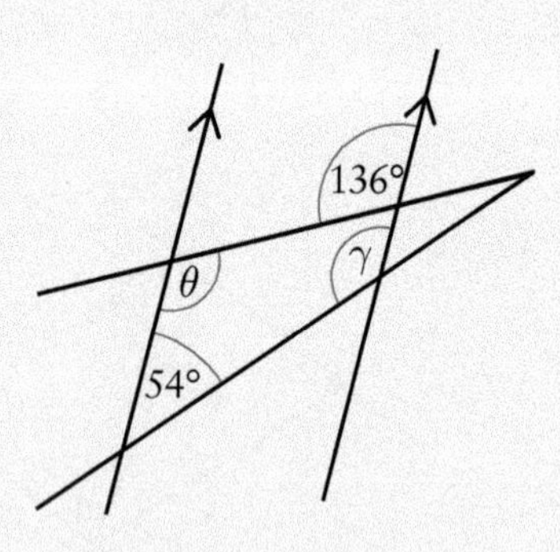

g

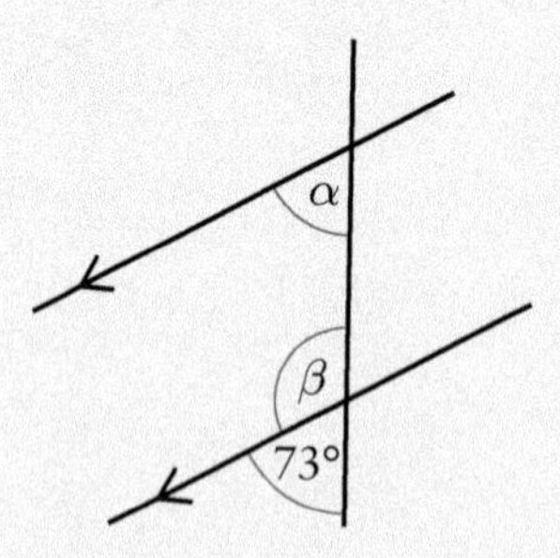

h

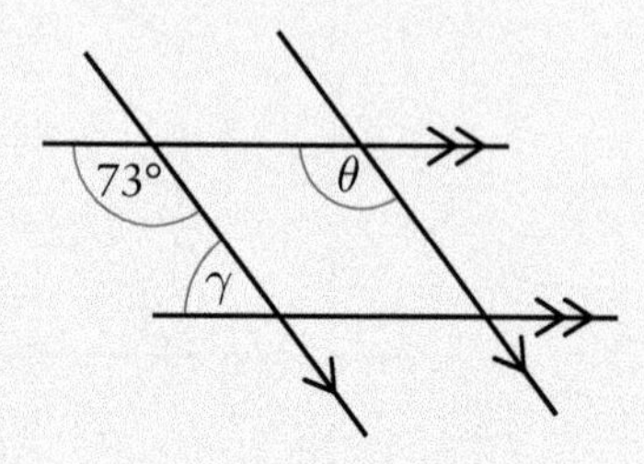

i

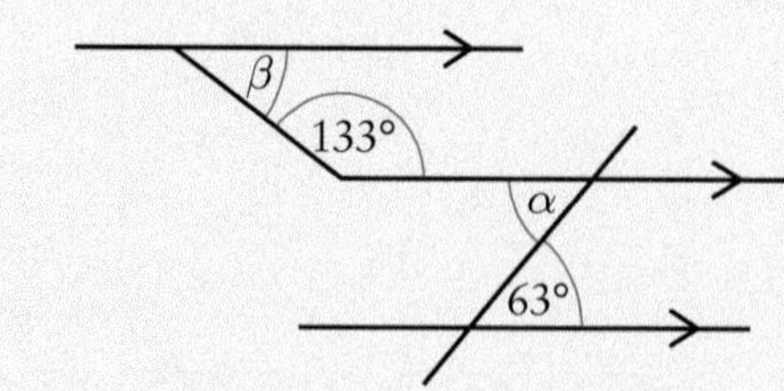

9 For each diagram, find the value of the variable.
You must state your reasons.

a

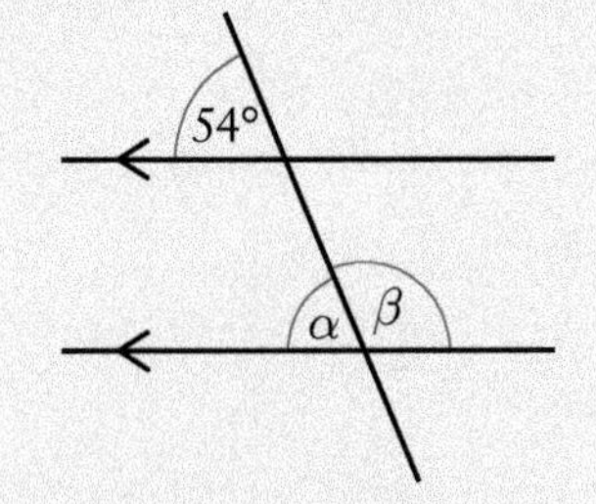

b

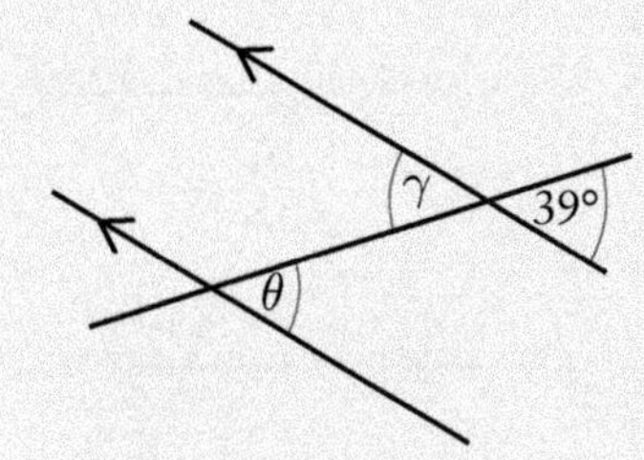

c

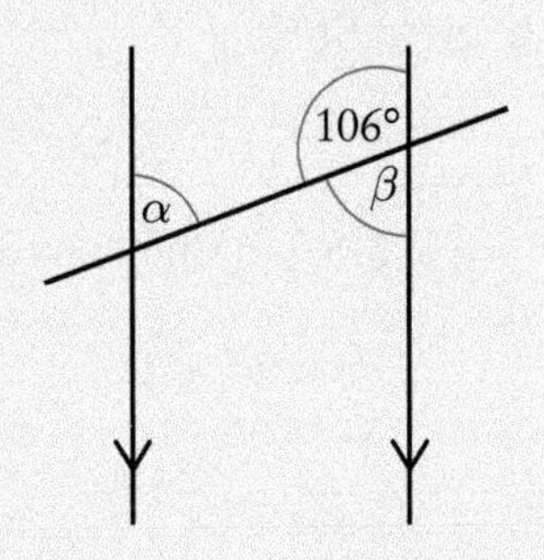

d

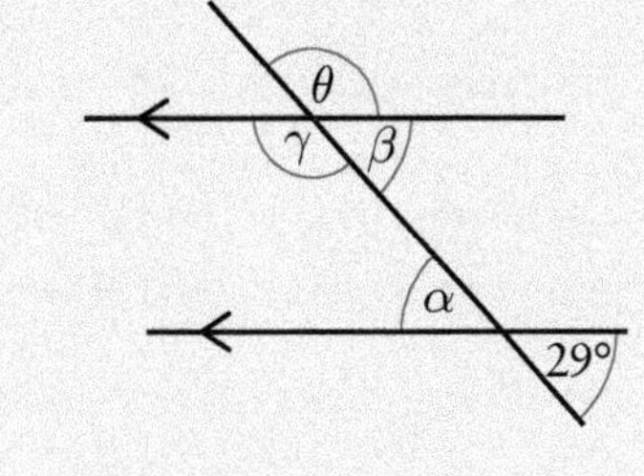

e

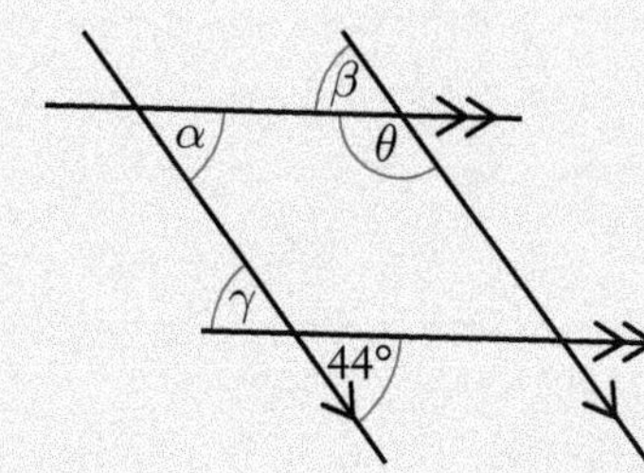

f

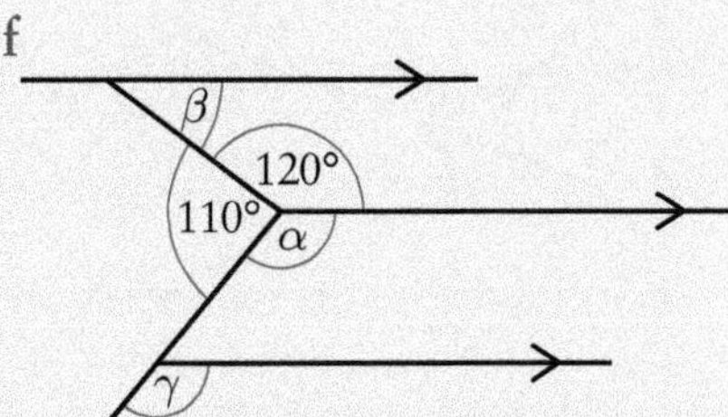

10 Match the pairs of alternate angles. Explain why they are alternate.

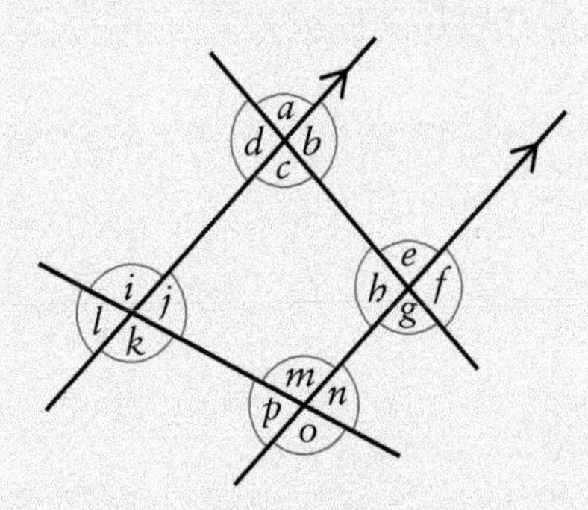

11 Sanura is completing a question in her mathematics lesson.
She says, '*ABC* is a corresponding angle with 100°'.
Draw a diagram that would match the question she has just answered.

12 Using the diagram in Question 10, match the pairs of corresponding angles. Explain why.

13 Create a question involving an angle of 20° and the co-interior angle *ABC*.

14 Max looks at this diagram.

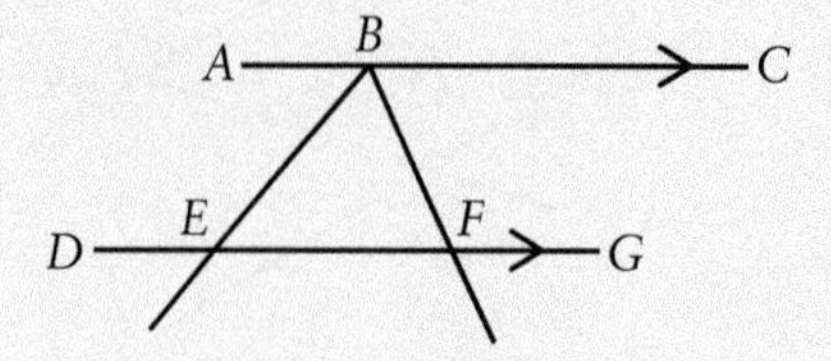

They say, 'Angle *EBF* is co-interior with angle *BEF*'.

Are they correct? Explain your answer.

15 Carlos is asked to work out the size of angle *DEB* in this diagram, giving reasons for his answer.

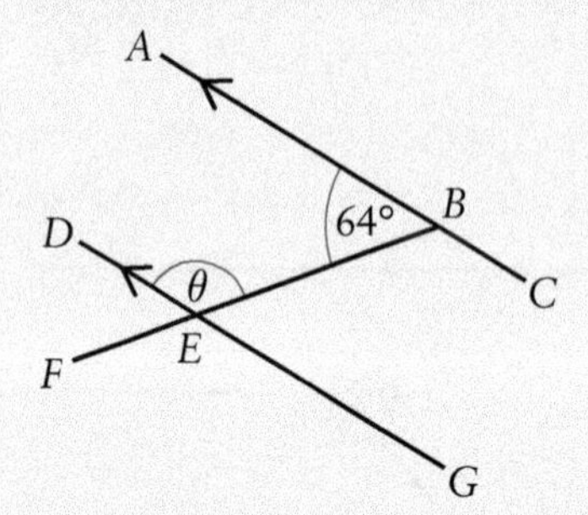

He writes down these steps.

Step 1: Angle DEF = 64° as they are alternate angles.

Step 2: $180 - 64 = \theta$ as angles on a straight line add up to 180°.

Step 3: $\theta = 116°$.

Is Carlos correct? Explain your answer.

16 Zahra is given this diagram.

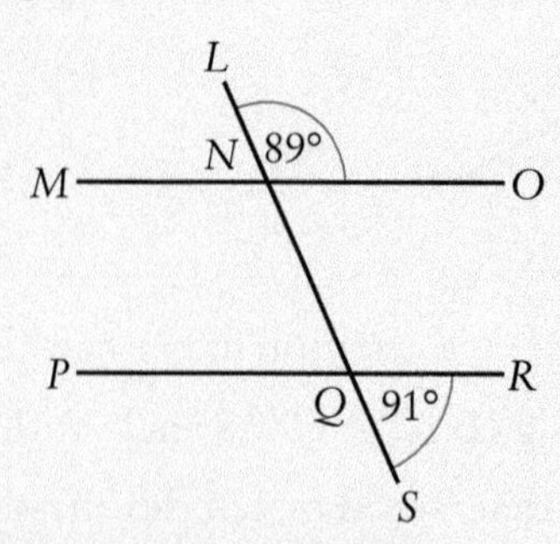

She says, 'the lines *MO* and *PR* are parallel'.

Is she correct? Explain your answer.

17 Copy the diagram or draw one similar with a pair of parallel lines cut by a transversal.

Challenge Q17

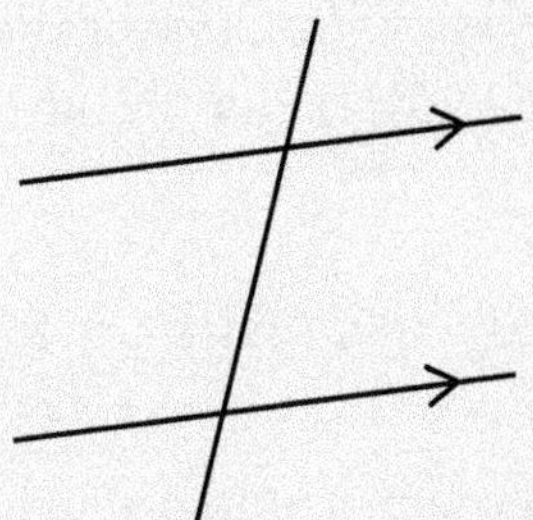

Can you use your diagram to write down five questions about the angles in your diagram, using what you have learned in this chapter? You should add your own labels to your diagram.

Swap diagrams with someone else and answer each other's questions. Did anyone in the group write more difficult questions? If so, how did they make their questions more difficult?

Investigation 6.1

Draw a pair of parallel lines.

Draw another pair of parallel lines that cross your first pair of parallel lines to form a four-sided shape. Label the four angles inside the shape a, b, c and d. For example:

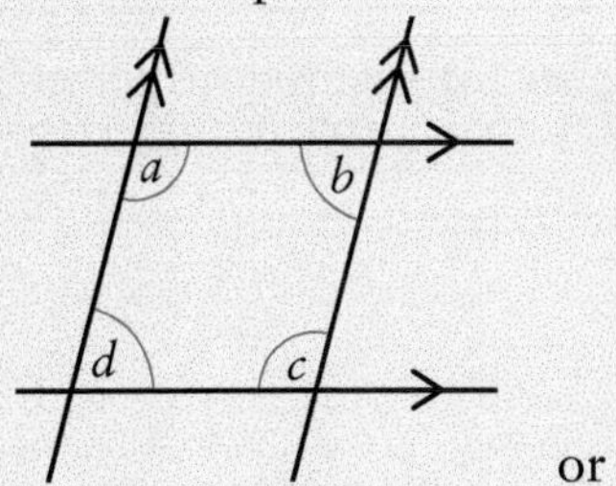

or

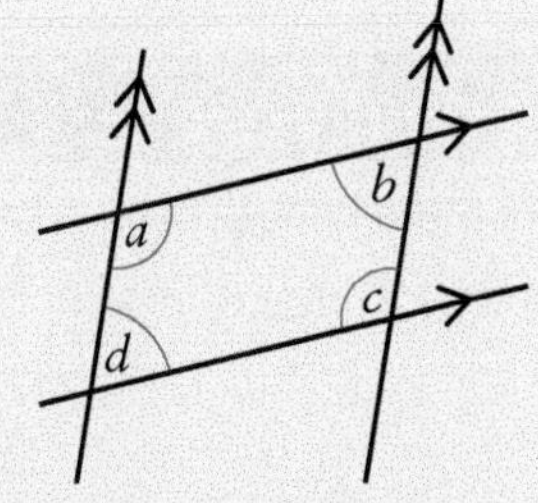

Measure angle a. Use your measurement and knowledge of angles on parallel lines to work out angles b, c and d.

What do you notice about the four angles inside the shape?

Repeat this with different pairs of parallel lines. What do you notice?

Using the properties of the shape, research the name of this shape.

A special case of this shape is used at the entrance to the Louvre Museum in Paris.

The entrance to the Louvre Museum is a pyramid made of metal and glass. A total of 603 of the glass panels have opposite parallel sides and the sides are all equal in length, research the name of this shape. One of the angles in the four-sided shape is 114°. Work out the other angles in the shape.

6.3 Reasoning involving angles in a triangle

You have previously used the fact that the angles in a triangle total 180°. You can check this by drawing a triangle, measuring each angle and adding them together.

Explore 6.4

Draw a triangle and an exterior angle and cut them out like this:

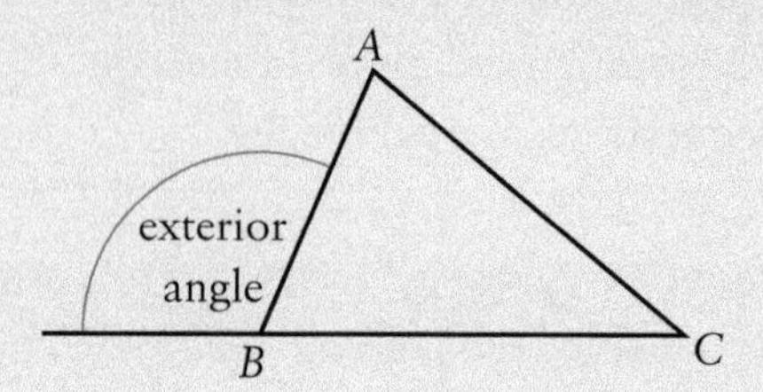

Tear off angles BAC and ACB and place them over your exterior angle.
What do you notice about the exterior angle and the two interior angles?
Repeat this for other triangles. Does the same happen? Can you explain why?

Worked example 6.4

a Find the value of angle a, giving reasons for your answer.

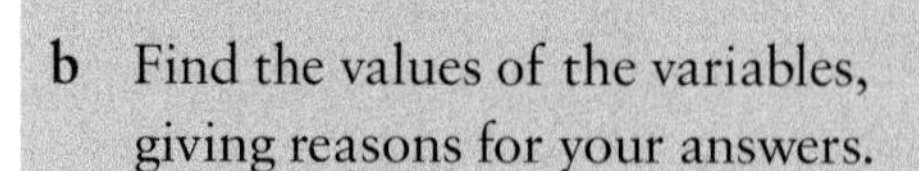

b Find the values of the variables, giving reasons for your answers.

c Work out the value of d, giving a reason for your answer.

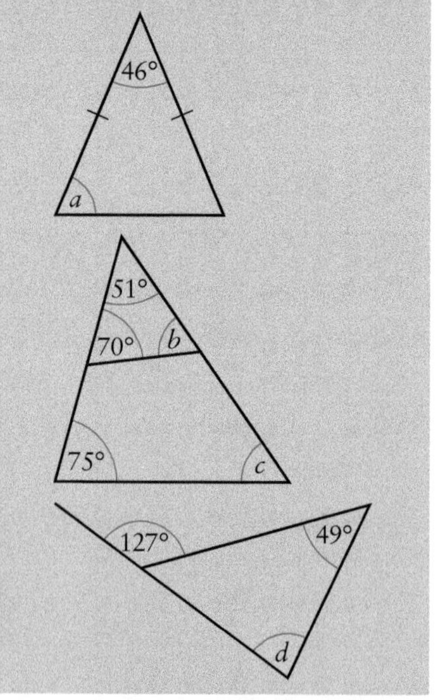

Solution

a Understand the problem

We need to work out the value of a.

Make a plan

We can use the fact that the angles in a triangle total 180° and this is an isosceles triangle, so the two angles at the base of the two equal sides are equal.

Carry out the plan

The angles in a triangle total 180° and in an isosceles triangle, the two angles at the base of the two equal sides are equal:

$a + a + 46 = 180$

$2a + 46 = 180$

$2a = 134$

$a = 67°$

Look back

Does the answer work?

$67 + 67 + 46 = 180$

b Understand the problem

We need to work out the values of b and c. There are two triangles in the diagram so two calculations. The smaller triangle has angles 51°, 70° and b. The larger triangle has angles 51°, 75° and c.

Make a plan

We can use the fact that the angles in a triangle total 180° to work out the values of b and c.

Carry out the plan

The angles in a triangle total 180°:

$b + 51 + 70 = 180$

$b = 59°$.

The angles in a triangle total 180°:

$c + 51 + 75 = 180$

$c = 54°$.

Look back

Does the answer work?

$59 + 51 + 70 = 180$ and $54 + 51 + 75 = 180$

Reminder

Angles in a triangle sum to 180°.

c **Understand the problem**

We need to work out the size of the angle labelled d.

Make a plan

We can use the fact that the exterior angle of a triangle is equal to the sum of the two interior angles at the other two vertices of the triangle.

Carry out the plan

$d + 49 = 127$

The exterior angle of a triangle is equal to the sum of the two interior angles at the other two vertices of the triangle.

$d = 78°$

Look back

Does the answer work?

$78 + 49 = 127$

Reflect

Can you explain why the angles in a triangle sum to 180°?

Look at your results from the Explore activity. Does the same happen if you use any two angles in the triangle? Can you explain why?

Practice questions 6.3

1 For each diagram, find the value of the variable. You must state your reasons.

a 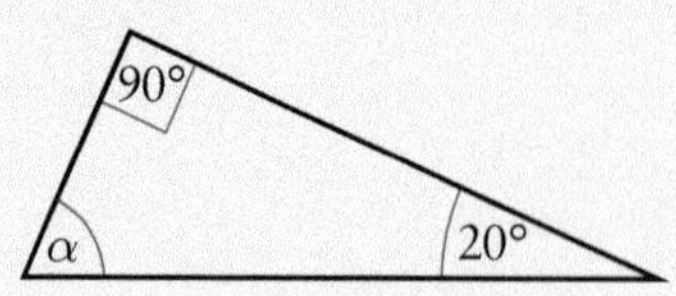

b 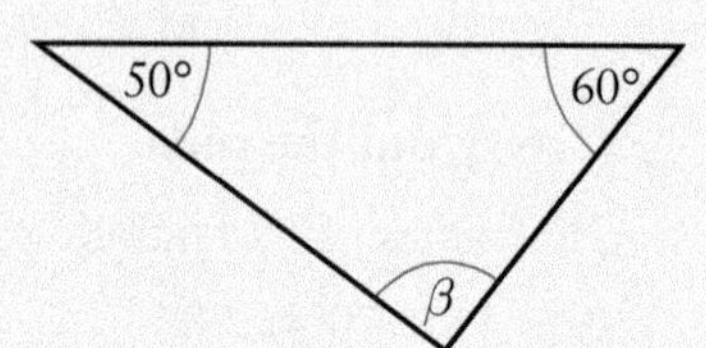

c 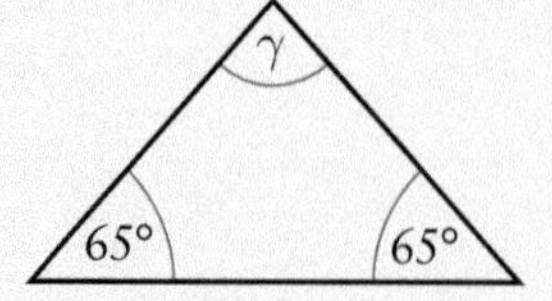

d 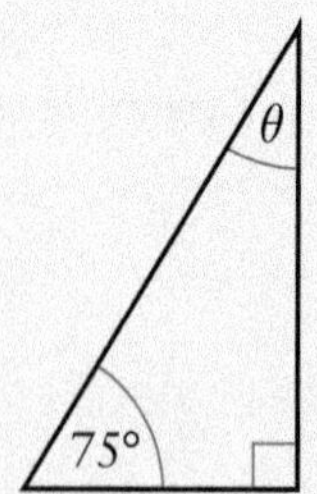

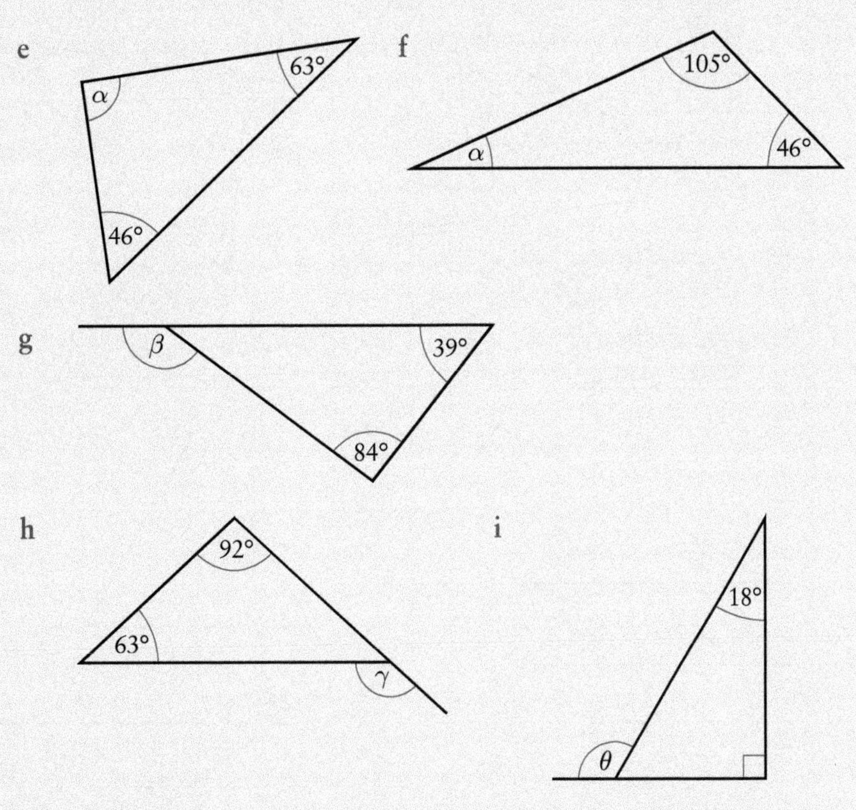

2 For each diagram, find the value of the variable. You must show your working.

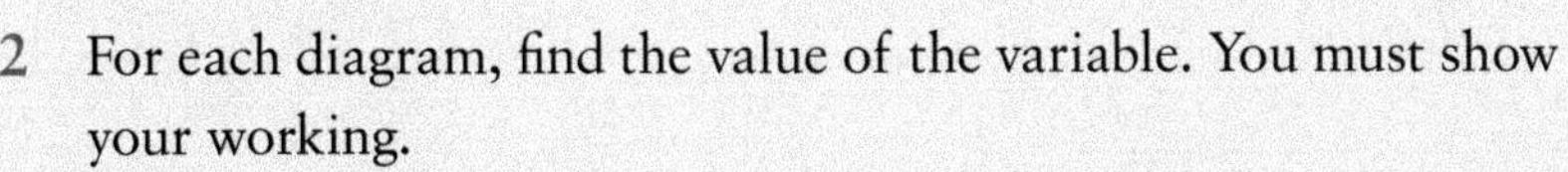

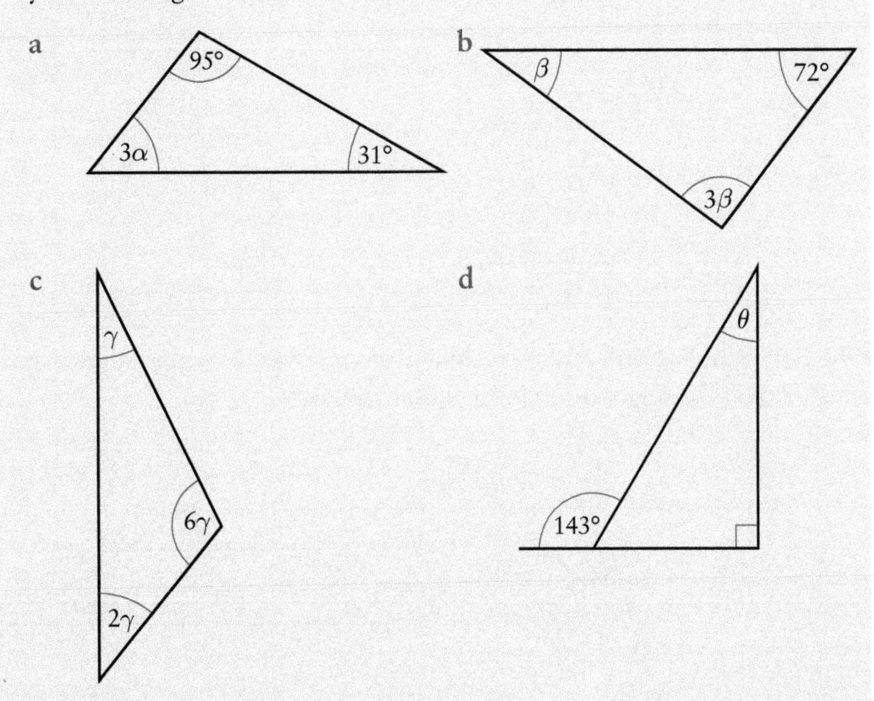

e

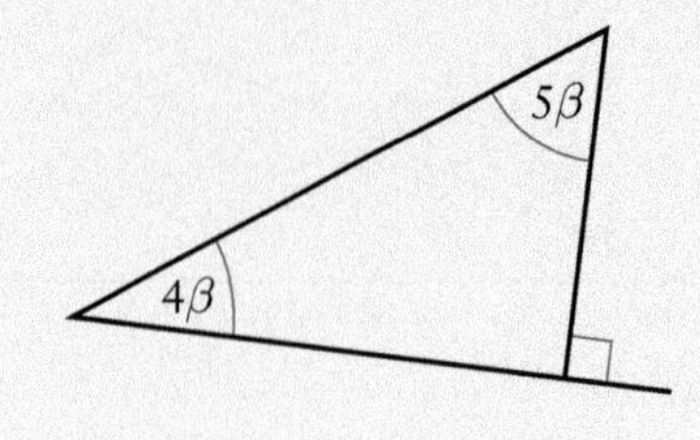

f

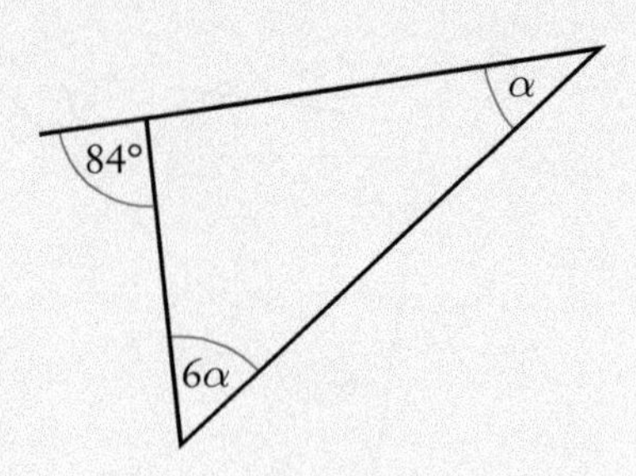

3 Work out the size of all the angles in each of these triangles. Show your method.

a

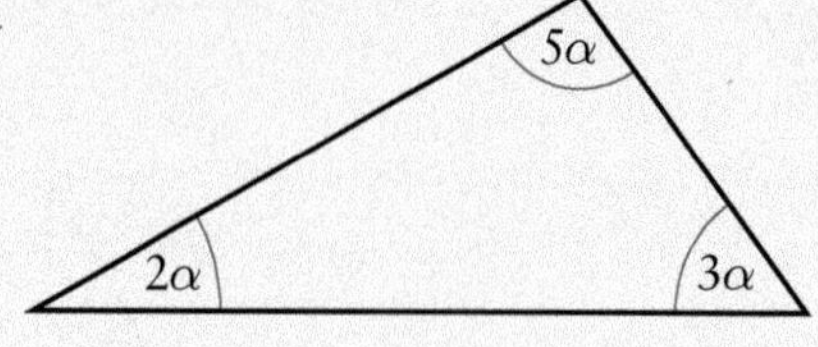

b

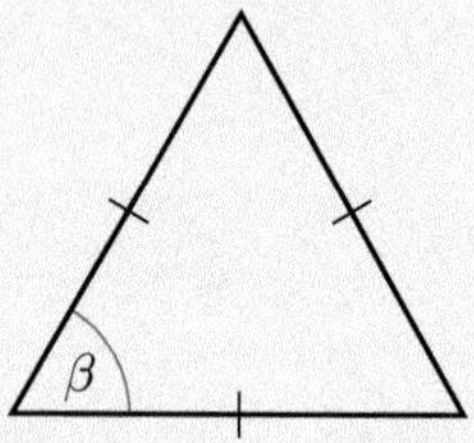

c

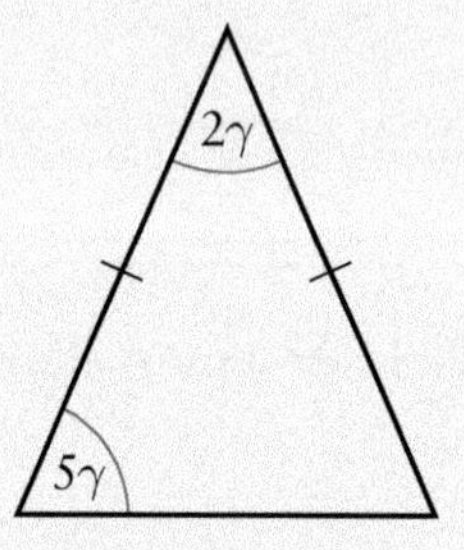

d

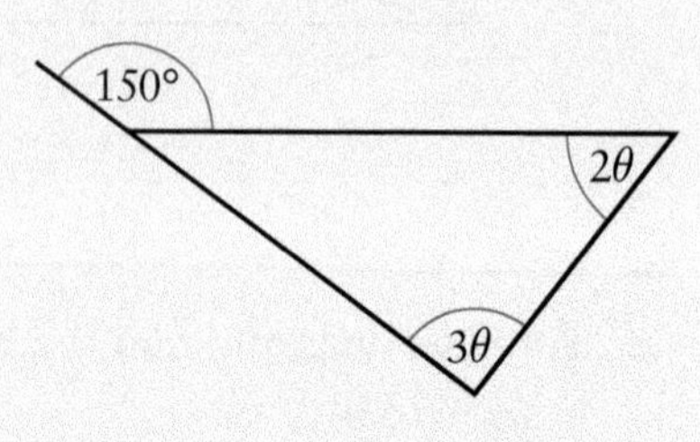

e

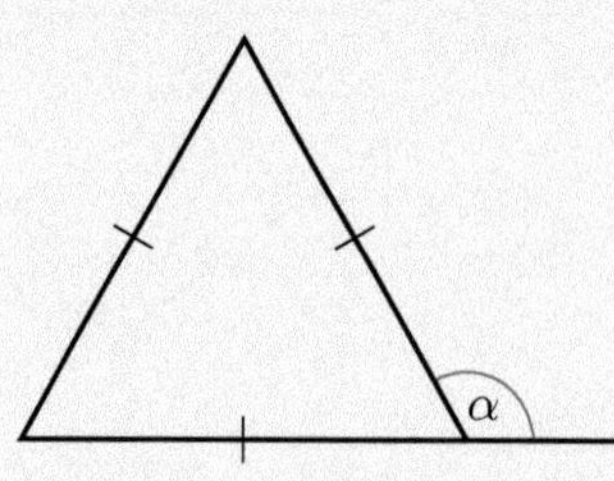

f

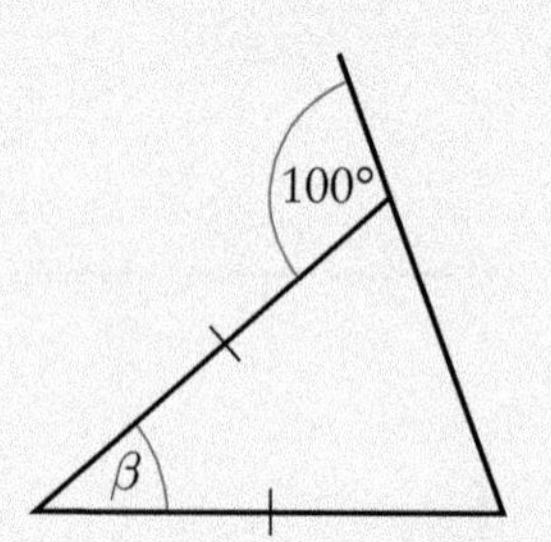

4 For each diagram, find the value of the variable. You must show your method, stating any angle facts that you use.

a

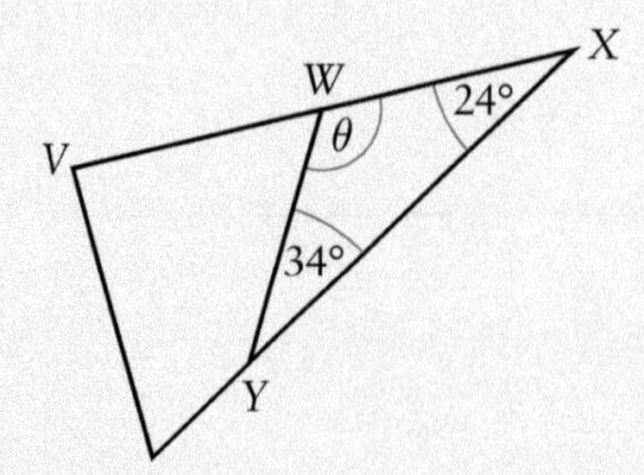

b

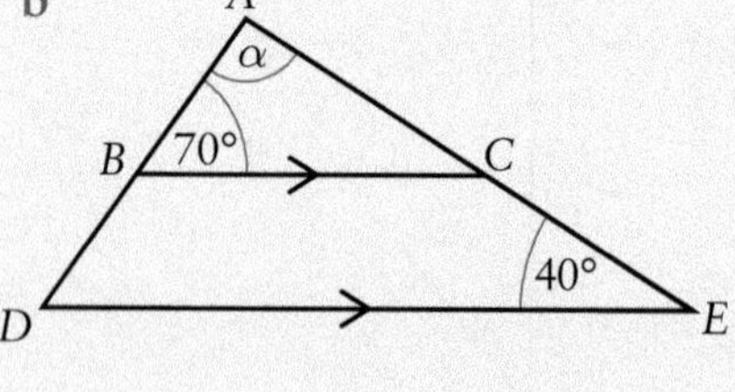

c

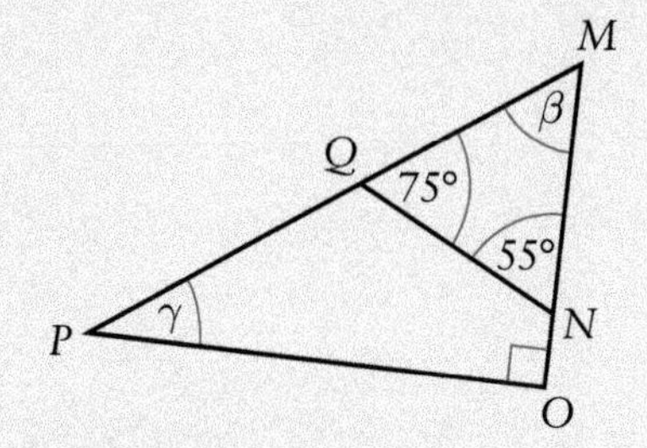

d

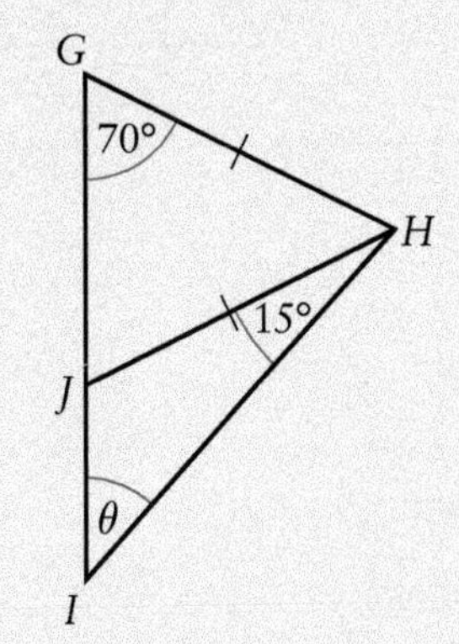

e

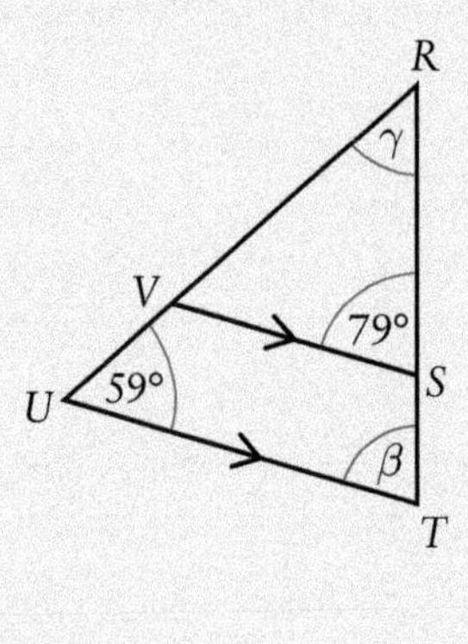

5 For each diagram, find the value of the variable. You must show your method, stating any angle facts that you use.

a

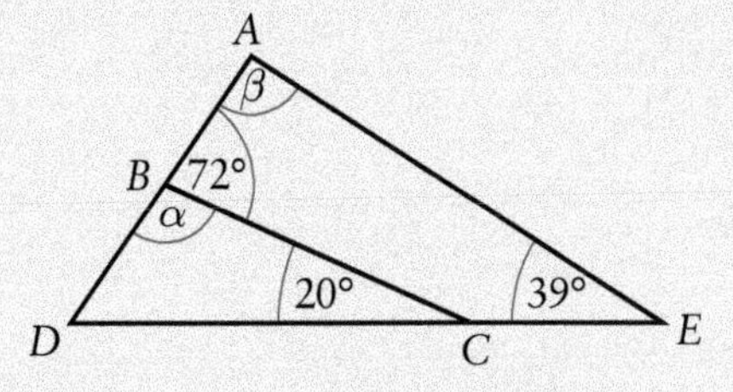

b

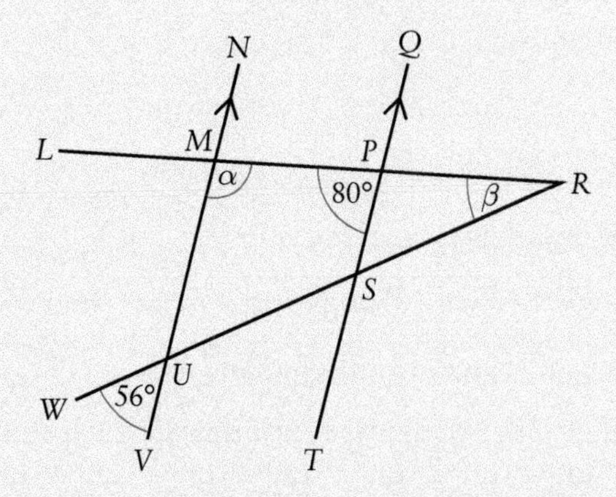

c

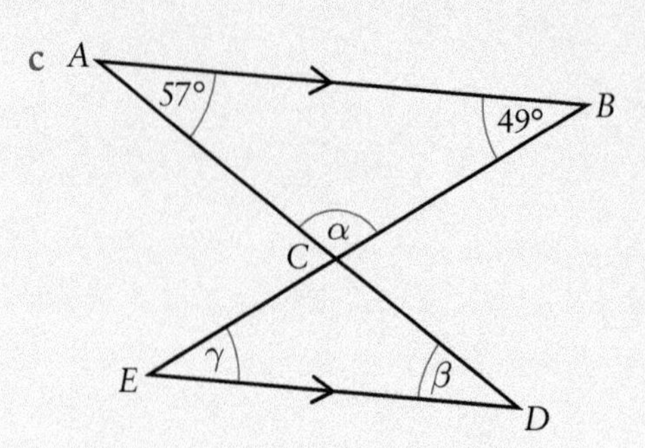

d

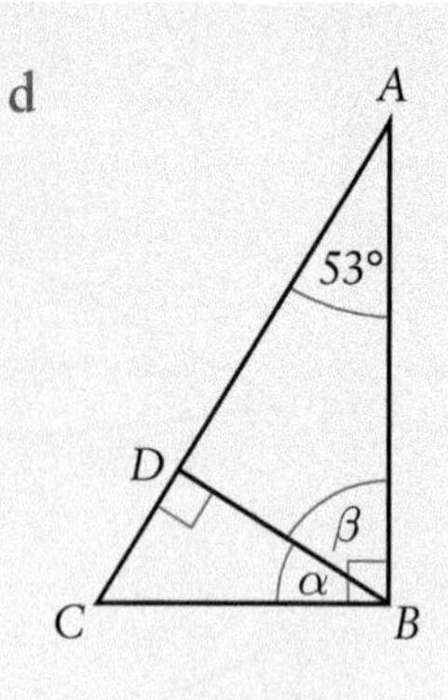

e

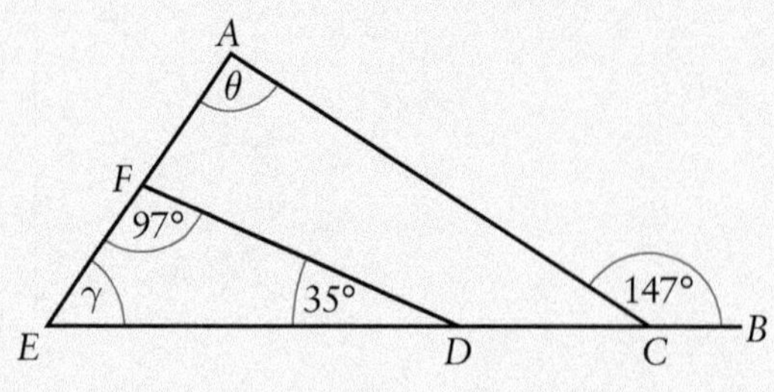

f

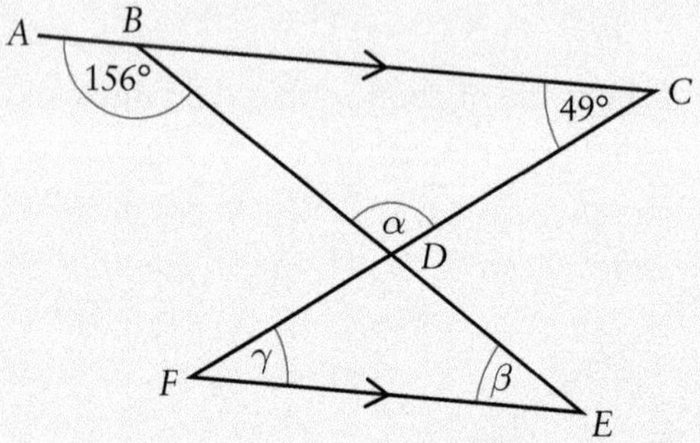

6 State whether these statements are true or false.
Give reasons for your answers.

a The three angles in a triangle are 40°, 100° and 50°.

b The three angles in a triangle are 60°, 30° and 40°.

c An equilateral triangle has an angle of 50°.

d An isosceles triangle has two 50° angles and one 80° angle.

e The three angles in a triangle are 70°, 80° and 30°.

f An isosceles triangle has two angles measuring 80° each and one angle measuring 50°.

g The three angles in a triangle are 24°, 48° and 108°.

h An equilateral triangle has an angle of 60°.

7 For each diagram, find the value of the variable. You must show your method, stating any angle facts that you use.

a

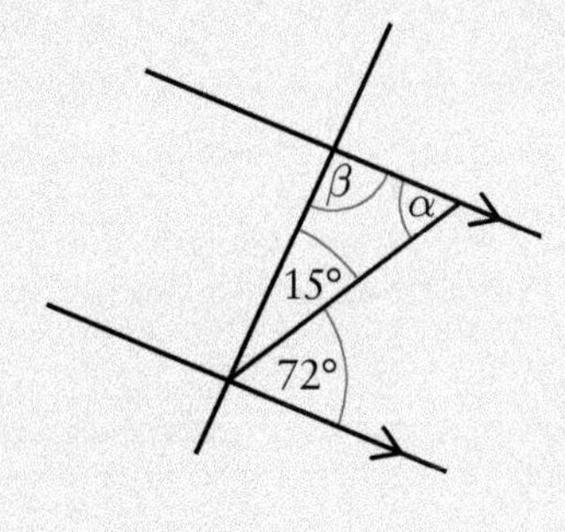

b

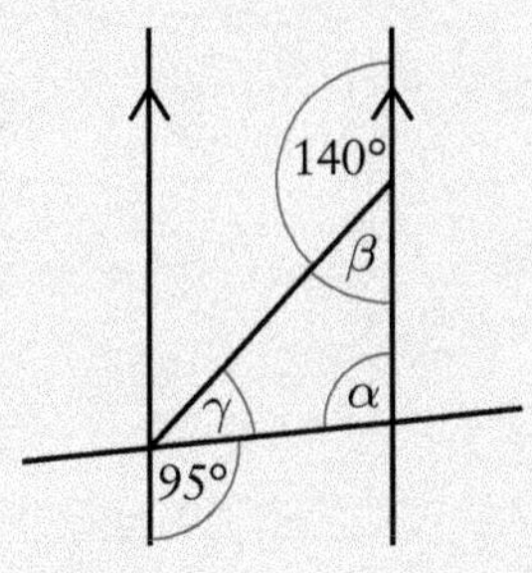

c

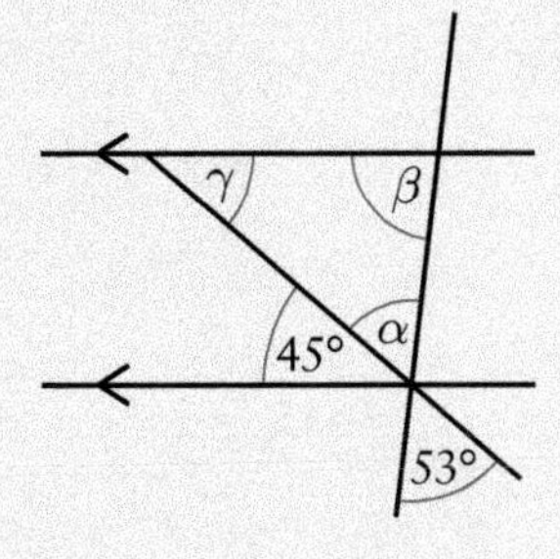

8 Lorenzo says he can draw two different diagrams for isosceles triangles that have an angle of 70°. Lorenzo is correct. Sketch his two different triangles.

9 In this diagram the α and β are in the ratio $\alpha : \beta = 5 : 6$

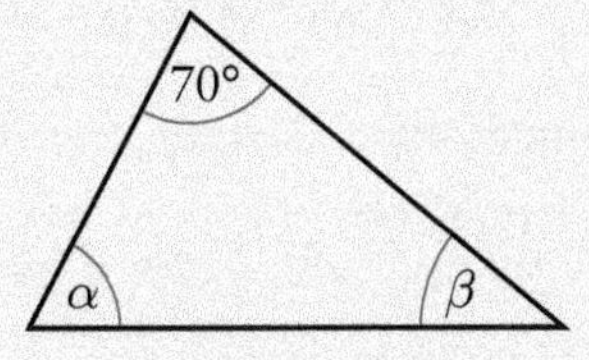

Work out the values of α and β. Show your method.

10 Here is a diagram of a triangle with angles α, β and γ.

Challenge Q10

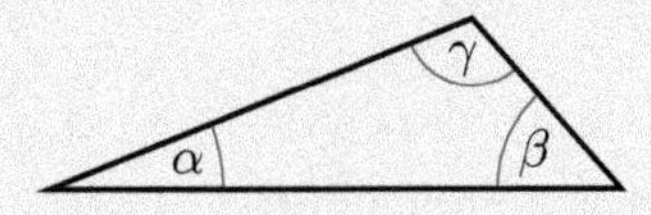

β is twice the size of α and γ is three times the size of β.

Work out the size of each angle in the triangle. Show your method.

11 Use the diagram to decide whether these statements are true or false. Give reasons for your answers.

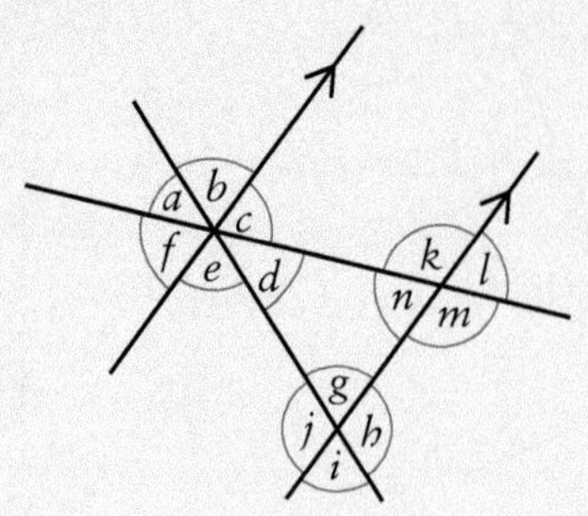

a $c = f$ b $d + g + n = 180°$ c $d = m$

d $g = k$ e $c = n$ f $c + d = h$

g $e = g$ h $n = l$ i $a + b + c = k + l$

12 Jamal is given this diagram.

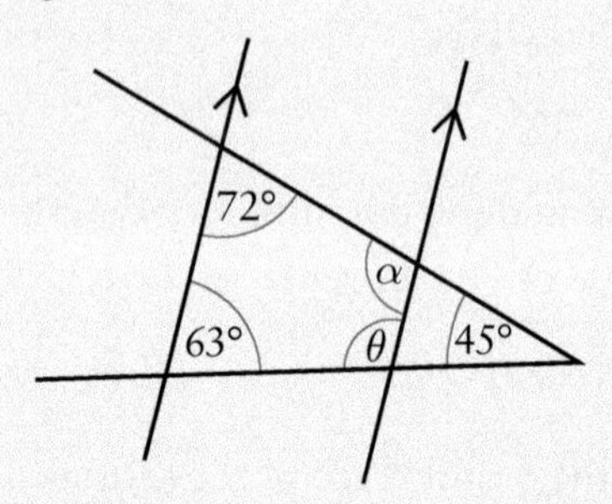

He writes, '$\alpha = 72°$ and $\theta = 63°$' Is he correct? Explain your answer.

13 Write down the value of the variable. You must show your method.

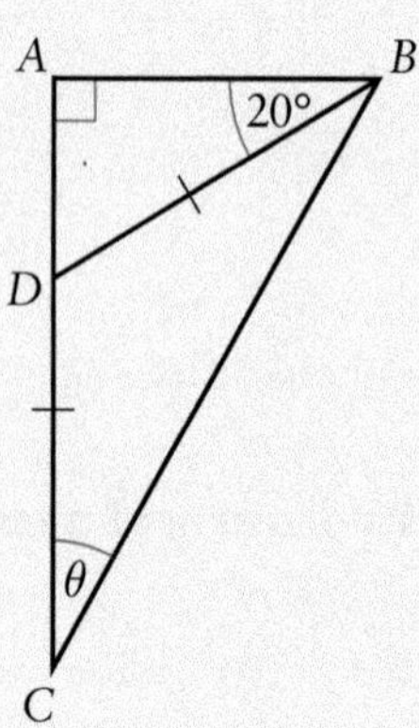

Challenge Q14

14 The diagram shows a triangle with angles of size α, β and γ.

β is 50% of the size of α and γ is 30% of the size of α.

Write down the value of each angle.

You must show your method.

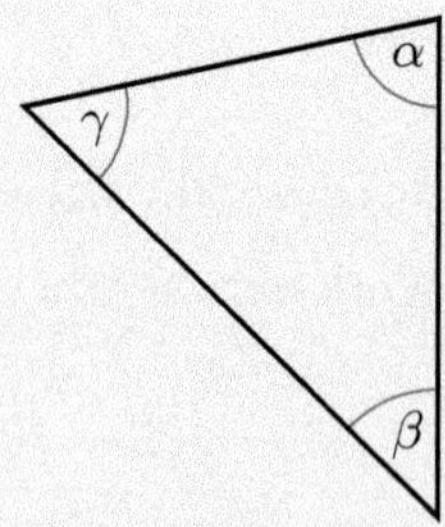

Investigation 6.2

Using what you have learned about angles so far in this chapter can you copy the diagram and label as many angles on it as you possibly can. Can you give reasons for your angles?

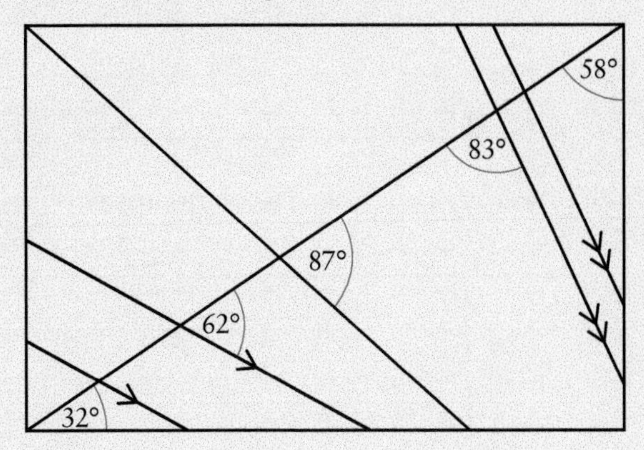

Can you create your own rectangle with missing angles in it? Swap your rectangle with someone else and try to complete each other's diagrams.

Thinking skills

Social skills

Communication skills

6.4 Angles in a quadrilateral

A quadrilateral is a shape with four straight sides.

Explore 6.5

Draw any quadrilateral and cut it out.

Tear the vertices off your quadrilateral.

Fit the angles together so the vertices all touch like this:

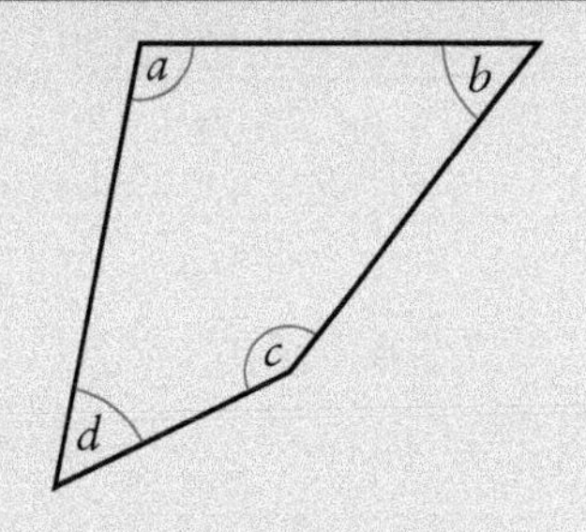

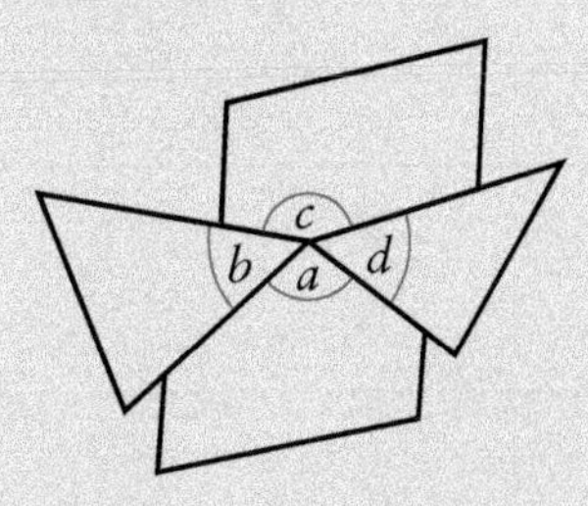

What do you notice? What does this tell you about the angles in your quadrilateral?

Repeat with at least two more quadrilaterals.

Compare your results with others in your group.

Worked example 6.5

a Find the value of α, giving a reason for your answer.

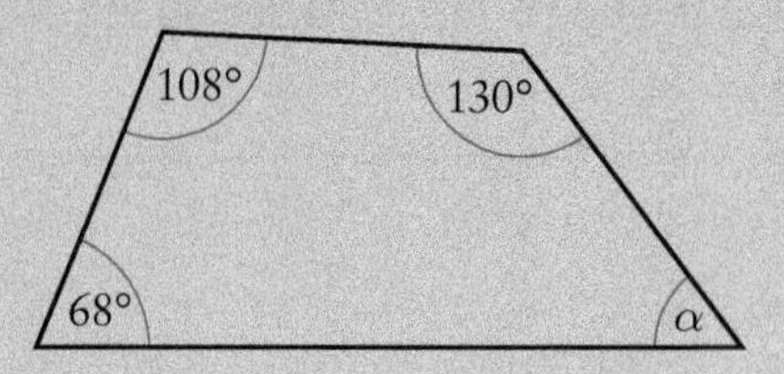

b Find the value of β, giving a reason for your answer.

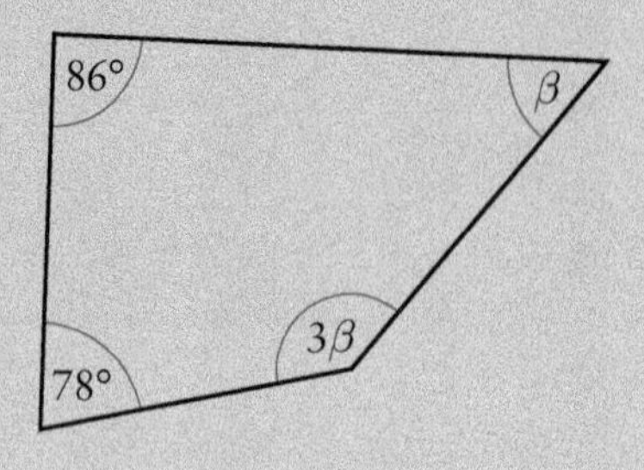

Solution

a **Understand the problem**

We need to work out the value of α.

Reminder

Angles in a quadrilateral sum to 360°.

Make a plan

We can use the fact that the angles in a quadrilateral total 360°.

Carry out the plan

$\alpha + 68 + 108 + 130 = 360$ as the angles in a quadrilateral total 360°.

$\alpha = 54°$

Look back

Does the answer work?

$54 + 68 + 108 + 130 = 360$

b **Understand the problem**

We need to work out the value of β. There are two unknown angles in the diagram.

Make a plan

We can use the fact that the angles in a quadrilateral total 360° to write and solve an equation to work out the value of the variable.

Carry out the plan

The angles in a quadrilateral total 360°:

$\beta + 3\beta + 86 + 78 = 360$

$4\beta + 164 = 360$

$4\beta = 196$

$\beta = 49°$

Look back

Does the answer work?

$49 + 3 \times 49 + 86 + 78 = 360$

Reflect

An alternative to the Explore activity would be to draw a quadrilateral, measure the angles and add them together. What may be a problem with this method of seeing what the angles in a quadrilateral total? You could try it to see what happens.

Can you explain why the angles in a quadrilateral sum to 360°?

Practice questions 6.4

1 Work out the value of the variable for each of these. You must show your method.

a
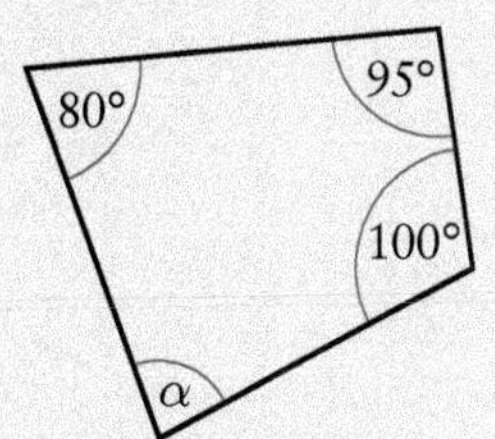

b
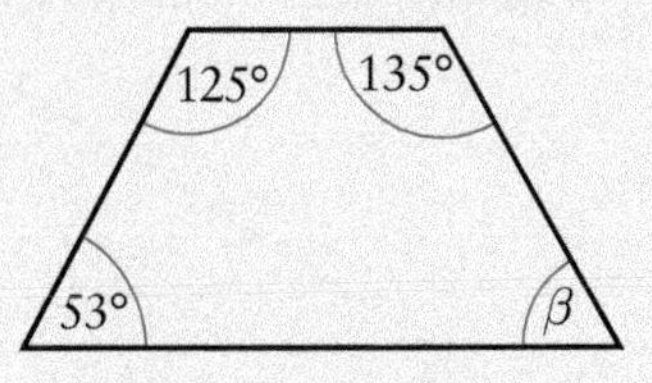

c
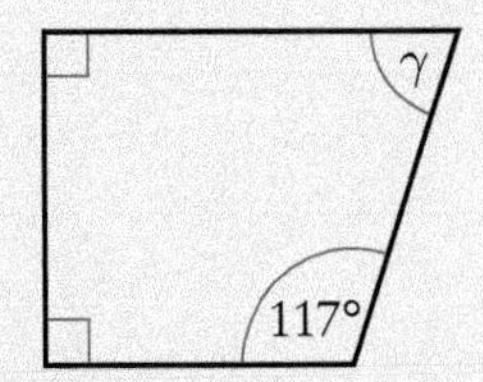

d
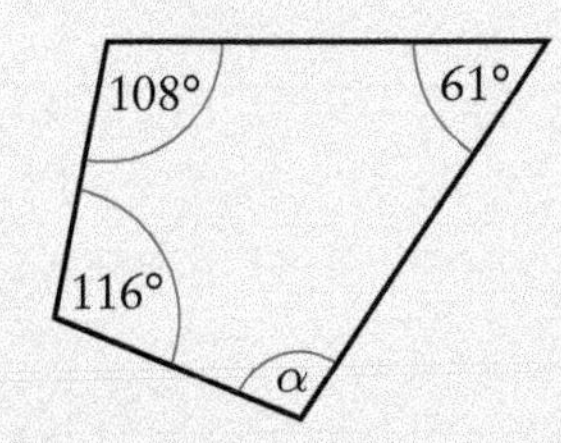

e

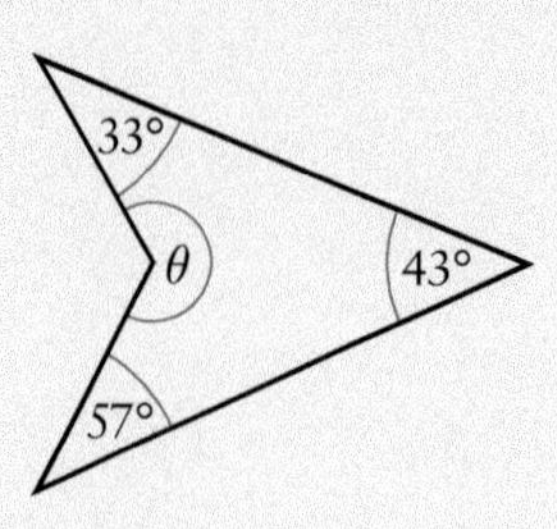

f

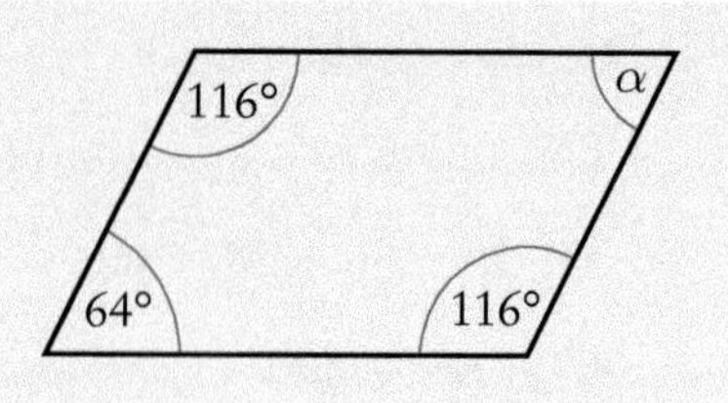

2 Work out the value of the variable for each of these. You must show your working.

a

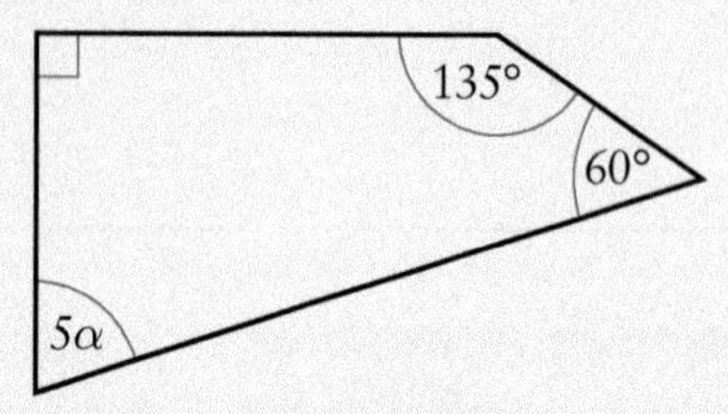

b

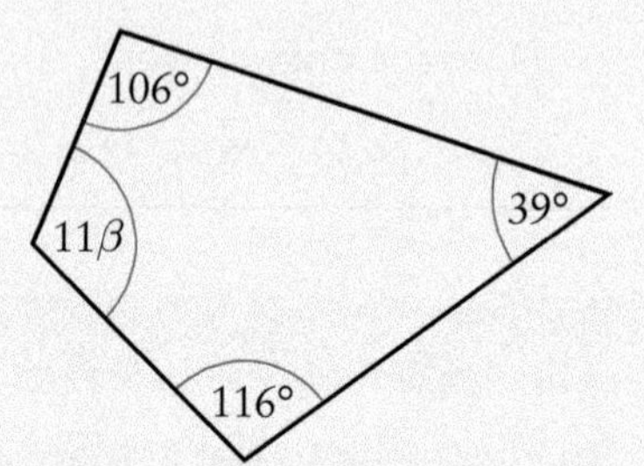

c

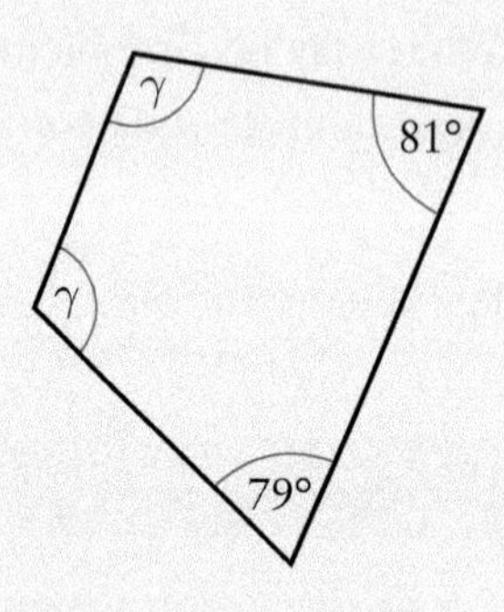

d

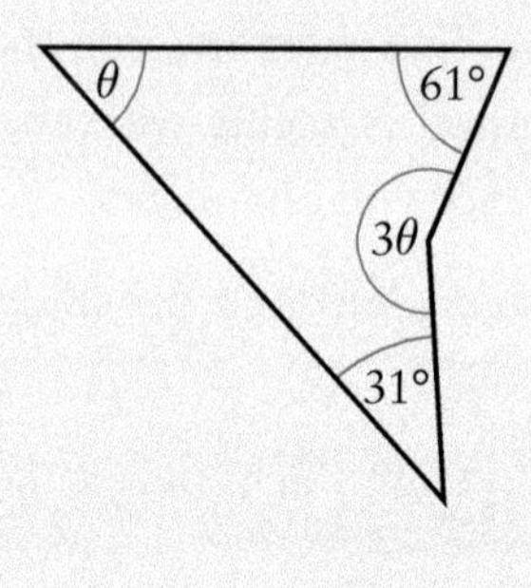

3 Work out the size of each angle in these diagrams. You must show your method.

a

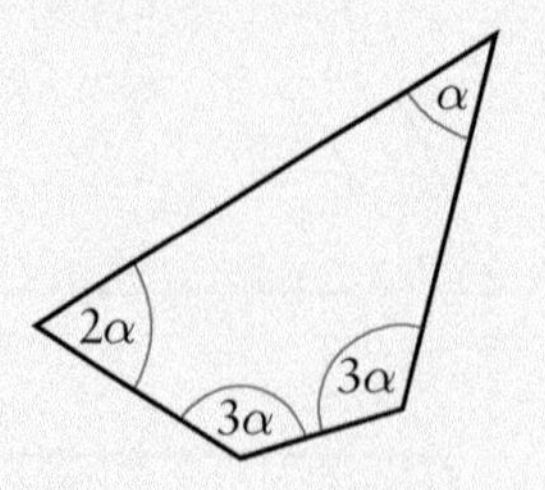

b

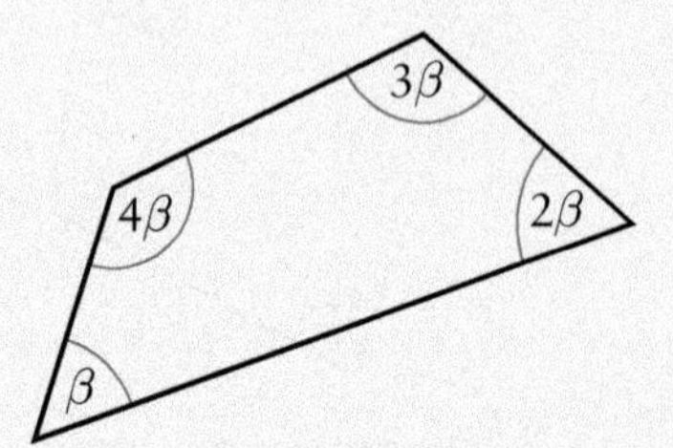

c

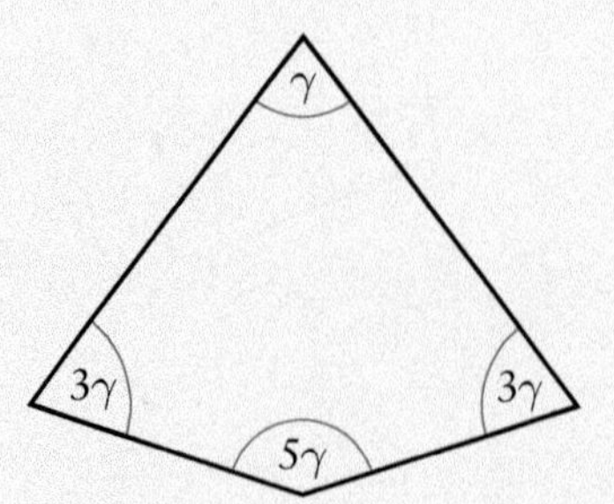

d

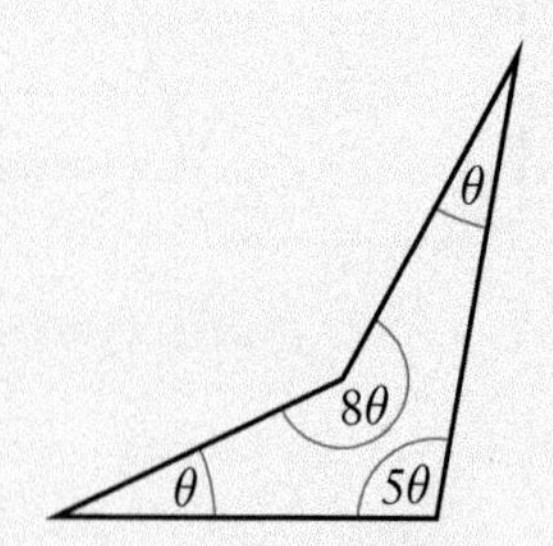

4 Work out the value of the variable in each diagram. You must show your method.

a

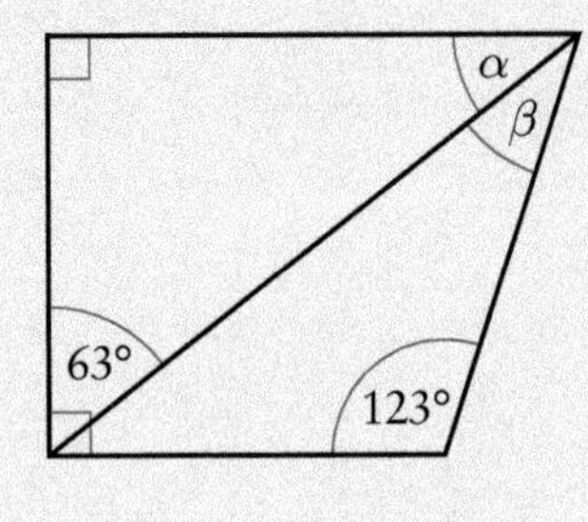

b

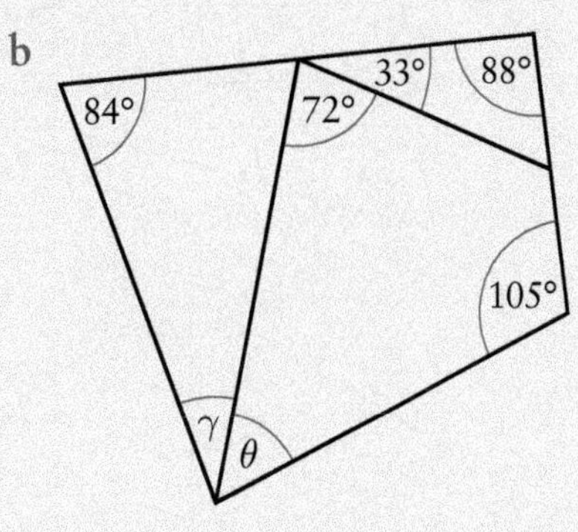

c

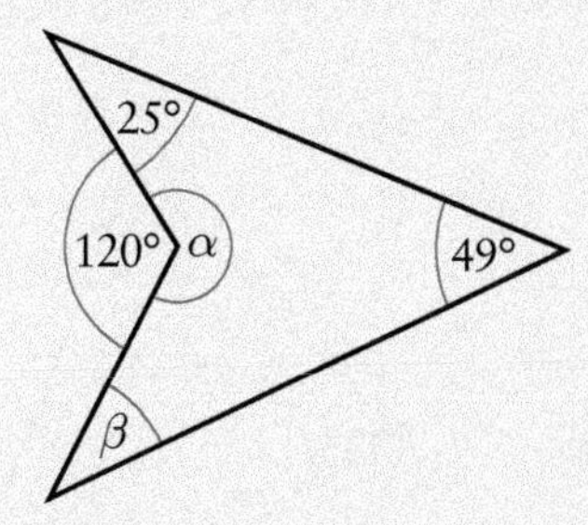

d

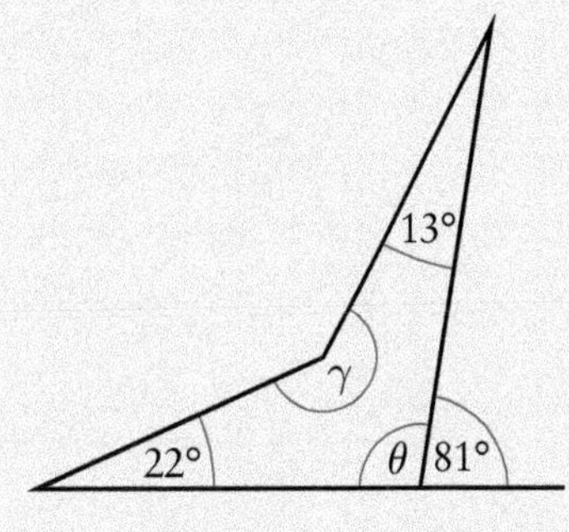

e

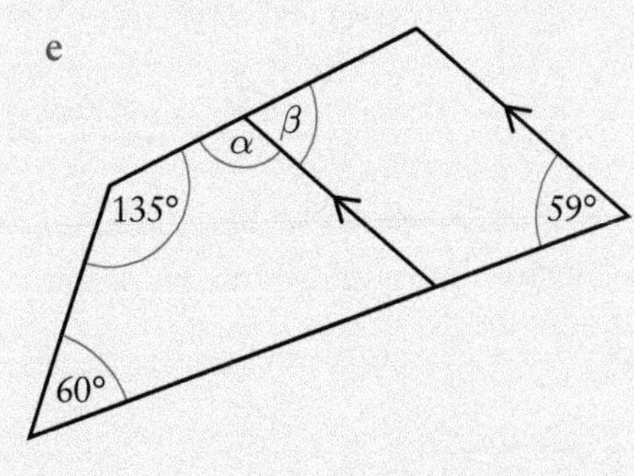

f

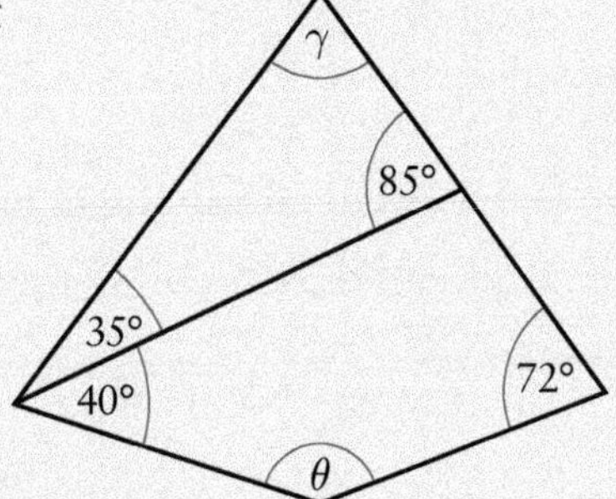

5 Work out the value of the variable in each diagram. You must show your method.

a

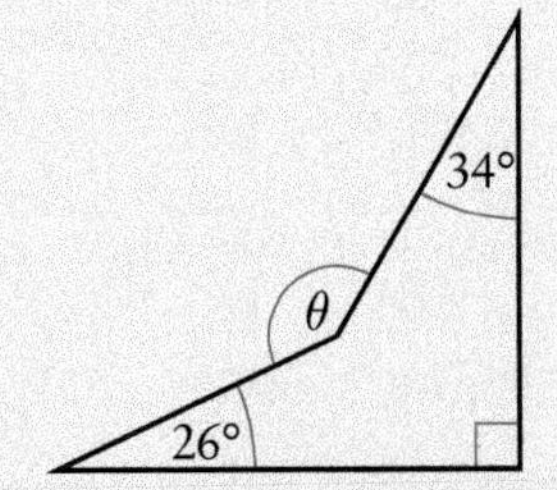

b

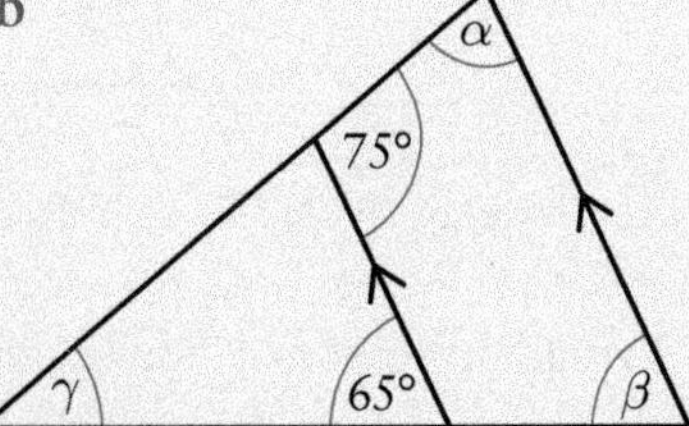

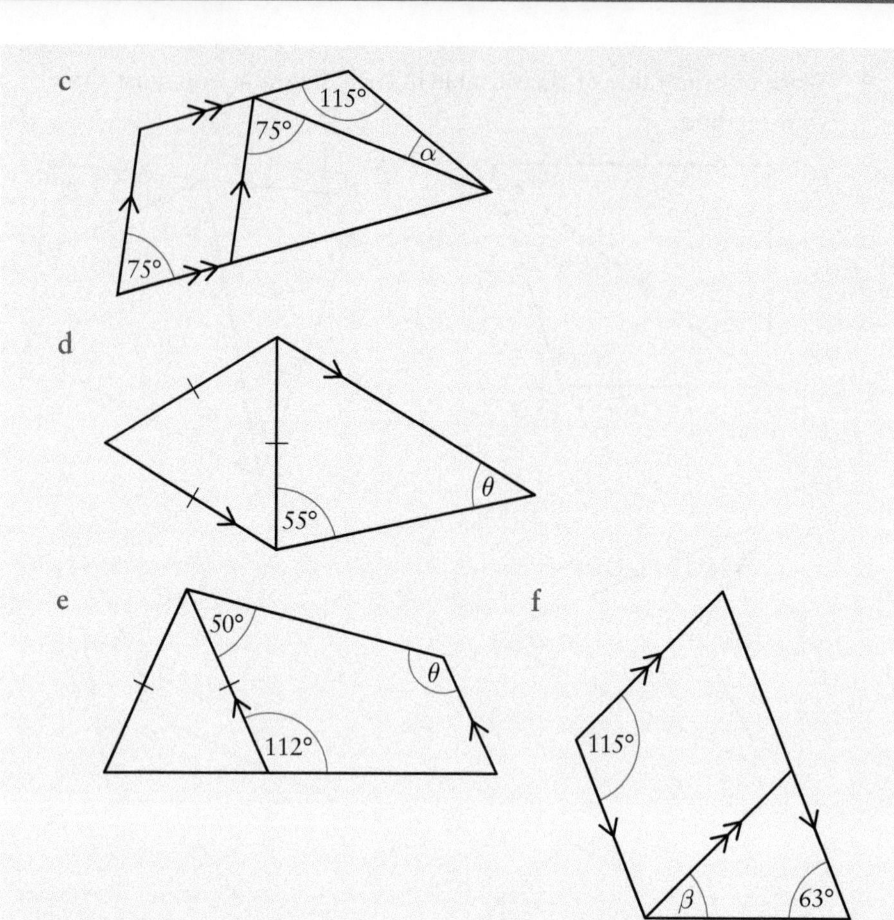

6 A quadrilateral has angles in the ratio 4 : 5 : 12 : 15

Work out the sizes of all the angles. You must show your working.

7 Arlo is given this problem to solve:

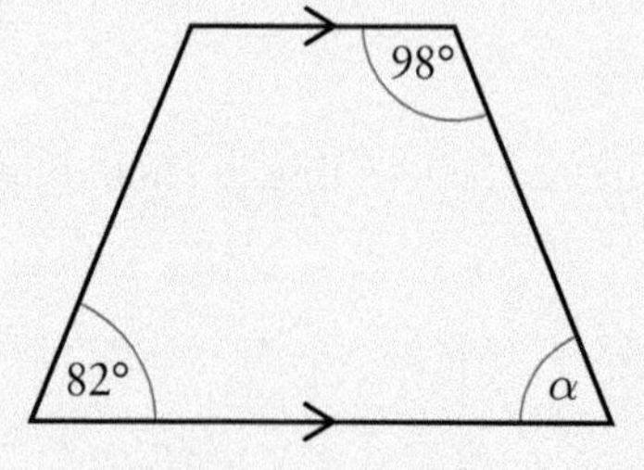

He writes, '$\alpha = 82°$'. Is Arlo correct? Explain your answer.

8 A quadrilateral is made by joining two identical, equilateral triangles.

Ela says, 'this means all four angles in the quadrilateral must be equal'.

Is she correct? Explain your answer.

9 The diagram shows the quadrilateral $ABCD$.

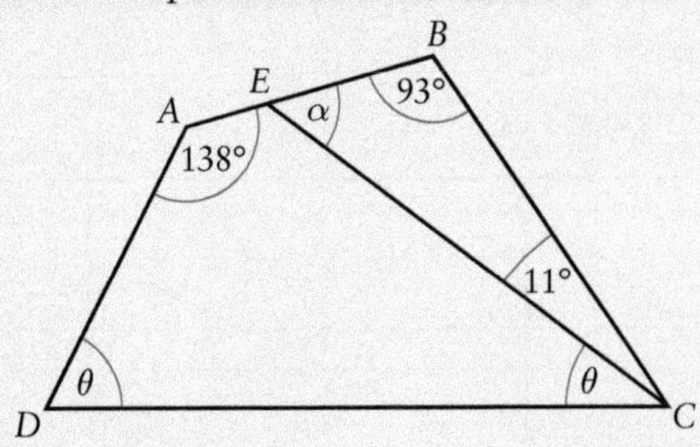

Work out the value of the variables. Show your method.

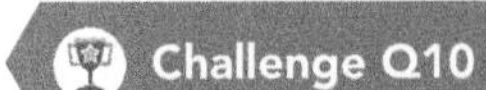

10 Anika has a quadrilateral with angles in the ratio $1:2:3:4$
She says, 'the difference between the largest and the smallest angles is 108°'.
Is Anika correct? Explain your answer.

Activity 6.1

Challenge

Look at these triangle and quadrilateral jigsaw pieces.

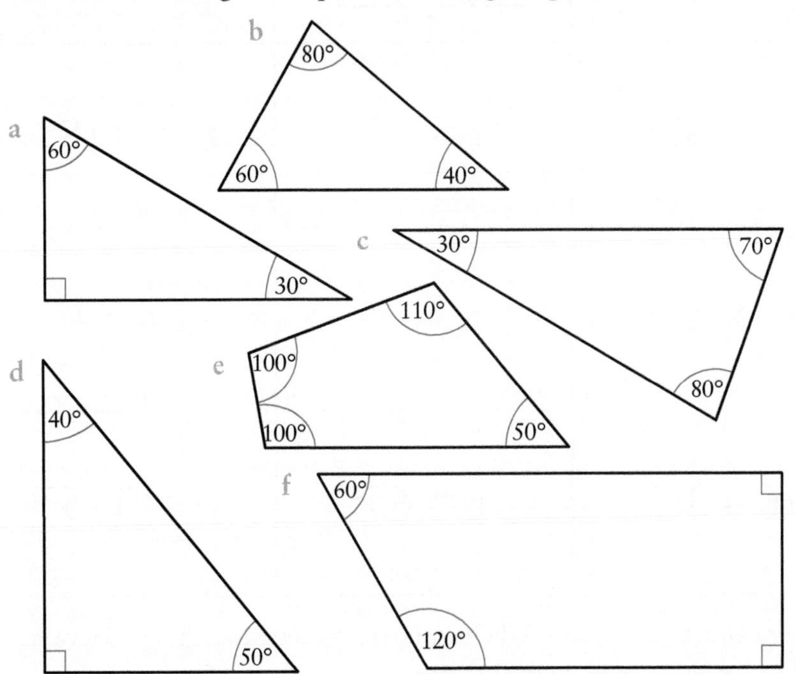

The pieces can be placed together to make a rectangular jigsaw. Trace and cut out the jigsaw pieces. Arrange them so that they make a rectangle. You can use the angle facts covered in this chapter to help you.

Make your own rectangular jigsaw. Start by drawing a rectangle and then draw your triangles and quadrilaterals on it. Write the angles on your shapes and then cut them out. Swap your jigsaw with someone else and see if you can put them back together.

 Activity 6.2

Angle snap game

Copy the following onto blank cards.

$11a$, $7a$	a, a, a, a, a	a, $62°$
$62°$, a	a, a, a, a, a	$65°$, a
$115°$, a	$65°$, $120°$, a	$78°$, $3a$, $94°$, a
$a = 62°$	$a = 55°$	$a = 118°$
$a = 47°$	$a = 5°$	$a = 54°$
$a = 36°$	$a = 65°$	$a = 115°$

Use your cards to play Snap. When a player correctly matches a diagram with the value of *a*, they must also give a correct reason for the angle.

Alternatively, place the cards face down. Players take turns to turn two cards over. If the cards match and the player gives a correct reason for the value of a, then the player keeps the cards and has another go. If the cards do not match the player turns them face down again. Players take turns until all the cards are paired. The winner is the player with the most cards at the end of the game.

You can create your own set of cards.

Investigation 6.3

Thinking skills

Social skills

Communication skills

Draw a rectangle and cut it out.

Using one cut, cut your rectangle into two quadrilaterals. For example:

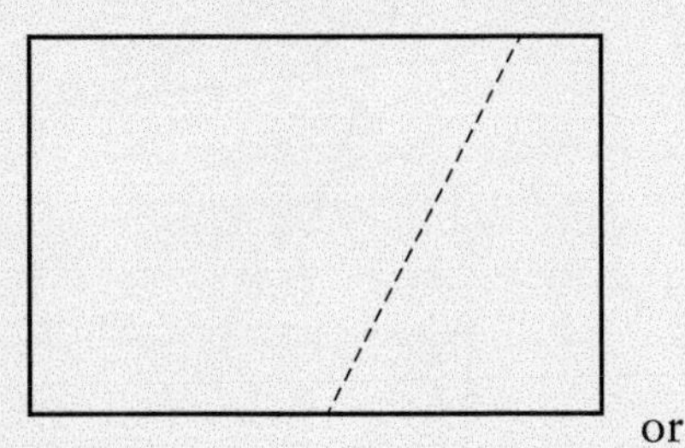

or

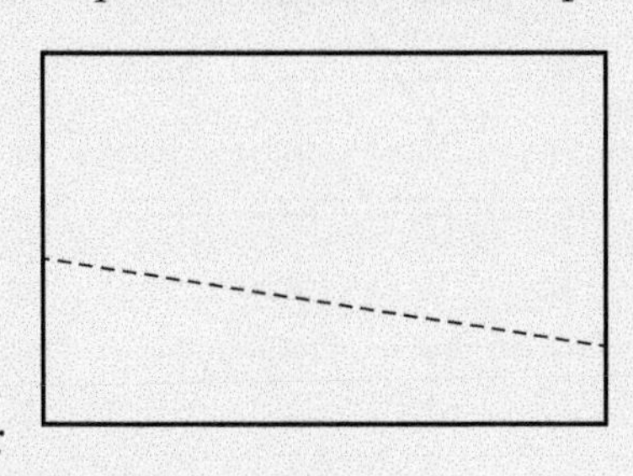

Repeat this process for other rectangles. What do you notice?
Will this always happen? Explain your reasoning.

Self-assessment

- I can reason with adjacent angles in right angles.
- I can reason with adjacent angles on straight lines.
- I can identify vertically opposite angles.
- I can reason with angles at a point.
- I can identify alternate, corresponding and co-interior angles.
- I can reason using the angles in a triangle.
- I can calculate missing angles in a quadrilateral.

? Check your knowledge questions

1 For each of the following, write down the value of the variable.
Give a reason for each answer.

a

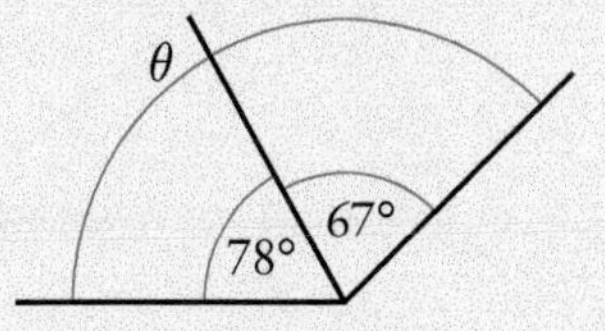

b

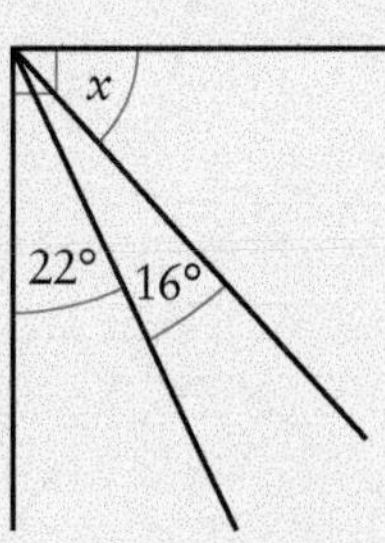

c

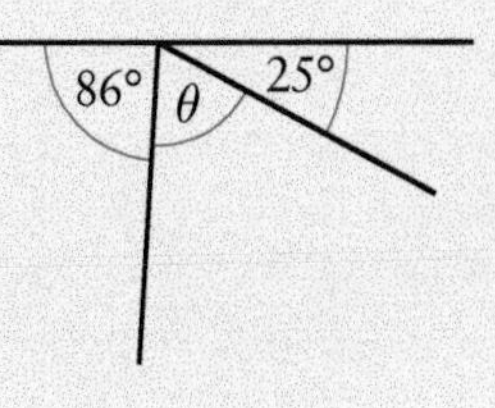

d

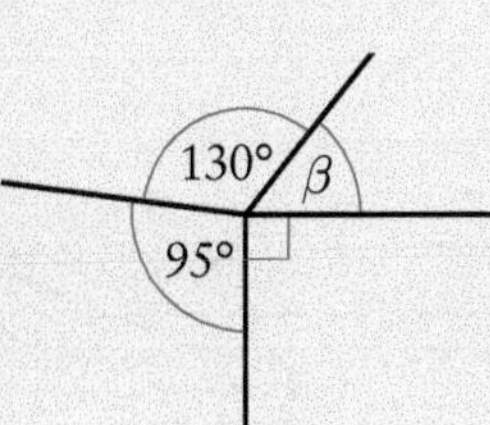

e

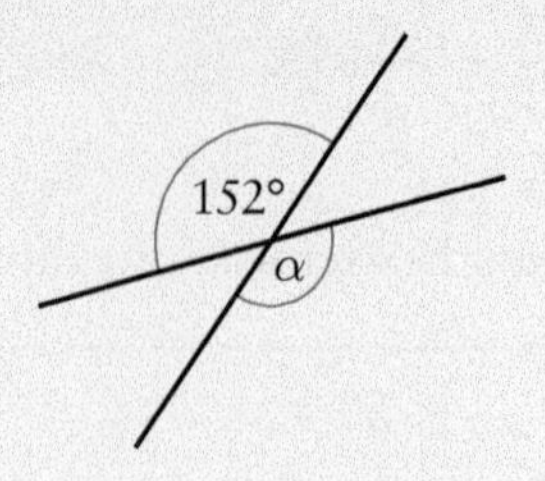

f

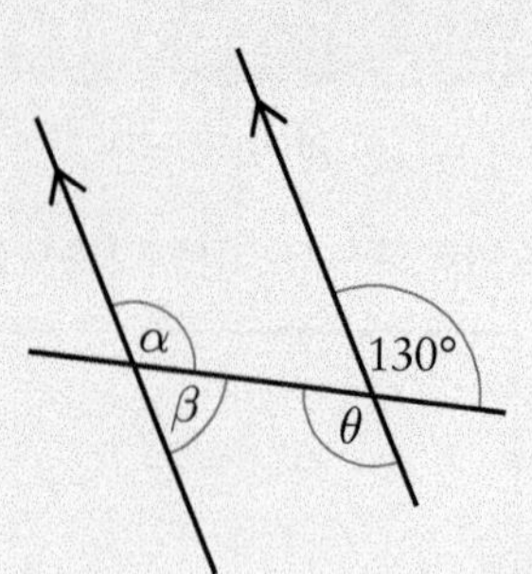

g

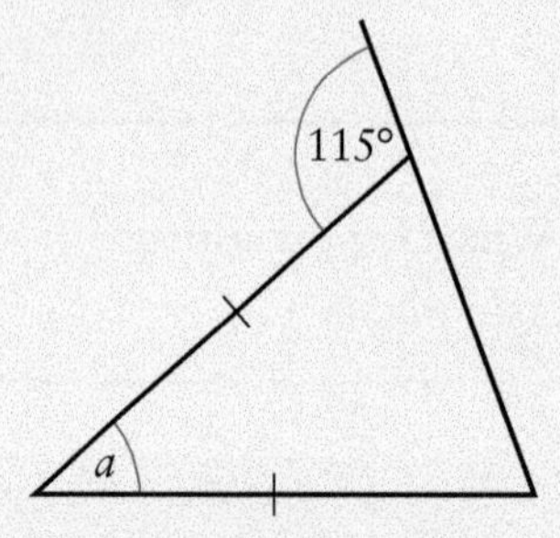

h

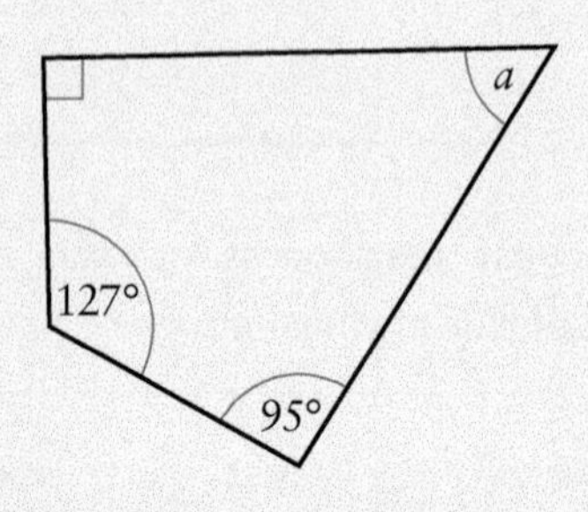

i

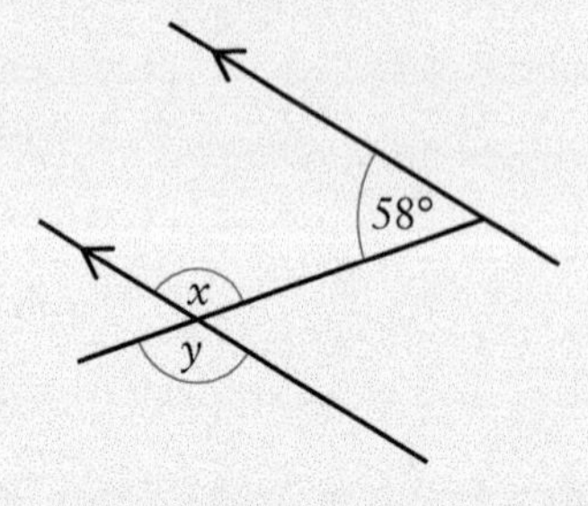

2 Write down the value of the variable in each of these.
Show your working.

a

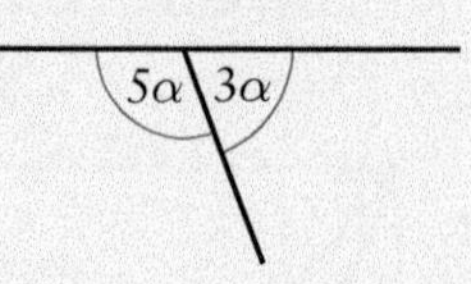

b

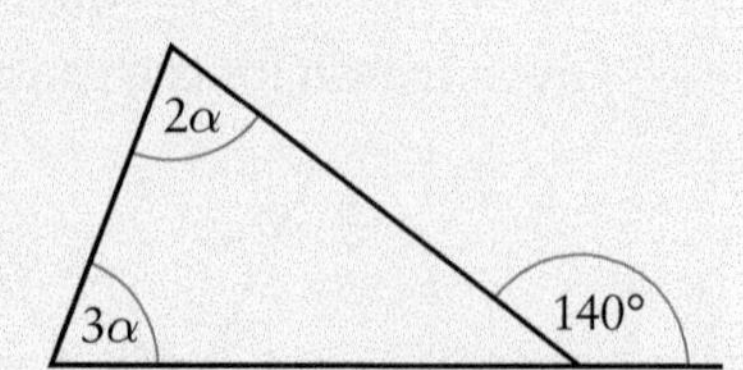

c

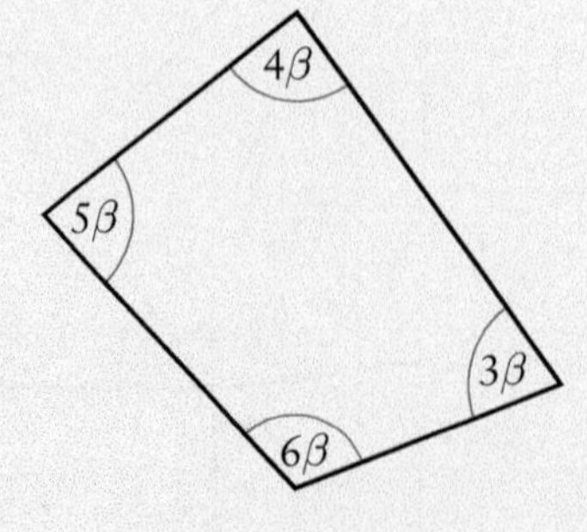

3 Here is a diagram of a right-angle.

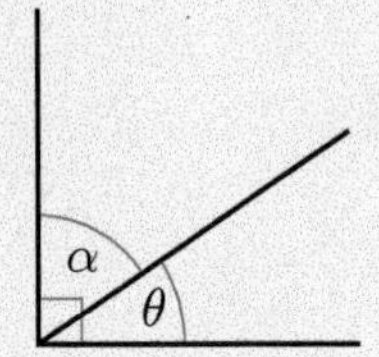

α is 1.5 times the size of θ.

Work out the values of α and θ. Give reasons for your answers.

4 In the diagram the values of α and β are in the ratio $\alpha : \beta = 19 : 5$. Work out the size of both angles. You must show your working.

Challenge Q4

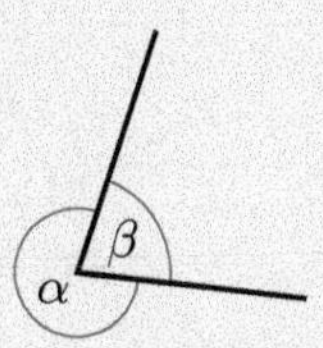

5 Draw a diagram showing an angle of 35° and co-interior angle a. Write the value of a on your diagram.

6 Work out the variables in these diagrams.
Give reasons for your answers.

a
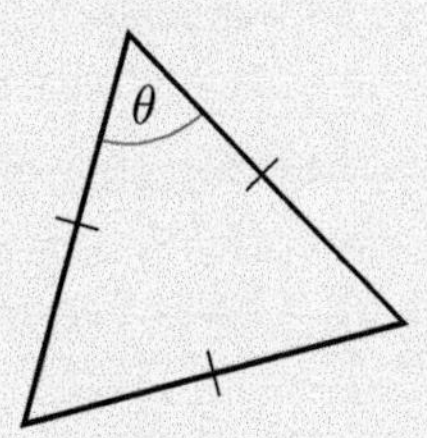

b
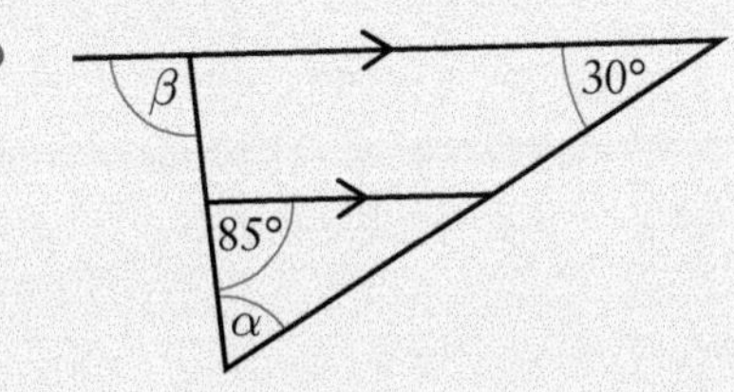

c
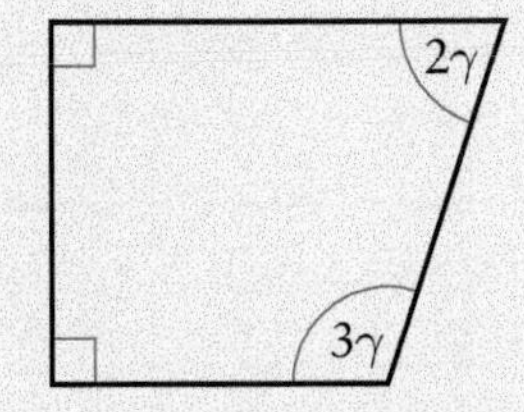

7 Work out the values of the variables in the diagram.
State reasons for your answers.

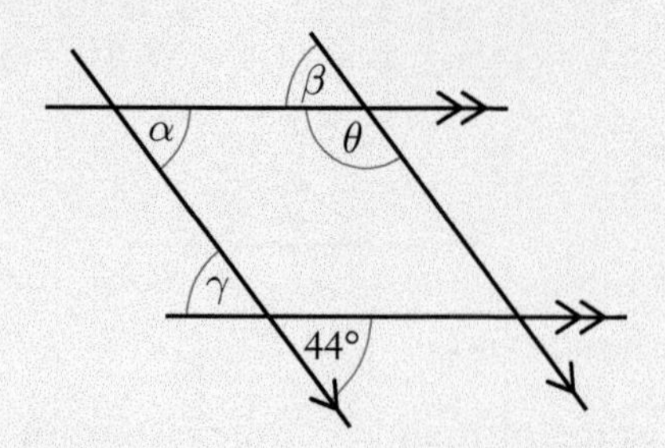

Challenge Q8

8 Use the diagram to work out the size of angle α and θ.
Give reasons for your answers.

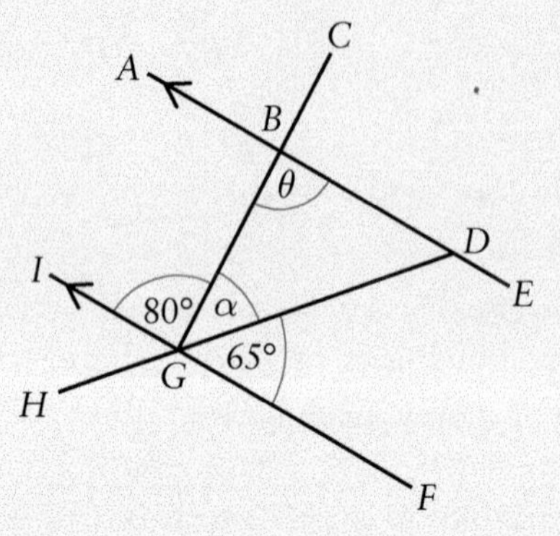

7 Circles

7 Circles

KEY CONCEPT

Relationships

RELATED CONCEPTS

Approximation, Generalisation, Models, Space

GLOBAL CONTEXT

Orientation in space and time

Statement of inquiry

Generalising relationships between shapes helps us to model real-life spacial patterns.

Factual

- How can we measure different parts of a circle?

Conceptual

- How accurate are measurements of length? Why do all measurements of length involve rounding?

Debatable

- Do irrational numbers occur in the natural world? Do perfect circles occur in the natural world?

Do you recall?

1 Use a pair of compasses to draw a circle with a radius of 4 cm.

2 The table shows the species of birds in a garden one day.

Colour	Frequency	Angle
Pigeon	9	90°
Sparrow	17	170°
Magpie	6	60°
Robin	4	40°

Draw a pie chart for this data.

3 Calculate the size of the angle labelled x in this diagram:

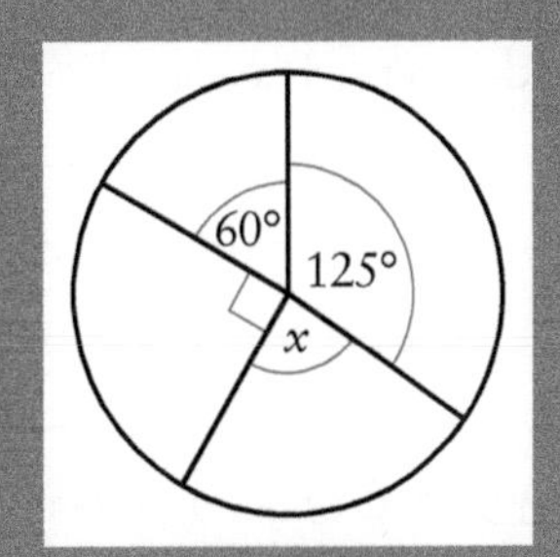

4 Give the mathematical name for each triangle:

a 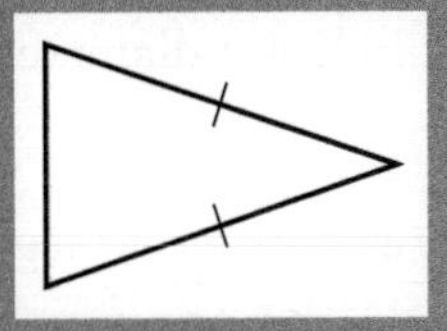

b 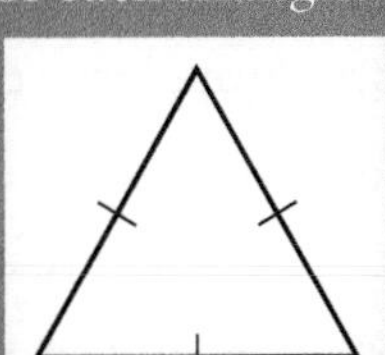

c 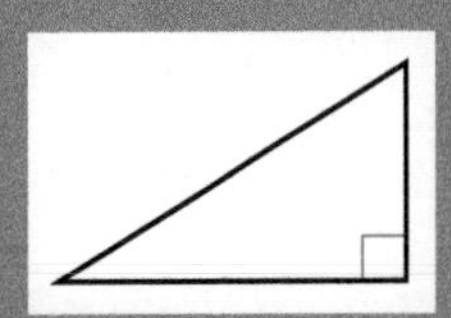

5 Estimate the answer to each calculation by rounding values to 1 significant figure:

a 3.7×40 b 9.2×8.7 c 4.5176×21.38

6 Solve to find the value of x:

a $12 = 4x$ b $2x^2 = 50$

7 Use the equation $y = 2x^2$ to find the value of y when $x = 3$ and the value of x when $y = 14$
Give your answers to 1 decimal place where appropriate.

8 Calculate the area of each shape.

a 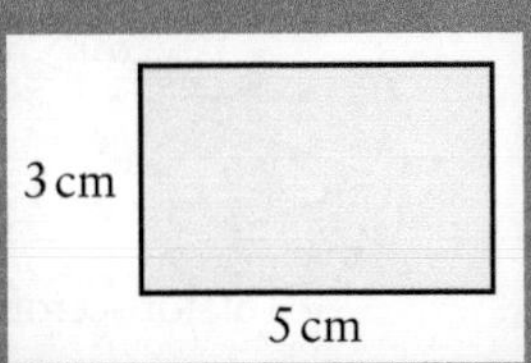

b 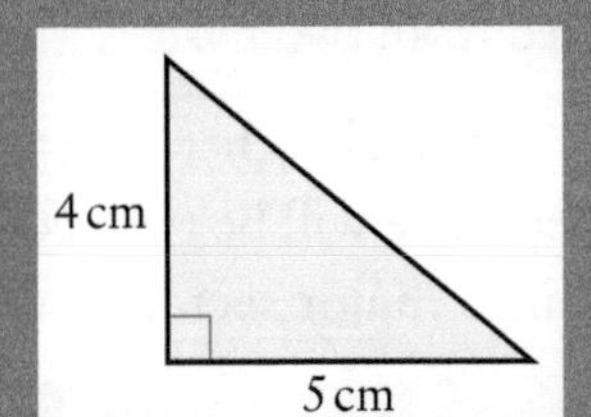

c 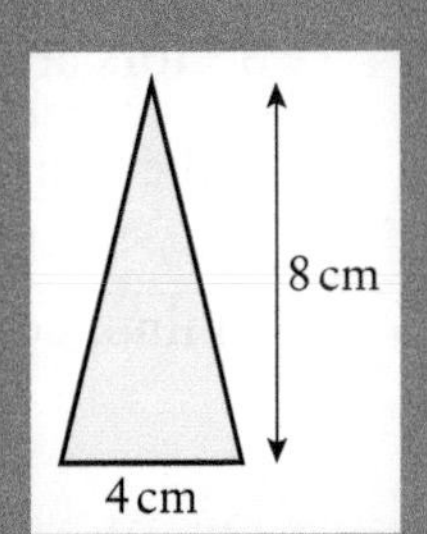

Hint

Radius and diameter are also used to mean the length of the radius and the length of the diameter. For example, you may read: 'a circle with centre O and radius 5', which means that the length of the radius is 5 units.

Hint

The plural of radius is radii.

When you draw a circle using compasses, the distance you open your compasses is the radius of the circle.

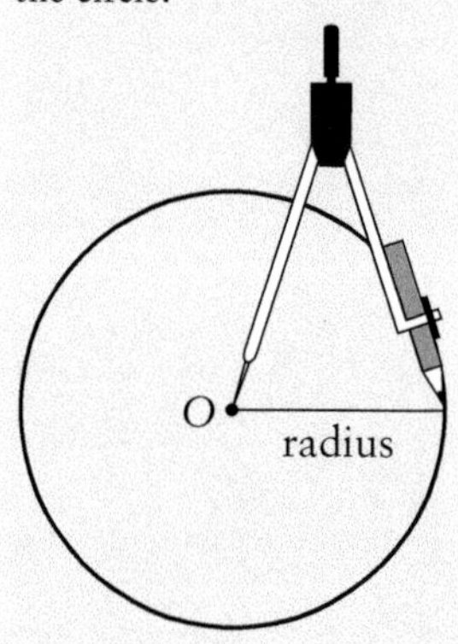

7.1 Parts of a circle

A **circle** is the set of all points that are a fixed distance from a centre.

The distance around a circle is called the **circumference**.

A **radius** of a circle is a straight line segment from the centre to any point on the circumference.

A **diameter** of a circle is a straight-line segment through the centre that touches two points on the circumference.

The diameter is twice the radius.

Here is a circle with a diameter and radius, shown with two other lines.

- A **tangent** to a circle is a straight line which touches the circle at only one point.
- A **chord** of a circle is a straight-line segment whose endpoints both lie on the circle.

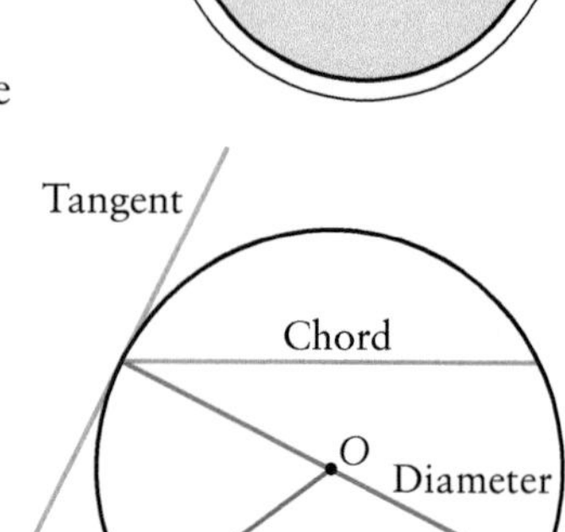

Note: a chord that passes through the centre of a circle is a diameter.

Circles can also be divided into parts:

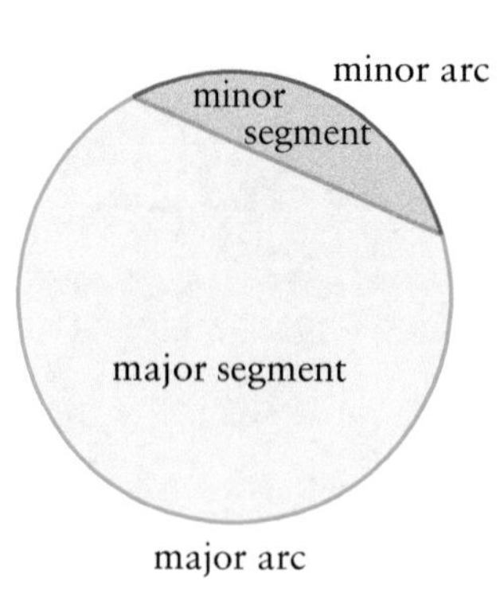

A chord divides a circle into two **segments**, a **minor segment** and a **major segment**. If the chord is a diameter, then the two segments are equal in size and we call them both semicircles.

Likewise, the endpoints of the chord define two **arcs**, one **minor** and one **major**, that are both part of the circumference.

Sectors are created from two radii.

If the radii are at any angle other than 180° to each other, then there will be two sectors: a **minor sector** and a **major sector**.

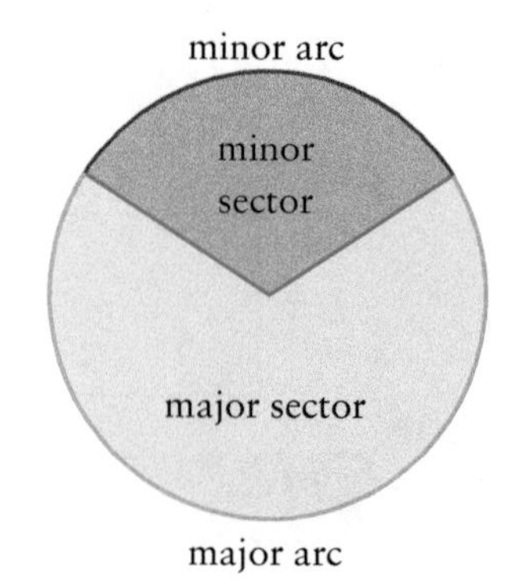

Explore 7.1

Can you draw a circle with 10 equal sectors?

How can you draw a regular decagon inside a circle?

Investigation 7.1

Research how the Russian artist Kandinsky used circles in his paintings. Are they perfect circles?

A few tips:

- You can make accurate drawings of regular polygons by constructing a circle first.
- An **inscribed polygon** is a polygon constructed inside a circle. All the vertices of the polygon are on the circle.
- **Concentric** circles have the same centre.

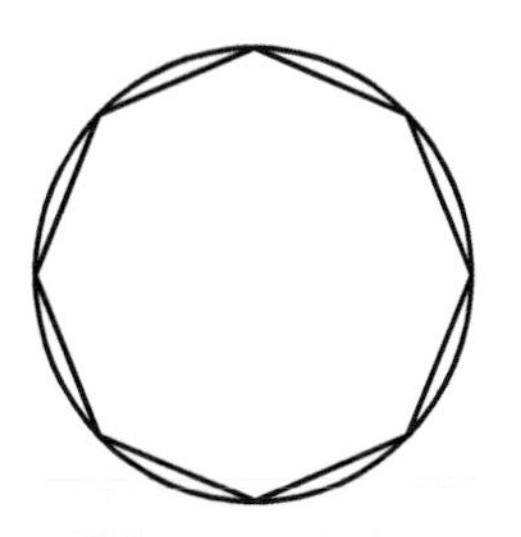

Worked example 7.1

Draw two concentric circles with radii 4 cm and 7 cm.

Solution

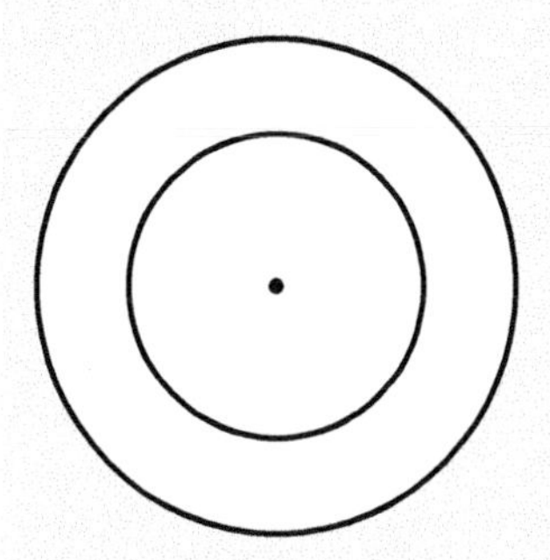

Draw a circle with radius 4 cm.

Open the compasses to 7 cm.

Place the compass point on the centre of the 4 cm circle.

Draw another circle.

Reflect

What is the distance between the two circles?

How could you draw two concentric circles with a distance of 5 cm between them?

Connections

Drawing a perfect circle freehand is probably impossible. According to the Italian Renaissance artist Vasari, the artist Giotto could draw a perfect freehand circle with paintbrush and ink, which proved his skill as a painter.

Connections

A ring shape, like the one in the Worked example, is called an annulus. In an annular solar eclipse the moon passes in front of centre of the sun, so the visible light from the sun forms an annulus.

Communication skills

Self-management skills

Investigation 7.2

Two concentric circles do not meet – they do not intersect or touch each other.

How many ways can two circles meet or not meet?

Practice questions 7.1

1 Name the part of the circle shown in dotted red:

a 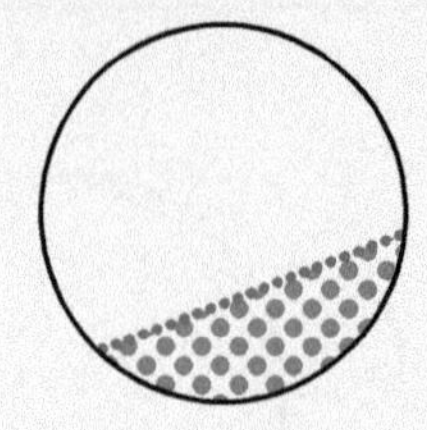
b 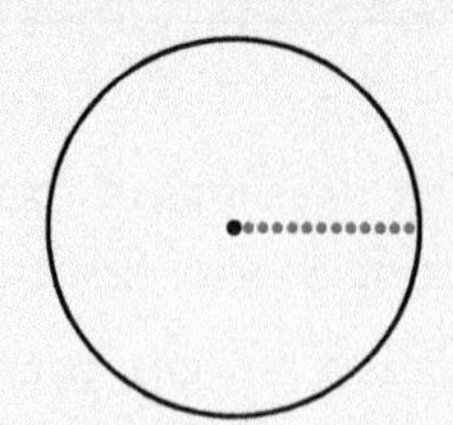
c 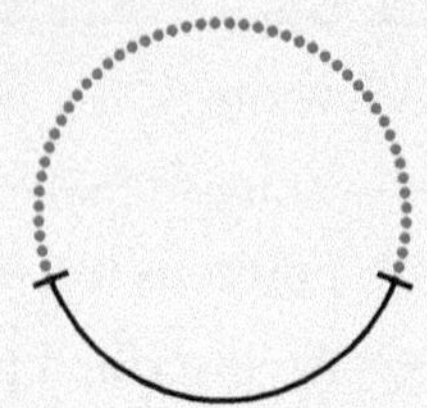

d 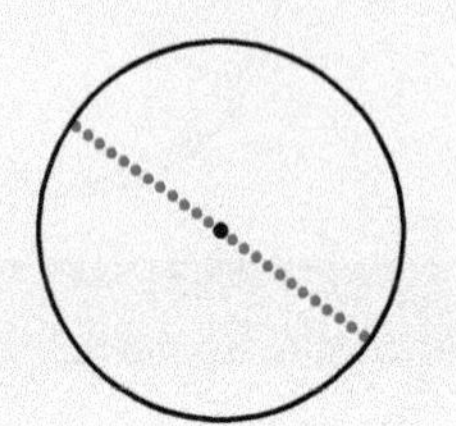
e 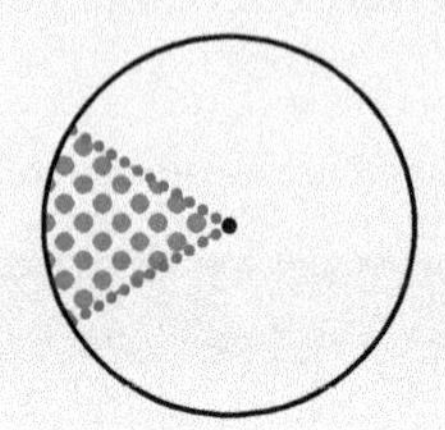
f 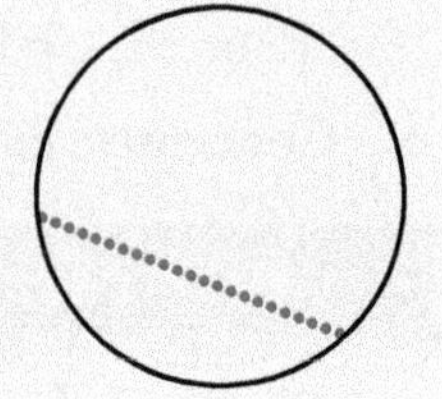

2 Name the circle features indicated in the pictures.

a 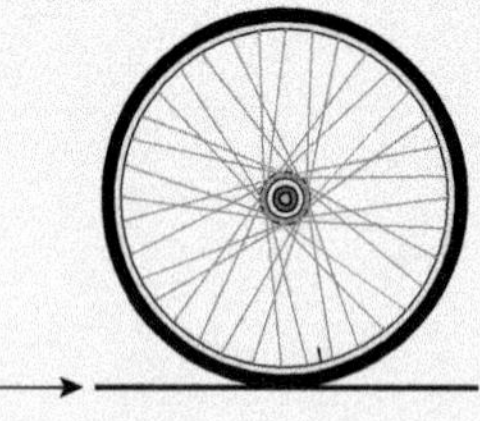
b 
c 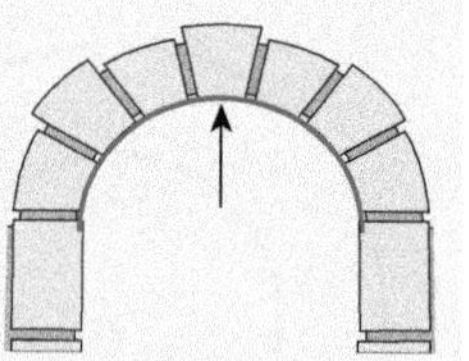

3 In the diagram, name the features of the circle marked with lines:

a AB b OC

c BC d DE

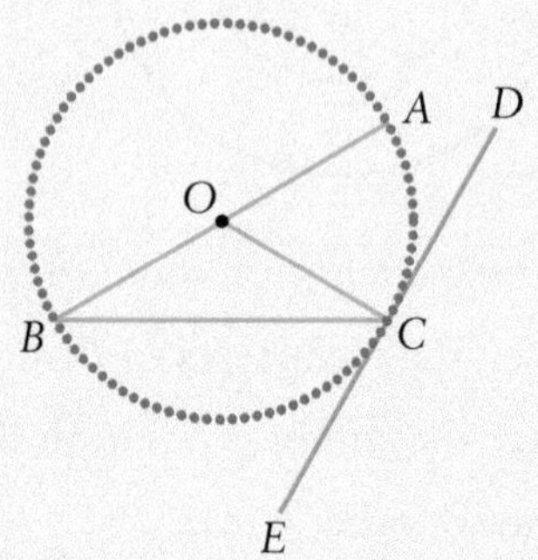

4 Is the statement below true?

'In a pie chart, sets of data are represented in segments of a circle.'

If it is false, write a true statement.

5 Draw a circle with radius 7 cm.

6 Draw a circle with diameter 11 cm.

7 The diagram shows an inscribed regular nonagon. All nine sectors are congruent.

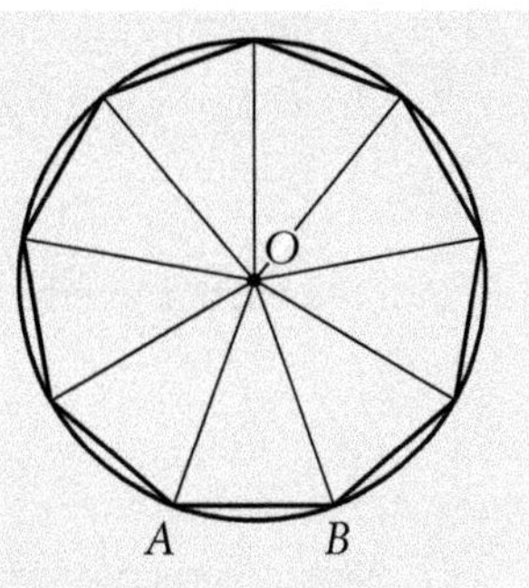

   a Work out the angle at the centre for each sector.

   b What type of triangle is AOB? Explain how you know.

8 Make an accurate drawing of an inscribed regular octagon.

9 Draw a sector of a circle with an angle measuring:

   a 50° b 125°

10 Make an accurate drawing of this shape. Explain your steps.

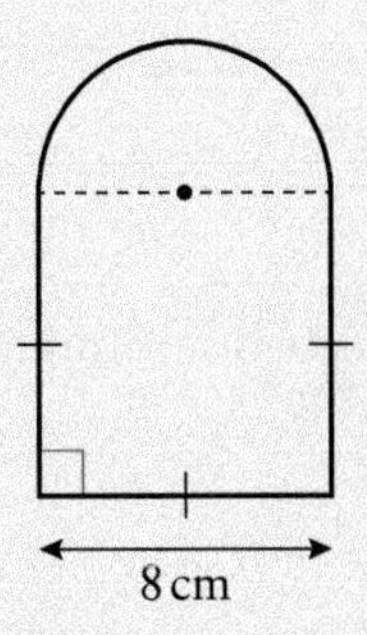

11 Make an accurate drawing of this diagram. Explain your steps.

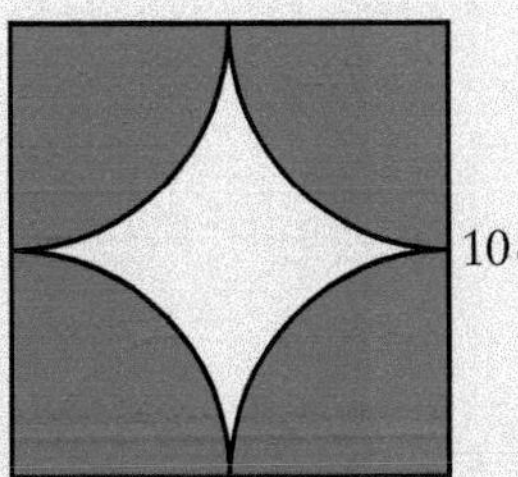

12 The diagram shows an equilateral triangle inscribed inside a circle, with another circle drawn inside the triangle. Make an accurate drawing of this diagram and explain your steps.

Challenge Q12

Your drawing can be larger than the one given.

13 Make an accurate drawing of a hexagon. Show that a hexagon can be divided into six equilateral triangles.

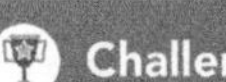

7.2 Circumference of a circle

You can measure the circumference of a coin by rolling it along a ruler:

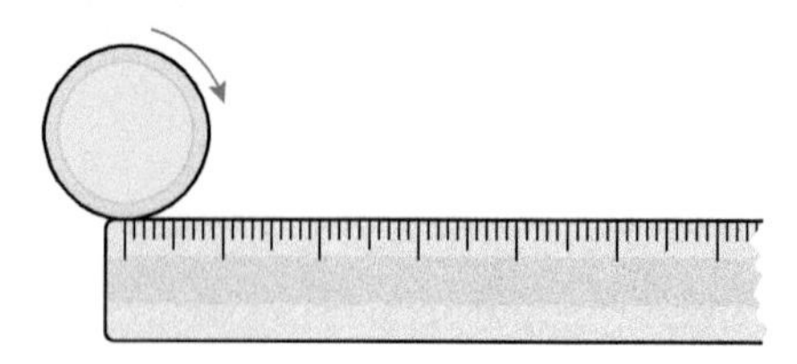

You can measure the circumference of a circle using a tape measure, or some string:

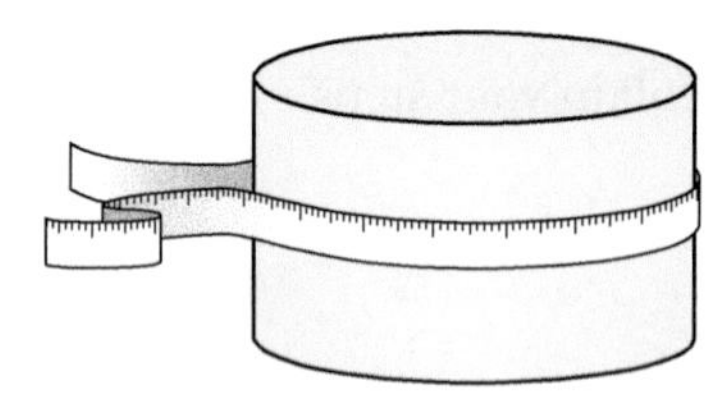

You can measure the diameter of a circle using a ruler and two set squares:

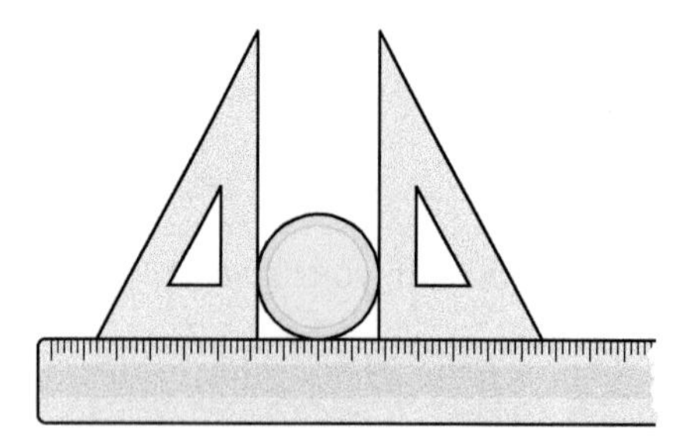

Could you measure a circular **hole** using any of these methods? If not, how could you do it?

Hint

You can draw circles with different radii using dynamic geometry software such as GeoGebra. The software can give you the circumference of the circle accurate to 1 decimal place.

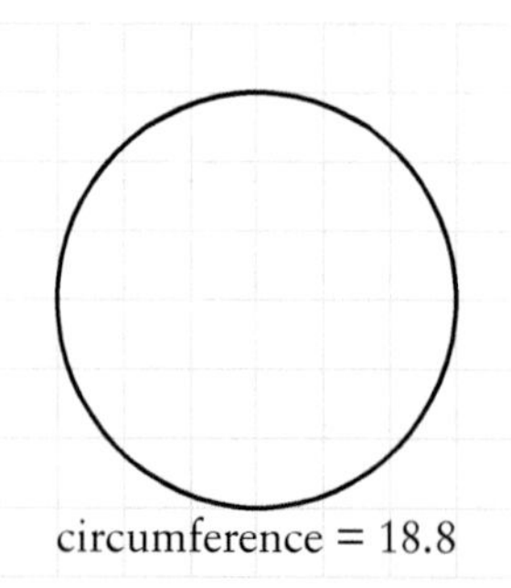

Explore 7.2

How could you find the circumference of a circle with a 1 cm diameter? With a 10 cm diameter? What is the relationship between the circumference of a circle and its diameter?

Worked example 7.2

A bicycle wheel has diameter 60 cm.

a Estimate the circumference of the wheel.

b Estimate how far the bicycle wheel rolls in 1 full turn.

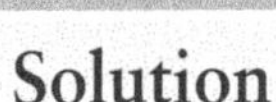

Solution

a You may know by now that the circumference is roughly three times the diameter.

$60 \times 3 = 180$ cm

Estimate for circumference is 180 cm

b Therefore, the bicycle wheel rolls approximately 180 cm in one full turn.

Reflect

How could you estimate the circumference of a circle from its radius?

For every circle, the ratio $\frac{\text{circumference}}{\text{diameter}} = 3.1415926\ldots$

This decimal number continues forever and has no repeating patterns in its digits.

The Greek letter π (pronounced 'pi') is used to represent this number.

$\frac{\text{circumference}}{\text{diameter}} = \pi$ or circumference $= \pi \times$ diameter

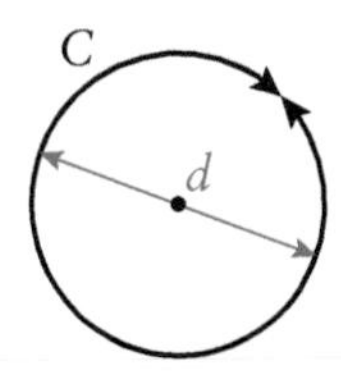

$C = \pi d$

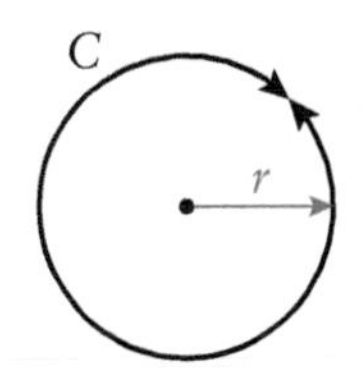

diameter $= 2 \times$ radius

$C = 2\pi r$

You can use the π button on your calculator in circumference calculations.

Fact

A rational number can be written as a finite or recurring decimal, or as a fraction or ratio. The value of π is called an irrational number, because it cannot be written exactly as a decimal or a fraction.

Worked example 7.3

a Calculate the circumference of this circle, to 1 decimal place (d.p.).

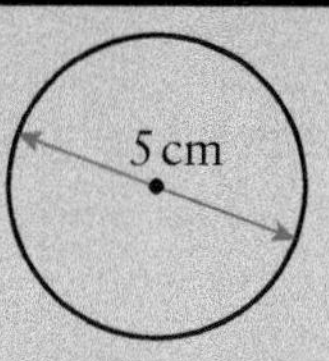

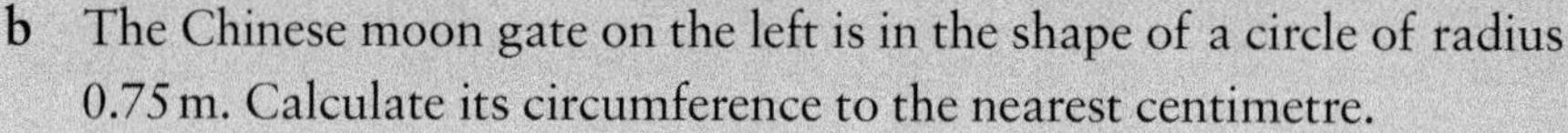

b The Chinese moon gate on the left is in the shape of a circle of radius 0.75 m. Calculate its circumference to the nearest centimetre.

Solution

a Use the formula $C = \pi d$ and the π key on your calculator.

$C = \pi d$

The diagram shows that $d = 5$ cm, so:

$C = \pi \times 5$

$= 15.7$ cm (1 d.p.)

b Use the formula $C = 2\pi r$

We are given that $r = 0.75$ m, so:

$C = 2 \times \pi \times 0.75$

$= 4.71$ m (2 d.p.)

Reflect

How do you decide which formula to use to calculate the circumference?

Research skills

Investigation 7.3

The Greek mathematician, Archimedes, used a geometric approach to calculate π.

He used inscribed and circumscribed polygons, like this:

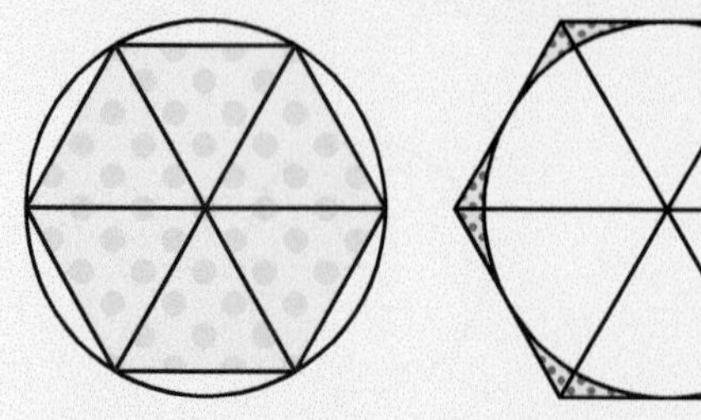

inscribed polygon **circumscribed polygon**

The two circles are the same size.

Perimeter of inscribed polygon	<	circumference of circle	<	perimeter of circumscribed polygon

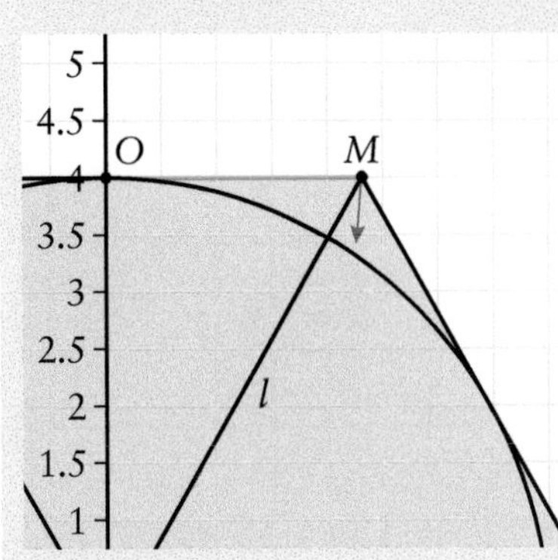

By increasing the number of sides of the polygon, Archimedes got closer and closer estimates of the circumference of the circle, which he then used to calculate π.

Research Archimedes' method, and find an animation showing how his method worked.

Practice questions 7.2

1 Work out the circumference of each circle to 1 decimal place.

a Circle with a diameter of 8 mm.

b Circle with a radius of 7 cm.

c Circle with a diameter of 20 mm.

d Circle with a diameter of 21 cm.

e Circle with a radius of 4.7 mm.

f Circle with a diameter of 12.1 mm.

2 Estimate the circumference of each circle:

a

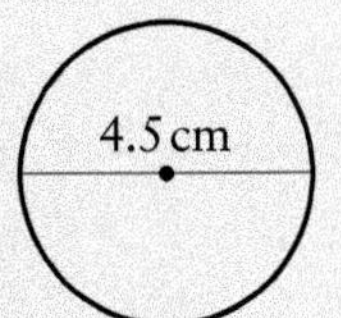

b

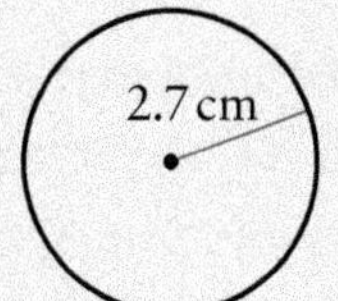

Hint Q2

Round all values in the formula to 1 significant figure first.

3 Measure the diameter of each circle and then calculate its circumference to 1 decimal place.

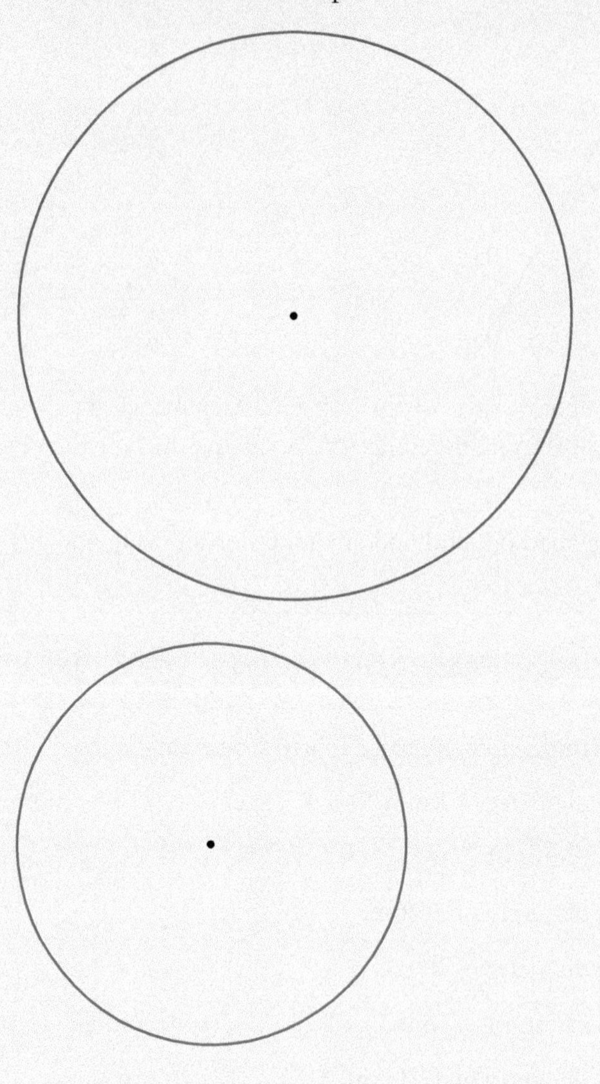

Hint Q4

A revolution of the wheel is one full turn.

4 A tractor has wheels of diameter 120 cm. How many metres does the tractor travel in one revolution of its wheels?

5 A Ferris wheel has a diameter of 126 metres. It has 60 passenger cars and takes 25 minutes to complete one revolution.

How far does each passenger car travel in one revolution?

6 The radius of car tyre is 35 cm. How many revolutions does this car wheel make in a 1 km journey? Give your answer to the nearest whole number.

7 Noor bakes a circular cake of diameter 20 cm. She wants to put a ribbon around the cake. The ends of the ribbon must overlap by 2 cm.

What length of ribbon does she need? Give your answer to 1 decimal place.

8 The circumference of a circle is 24 cm. Calculate its diameter. Give your answer to 1 decimal place.

9 The circumference of a circle is 1.4 m. Calculate the radius of the circle in centimetres to 2 decimal places.

10 For an art project, Leo uses 1 m of wire to make a circle. What is the radius of his circle, to the nearest mm?

11 Here are two concentric circles.

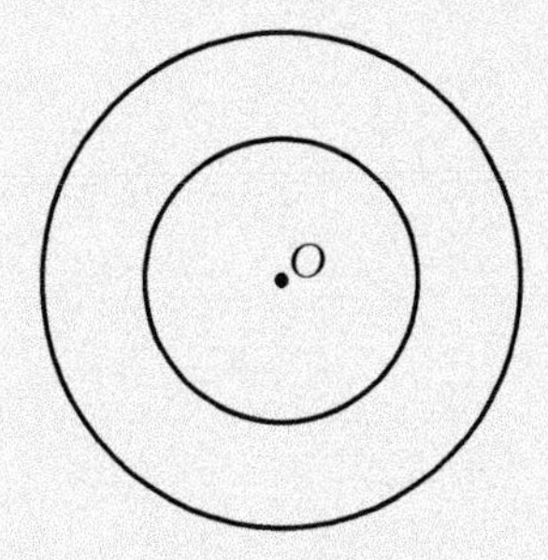

The diameter of the inner circle is 5 cm. The distance between the two circles is 2 cm. Work out the circumference of the outer circle.

12 The circumference of this circle is equal to the perimeter of this square.

What is the side length of the square?

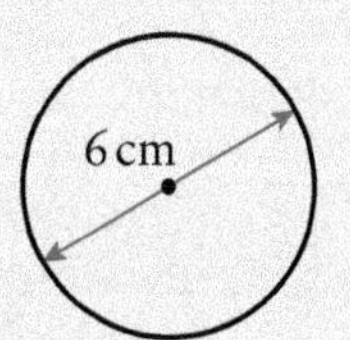

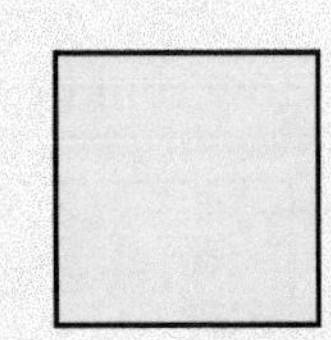

13 A square has integer side lengths.

Is it possible to draw a circle that has exactly the same perimeter as the square? Explain your answer.

14 The Earth is a sphere of radius 6371 km.

a Calculate the circumference of the Earth at the Equator, to the nearest kilometre.

b A satellite orbits the Earth above the Equator, at a height of 200 km above the Earth's surface. How far does the satellite travel in one orbit?

 Research skills

Investigation 7.4

Create a spreadsheet to calculate the distance around the equator of a planet, given the radius of the planet in km.

Research the data you need, and calculate the distance around the equator for each planet in the solar system.

7.3 Area of a circle

 Thinking skills

Explore 7.3

This circle is drawn on a 1 cm grid. How could you estimate its area?

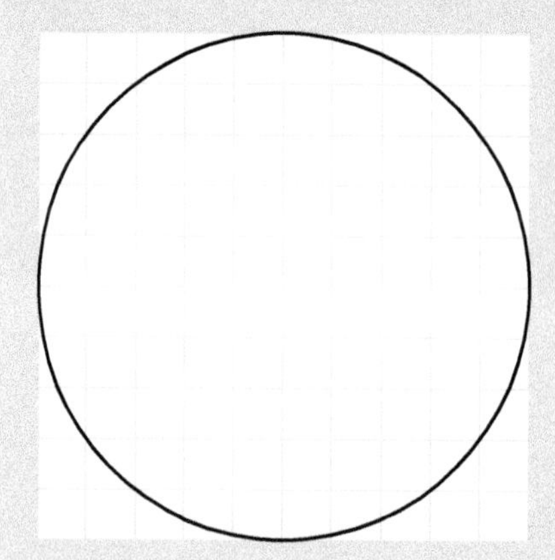

How could you estimate the area of the circle using the diagram below?

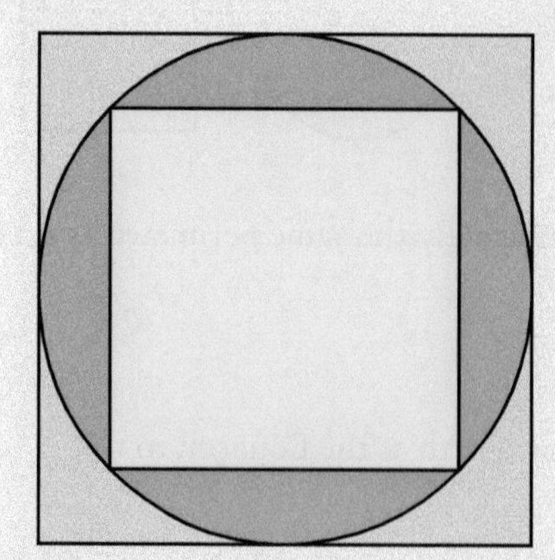

Hint

You can draw circles with different radii using dynamic geometry software. The software can give you the area of the circle accurate to 1 decimal place.

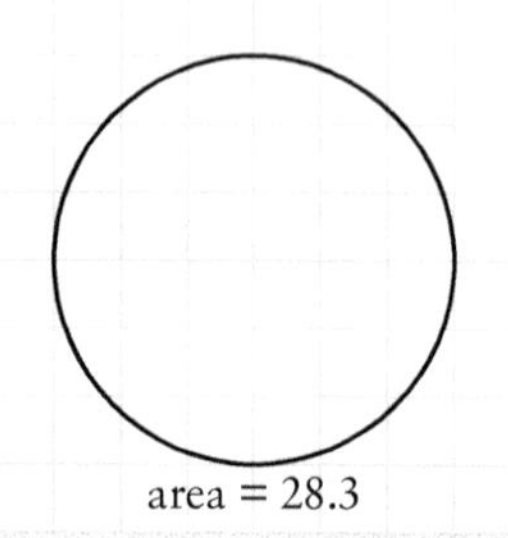

Investigation 7.5

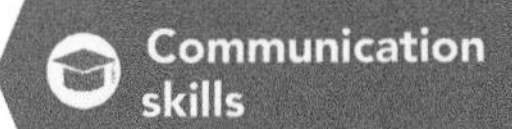

Communication skills

Draw a large circle using a pair of compasses.

Measure and draw 12 equal sectors inside your circle.

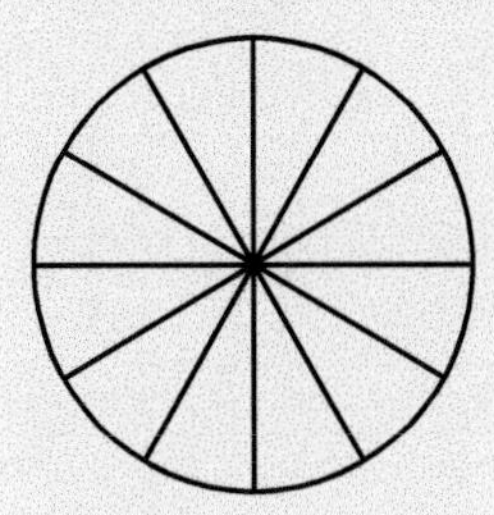

Cut out all the sectors and arrange them like this:

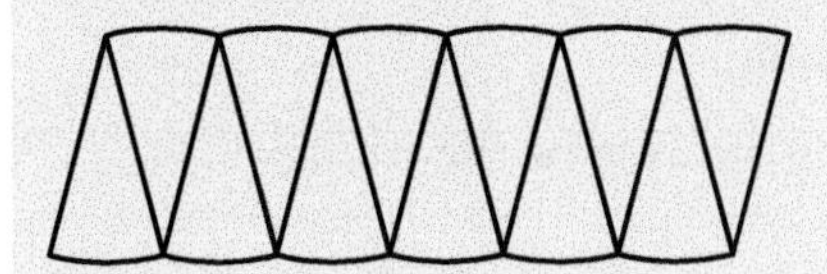

This shape is roughly a parallelogram.

The height of the parallelogram is approximately r, the radius of the circle.

How does the base of the parallelogram relate to the circumference of the circle?

Write an expression for the base of the parallelogram.

Use your expressions to write a formula for the area of a circle.

Hint

You can search online for an animation of arranging sectors of a circle into a parallelogram to find the area.

Investigation 7.6

Communication skills

Here is a circle of radius r, made of concentric rubber rings with no gaps between them.

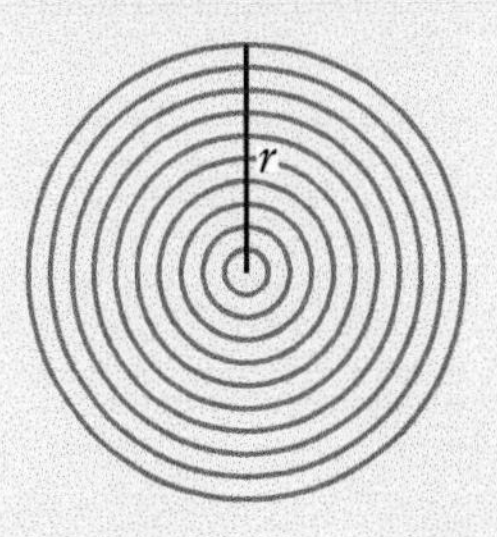

If you cut the circle along its radius and flatten the rubber rings, they would make a triangular pile:

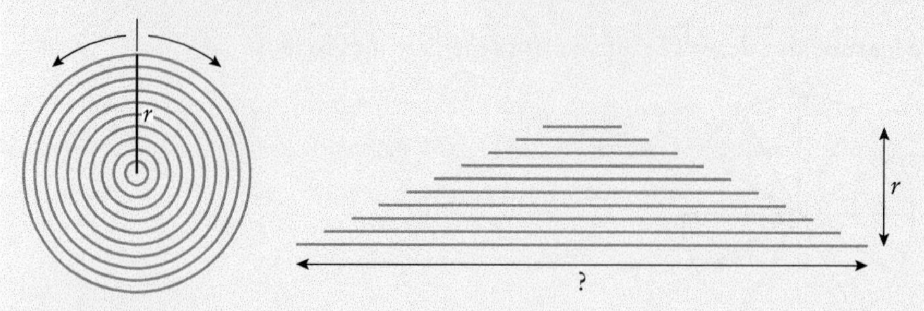

The height of this triangle is r.

The base of the triangle is the length of the longest ring.

Write an expression for the base of the triangle.

Use your expression to write a formula for the area of the triangle.

Reflect

The two Investigations used different methods to find a formula for the area of a circle. Which did you prefer and why?

Fact

The formula for the area of a circle of radius r is:

Area = $\pi \times \text{radius}^2$

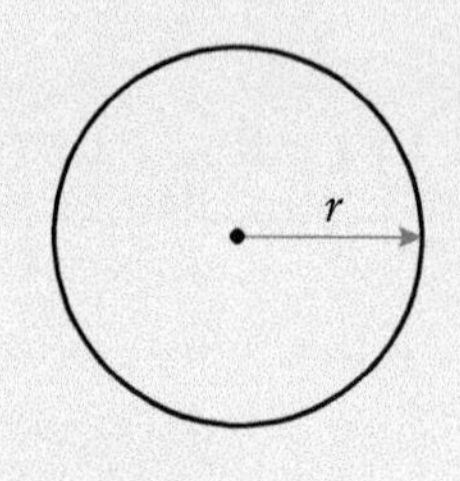

$A = \pi r^2$

Worked example 7.4

Calculate the area of this circle. Give your answer to 2 decimal places.

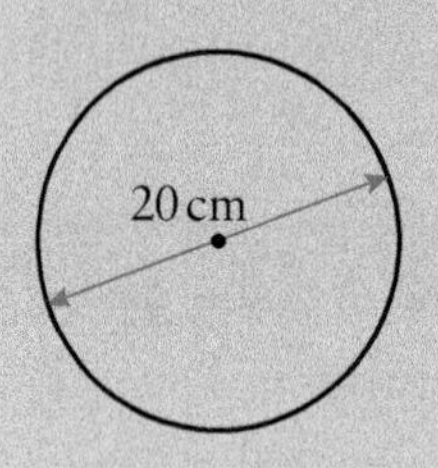

Solution

Use the formula:

$A = \pi r^2$

The diagram tells us that $d = 20$ cm.

The radius is half the diameter, so:

$r = 20 \div 2 = 10$

Substituting r into the formula, $A = \pi r^2$

$A = \pi \times 10^2$

$\quad = 314.16$ cm (2 d.p.)

Reflect

The formulae for the circumference and the area of a circle both involve π.

How can you remember which is which?

Worked example 7.5

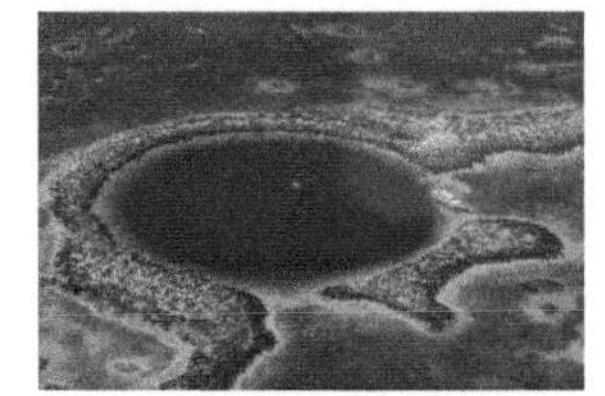

The Great Blue Hole in Belize's Lighthouse Reef is approximately a circle with circumference 999 metres. Calculate an estimate of its area.

Give your answer to the nearest whole number.

Solution

Make a plan

Use the circumference of the circle to find the radius.

Use the radius to calculate the area of the circle.

As the Great Blue Hole is not exactly a circle, this will be an estimate of its area.

Carry out the plan

Write down the formula:

$C = 2\pi r$

Substitute for C:

$999 = 2\pi r$

Divide both sides by 2π:

$\frac{999}{2\pi} = r$

$r = 158.9957\ldots$

Use the exact value for r, from your calculator, in the formula:

$A = \pi r^2$

$= \pi \times (158.9957\ldots)^2$

Round the answer on your calculator, and include the units:

$A = 79\,418\,\text{m}^2$

Look back

Check by estimation that your answer is sensible:

$\pi \approx 3$, so $\frac{999}{2\pi} \approx \frac{999}{6} = 166.5$ which is close to the calculated value of r.

Why is the estimated value larger than the actual value?

Reflect

How would you answer a question where you were given the area and asked to find the circumference of a circle?

Fact

The Rhind Mathematical Papyrus, in the British Museum in London, contains examples of Egyptian mathematics transcribed by a scribe called Ahmes in 1550 BCE.

It shows solutions to many problems, including calculating the area of a circle using an approximation of π.

It also states that the ratio $\frac{\text{area of circle}}{\text{area of circumscribing square}} = \frac{64}{81}$

Use dynamic geometry software to verify this ratio.

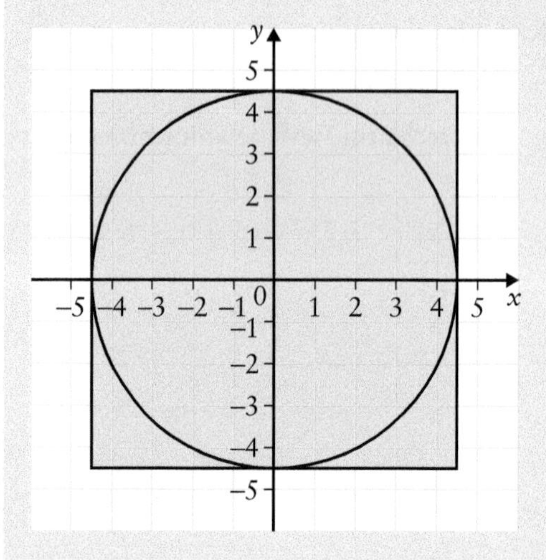

Practice questions 7.3

1 Work out the area of each circle. Give your answers to 1 decimal place.

a Circle with a radius of 7 cm.

b Circle with a diameter of 2 cm.

c Circle with a diameter of 8 m.

d Circle with a radius of 5 m.

e Circle with a diameter of 7 mm.

f Circle with a radius of 2.7 cm.

g Circle with a diameter of 21.6 cm.

h Circle with a diameter of 11 cm.

2 Measure each circle and calculate its area. Give your answers to 2 decimal places.

a

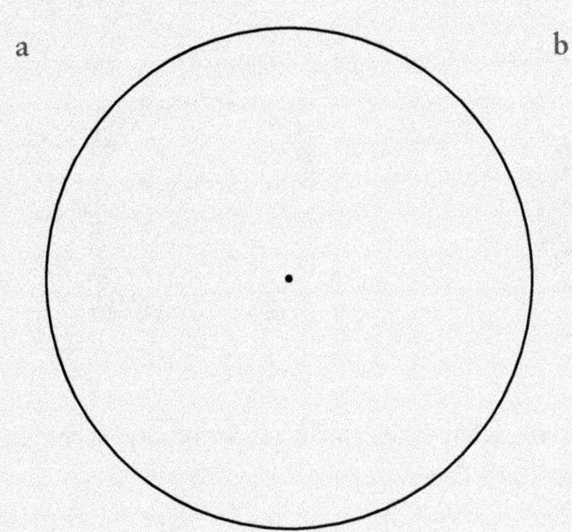

b

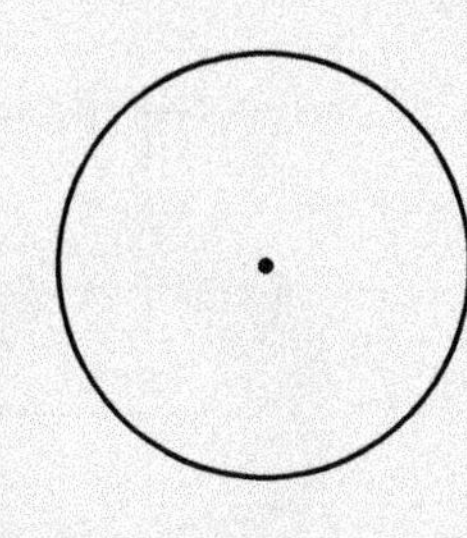

3 Work out an estimate for the area of a circle with diameter 4.2 m.

4 A circular rug has radius 1.2 m. Work out:

a the area

b the circumference of the rug.

Give your answers to 2 decimal places.

5 The light from a street lamp shines over a circular area on the ground. The circle extends 2 m from the street lamp in all directions. Work out the area of the circle of light.

6 This circular courtyard, at the Alhambra in Granada, Spain, has diameter 30 metres. Work out the area of the courtyard, to the nearest square metre.

7 A circle has an area of $30\,cm^2$. Work out the radius of this circle to 1 decimal place.

8 A circle has an area of $425\,cm^2$. Work out the diameter of this circle to 1 decimal place.

9 A circular pond has area $12\,m^2$. Work out the radius of the pond, to the nearest centimetre.

10 Max has a rectangular garden 5 m by 3.5 m. Can he fit a circular patio with area $10\,m^2$ in his garden? Show working to explain.

11 Work out the shaded area:

a

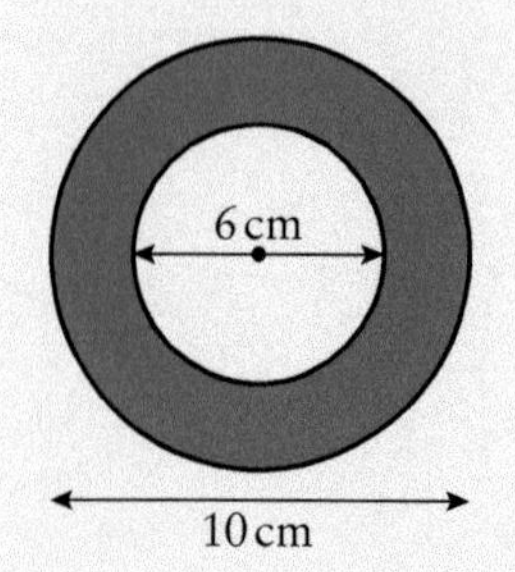

b

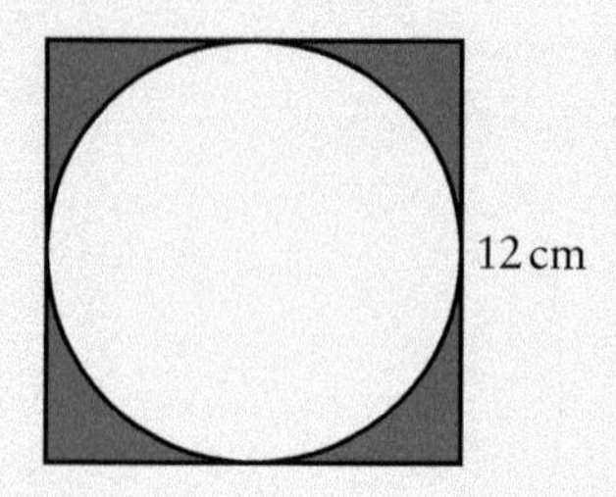

12 This circle and this triangle have the same area. Work out the length of the base of the triangle, to 1 decimal place.

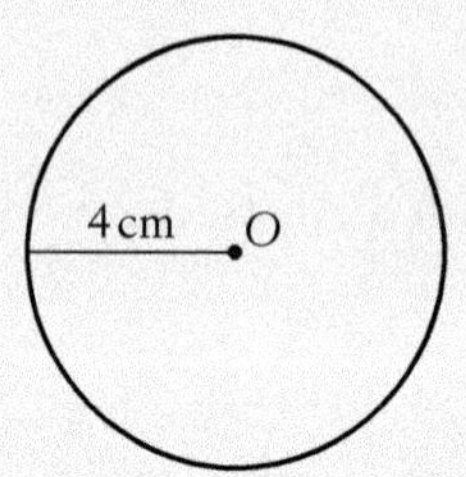

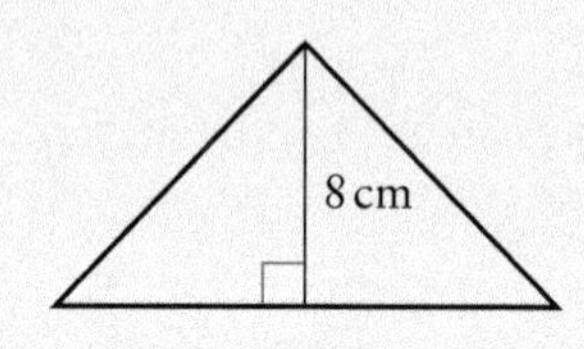

Challenge Q13

13 How many paper circles of diameter 5 cm can you cut from an A3 sheet of paper (297 mm × 420 mm)?

Challenge Q14

Self-management skills

14 Investigate the effect of doubling the radius of a circle on its circumference and its area.

7.4 Semicircles and quadrants

A **semicircle** is half a circle:

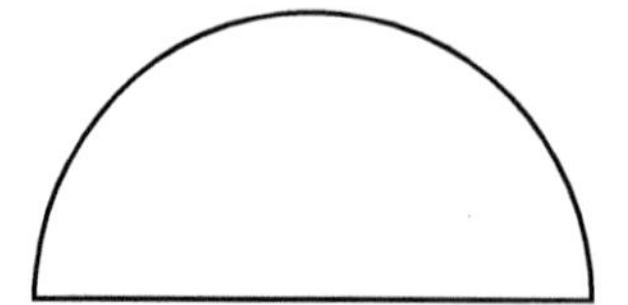

A **quadrant** is a quarter circle:

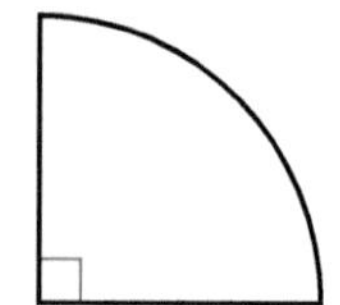

Explore 7.4

a How could you work out the area of these semicircles?

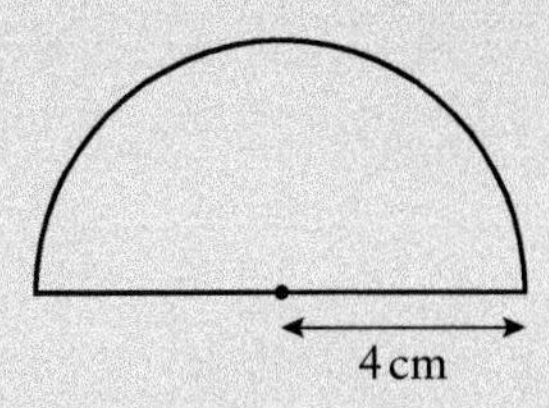

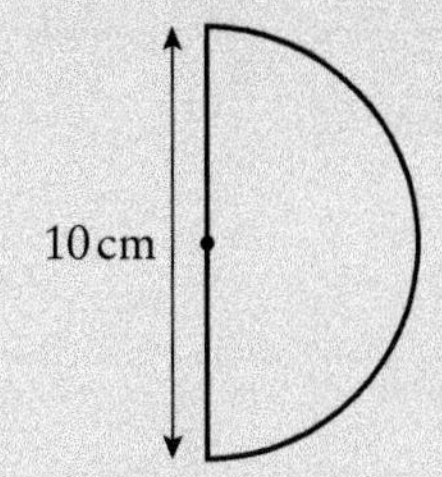

b Can you write an expression for the area of this semicircle?

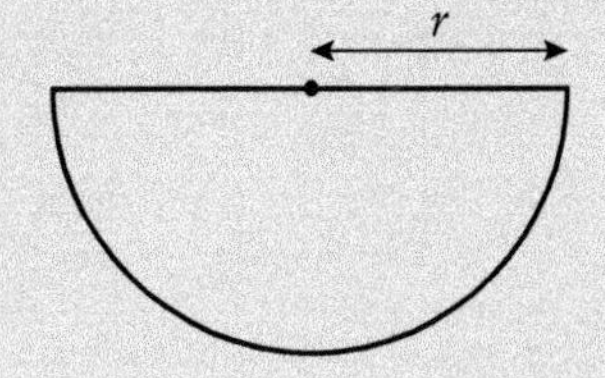

c Can you write an expression for the area of this quadrant?

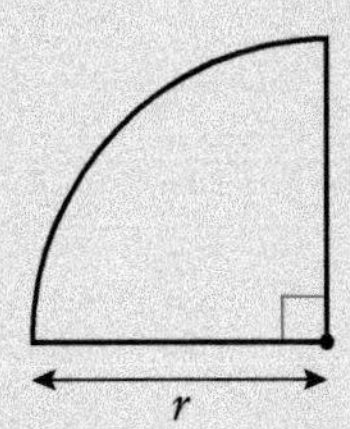

Explore 7.5

a How could you work out the perimeter of these semicircles?

i

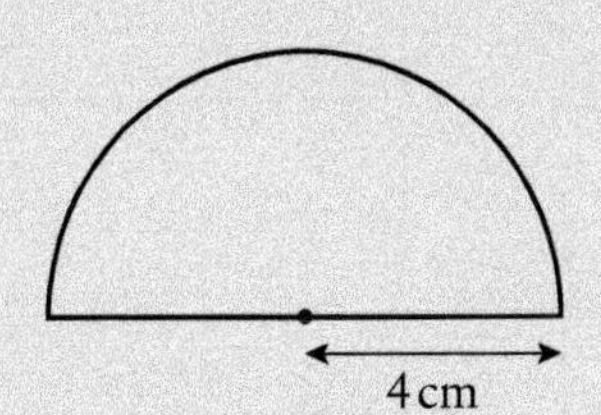

ii

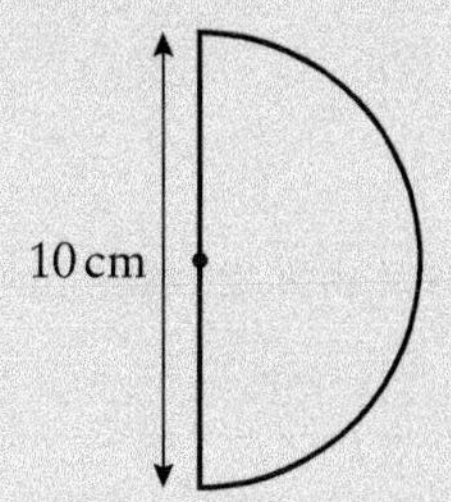

b How could you work out the perimeter of this quadrant?

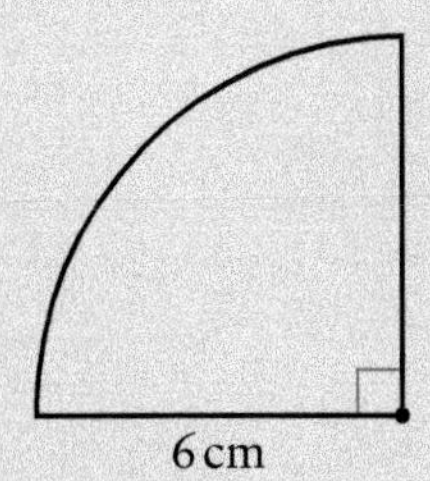

 Worked example 7.6

Given that the height of the door in the picture is 2.8 m and the width is 2.2 m, calculate its area and perimeter. You can assume that the curved part makes a semicircle.

Solution

Make a plan

Split the shape into a rectangle and a semicircle.

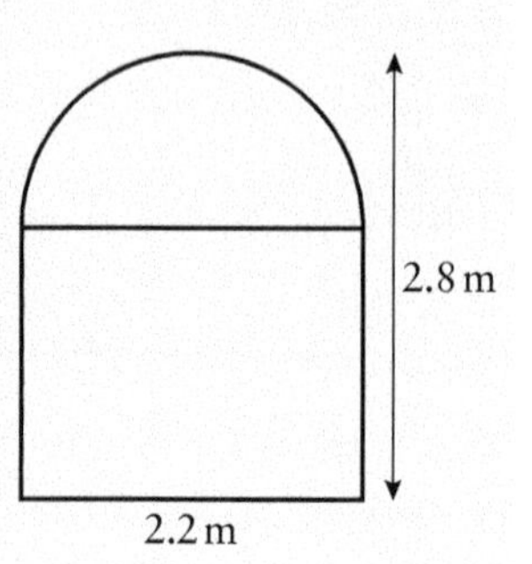

Find the radius of the semicircle and the height of the rectangle.

Find the two areas separately, then add them together.

Split the perimeter into straight edges and an arc. Find these lengths separately, then add them together.

Carry out the plan

diameter of semicircle = 2.2 m

So, radius of semicircle = 2.2 ÷ 2 = 1.1

Area of semicircle $= \frac{1}{2}\pi r^2 = \frac{1}{2}\pi \times 1.1^2$

$= 1.90066\ldots \text{ m}^2$

height of rectangle = 2.8 − 1.1 = 1.7 m

area of rectangle = 2.2 × 1.7 = 3.74 m²

1.7 m

2.2 m

total area = 1.90066... + 3.74... = 5.64066...

$= 5.64 \text{ m}^2$ (2 d.p.)

Work out the total length of the straight edges:

total length of straight edges = 2.2 + 1.7 + 1.7 = 5.6 m

Work out the length of the semicircular arc:

arc length of semicircle $= \frac{1}{2}$ circumference of whole circle

arc length $= \frac{1}{2}\pi d$

$= \frac{1}{2}\pi \times 2.2$

$= 3.4557\ldots$ m

total perimeter = 5.6 + 3.4557…

= 9.0557…

= 9.06 m (2 d.p.)

Look back

The area of the semicircle is less than the area of the rectangle, which seems sensible.

Reflect

Why don't you calculate the perimeter of the semicircle in this example?

What is the difference between arc length and perimeter of a sector?

Practice questions 7.4

1 Calculate the area of each of these sectors. Give your answers to 1 decimal place.

a 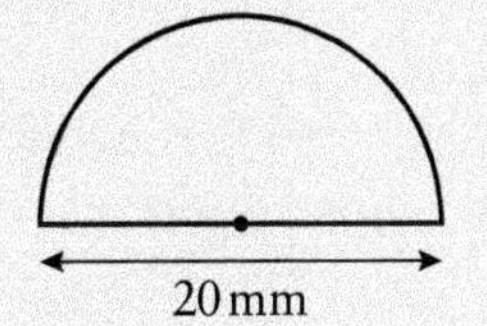

b 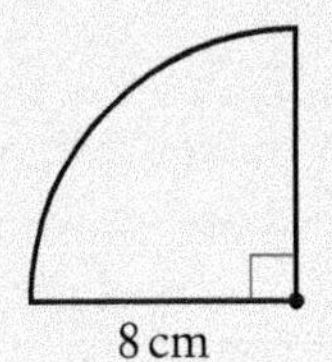

c 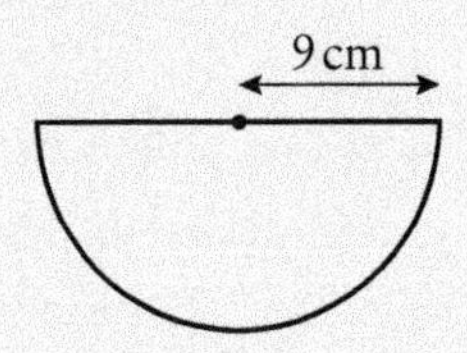

2 Calculate the length of the arc for each of these sectors. Give your answers to 1 decimal place.

a 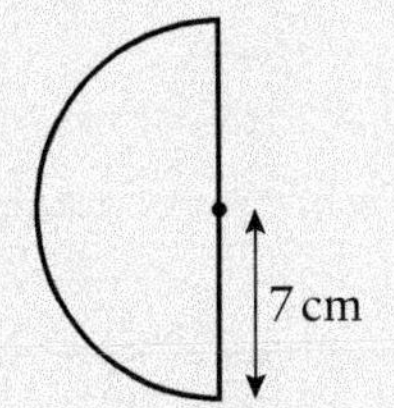

b 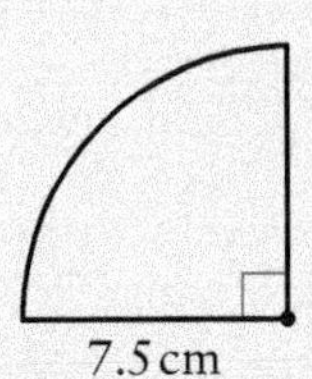

c 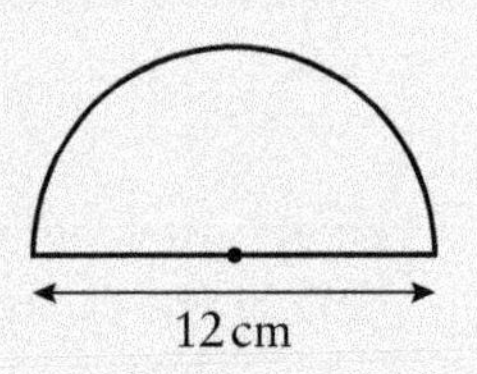

3 Calculate the perimeter of each of these sectors, to 2 decimal places.

a 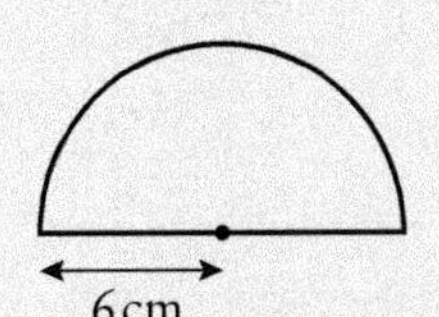

b 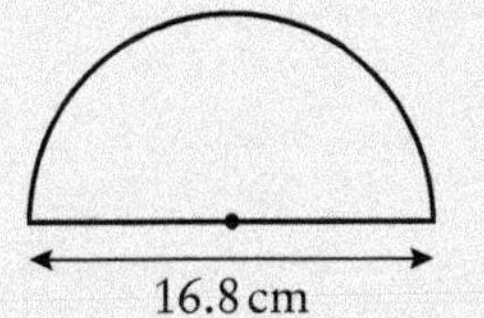

c 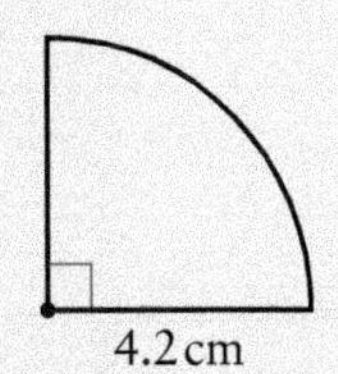

4 The base of the semicircular window on the right measures 1.8 m. Calculate the area of the window.

5 For the shape below, work out:

a the area

b the perimeter.

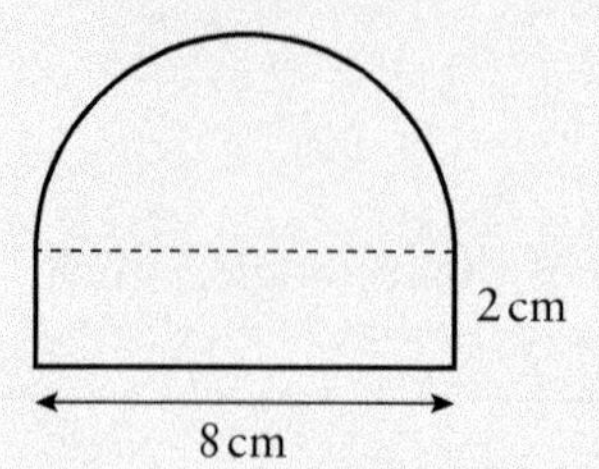

6 Calculate the perimeter and area of this shape.

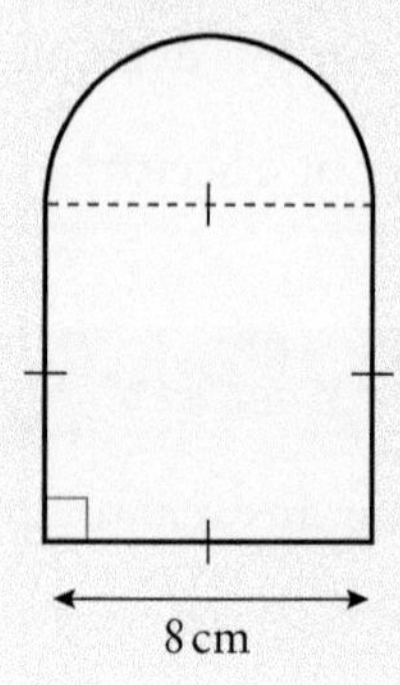

7 This skateboard shape is made from a rectangle and two semicircles. Calculate the area of the skateboard, to the nearest square centimetre.

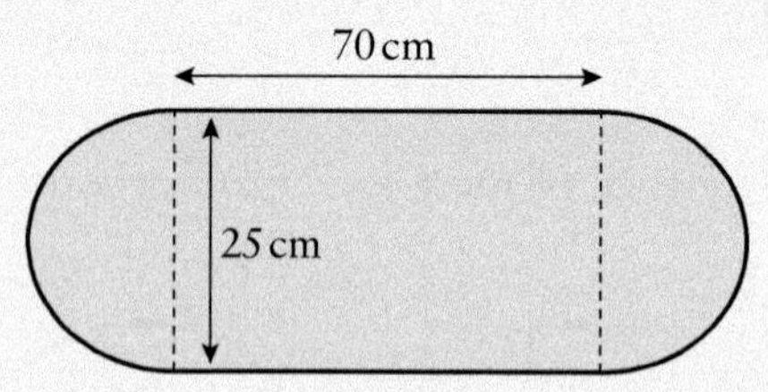

8 Here is the plan of a five-a-side football pitch. What percentage of the area of the pitch is inside the two penalty areas?

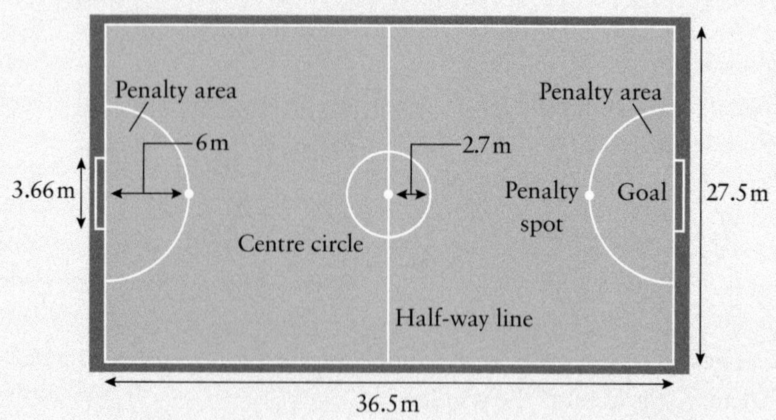

9 A sign for an ice cream shop is made from a semicircle and an isosceles triangle, as shown.

The sign is made from metal. 1 m^2 of the metal costs \$15. Calculate the cost of the metal for the sign.

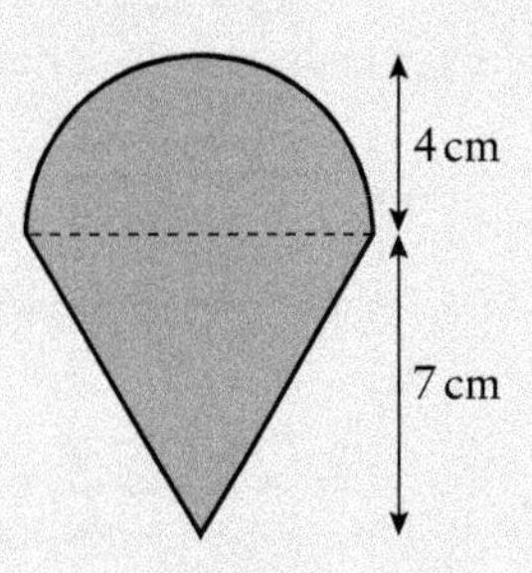

10 The diagram shows the plan of a race track.

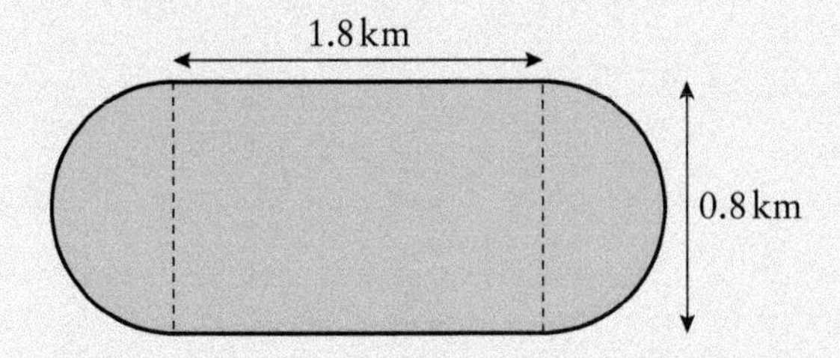

a Find the length of one lap.

b A motorcycle race is 12 laps of the track. Find the distance of the race, to the nearest kilometre.

11 The minute hand of a clock is 8.5 cm long.

a How far does the tip of the minute hand travel in half an hour?

The ratio of the length of the minute hand to the length of the hour hand is 5 : 3

b How far does the tip of the hour hand travel in three hours?

12 Find the area and perimeter of this shape made from semicircles:

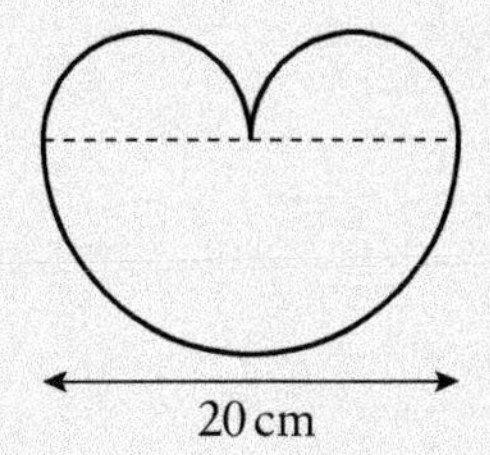

13 Here is a design for a brooch. The brooch has silver wire around its perimeter. Find the length of silver wire needed.

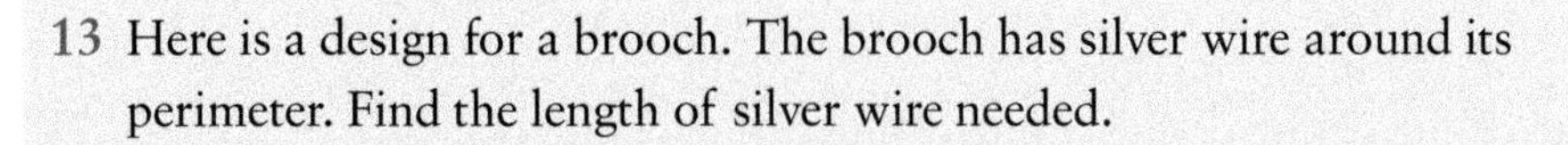

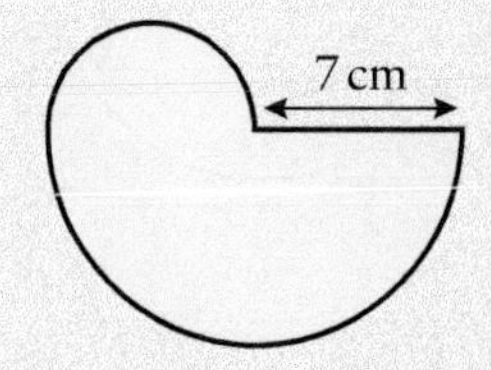

14 Calculate the shaded area.

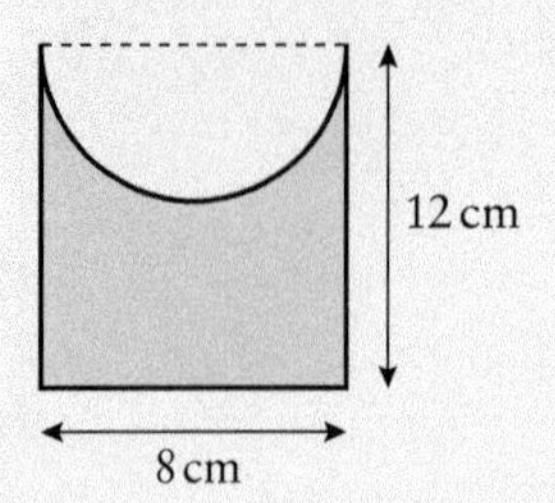

Challenge Q15

15 This circle has diameter 2.4 m.

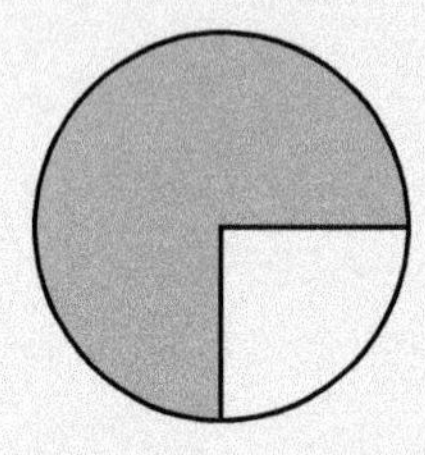

Find the shaded area.

Challenge Q16

16 In each diagram, the square has side length 12 cm. Show that the shaded areas in the two diagrams are equal.

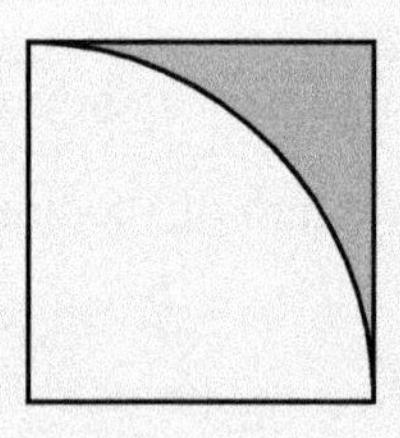

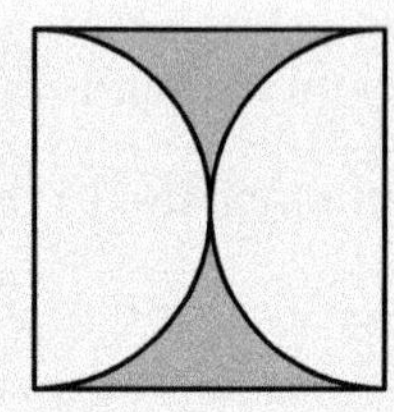

Challenge Q17

17 Find the perimeter of the shaded part of this 10 cm by 10 cm square.

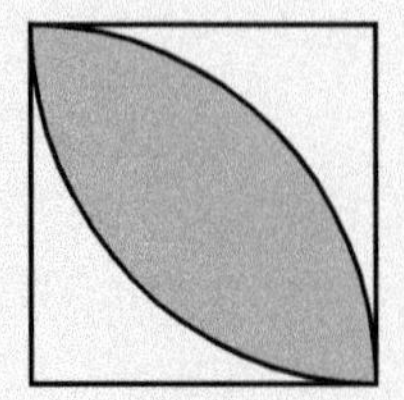

Challenge Q18

18 The diagram shows the plan of a running track with two lanes.

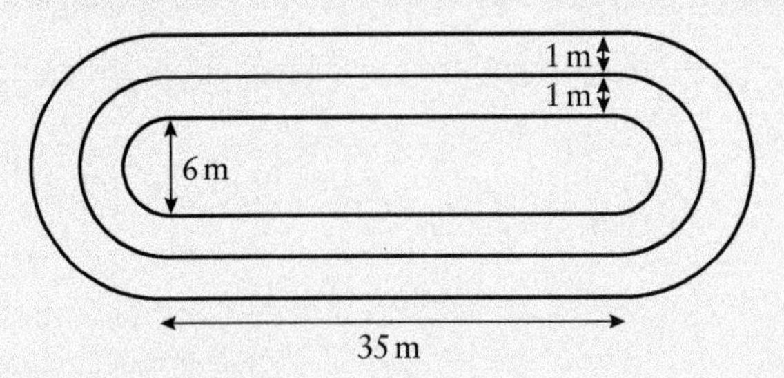

Each lane is one metre wide.

a Which is shorter: a lap around the outside edge of a lane, or a lap around the inside edge? Show working to explain.

Two athletes run around the track, one in each lane. Both run around the inside edge of their lane.

b How much farther does the athlete in the outer lane run, compared with the athlete in the inner lane?

Investigation 7.7

Investigate the dimensions and starting positions for a 400 m track, such that each athlete in each lane runs exactly 400 m.

7 Circles

Self-assessment

- I can name the parts of a circle.
- I can use compasses to draw a circle of given radius.
- I can use compasses to draw a circle of given diameter.
- I can draw and label a diameter, radius, sector, segment, chord and arc on a circle.
- I can identify a major segment and minor segment.
- I can identify a major sector and minor sector.
- I can draw a regular polygon inside a circle.
- I can draw concentric circles.
- I can draw a sector with a given angle.
- I can make accurate drawings of diagrams involving circles and part circles.
- I can estimate the circumference of a circle.
- I can calculate the circumference of a circle given its radius or diameter.
- I can solve problems involving the circumference of circles.
- I can calculate the diameter or radius of a circle, given its circumference.
- I can estimate the area of a circle.
- I can calculate the area of a circle given its radius or diameter.
- I can solve problems involving the area of circles.
- I can calculate the diameter or radius of a circle, given its area.
- I can calculate the area of semicircles and quadrants.
- I can calculate the perimeter of semicircles and quadrants.
- I can calculate the area and perimeter of shapes made from circles, semicircles, quadrants and other 2D shapes.

? Check your knowledge questions

1 Name the parts of the circle shown in dotted red in these diagrams.

a 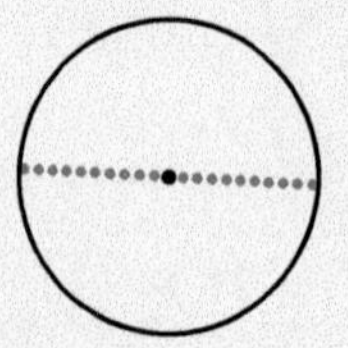

b 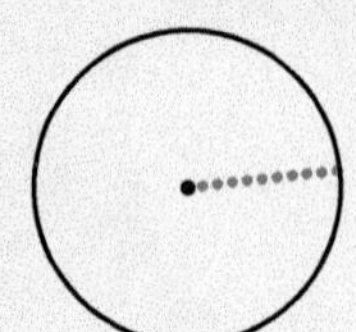

c 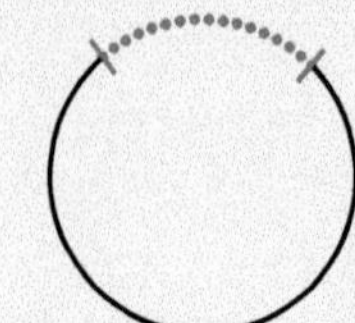

d 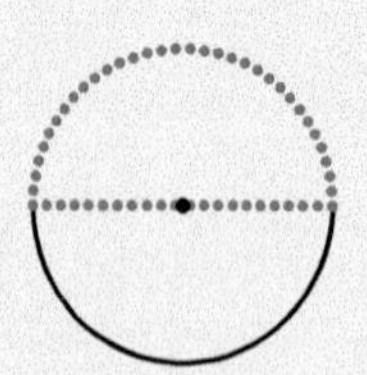

e 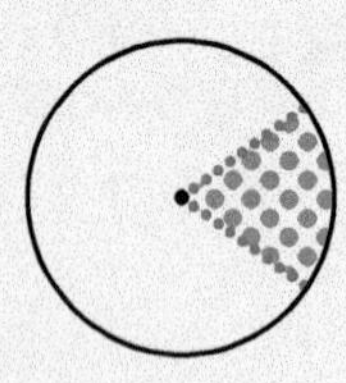

2 Draw a circle of diameter 12 cm.

3 Draw two concentric circles with radii 4 cm and 2.5 cm.

4 Make an accurate drawing of an inscribed octagon.

5 Draw a sector of a circle with radius 6 cm and angle 70°.

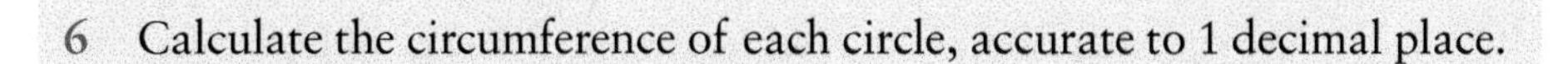

6 Calculate the circumference of each circle, accurate to 1 decimal place.

a

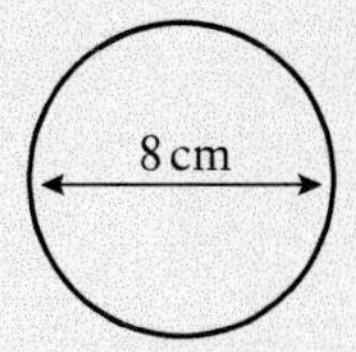

b

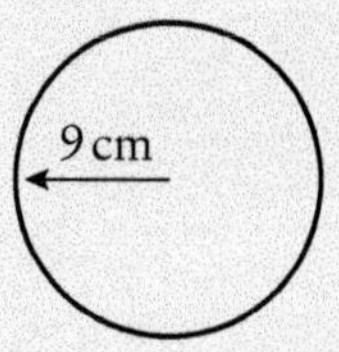

7 Calculate the area of each circle, accurate to 1 decimal place.

a

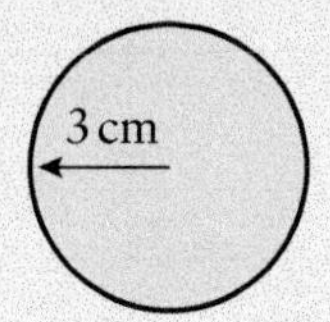

b

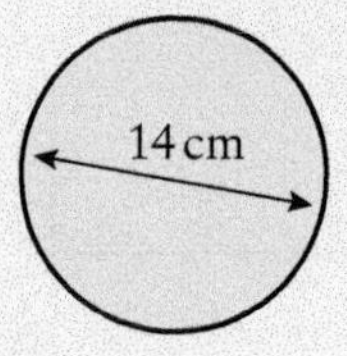

8 Estimate the area and circumference of this circle:

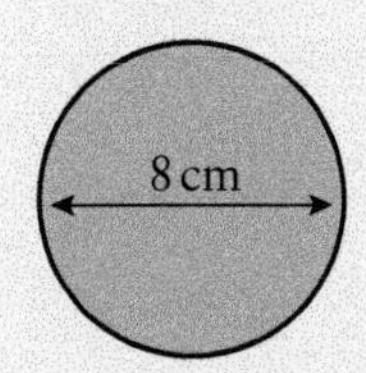

9 A circle has circumference 65 cm. Work out its radius, to the nearest mm.

10 A circle has area 100 cm^2. Work out its radius, to the nearest mm.

11 Find the perimeter of each shape, to 1 decimal place.

a

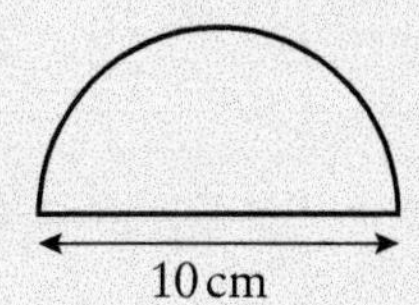

b

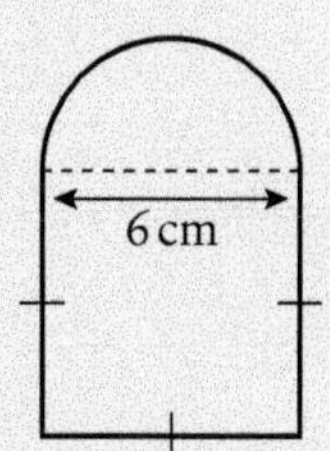

c

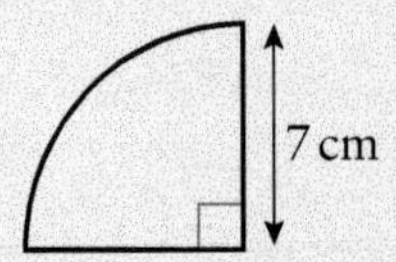

d

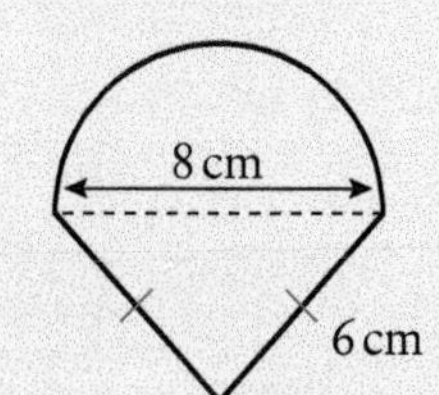

12 Calculate the area of each shaded region, to 2 decimal places.

a

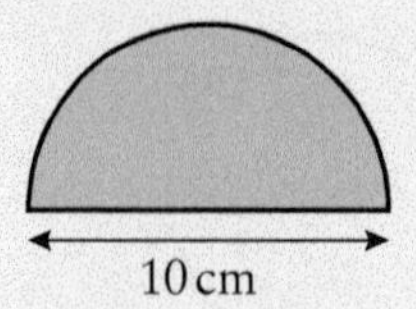

b

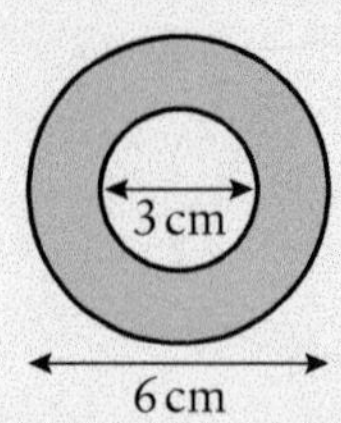

c

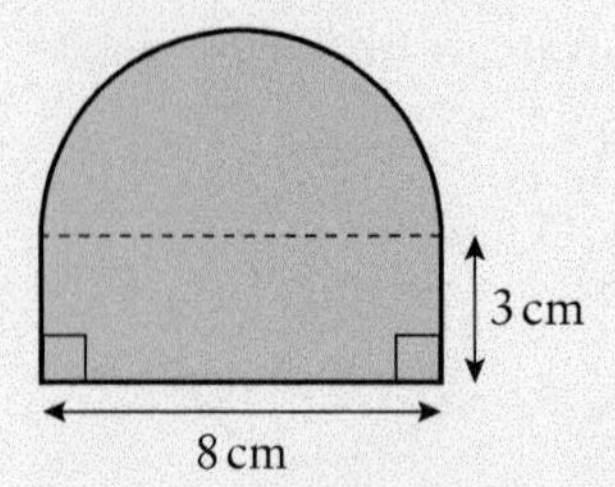

d

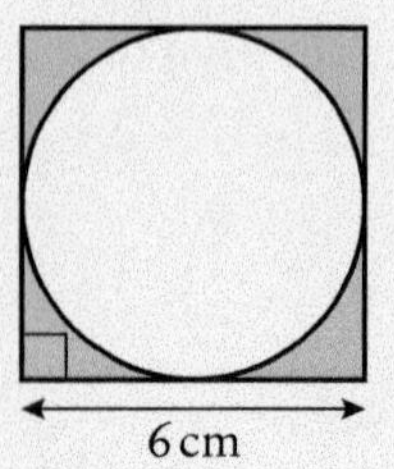

13 A trundle wheel measures a distance of 1 m with every revolution. What is its radius, to the nearest mm?

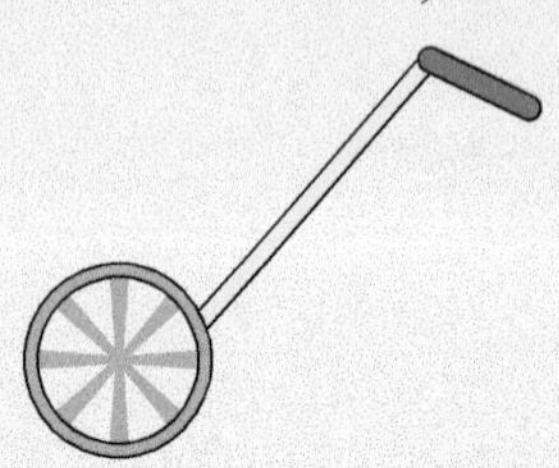

8 Congruent shapes and constructions

8 Congruent shapes and constructions

KEY CONCEPT

Logic

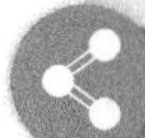

RELATED CONCEPTS

Equivalence, Patterns, Representation, Space, Systems

GLOBAL CONTEXT

Orientation in space and time

Statement of inquiry

Logical systems can be used to represent equivalent shapes in space.

Factual

- What are congruent shapes?

Conceptual

- How can you test if shapes are congruent?

Debatable

- Do congruent shapes make a structure stronger?
- Where can you find congruent shapes?

Do you recall?

1 Look at the diagram below and answer the questions.

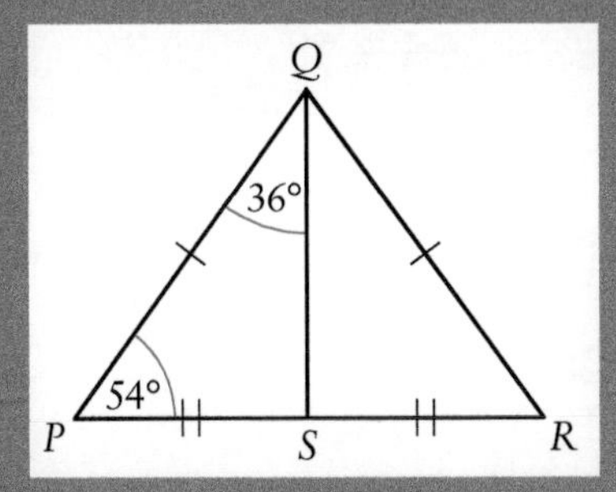

a Which side of triangle RQS is congruent to PQ?

b What length is equal to PS?

c Which angle is congruent to $\angle QPR$?

d What is the size of each of the angles $\angle QRP$, $\angle QSP$ and $\angle RQS$?

2 In the diagram below A and B are the centres of the circles.

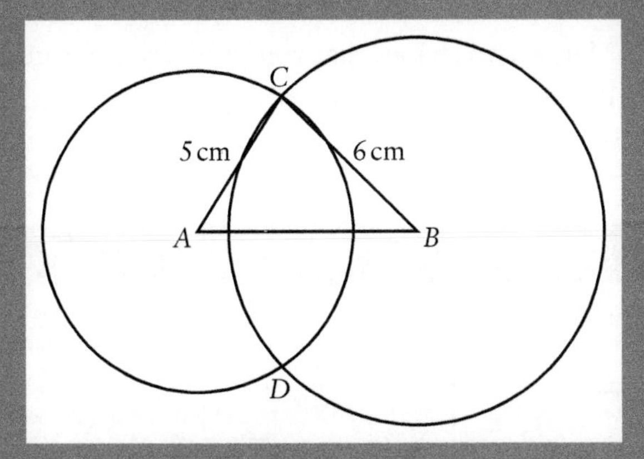

a How far is C from A?

b How far is C from B?

c Which point or points are 5 cm from A and 6 cm from B?

d Describe how you could find a point 4 cm from A and 7 cm from B.

8.1 Congruent shapes

Explore 8.1

The diagram shows four pairs of shapes. Three of the pairs are congruent; one pair is not. Discuss with others what you think congruency might mean. Are you confident with your decision or would you like to see one or more pairs of shapes? Draw two shapes that you think are congruent and see if others agree. Research the word congruency (in its mathematical context) to see if you are correct.

Worked example 8.1

Which pairs of these figures are congruent?

Solution

Understand the problem

Pair the figures that are the same shape and size.

Make a plan

Measure the sides and angles of each figure and then compare them with each other to see which are the same.

Carry out the plan

A and G are congruent.

The angles and sides in figures A and G are equal.

E and J are congruent. E and J are squares of the same size.

F and H are congruent. The angles and sides in triangles F and H are equal.

Look back

Are the pairs of figures congruent and are there any other pairs of congruent figures? Yes, the pairs of figures are congruent as the sides and angles measure the same in each pair. There are not any other pairs of congruent figures as they are either different shapes or have different measurements.

Reflect

Look at the Worked example. Can you explain why triangle I is not congruent to triangles F and H, or why figure C is not congruent to figure G?

Draw a rectangle on squared paper. Reflect your rectangle, translate your rectangle and rotate your rectangle. Can you explain why all of the rectangles you have drawn are congruent?

Practice questions 8.1

1 The two figures here are congruent.

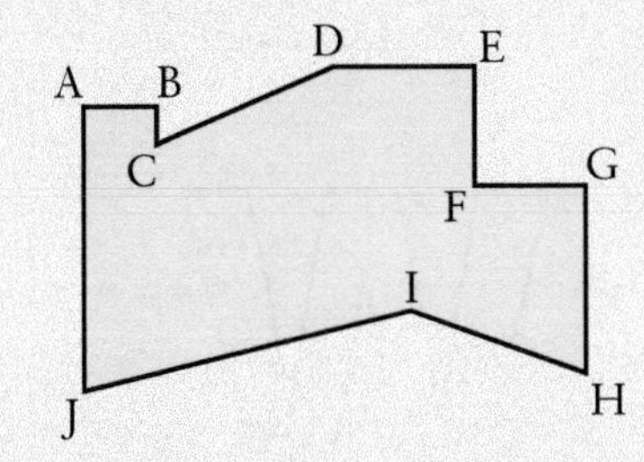

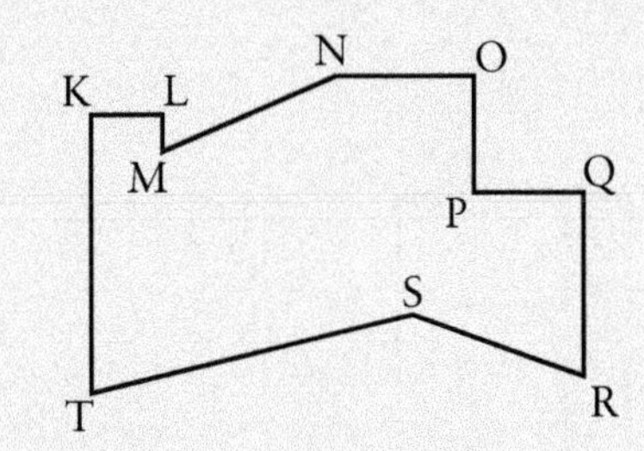

a Use them to write down the angle that matches with:

i $\angle CDE$ ii $\angle GHI$ iii $\angle HIJ$

b Use them to write down the side that matches with lines:

i AJ ii CD iii HI

c Gaia says, 'The area of the two shapes is the same'. Is she correct? Explain your answer.

2 Make four copies of this regular hexagon.

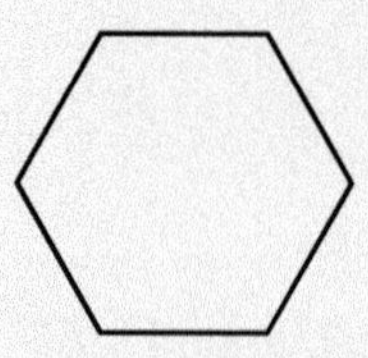

a Divide your first hexagon into two congruent figures.

b Divide your second hexagon into three congruent figures.

c Divide your third hexagon into six congruent figures.

d Divide your fourth hexagon into four congruent figures.

e Look back at parts **a** to **d**. Are there any other ways of dividing the hexagon into these numbers of congruent figures?

3 Copy each of these figures. For each figure, use one line to divide them into two congruent figures.

a
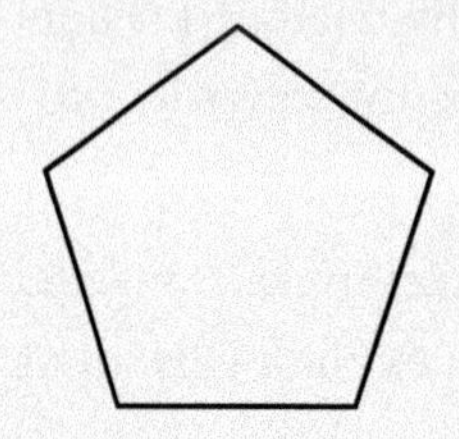

b
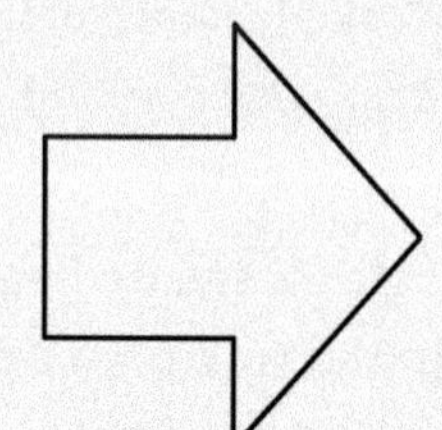

c

d
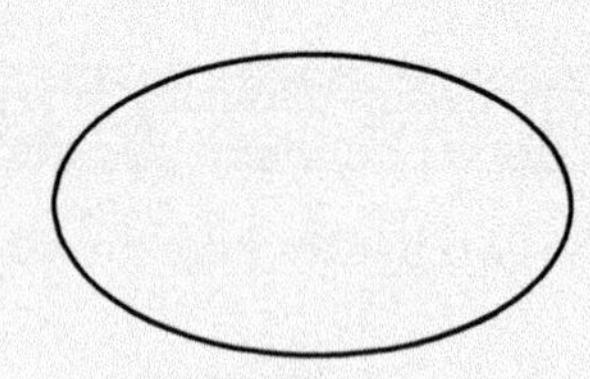

e
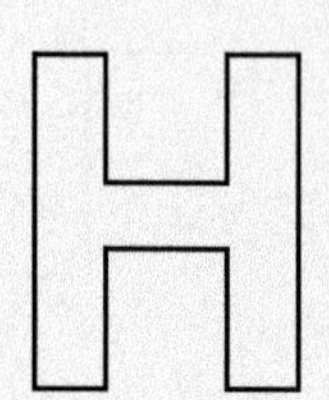

f
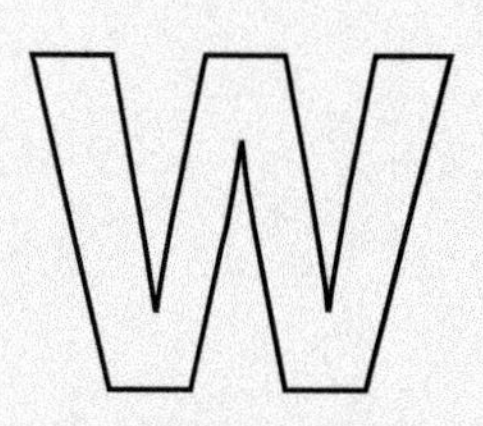

g
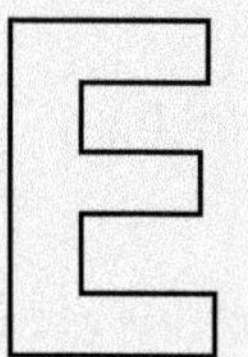

h
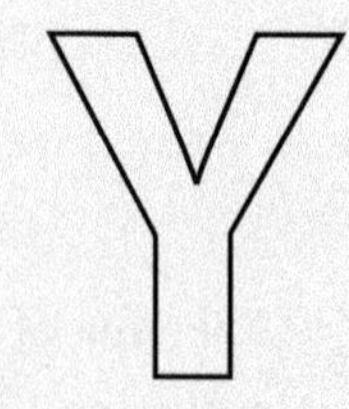

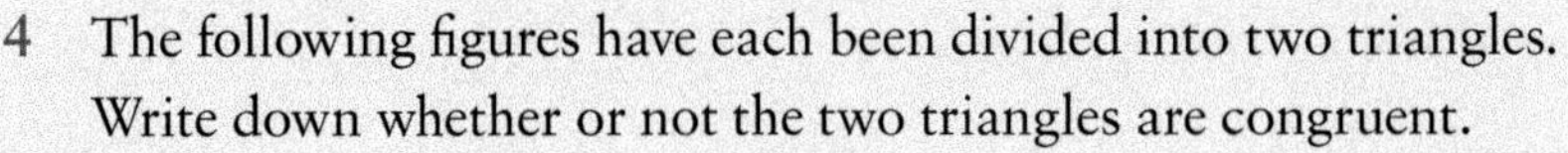

4 The following figures have each been divided into two triangles. Write down whether or not the two triangles are congruent.

a

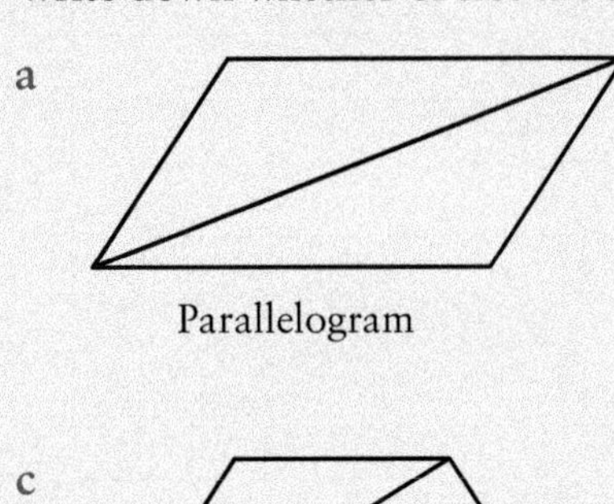

Parallelogram

b

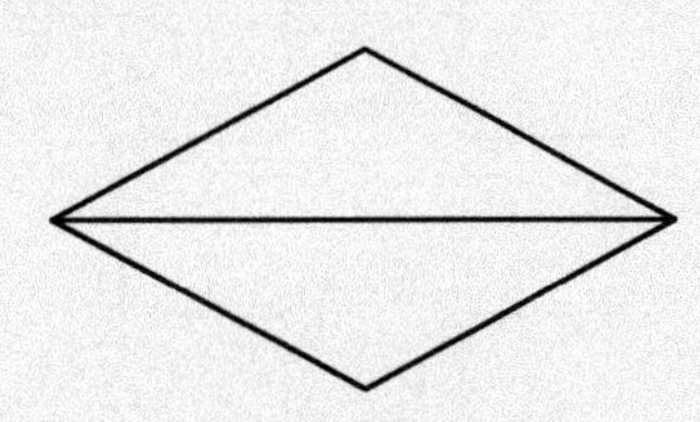

Rhombus

c

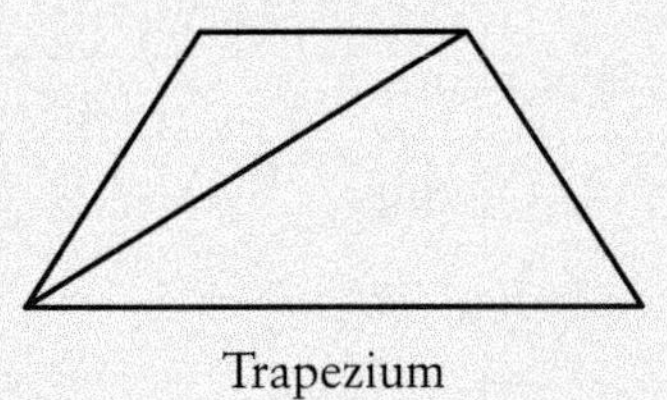

Trapezium

d

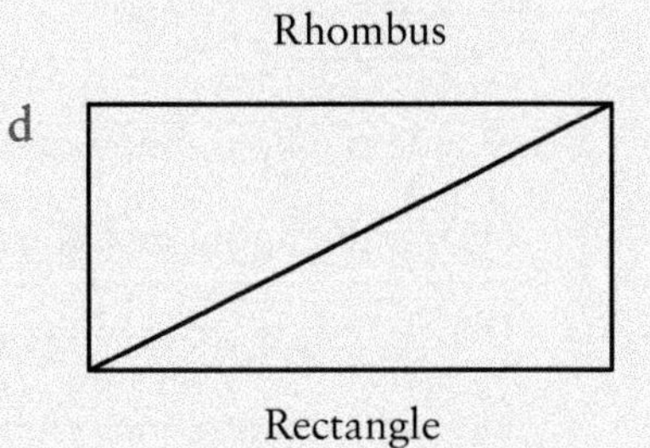

Rectangle

e

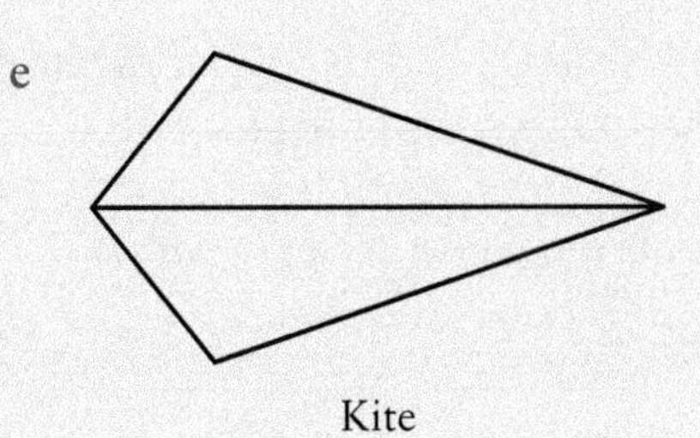

Kite

f

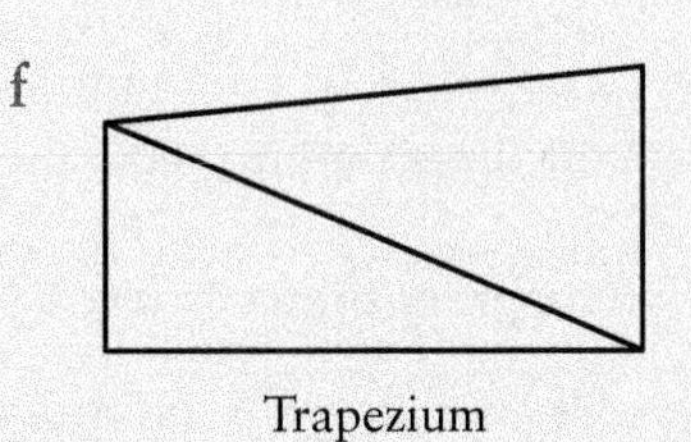

Trapezium

g

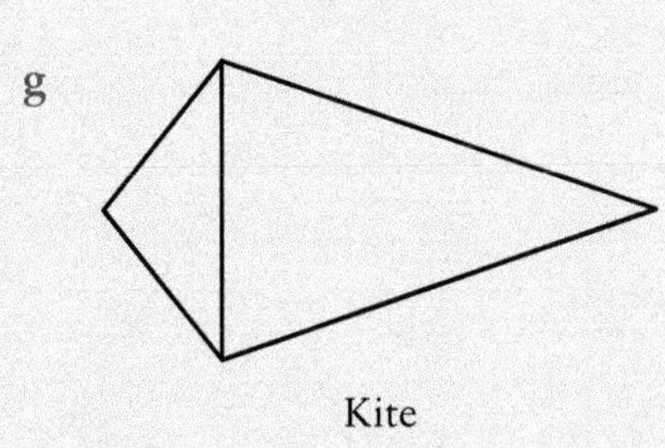

Kite

h

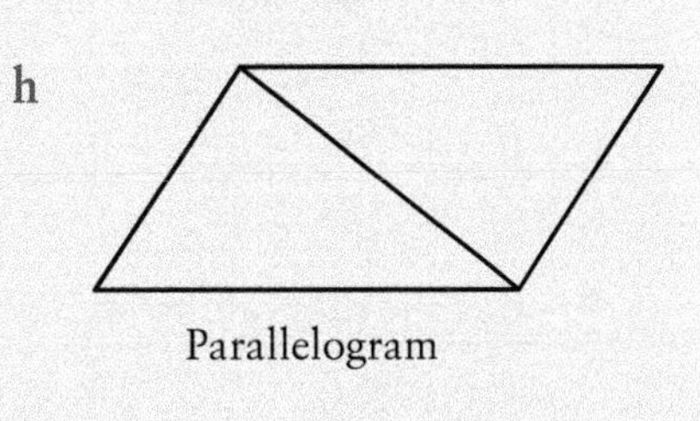

Parallelogram

5 The diagram shows three figures.

A

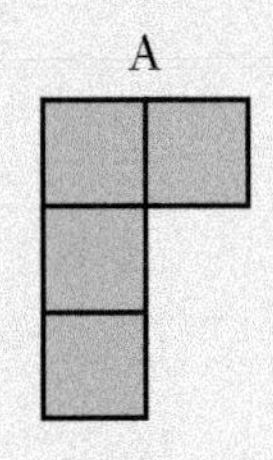

B

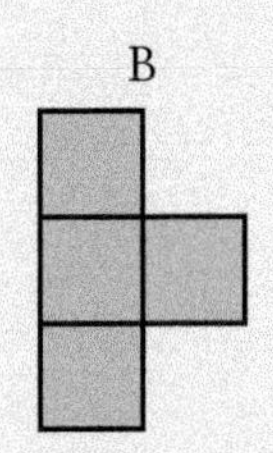

C

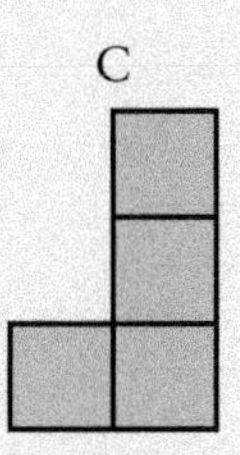

a Write down the area of each figure.

b Which two figures are congruent?

c Do figures A and B have the same area? The same perimeter?

d Are figures A and B congruent? Explain your answer.

6 Copy this diagram four times.

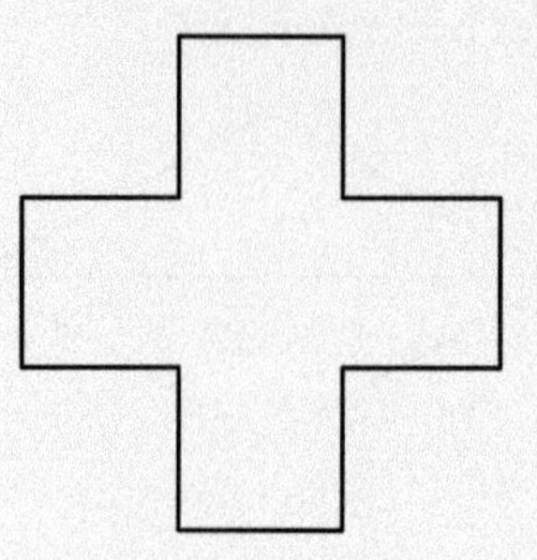

a Divide the figure into two congruent figures.

b Divide the figure into four congruent figures.

c Divide the figure into five congruent figures.

d Divide the figure into eight congruent figures.

e Look back at parts **a** to **d**. Are there any other ways of dividing the diagram into these numbers of congruent figures?

7 The diagram shows two congruent triangles *ABC* and *CDE*.

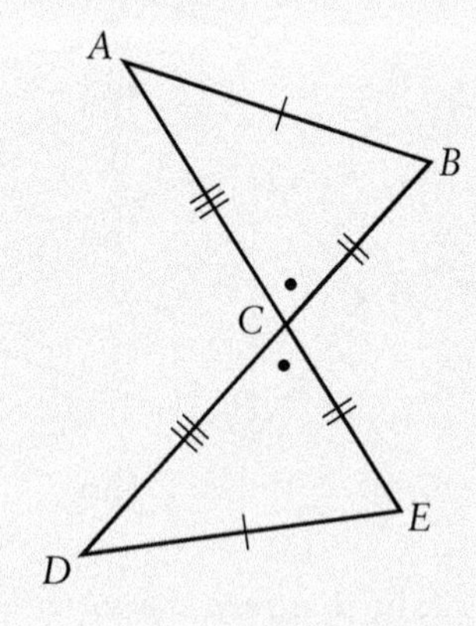

a Use the diagram to write down the angle that matches with:

i $\angle CAB$

ii $\angle ABC$

iii $\angle ACB$

b Use the diagram to write down the side that matches with lines:

i *AB*

ii *BC*

iii *AC*

8 Here are two circles:

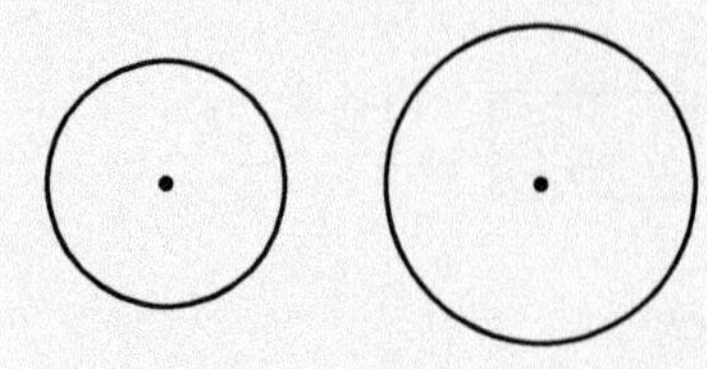

a Are these circles congruent? Explain your answer.

b What conditions must be satisfied for two circles to be congruent?

9 Idris looks at figure A and then draws figure B.

Figure A

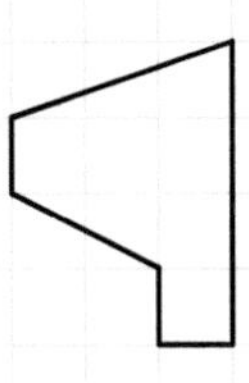

Figure B

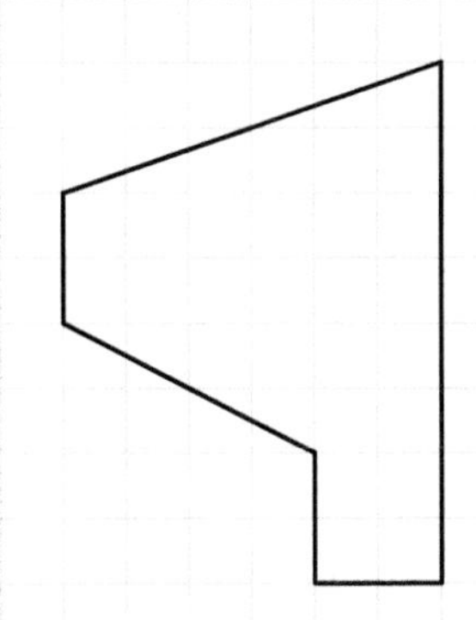

Idris says, 'Figure B is the same as figure A, so it is congruent'.

Kiara says, 'Figure B is not congruent to figure A'.

Who is correct? Explain your answer.

10 Opel says, 'All squares are congruent'. Is she correct?
Explain your answer.

11 Identify the pairs of congruent figures in each of these diagrams.

a

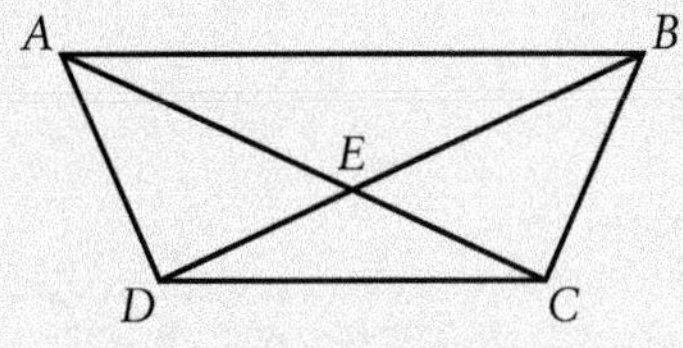

Isosceles trapezium

b

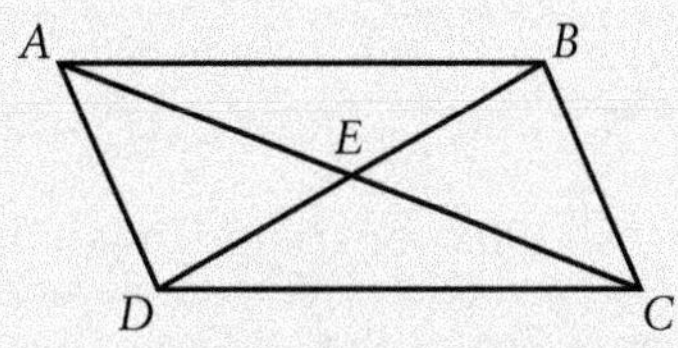

Parallelogram

Explore 8.2

The pictures show two different rollercoaster cars.

Why do you think rollercoaster cars are congruent figures?

 Thinking skills

 Communication skills

 Research skills

Investigation 8.1

A tessellation is a tiling pattern using one or more congruent shapes.

Here is an example of a tessellation using one shape:

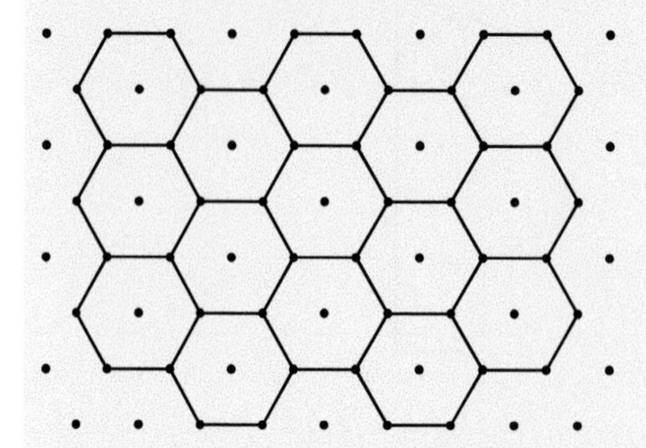

Here is an example of a tessellation using two shapes:

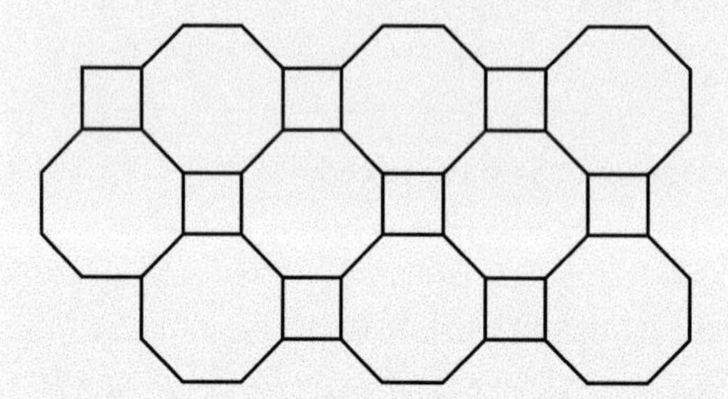

Draw some patterns using triangles that tessellate on their own.

Do all types of triangle tessellate? Explain why.

Try investigating other shapes that tessellate on their own. What quadrilaterals tesselate?

What shapes do not tessellate?

Tessellations have been used in many art designs. One artist famous for using tessellations is M. C. Escher, born in the Netherlands in 1898. Research Escher to find out more about his artwork. Can you see how he used the idea of tessellations and developed this within some of his pieces of art?

Design your own tessellation artwork. First, start with a shape that tessellates on its own, for example, a rectangle. Translate small parts of the shape to its opposite side. Both the following examples start with a rectangle. The first example takes two triangles from each corner at the bottom of the rectangle and translates them to the top of the rectangle. The second example takes a segment from the left of the rectangle and translates it to the right of the rectangle.

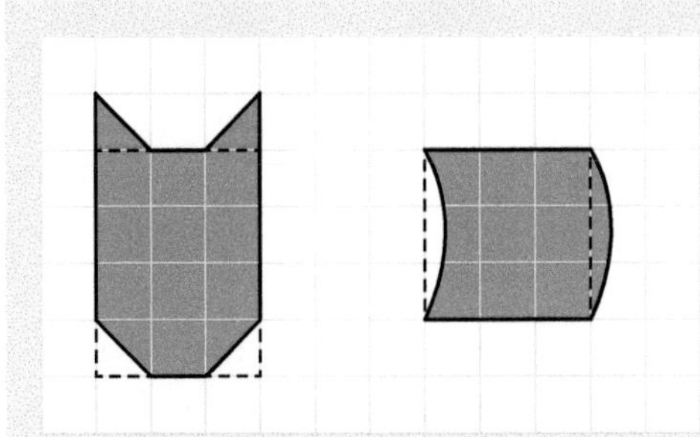

Tessellate your shape and then turn it into art. The first example above could be developed into a cat tessellation artwork:

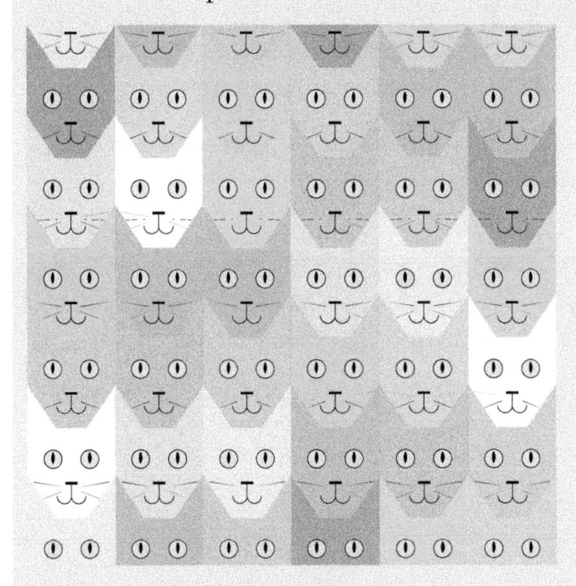

Here is more complicated cat tessellation using the style of graphic artist, M. C. Escher:

8.2 Basic constructions

8.2.1 Constructing triangles

Construction means to build something. Constructions in mathematics are linked to this as they are accurate drawings of shapes, angles and lines. When you are asked to construct you will need a ruler, a pencil and a pair of compasses.

Explore 8.3

Draw a triangle with sides 5 cm, 6 cm and 8 cm, as accurately as you can.

Swap your triangle with someone else and check each other's accuracy. Explain the steps you followed to draw your triangle.

Explore 8.4

Diane was asked to construct the triangle ABC where:

$AB = 7$ cm, $AC = 5$ cm and $BC = 4$ cm.

Here are the steps she took:

Step 1 A ——— B (7 cm)

Step 2 Draw an arc using compasses 5 cm from point A.

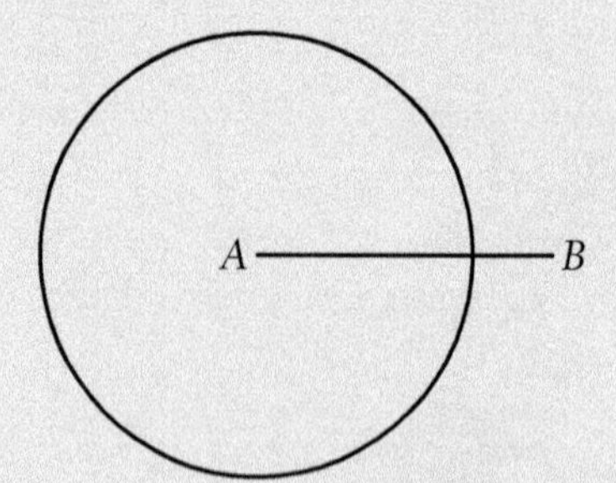

Step 3

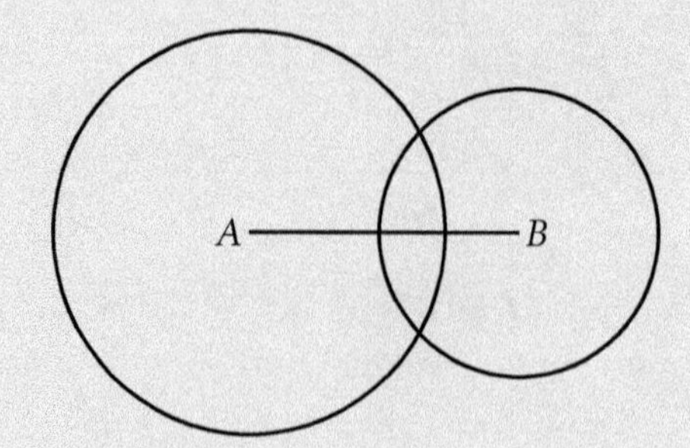

Hint

In construction, it is good practice to first make a hand drawn sketch of the figure you are about to construct. For example, for this Explore activity:

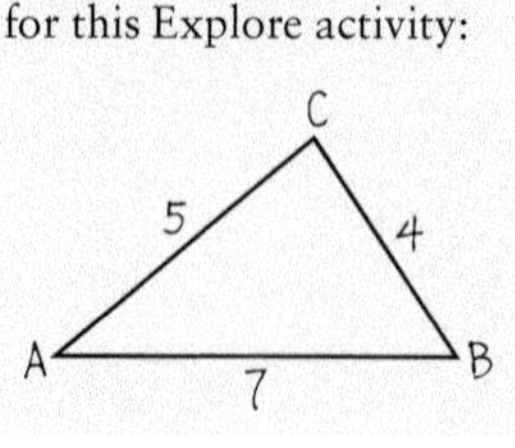

Step 4

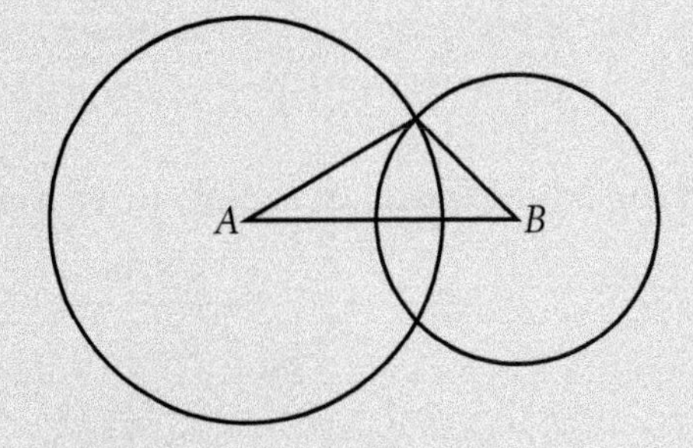

Describe what you think she did. How many such triangles are you able to find?

Reflect

When constructing a triangle, does it matter which side you choose as your first side? Can you give a reason for your answer?

Practice questions 8.2.1

1 Construct a triangle with sides of length:

a 5 cm, 6 cm and 7 cm

b 8 cm, 5 cm and 8cm

c 9 cm, 5 cm and 7 cm.

2 a Use compasses to construct a triangle that is congruent to triangle ABC.

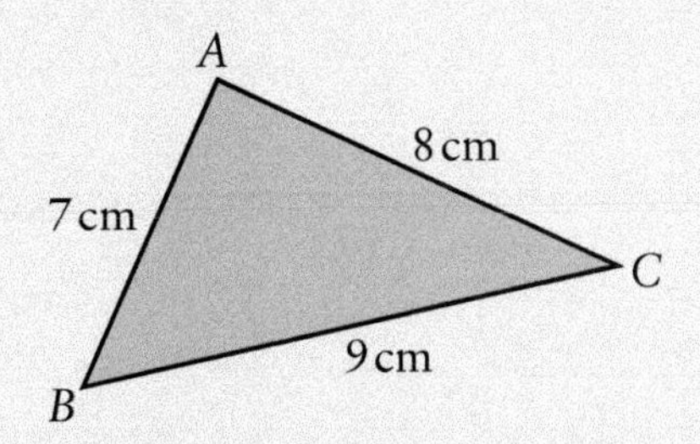

b Use compasses to construct a triangle with side lengths that are half of the side lengths of triangle ABC. Label your triangle PQR.

3 Using your triangles from Questions 2a and 2b, write down the matching pairs of angles and sides in these two congruent triangles.

4 Use a ruler and pair of compasses to accurately draw these triangles.

a

4 cm

4 cm   4 cm

b

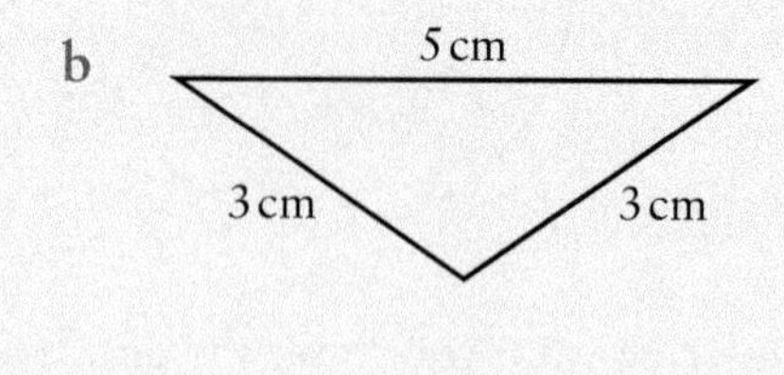

5 Construct an equilateral triangle with side length 6 cm.

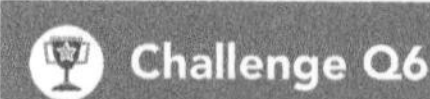
Challenge Q6

6 A tetrahedron is shown below.

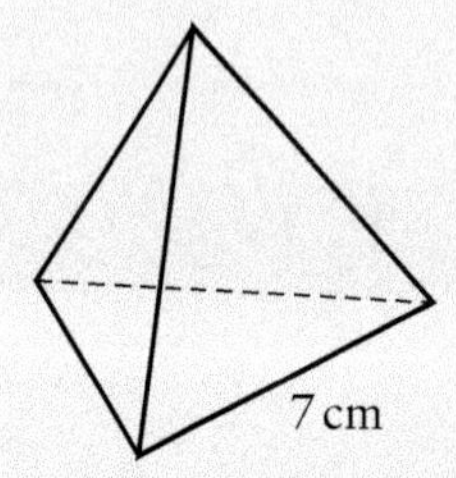

Construct an accurate drawing of a net of the tetrahedron.

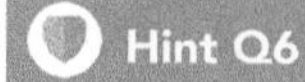
Hint Q6

A tetrahedron is a triangular-based pyramid. All four faces are equilateral triangles. The diagram shows a sketch of the net.

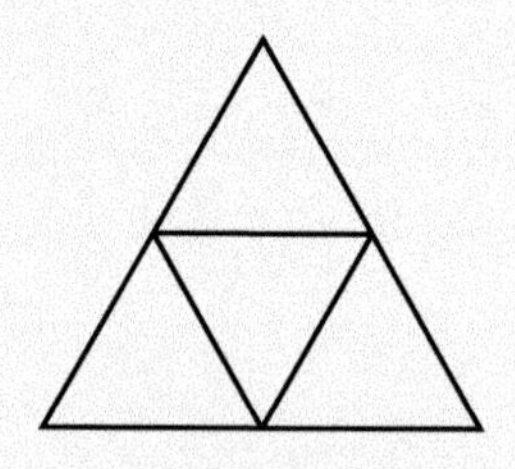

7 A triangle cannot be constructed with sides 5 cm, 7 cm and 15 cm or sides 6 cm, 7 cm and 13 cm. Explain why and state the conditions that must be satisfied in order to construct a triangle.

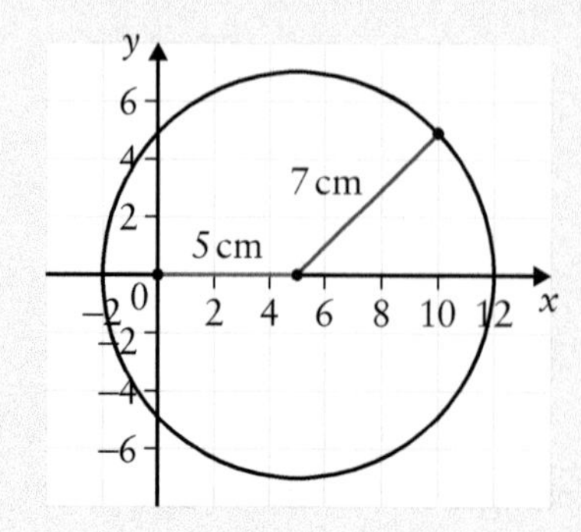

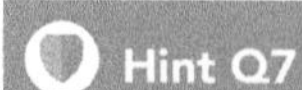
Hint Q7

First draw a 5 cm line segment, then draw a circle of 7 cm with its centre at one end of the line segment and then contemplate what a third side, of 15 cm, would look like.

Challenge Q8

8 Construct all possible isosceles triangles with side lengths 5 cm and 7 cm.

9 This tessellation uses a rhombus.

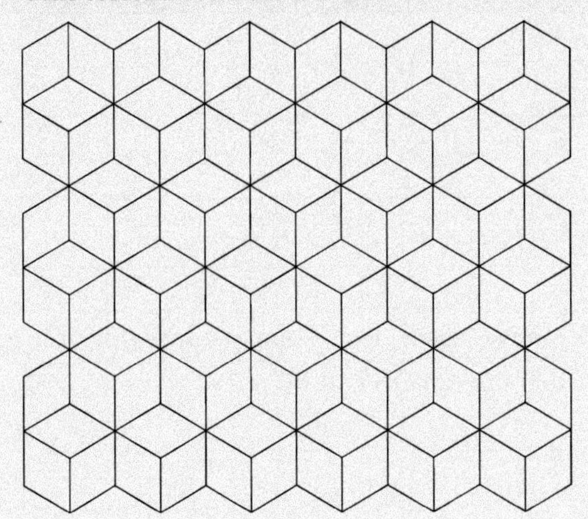

Each rhombus is constructed using two equilateral triangles with side of length 4 cm.

Using a ruler and compasses, construct one rhombus in the tessellation.

Challenge Q9

10 Draw a line with point A on the line like this:

Construct an angle of 60° with vertex at A.

8.2.2 Bisecting angles

To bisect means to cut exactly into two equal parts.

Explore 8.5

Draw these angles using a protractor.

40°, 160°, 72°, 135°.

Can you draw a line to divide your angles exactly in half?

Were any of the angles easier to divide into half than others? Give a reason for your answer.

Worked example 8.2

Bisect the angle ABC using only a pair of compasses and a straight edge.

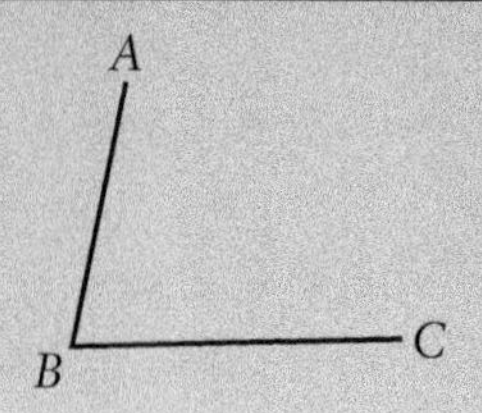

Solution

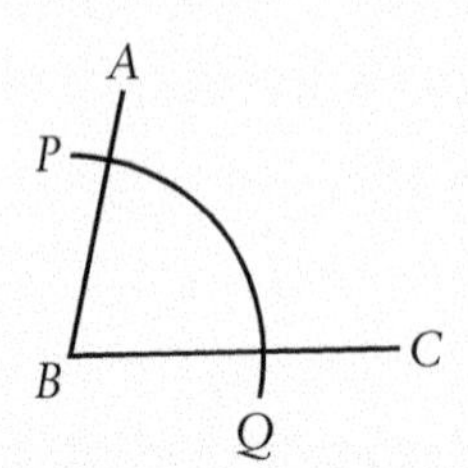

We put the tip of our compasses at B and draw an arc that crosses AB and BC.

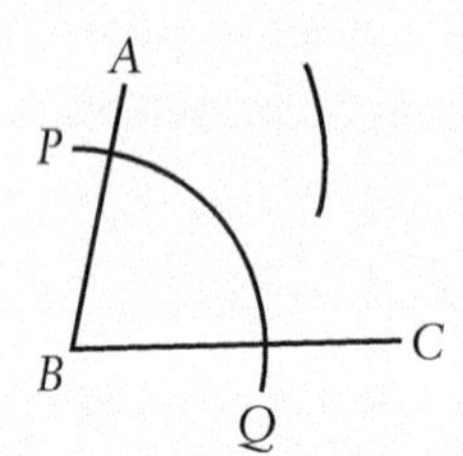

With the point of intersection P as the centre, we draw an arc.

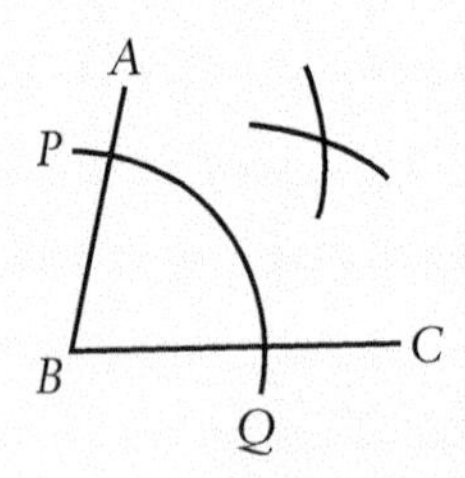

Without changing the setting of the compasses, we draw an arc with the point of intersection Q as the centre. The two arcs drawn from both points of intersection should cross. Call the point D.

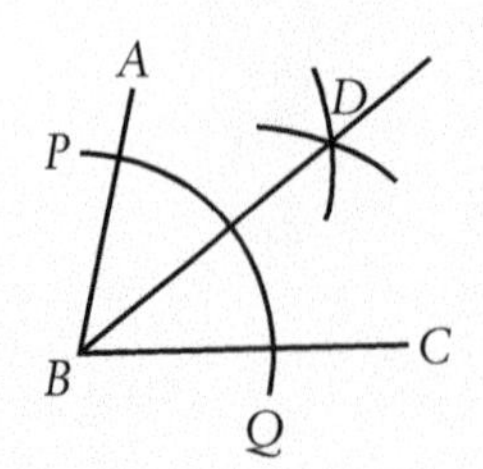

We draw a straight line from point B through the two intersecting arcs.

This line bisects the angle ABC.

Reflect

Can you explain why the line drawn in the worked example above bisects the angle ABC?

Which is more accurate: Constructing the angle bisector using compasses or using a protractor to split the angle into two halves?

Hint

Connect PD and QD.

Practice questions 8.2.2

1 Draw two different acute angles and bisect each of them.

2 Draw two different obtuse angles and bisect each of them.

3 Draw each angle using a protractor and construct its angle bisector using compasses and straight edge.

a 50° b 85° c 110° d 162°

4 Draw a scalene triangle with one obtuse angle. Bisect the obtuse angle in your triangle.

5 a Construct the triangle ABC:

b Bisect angle ABC.

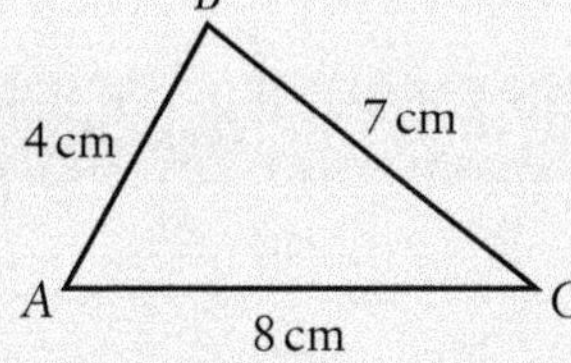

6 a Construct an equilateral triangle with side length 7 cm.

b Bisect an angle in your triangle.

c Without using a protractor, what is the size of the two angles you created in part **b**?

Investigation 8.2

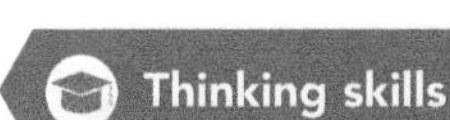

Construct an equilateral triangle with side length 10 cm.

Construct the angle bisector of each angle in your triangle.

What do you notice with your three angle bisectors?

Try this for isosceles and scalene triangles.

Does the same happen again?

Generalise your findings.

8.2.3 Perpendicular bisectors

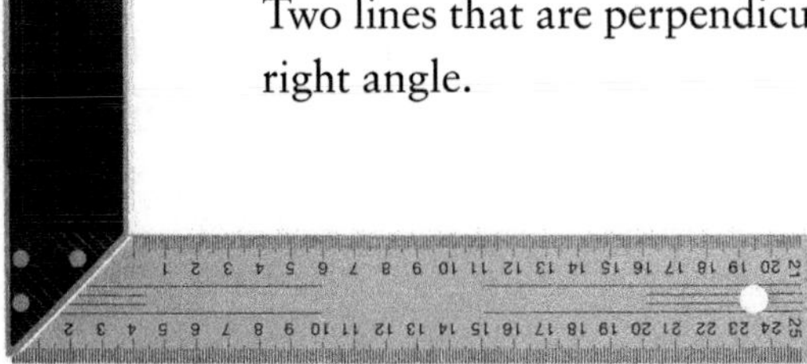

Two lines that are perpendicular to each other form a right angle.

Explore 8.6

Draw a line AB that is 9 cm long. Can you use a ruler and protractor to draw a line that is perpendicular to line AB and cuts the line AB in half? Swap your diagram with someone else's and check how accurate theirs is.

As with bisecting angles, perpendicular bisectors can be constructed more accurately using a pair of compasses. Draw another line AB that is 9 cm long. Can you figure out how to use a pair of compasses to construct the perpendicular bisector of your line?

Worked example 8.3

Bisect the line segment AB.

A ——————————— B

Solution

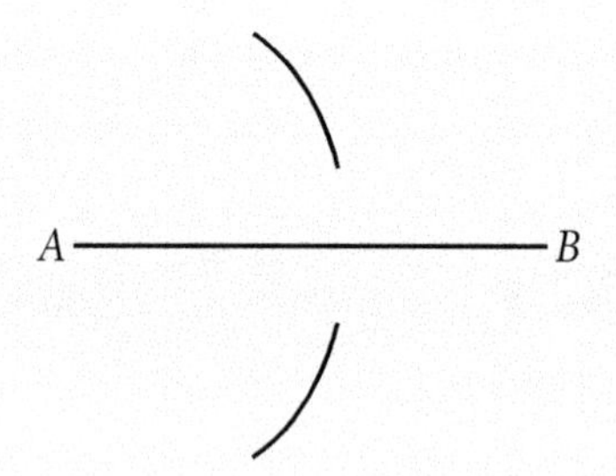

We open up our compasses to a radius of more than half of AB, and draw an arc above and below the line AB from point A.

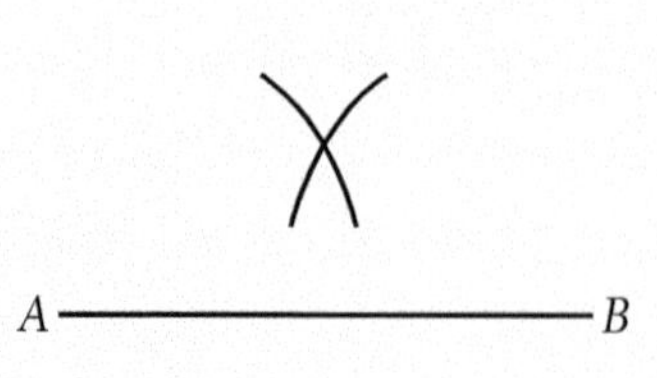

Without changing the setting of our compasses, we draw an arc above and below the line AB from point B.

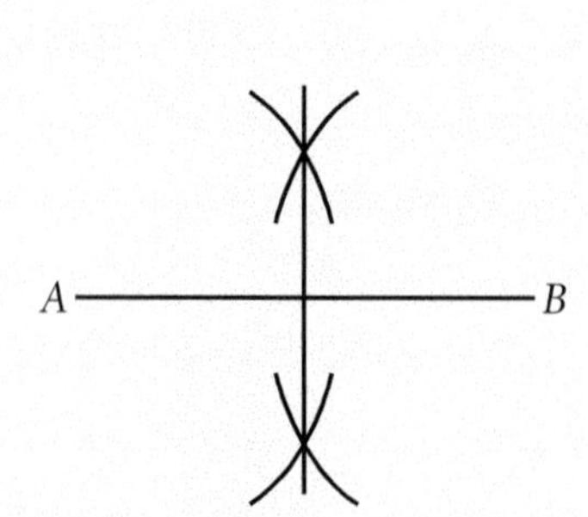

We draw a line passing through the intersecting arcs.

The line we have drawn bisects the line segment AB and forms a right angle with AB.

Reflect

Can you explain why the line constructed in Worked example 8.3 above bisects segment AB?

Can you explain why the compasses need to be set so the radius is set greater than half the length of the segment being bisected?

Practice questions 8.2.3

1 Draw a straight line segment 8 cm long.
Construct the perpendicular bisector of the line segment.

2 Draw each straight line segment and construct its perpendicular bisector.
a 10 cm b 7 cm c 12 cm d 6 cm

3 Draw a scalene triangle and construct the perpendicular bisector of one of its sides.

4 a Construct the triangle ABC, where $AB = 7$ cm, $BC = 6$ cm and $AC = 5$ cm.
b Construct the perpendicular bisector of the side AB.

5 a Draw a straight line segment 10 cm long and construct its perpendicular bisector.
b Construct the angle bisector of one of the right angles.
c Without using a protractor, what is the size of the two angles you created in part **b**?

6 Point P lies on the line AB.

Trace the diagram and construct a perpendicular line to the line AB from point P. Follow the steps in the diagrams below to help you.

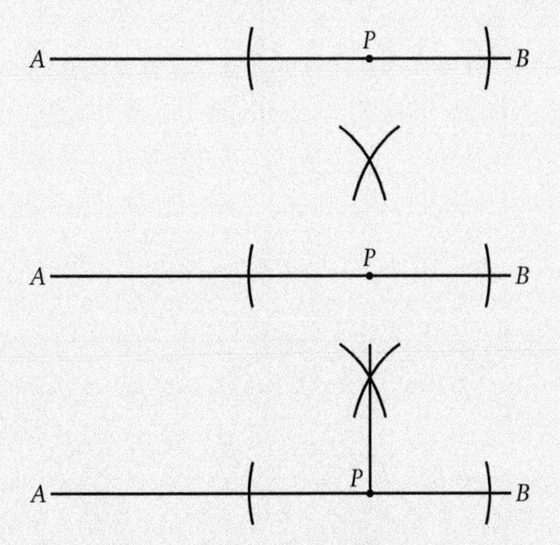

Hint Q6

The steps for constructing a perpendicular line to the line AB from point P is the same as bisecting the angle APB.

7 Copy each diagram. Construct a perpendicular line to the line AB from point P for each one.

a A ——— P ——— B

b A ——— P ——— B

8 Point P is above the line AB. Trace the diagram and construct a line perpendicular to AB from point P.

P

A ——— B

Follow the steps below to help you.

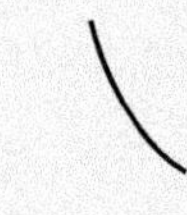

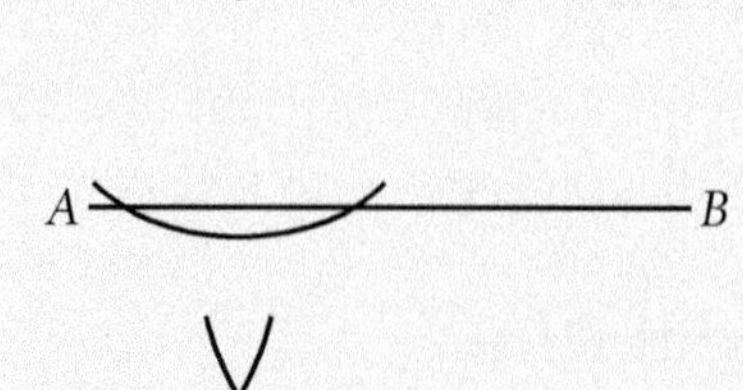

9 Copy each diagram.

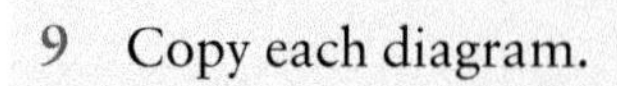

Construct a perpendicular line to the line AB from point P for each one.

a

b

10 The picture on the right shows a roof truss.

For this design of roof truss, the triangular outline of the roof truss is an isosceles triangle. The vertical piece of wood bisects the angle at the top of the roof truss. The other two interior pieces of wood are at right angles to the two equal sides of the roof truss.

Construct a scale drawing of a roof truss of this design with sides 6 m, 6 m and 10 m. Use a scale of 1 cm for 10 m.

You may complete this question using GeoGebra geometry.

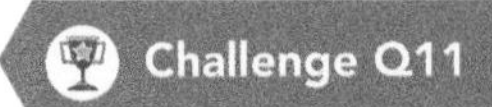

11 The picture on the right shows the steel frame of a commercial building. The outline of the windows are squares with length 2 metres.

Using ruler and compasses, construct an outline of the window using a scale of 1 cm for 0.25 m.

Investigation 8.3

Construct an equilateral triangle with side length 8 cm on paper or using GeoGebra.

Construct the perpendicular bisector of each side of your triangle.

What do you notice with your three perpendicular bisectors?

Try this for isosceles and scalene triangles.

Does the same happen again?

Generalise your findings.

Investigation 8.4

You are going to design and make your own bridge using popsicle (or lolly) sticks. You could use glue, elastic bands, sticky tape or Blu Tack® to join your popsicle sticks together.

Your bridge must span a gap of at least 30 centimetres and use congruent figures in its design.

First, sketch some designs. Choose which design you are going to construct.

Draw an accurate construction of your bridge using a ruler and compasses or GeoGebra geometry before building it.

What problems did you come across whilst designing and building your bridge? How did you overcome these problems?

What is the greatest mass your bridge will hold? How could you make your bridge stronger?

Explore 8.7

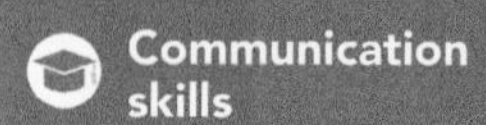

Using your knowledge of constructing triangles and perpendicular bisectors, can you construct these triangles? (Do not use a protractor.)

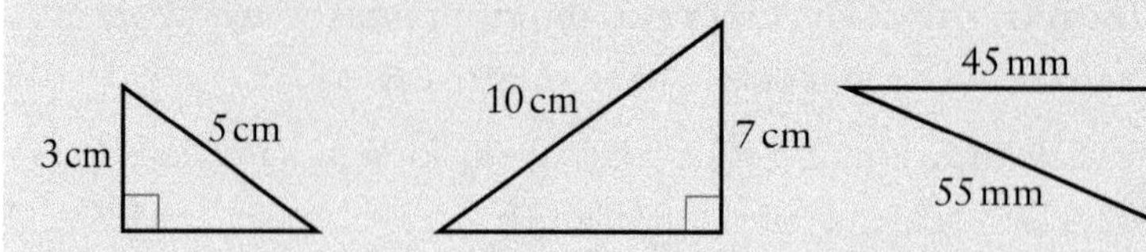

Explain your method.

Self-assessment

- I can identify congruent figures.
- I can construct triangles using a ruler and compasses.
- I can bisect angles.
- I can construct the perpendicular bisector of a line segment.
- I can construct a perpendicular line to a line segment from a point on the line.
- I can construct a perpendicular line to a line segment from a point above or below the line.

Check your knowledge questions

1 Copy each of these figures. For each figure, use one line to divide them into two congruent figures.

a

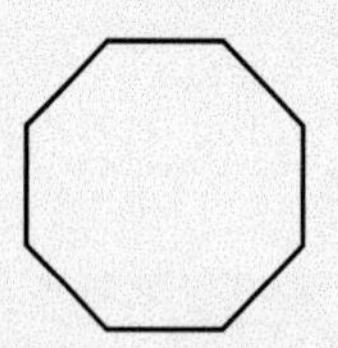

b

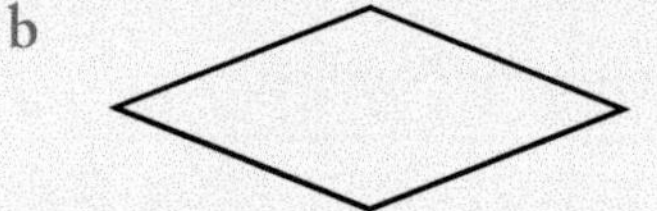

c

2 The diagram shows two congruent triangles ABC and CDE.

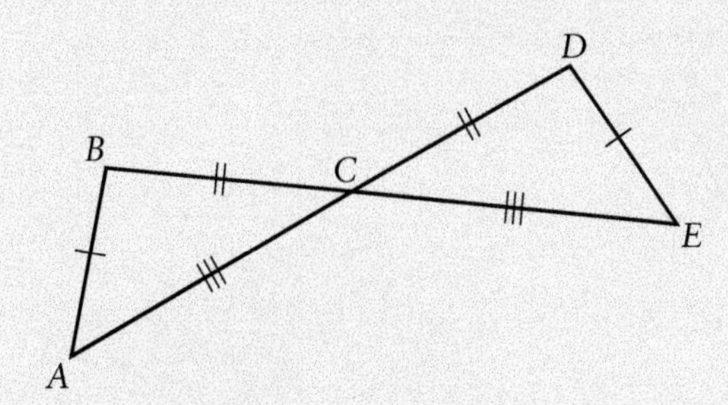

a Use the diagram to write down the angle that matches with

i $\angle CAB$ ii $\angle ABC$ iii $\angle ACB$

b Use the diagram to write down the side that matches with

i AB ii BC iii AC

3 A rectangle has area $20\,\text{cm}^2$. Using a ruler and compasses, construct a triangle with the same area as the rectangle.

4 Construct an equilateral triangle with side length 4 cm.

5 The diagram shows a triangle.

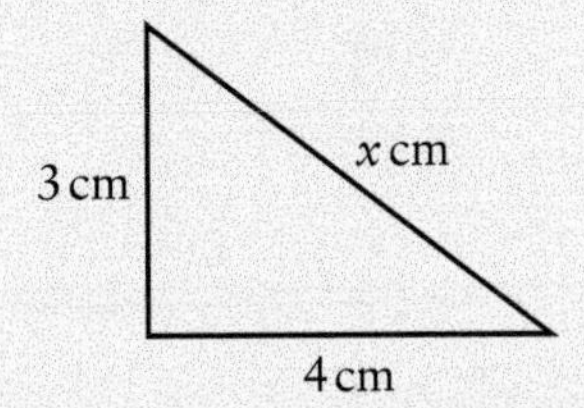

The side labelled x cm is 25% longer than the base of the triangle. Construct an accurate drawing of the triangle using a ruler and compasses.

6 Construct all possible isosceles triangles with side lengths 6 cm and 9 cm.

7 Draw an obtuse angle and bisect it.

8 Construct an isosceles triangle with one obtuse angle. Bisect the obtuse angle in your triangle.

9 Jay uses his ruler and compasses to bisect an angle. The diagram shows their work. Explain what mistake Jay has made.

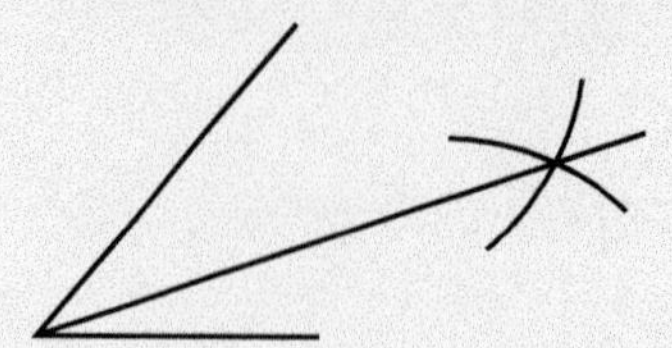

10 Use your triangle from Question 4 and construct the perpendicular bisector of one of the sides.

11 Copy each diagram. Construct a perpendicular line to the line AB from point P for each diagram.

a 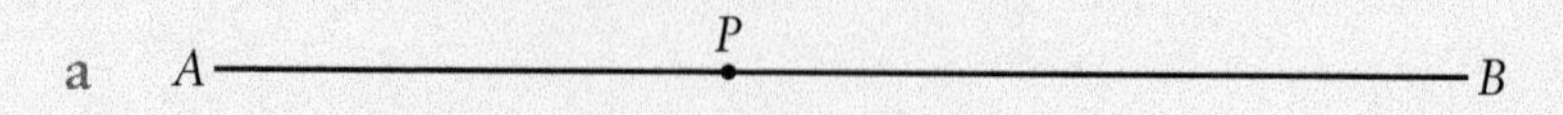

b 

Area and volume

9

9 Area and volume

KEY CONCEPT

Relationships

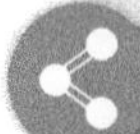

RELATED CONCEPTS

Models, Quantity, Representation, Space

GLOBAL CONTEXT

Orientation in space and time

Statement of inquiry

Identifying relationships between 2D and 3D representations of shapes helps us to quantify real-life models.

Factual

- What is surface area and what is volume?
- How do you find the volume and surface area of 3D objects?

Conceptual

- How are dimensions and surface area and volume related?
- How are different measurements related?

Debatable

- How do artists and architects exploit the different appearance of 3D shapes from different viewpoints in their designs?
- In designing objects, sometimes there is a conflict between form and functionality. Which should be given priority?

Do you recall?

1 Calculate the area of each shape:

a 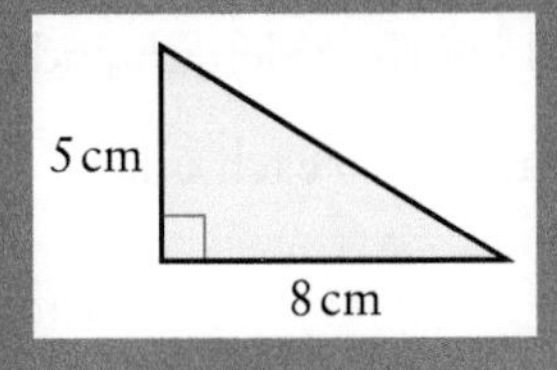

b 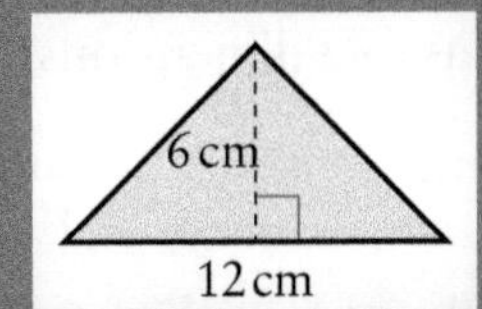

c 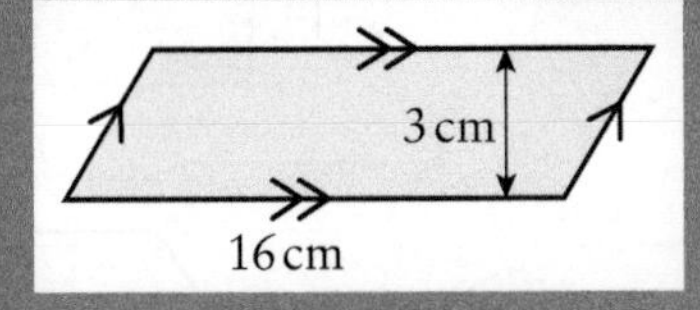

2 Give the mathematical name for each of these 3D shapes:

a 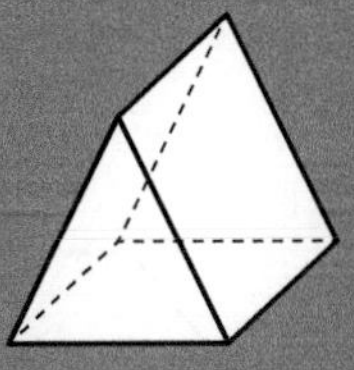

b 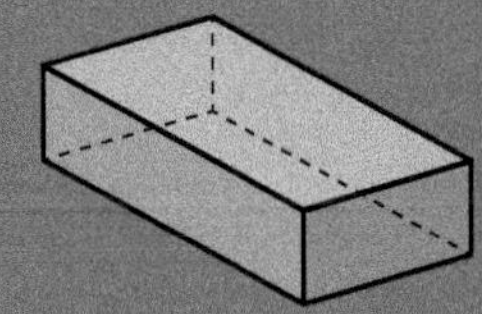

c 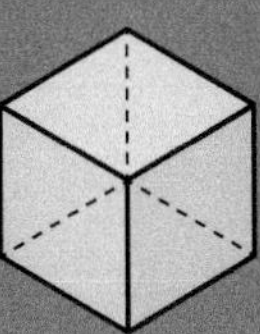

d 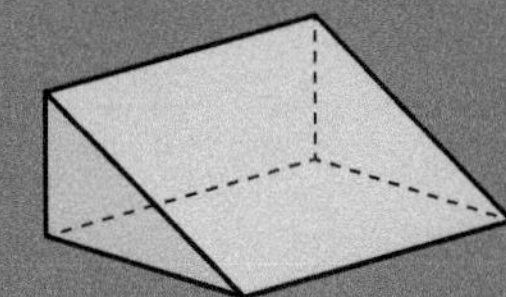

e 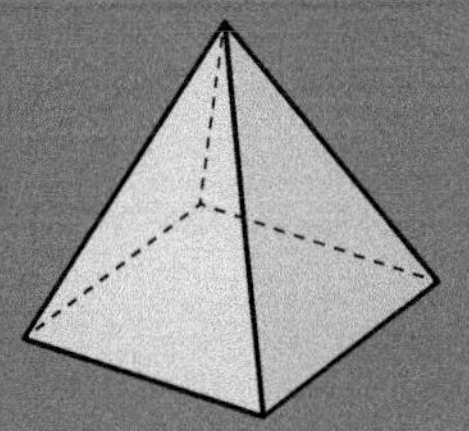

3 Calculate the volume of each 3D shape:

a 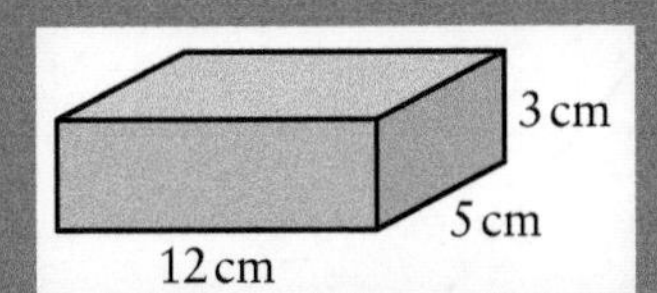

b 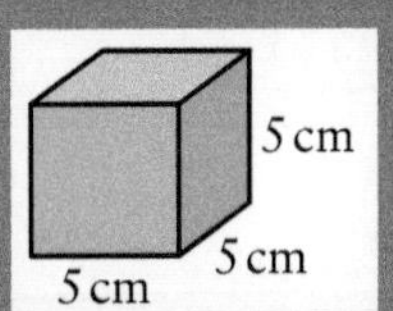

4 Solve these equations for x:

a $25x = 275$

b $7x + 15 = 141$

Area and volume

9.1 Areas of 2D shapes

A **2D shape** has two dimensions – length and width, or length and height.

These two dimensions are measured at right angles to each other.

The length and width of this rectangle are measured at right angles to each other.

length

width

The height h of this parallelogram is measured at right angles to the length b of its base.

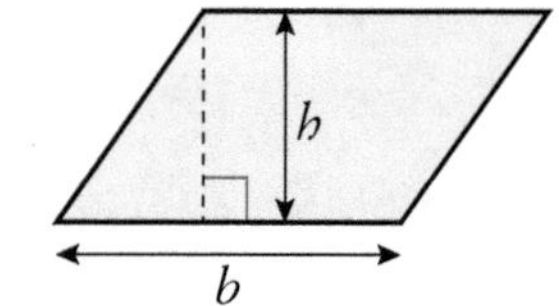

Fact

The plural of trapezium is **trapezia**.

In some countries, a trapezium is called a 'trapezoid', but in some countries a trapezoid can also mean a quadrilateral with no parallel sides. Make sure you know the term and definition used in your country.

A **trapezium** is a quadrilateral with exactly two parallel sides.

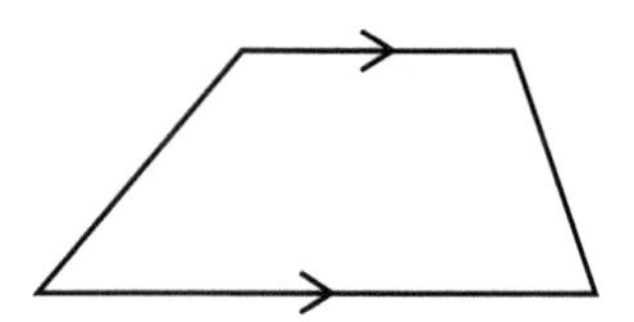

In an isosceles trapezium, the two non-parallel sides are the same length.

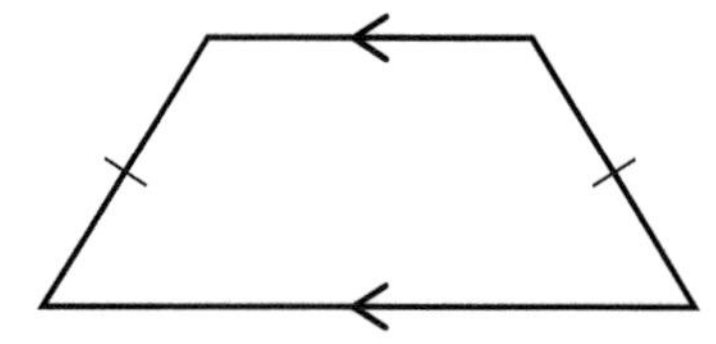

Explore 9.1

On a piece of paper, draw two congruent scalene triangles similar to the one below and cut them out.

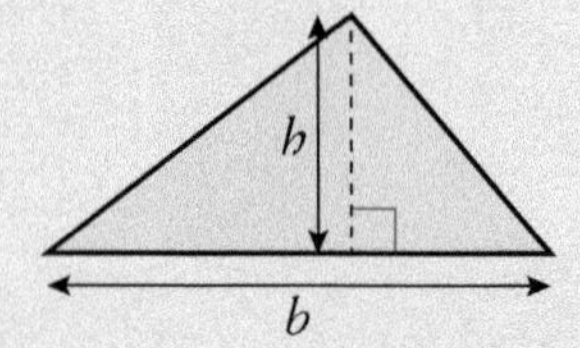

Try to arrange your two triangles to make a parallelogram.

Can you find a formula for the area of the parallelogram you just made?

Does this work for any two identical triangles?

Explore 9.2

On a separate piece of paper, draw two congruent trapezia similar to the one below and cut them out.

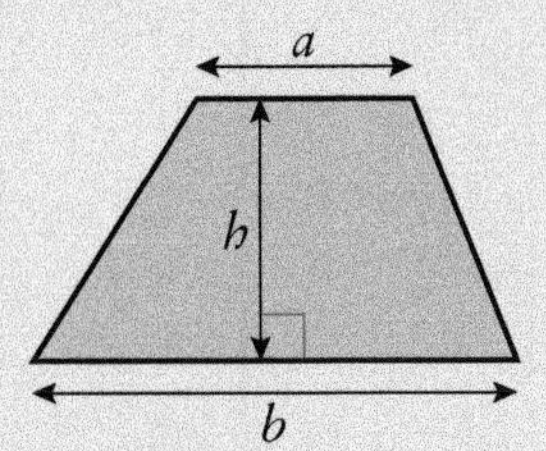

Try to arrange your two trapezia to make a parallelogram. What is the area of this parallelogram?

Can you find a formula for the area of a trapezium and justify your findings?

Does this work for any two identical trapezia?

Hint

You could draw and arrange your congruent shapes using dynamic geometry software.

Worked example 9.1

The building below has two sides in the shape of a trapezium. The other two sides are rectangular. A window-cleaning company needs to estimate the total area of the sides so that they can calculate the cost of cleaning the sides of the building. The taller part of the building is 90 metres. The shorter part is 60 metres. The base of the building is 120 m by 30 m. Help the company with their estimate.

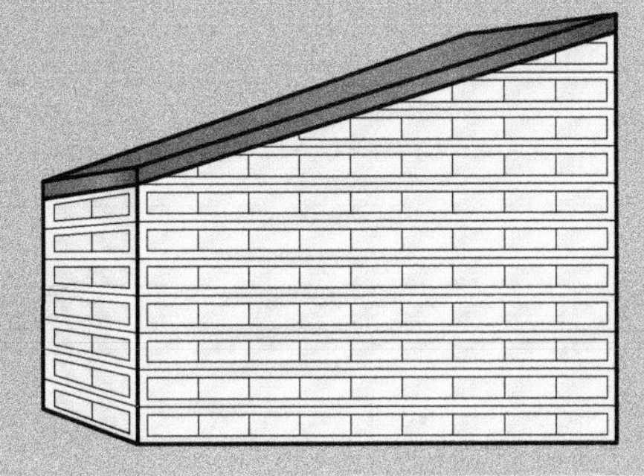

Solution

We need to calculate the areas of the two trapezium-shaped sides and the two rectangular-shaped sides.

Make a plan

Sketch the shapes of the sides and use the area of a rectangle and area of a trapezium formulae to make the calculations.

Carry out the plan

Here are sketches of the sides with dimensions as shown:

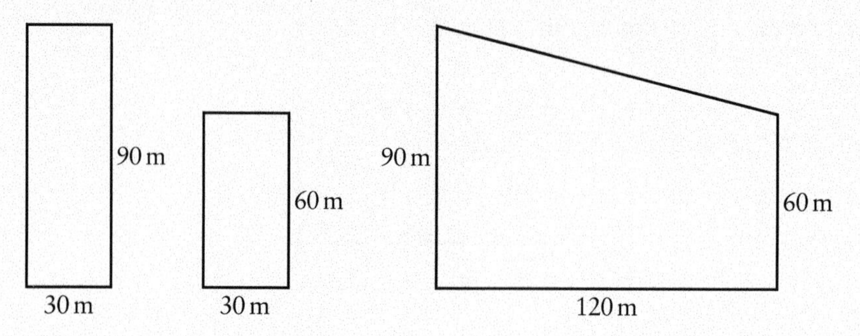

For the trapezium, label the parallel sides as:

$a = 60\,\text{m}$, $b = 90\,\text{m}$ and the height $h = 120\,\text{m}$.

Write the formula and substitute the values:

$$A = \frac{1}{2}(a + b)h$$

$$= \frac{1}{2}(60 + 90) \times 120 = 9000\,\text{m}^2.$$

Total area of trapezium sides is then $2 \times 9000 = 18\,000\,\text{m}^2$.

Make sure you include the units.

For the rectangles, label the sides as $l = 90\,\text{m}$ and $w = 30\,\text{m}$ for the taller side, and $l = 60\,\text{m}$ and $w = 30\,\text{m}$ for the shorter side.

$A_{Tall} = l \times w = 90 \times 30 = 2700\,\text{m}^2$

$A_{Short} = l \times w = 60 \times 30 = 1800\,\text{m}^2$

Thus, the total area of all sides is $18\,000 + 2700 + 1800 = 22\,500\,\text{m}^2$.

Reflect

Does it matter which side you label a and which you label b?

What mistake has this student made with the labelling?

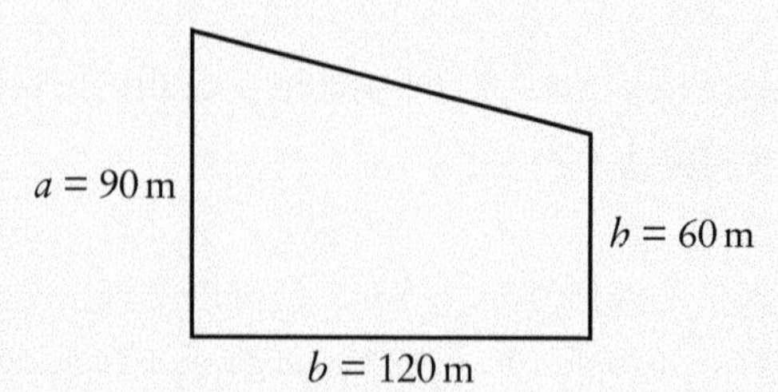

Another way of thinking about the area of a trapezium formula is:

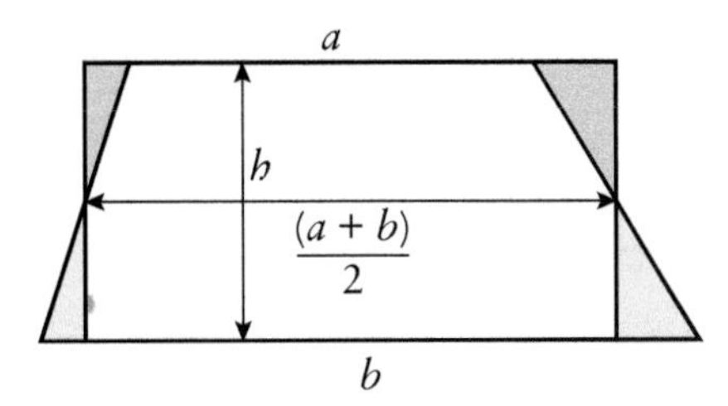

Find the mean length of the two parallel sides, then multiply by the height. Can you explain the connection between the formula and this diagram?

Connections

The mean of two values, a and b is $\frac{a+b}{2}$

Fact

The Temple of Kukulcán at Chichén Itzá in the Yucatán has the shape of a trapezium on all four sides.

Worked example 9.2

Max cuts a triangle from each end of this rectangle and rearranges the shapes to make a trapezium.

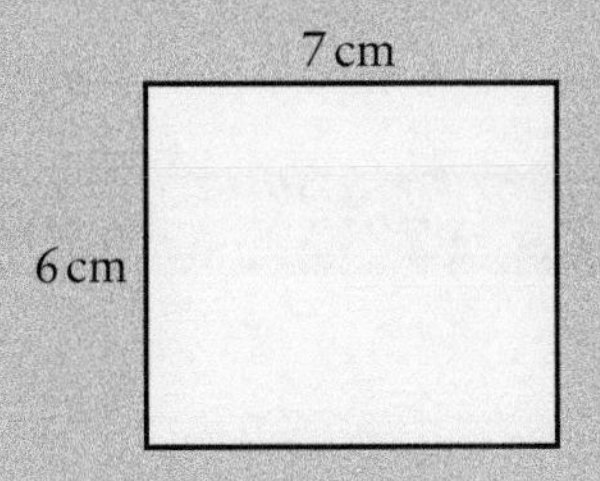

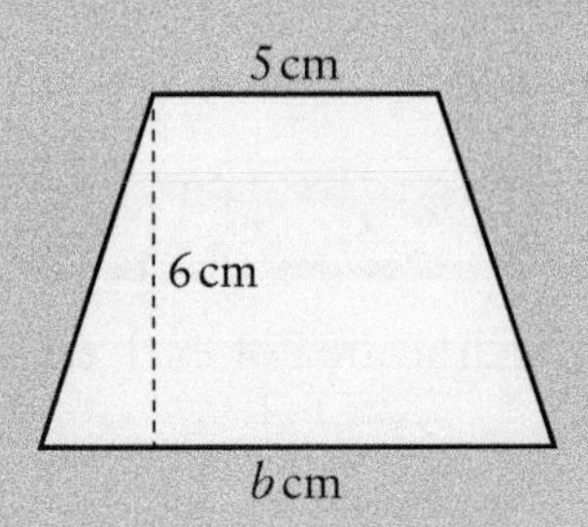

Find b, the length of its base.

Solution

Make a plan

Find the area of the rectangle. The trapezium has the same area.

We substitute the values we know into the area of a trapezium formula to give an equation.

We solve the equation to find the value of b.

Carry out the plan

Area of rectangle $= 6 \times 7 = 42\,\text{cm}^2$

Let the parallel sides be a and b, and the height h:

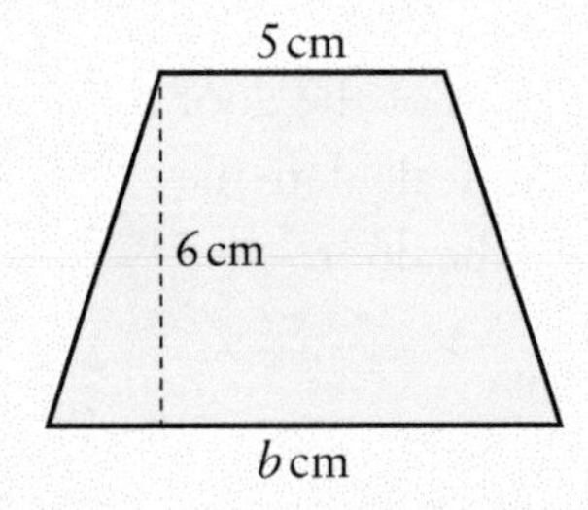

Write the formula:

$A = \frac{1}{2}(a + b)h$

Substitute the values:

$42 = \frac{1}{2}(5 + b) \times 6 = 3(5 + b)$

Expand the brackets:

$42 = 15 + 3b$

$27 = 3b$

Solve for b: $9 = b$

Looking back, if we substitute the values into the formula of the area of a trapezium, we have:

$\frac{1}{2}(5 + 9) \times 6 = 42$

Practice questions 9.1

1 Calculate the area of each trapezium.

a

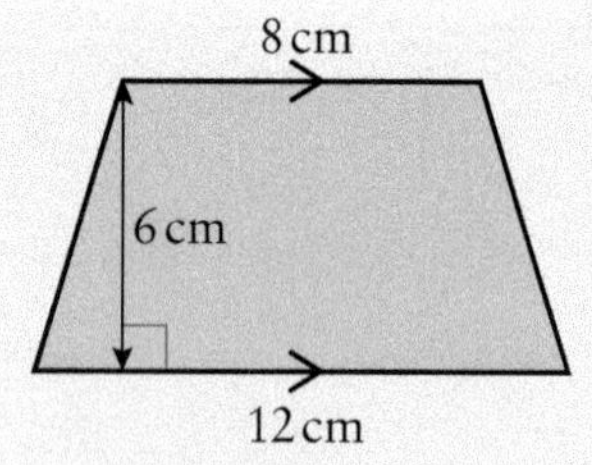

b

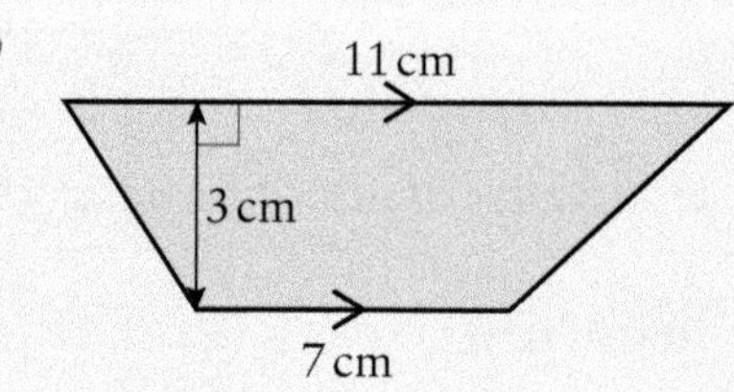

c

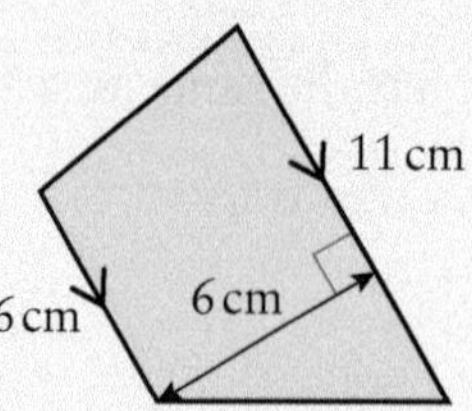

2 The two parallel sides of this door have lengths 90 cm and 40 cm. The width of the door is 75 cm. Calculate the area of the door.

3 This trapezium has been split into a triangle and rectangle.

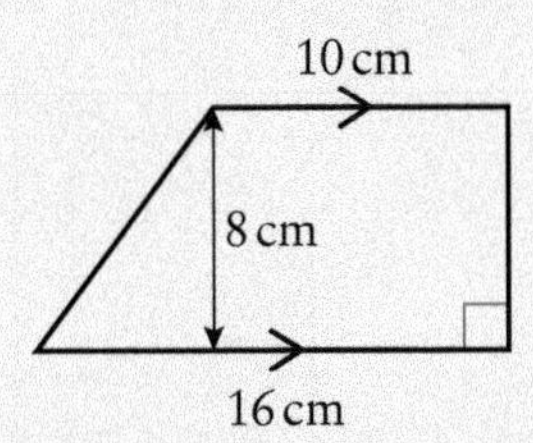

a Calculate the area of the triangle.

b Calculate the area of the rectangle.

c Find the area of the trapezium.

First find the base length of the triangle.

4 Calculate the area of this isosceles trapezium by splitting it into two triangles and a rectangle.

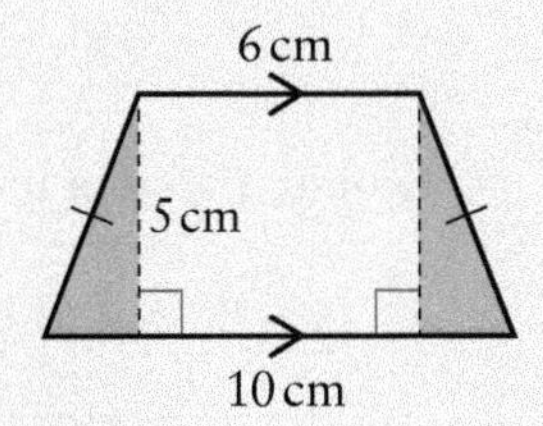

5 This Kuwait flag is a rectangle measuring 84 cm by 42 cm.

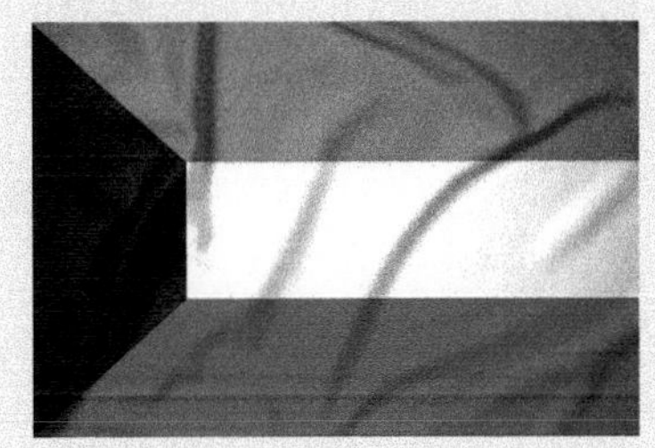

The central white rectangle is three quarters of the length of the flag and the heights of the green, white and red stripes are equal. Work out the area of each colour in the flag.

Fact

The Kuwait flag was used as the design for the world's largest kite. The kite was made by Abdulrahman Al Farsi and Faris Al Farsi and flown at the Kuwait Hala Festival in Flag Square, Kuwait City, Kuwait on 15 February 2005. It had an area of 1019 m^2.

6 The stage in a theatre is in the shape of an isosceles trapezium.

The dimensions of the stage are shown in the diagram:

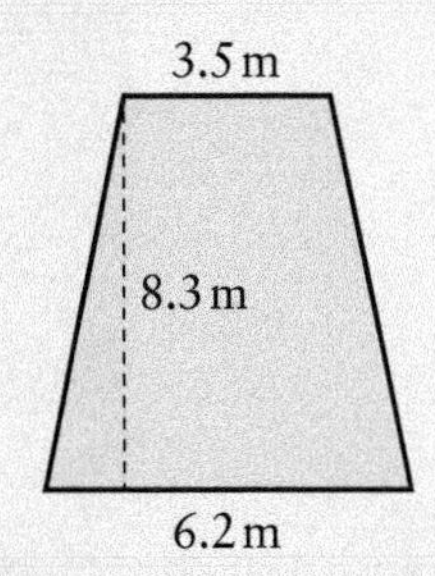

The stage floor is to be painted with varnish.

One tin of varnish covers 3 m^2 of floor.

How many tins of varnish are needed to paint the floor?

Hint Q6

Your answer must be a whole number of tins.

7 Here is a scale diagram of a field.

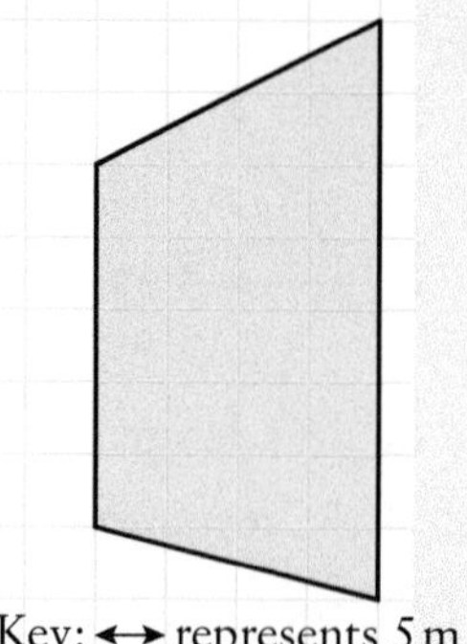

Calculate the area of this field.

8 Calculate the lengths labelled with letters:

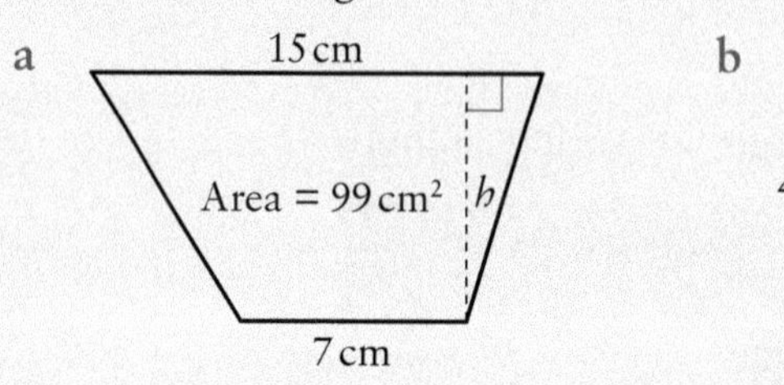

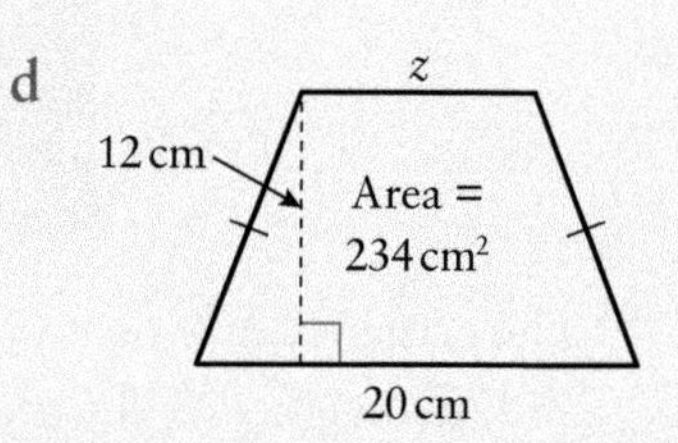

9 The front of the base of this trophy is a trapezium.

The width of the top the trapezium is 4 cm shorter than the width of the bottom of the trapezium. The height of the trapezium is 16 cm and it has area 320 cm².

Calculate the top and bottom widths of the trapezium.

10 Find the lengths of the two parallel sides in this trapezium:

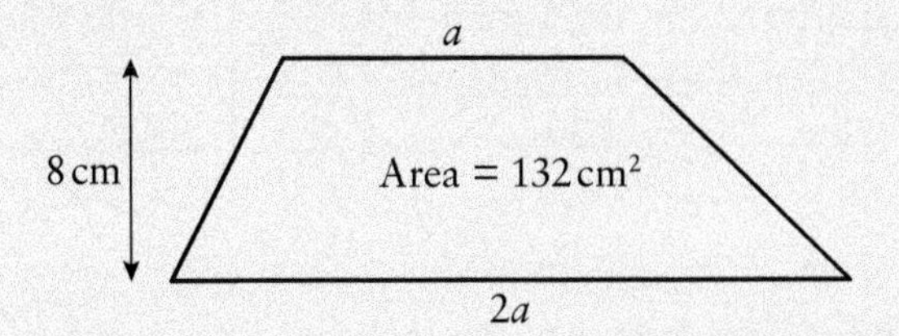

11 The diagram shows a trapezium of area 18 square units, drawn on a coordinate grid. Find the values of a and b.

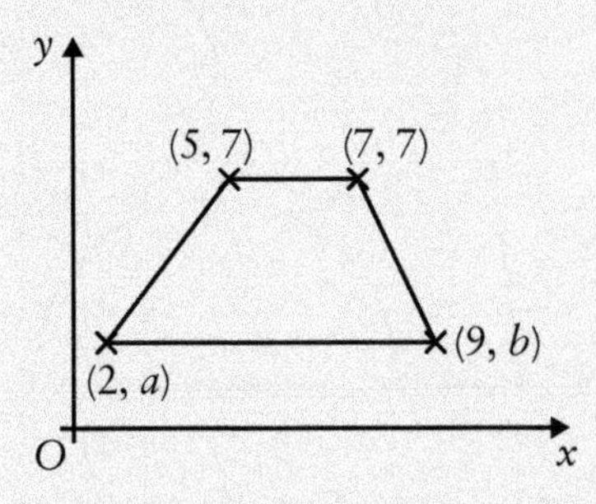

12 Calculate the area of this hexagon:

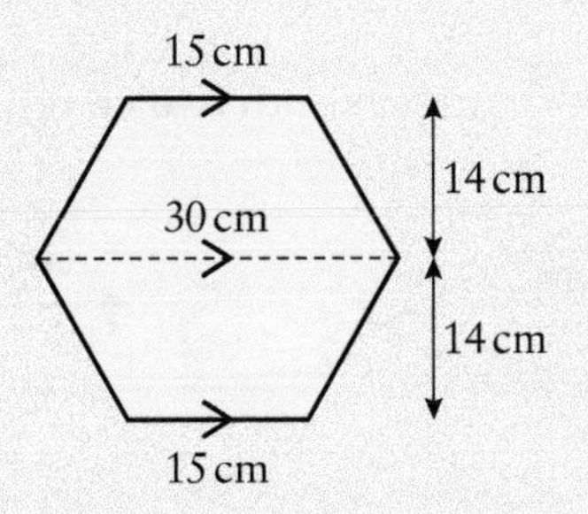

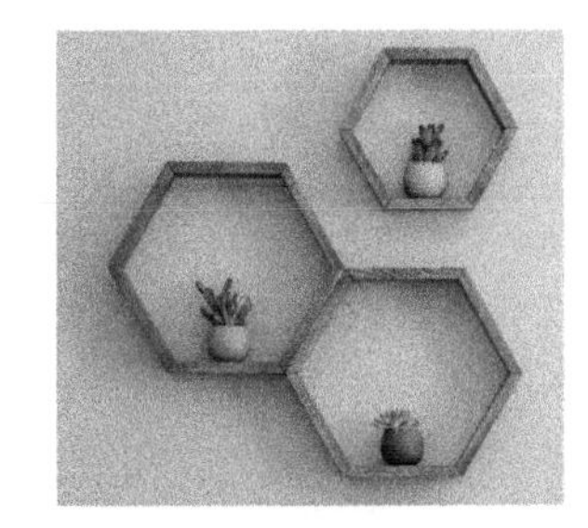

13 Calculate the area of each composite shape:

a

20 cm

9 cm

14 cm

b

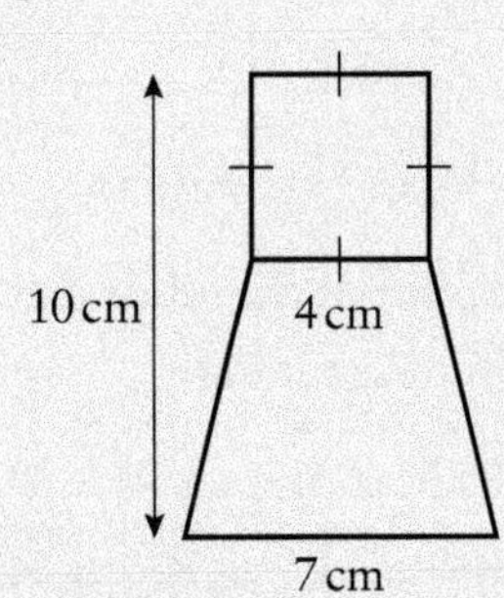

14 The diagram shows a quadrilateral, with a quadrant cut out at one vertex. Calculate the shaded area.

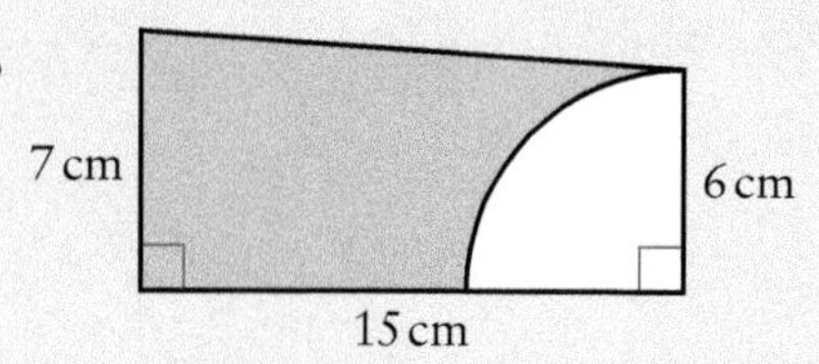

15 The diagram shows the plan view of a stage drawn on a centimetre grid. The scale is 1 cm represents 0.5 m.

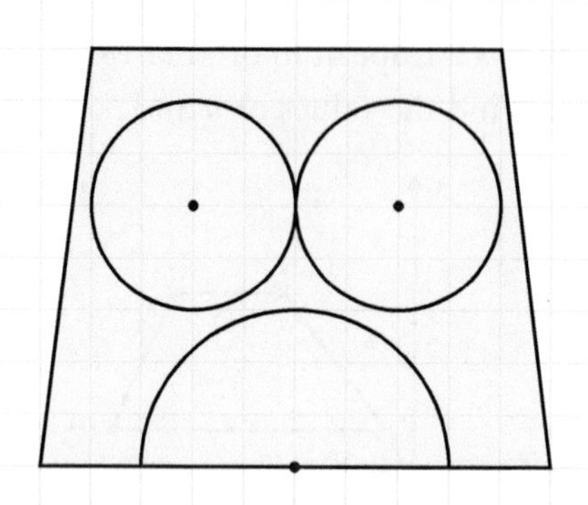

Three lights above the stage illuminate the two circular areas and the semi-circular area shown. What percentage of the stage area is lit by the three lights?

Challenge Q16

16 This rhombus has been split into two congruent triangles:

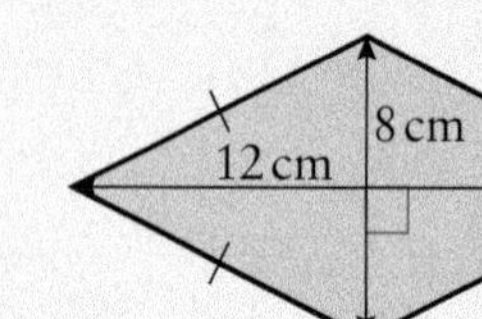

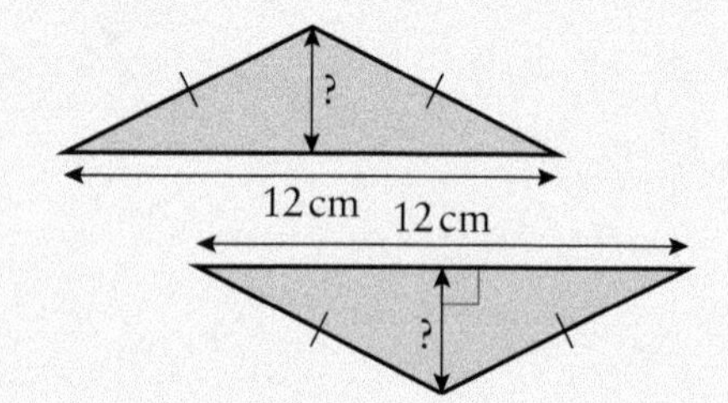

a Find the height of each triangle.

b Calculate the area of each triangle.

c Hence calculate the area of the rhombus.

d Split the rhombus into two congruent triangles along the 8 cm diagonal.

Repeat parts **a** to **c**. Does this give the same area?

e Write a formula for the area of a rhombus with diagonals of length x and y.

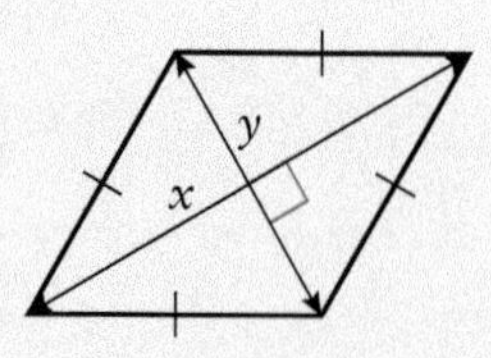

Challenge Q17a

17 a Split this kite into two congruent triangles and hence find its area.

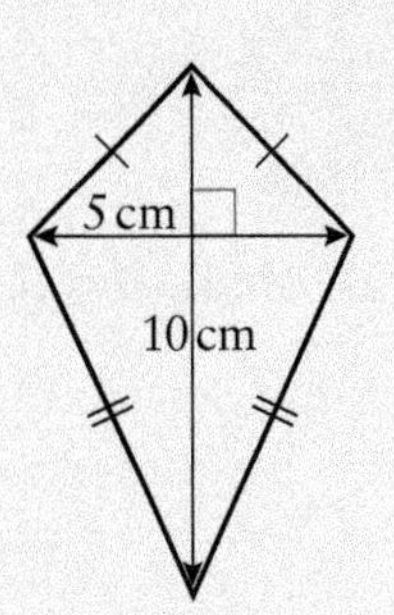

b Write a formula for the area of a kite with diagonals of length x and y.

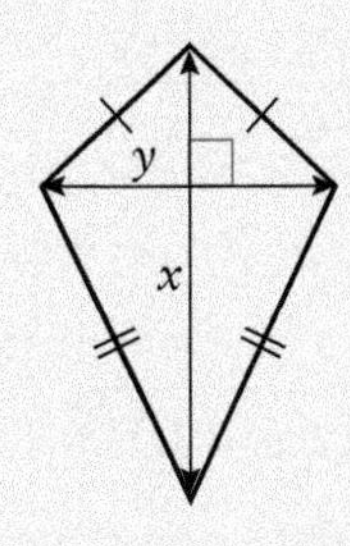

18 Calculate the area of each shape:

a

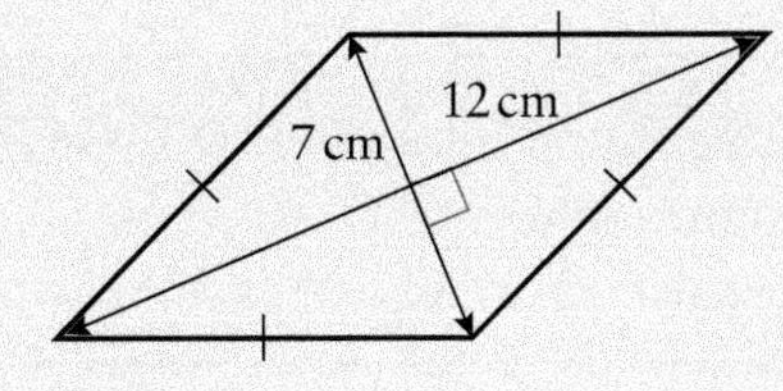

b

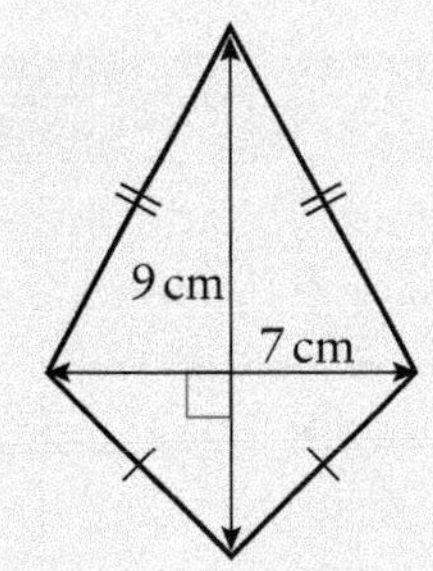

Challenge Q19

19 This window is a square of side 90 cm.

The area of the rhombus is $\frac{1}{27}$ of the total area of the window.

The width of the rhombus is 20 cm.

Calculate the height of the rhombus.

Challenge Q20

20 Sketch and label three different trapezia, each with area 36 cm^2.

 3D shapes

A **3D shape** has three dimensions – length, width and height.

You can describe a 3D shape by describing its faces, edges and vertices.

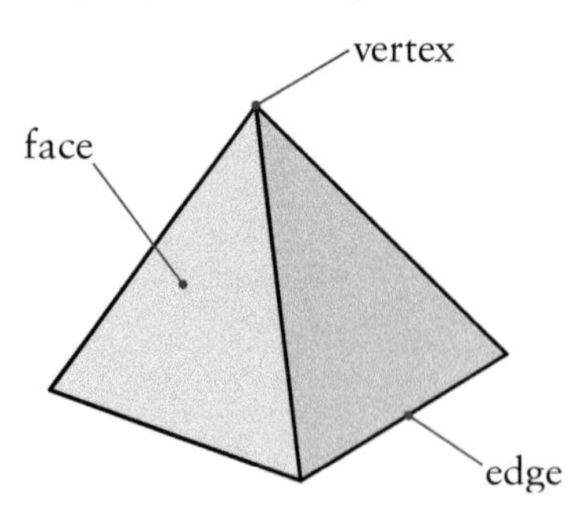

A **face** is a 2D surface of a shape.

An **edge** is where two faces meet.

A **vertex** is where edges meet. The plural of vertex is **vertices**.

 **Hint**

You may find it helpful to compile a table showing the numbers of faces, edges and vertices of 3D shapes.

 **Explore 9.3**

Can you find a relationship between the number of faces, edges and vertices on a 3D shape with flat surfaces?

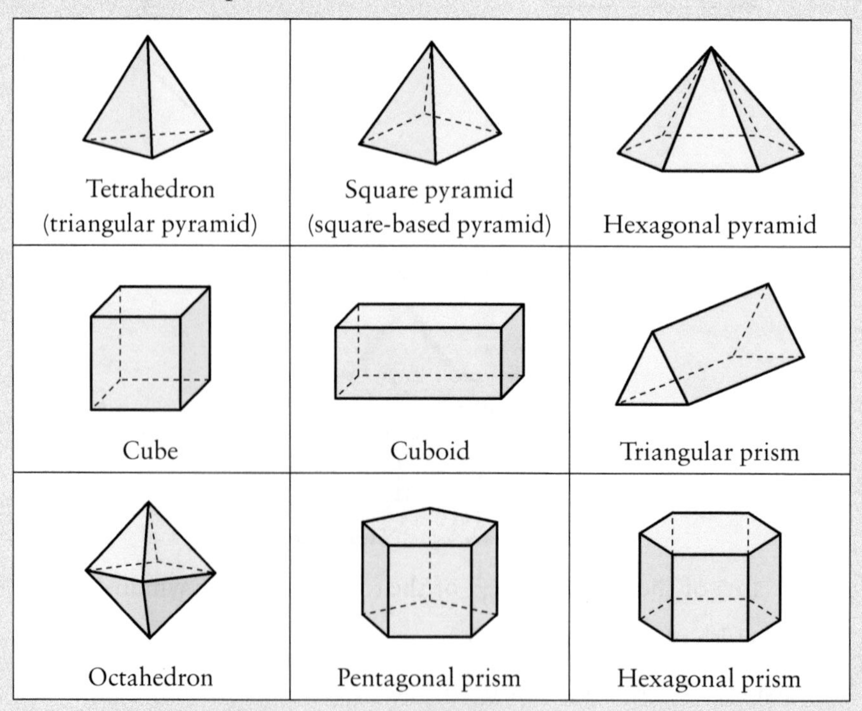

Does your relationship work for all 3D shapes?

 **Fact**

Euler's formula states the relationship between the number of faces, edges and vertices of a 3D shape. Leonhard Euler (1707 –1783) was a Swiss mathematician who wrote at least 92 books on mathematics, physics, astronomy, geography and logic. He also introduced some of the notation that mathematicians still use today.

A **prism** is a 3D shape that has:

- two parallel, congruent end faces
- the same cross-section all along its length.

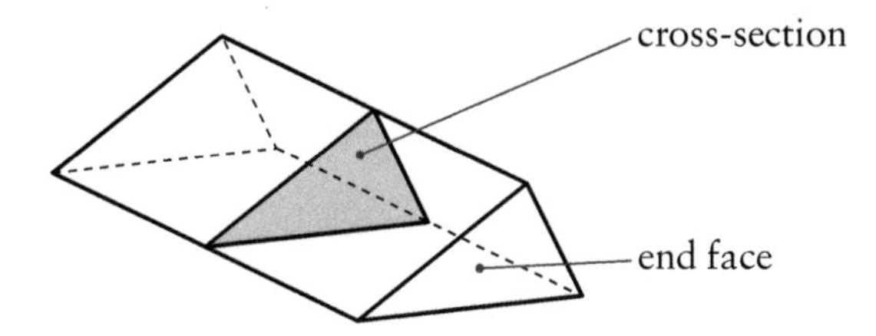

A cross-section is the shape you see when you cut a shape at right angles to its edge. In a prism, any cross-section is congruent to the end faces.

A cuboid is a prism with a rectangular cross-section.

You can draw prisms on isometric paper.

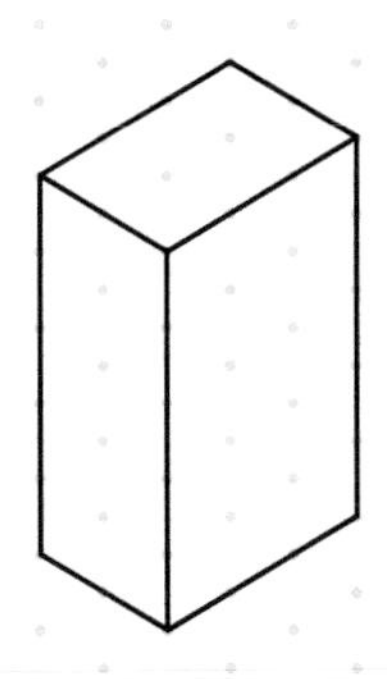

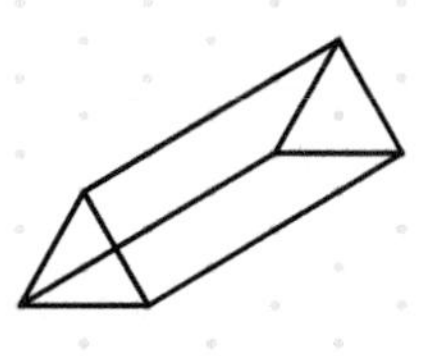

Explore 9.4

The diagram shows an L-shaped prism made of 1 cm cubes, drawn on isometric paper.

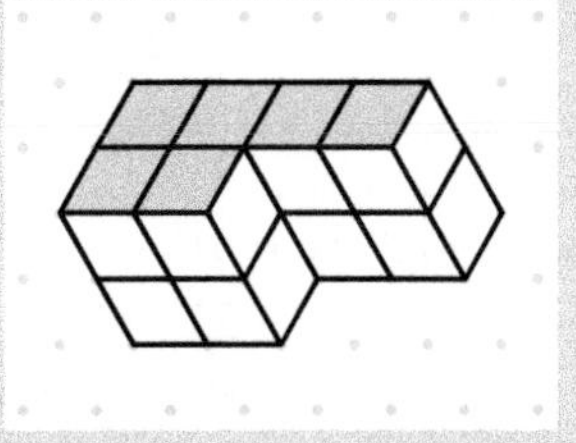

How can you find the volume of this prism?

Can you use the area of its cross-section to work out volume of a prism?

Fact

The formula for the volume of a prism is:

volume = area of cross-section × length

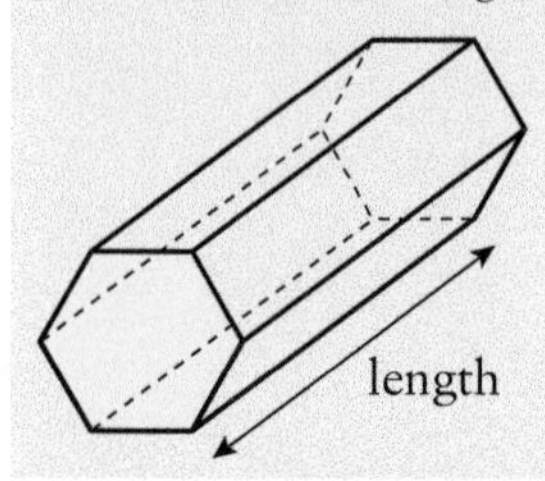

Worked example 9.3

Calculate the volume of this prism.

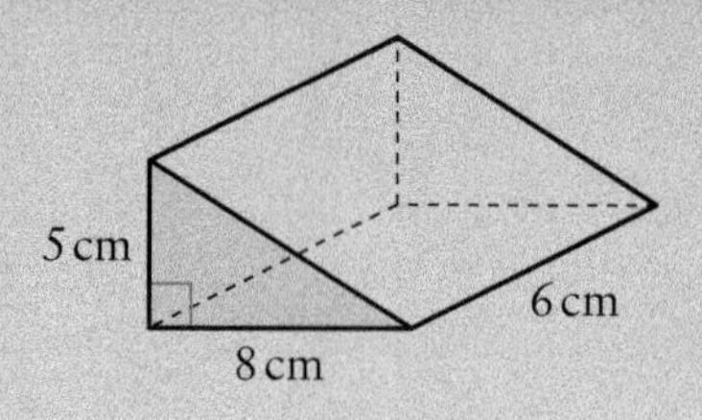

Solution

Make a plan

Identify the cross-section and calculate its area. Then multiply that area by the length.

Carry out the plan

The cross-section is a right-angled triangle.

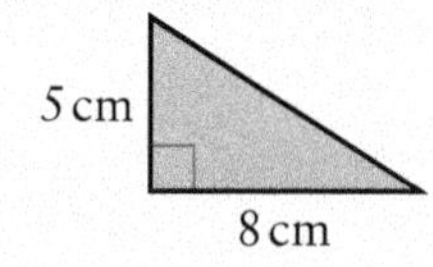

area of triangle = $\frac{1}{2}bh$

$A = \frac{1}{2}bh = \frac{1}{2} \times 8 \times 5 = 20\,\text{cm}^2$

volume = area of cross-section × length

volume = $20 \times 6 = 120\,\text{cm}^3$

Look back

This triangular prism is half of a cuboid:

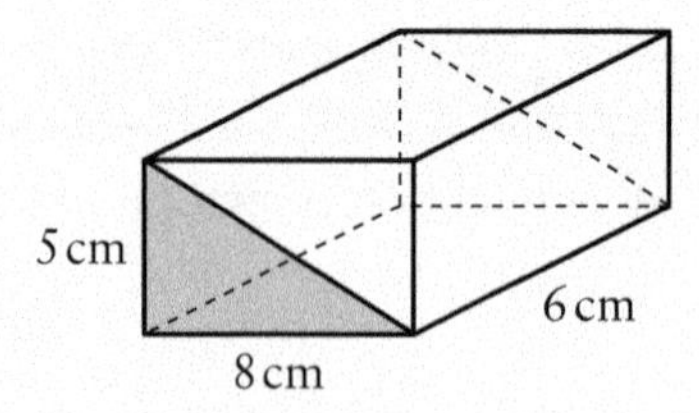

volume of cuboid = $lwh = 6 \times 8 \times 5 = 240\,\text{cm}^3$

Half the volume of the cuboid = $120\,\text{cm}^3$.

Hint

When you see a prism, first identify the shape of the cross-section.

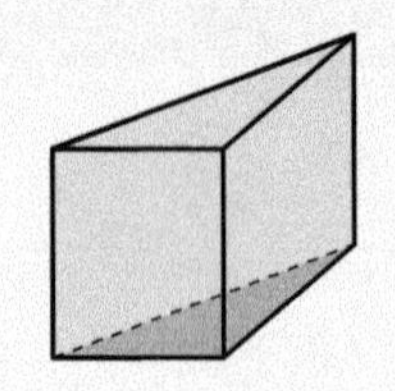

Reflect

Do we need two different methods for finding the volume of a cuboid?

volume of cuboid = lwh

volume of cuboid = area of cross-section × length

Practice questions 9.2

1 Which of these 3D shapes are prisms?

a 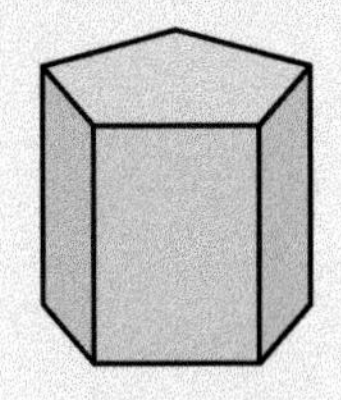

b 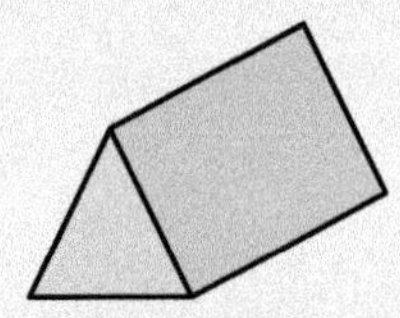

c 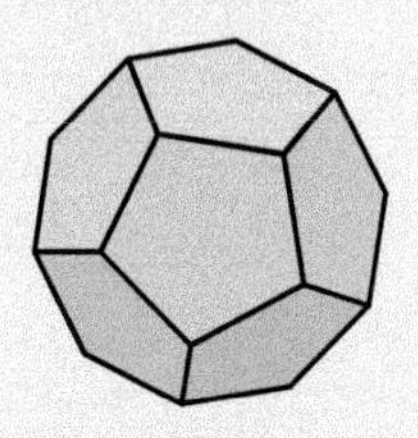

d 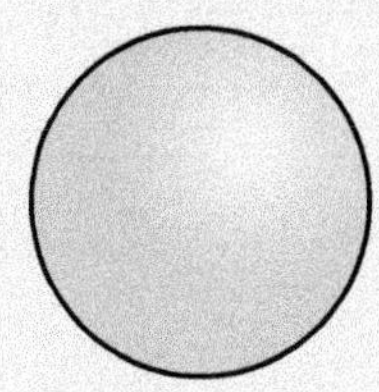

e 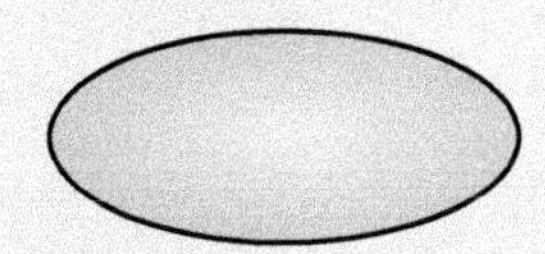

f 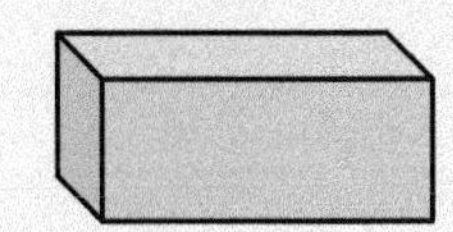

g 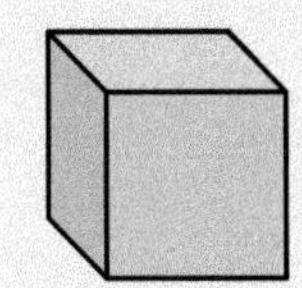

h 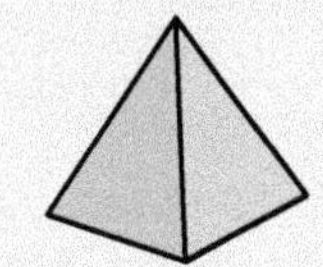

2 This diagram shows a cube of side 2 cm drawn on isometric paper.

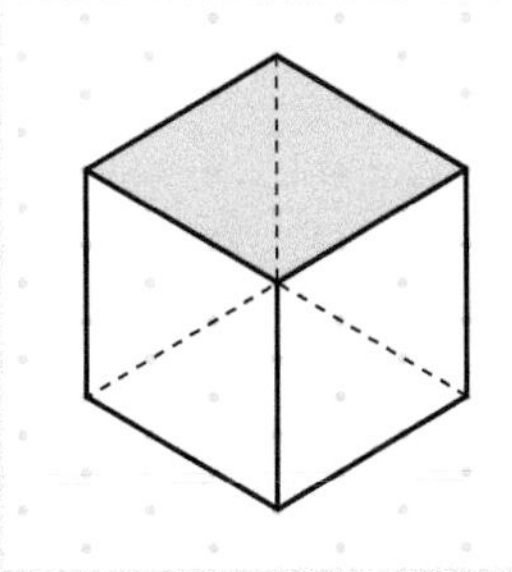

Draw the prisms below on isometric paper.

a 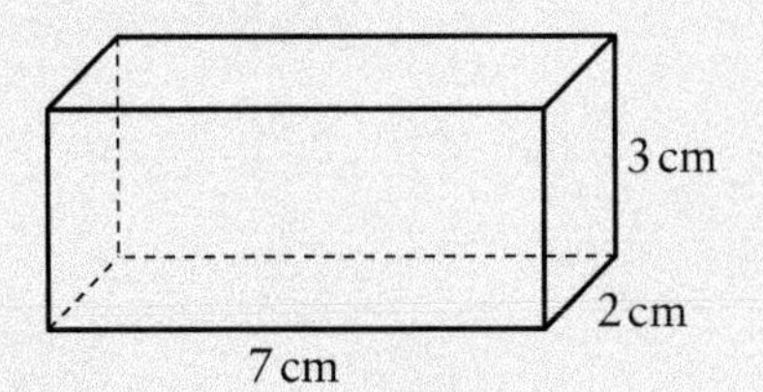

b 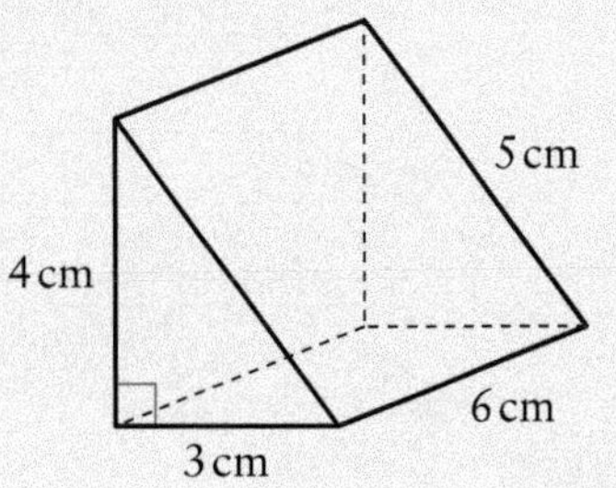

 **Hint Q2**

Find an online video tutorial on drawing prisms on isometric paper.

3 Sam makes this shape from cubes.

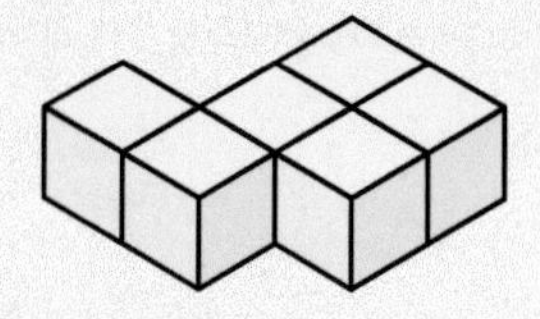

He adds two more layers to make a prism.

Draw Sam's prism on isometric paper.

Add shading to your drawing to give it a 3D effect.

4 This prism is rotated so it stands on its largest face.

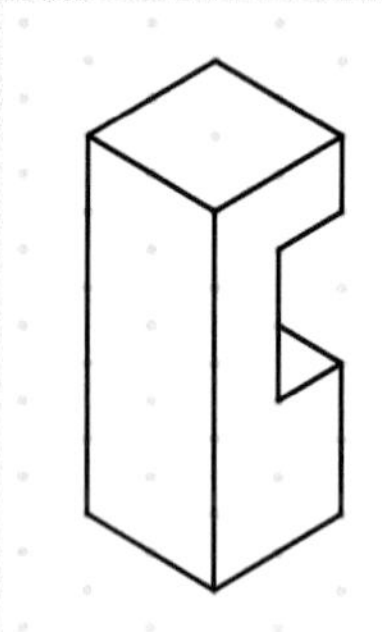

Draw the prism in its new position.

5 a Which 3D shape has three rectangular faces and two triangular faces?

b Describe the shapes of the faces, and the numbers of faces, edges and vertices of:

i a pentagonal prism

ii an octagonal prism.

6 Calculate the volume of each prism:

a

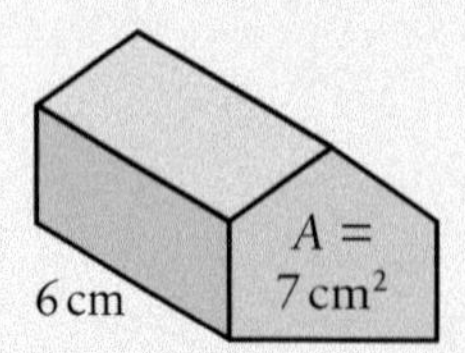

b

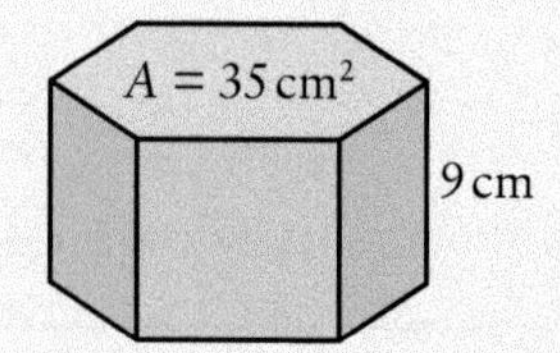

c

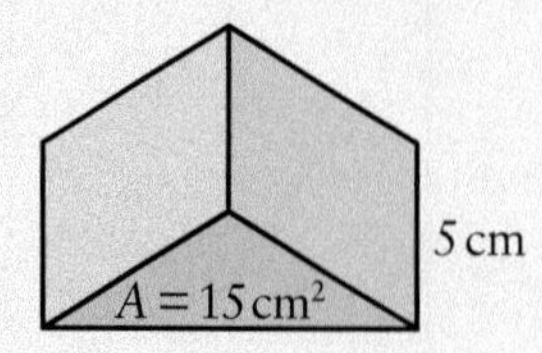

7 Calculate the volume of each triangular prism:

a

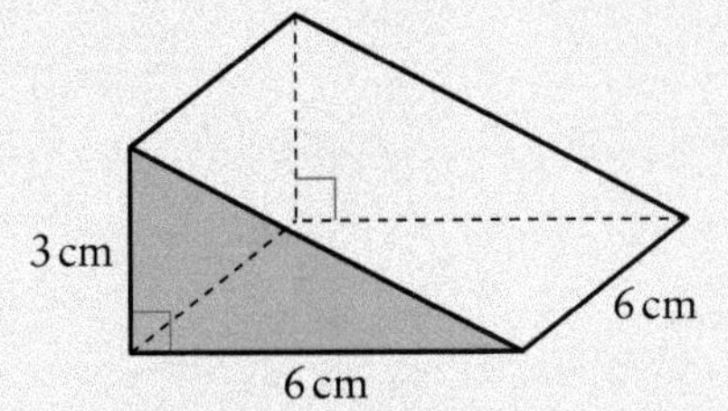

b

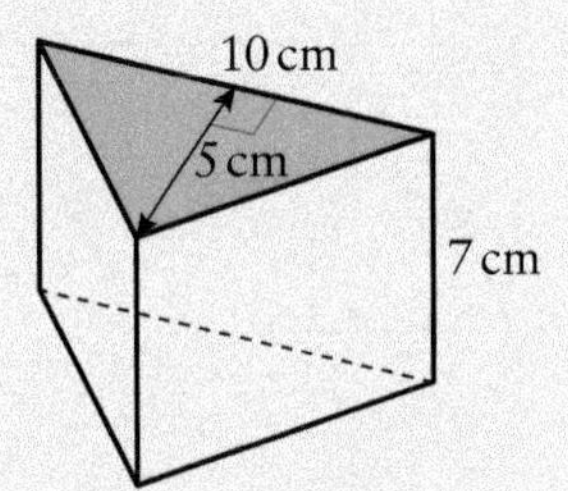

c

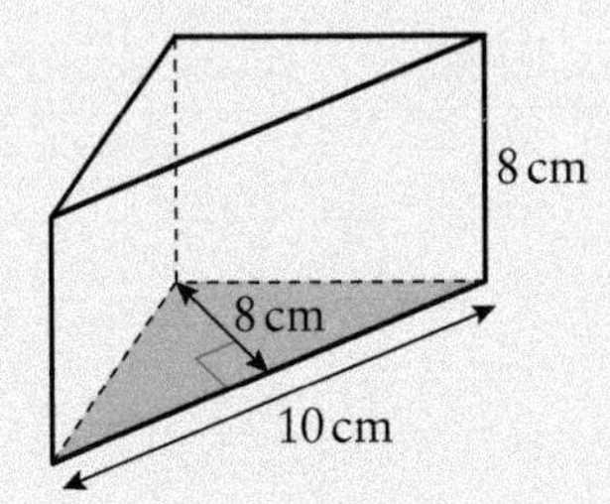

d

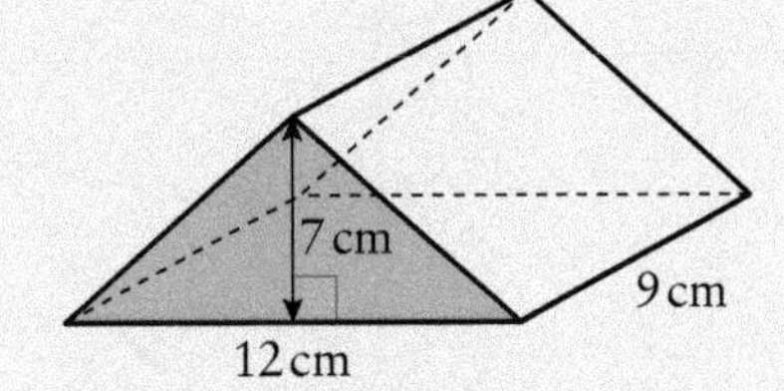

8 Calculate the volume of each prism:

a

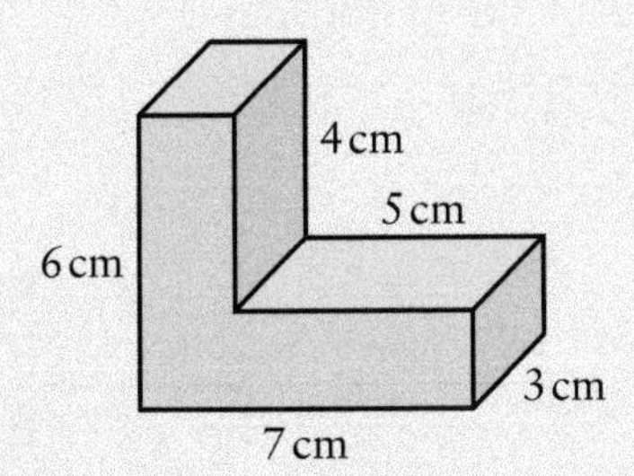

b

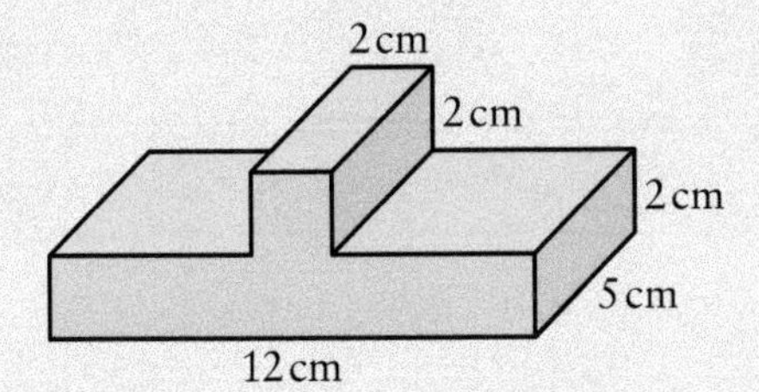

9 This cuboid has a triangular prism cut out through its centre.
Calculate the volume of the remaining shape.

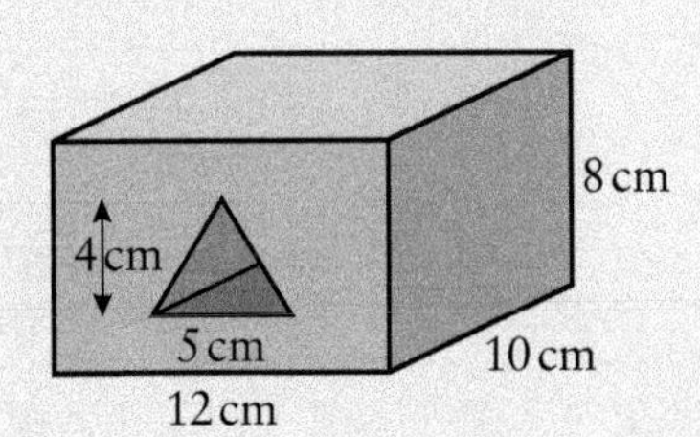

10 Calculate the volume of these prisms:

a

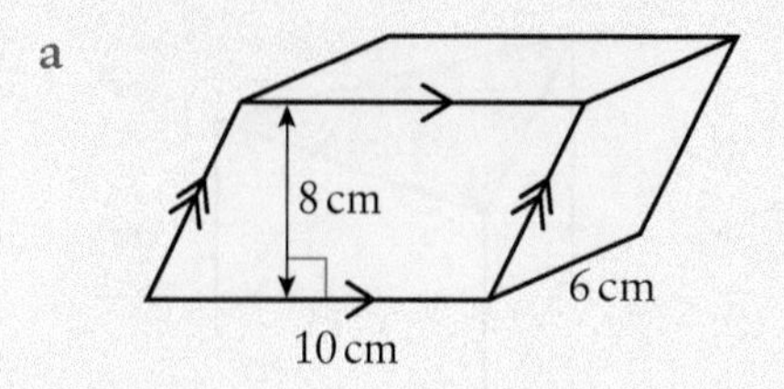

b

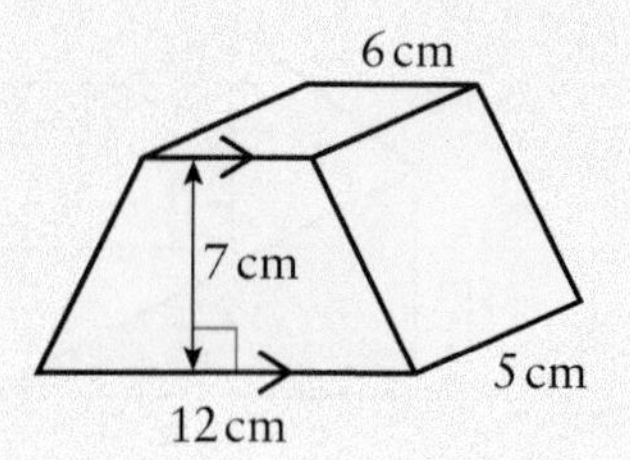

c

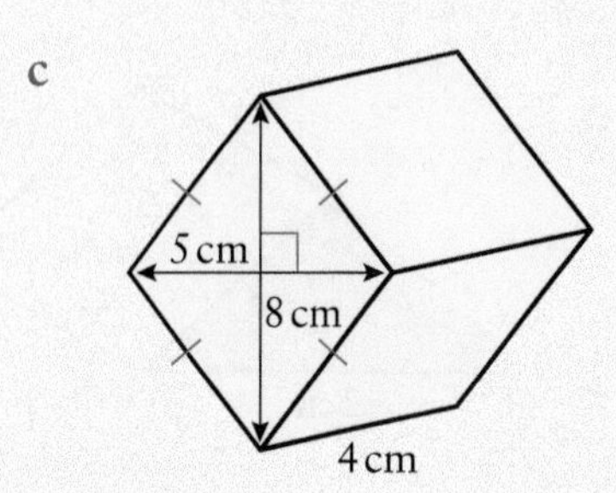

11 This marquee has maximum height of 3.4 m. The base is a rectangle 5 m by 25 m. The height of the side of the marquee is 2 m.

Calculate the volume of the marquee, in cubic metres.

12 This skip should only be filled level with the top.

The top of the skip is a rectangle 1.6 m by 1.2 m. The bottom of the skip is a rectangle 1.2 m by 1.2 m. The height of the skip is 1 m.

Calculate the volume of rubbish that can be put into this skip.

13 a This triangular prism has volume $112\,\text{cm}^3$. Calculate the length, l.

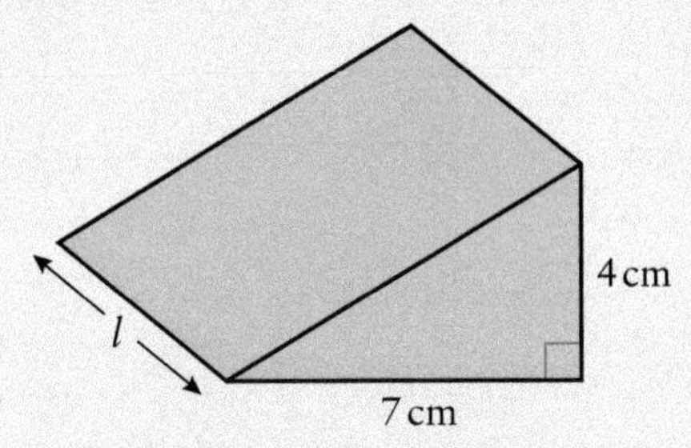

b This triangular prism has volume $15\,\text{cm}^3$. Calculate its height, h.

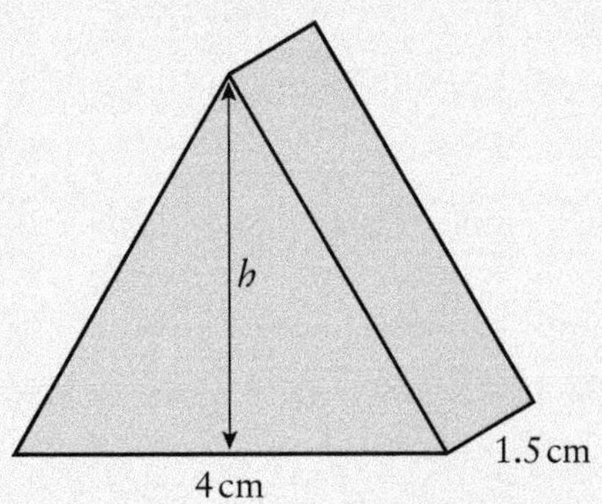

14 Which of these prisms has the larger volume? Show your working to justify your answer.

Prism A

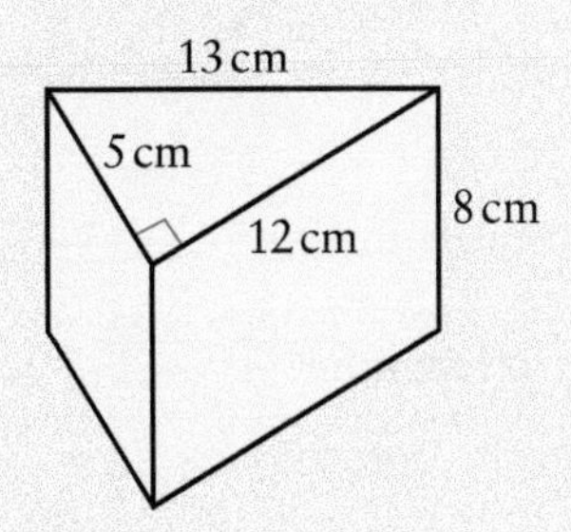

Prism B

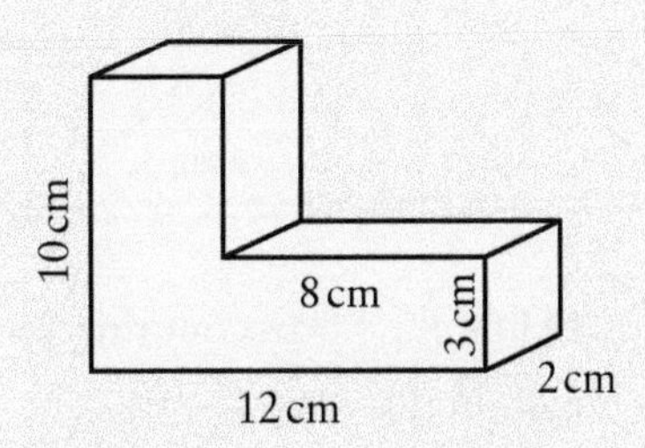

15 Calculate the volume of this 3D shape:

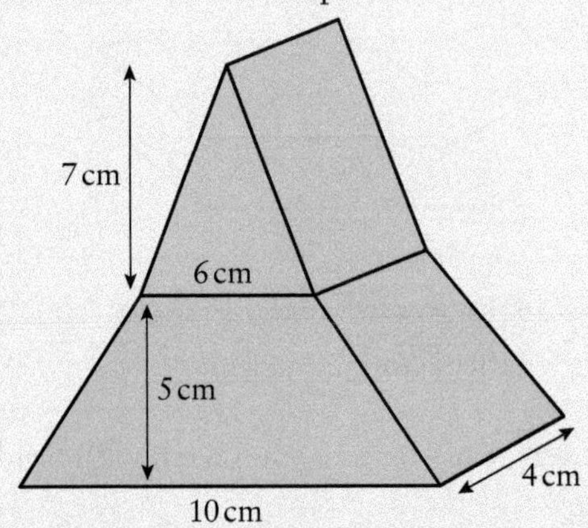

Challenge Q16

16 A 3D-printer makes this shape from plastic, 0.3 cm thick.

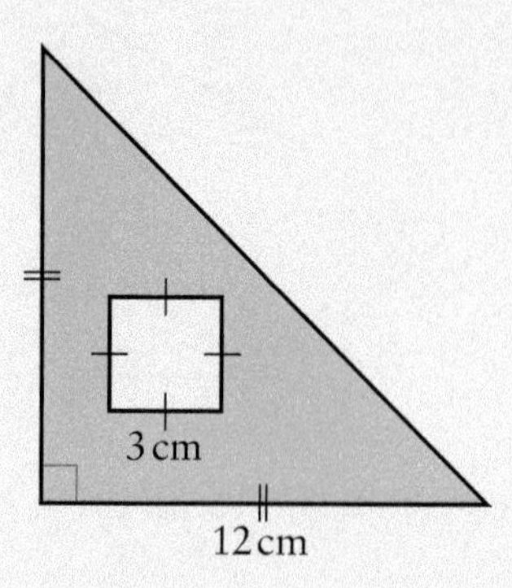

The shape is a right-angled isosceles triangle with a square-shaped hole in it.

Work out the volume of plastic in the finished shape.

Challenge Q17

17 A chocolate box is in the shape of a pentagonal prism.

The top of the box is a pentagon of area 20 cm^2. The height of the box is 4 cm.

The chocolates are in the shape of isosceles triangles:

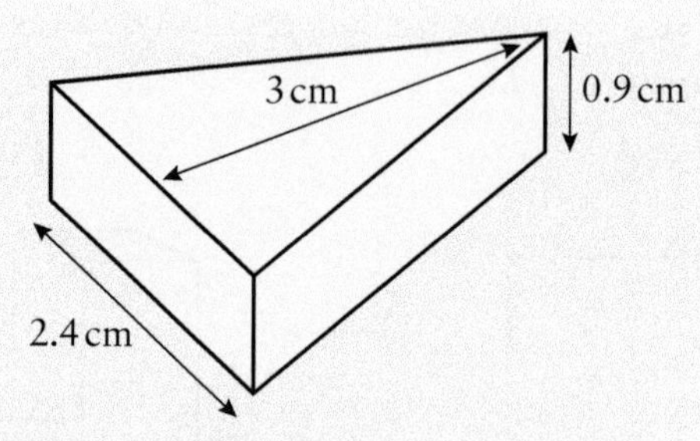

The chocolate box contains 15 of these chocolates.

Calculate the volume of empty space in the box.

Challenge Q18

18 This cuboid of modelling clay is shaped into a triangular prism.

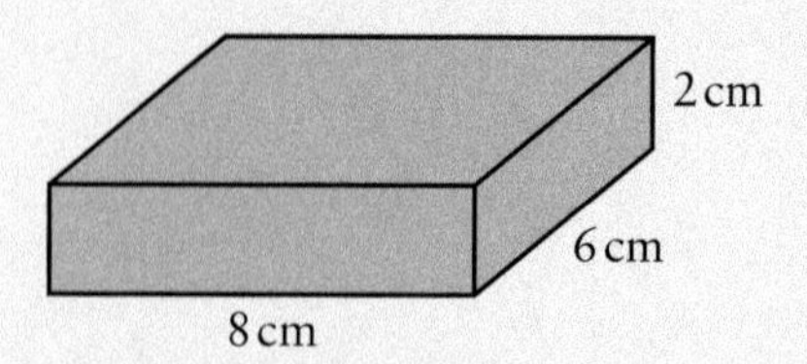

The base of the prism is a rectangle, 16 cm by 2 cm.

Find the height of the triangular prism.

Challenge Q19

19 How many different prisms can you sketch with height 8 cm and volume 120 cm^3?

9.3 Surface area

The surface area of a 3D shape is the total area of all its faces.

A net is a 2D shape that folds up to make a 3D shape.

Here is a cuboid and one of its possible nets.

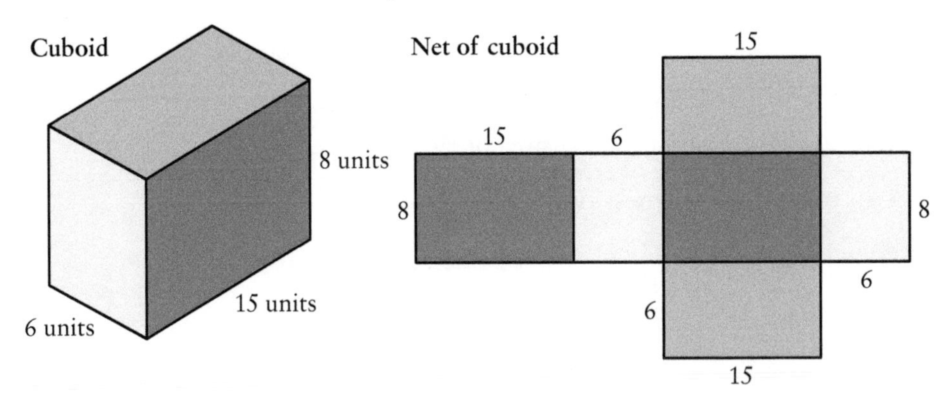

This is a triangular prism and one possible net.

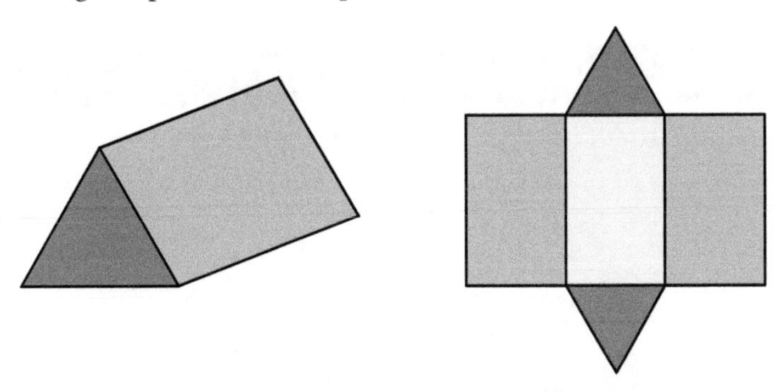

Investigation 9.1

How many possible nets of a cube are there?

Are all arrangements of six squares nets for a cube?

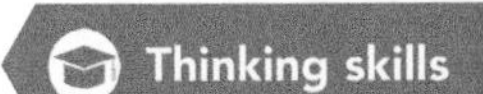

You can use a net to help you calculate the surface area of a 3D shape.

Explore 9.5

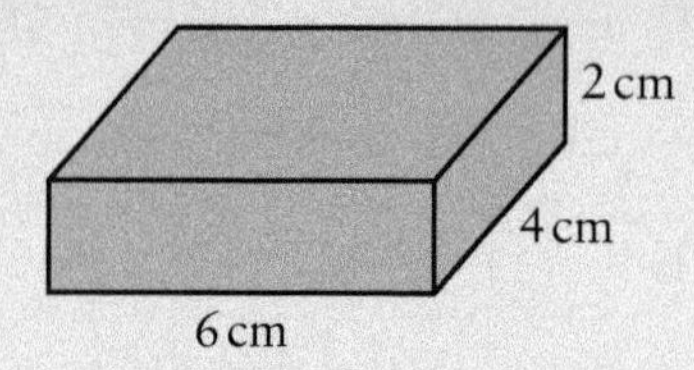

Can you sketch a net of this cuboid?

Can you calculate the surface area of this cuboid?

Can you write an expression for the surface area of this cuboid?

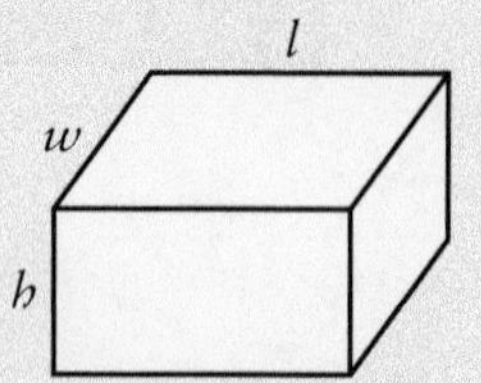

Worked example 9.4

Calculate the surface area of this triangular prism.

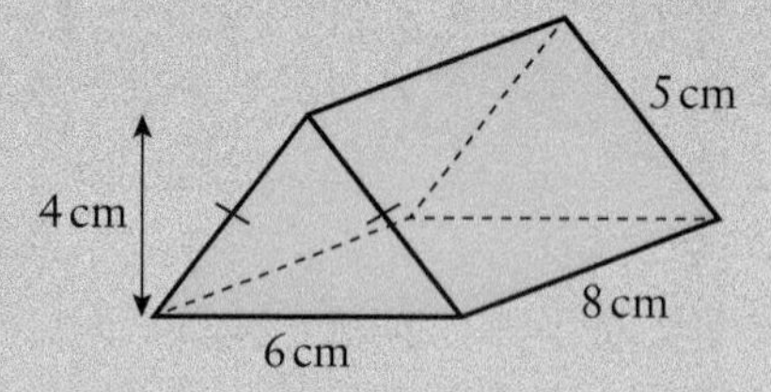

Solution

Understand the problem

The surface area is the total area of all five faces of the prism.

Make a plan

Sketch a net and label the measurements.

Calculate the area of each of the five faces, then add these areas together.

Carry out the plan

Sketch the net:

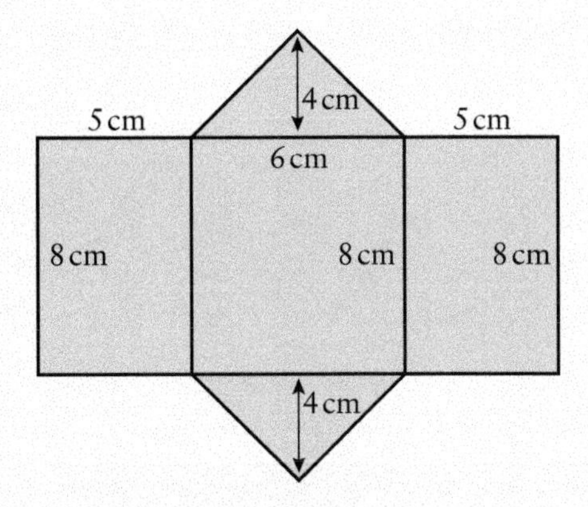

area of triangle $= \frac{1}{2}bh = \frac{1}{2} \times 6 \times 4 = 12\,\text{cm}^2$

Work out the area of each different sized face:

area of 5×8 rectangle $= 40\,\text{cm}^2$

area of 6×8 rectangle $= 48\,\text{cm}^2$

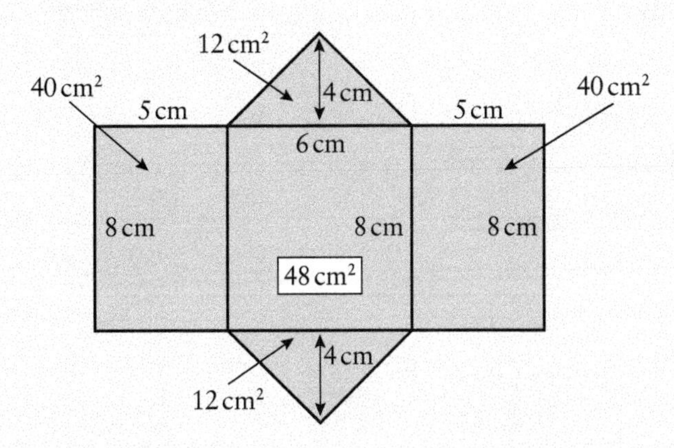

Add together the areas of all the faces:

total area $= 12 + 12 + 40 + 40 + 48 = 152\,\text{cm}^2$.

Reflect

What would be the same and what would be different in your method for calculating the surface area of this prism compared with the prism in the example?

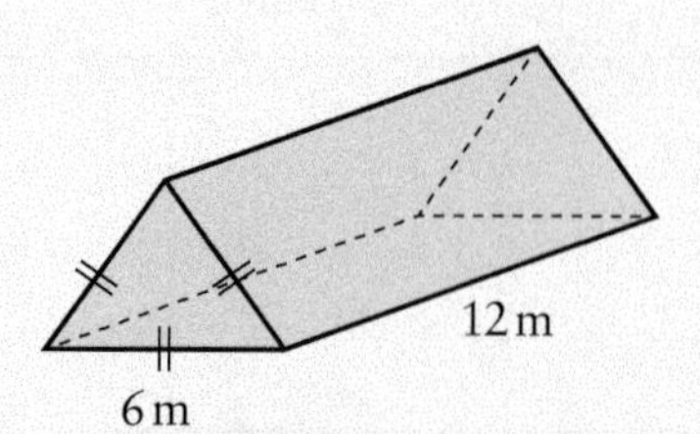

Practice questions 9.3

1 Sketch the 3D shape for each net.

a
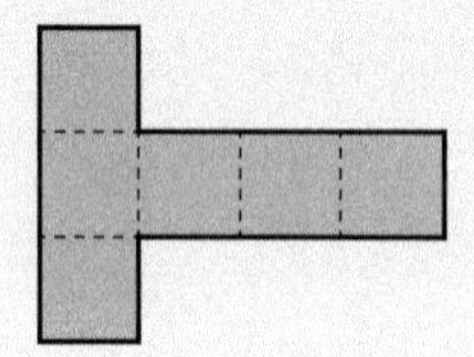

b
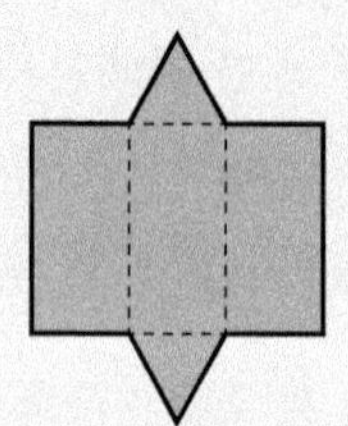

c
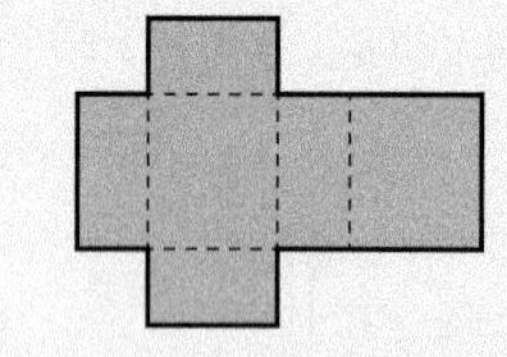

d
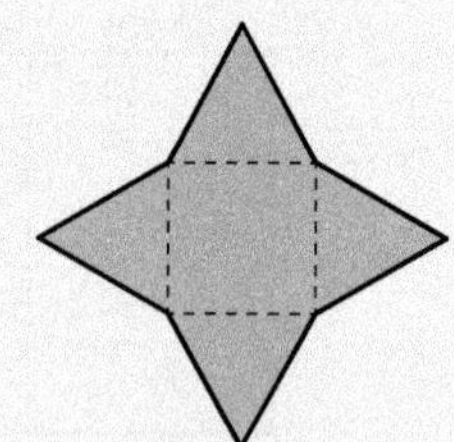

e
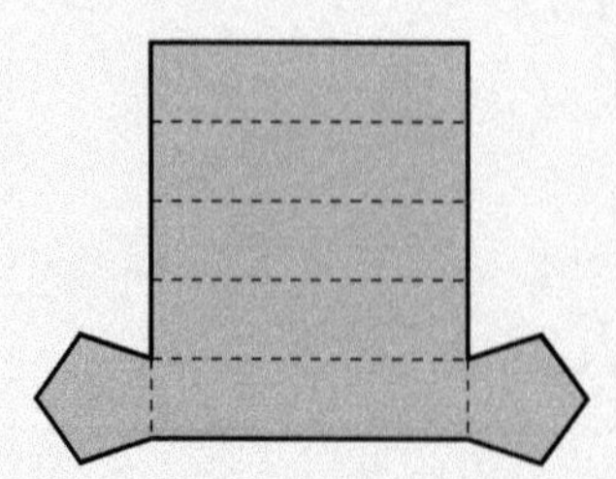

f
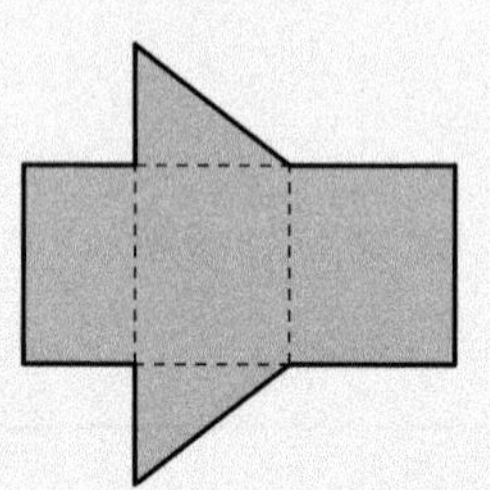

2 Sketch a net of each shape. Label the measurements you know.

a
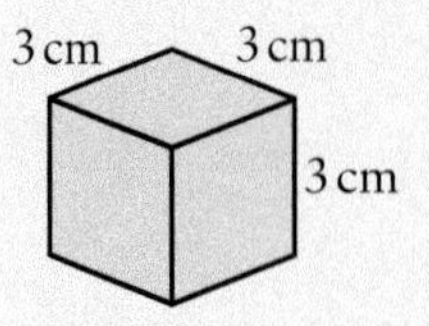

b
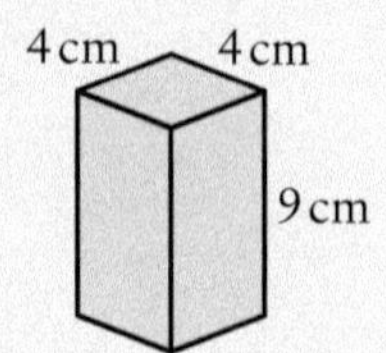

c
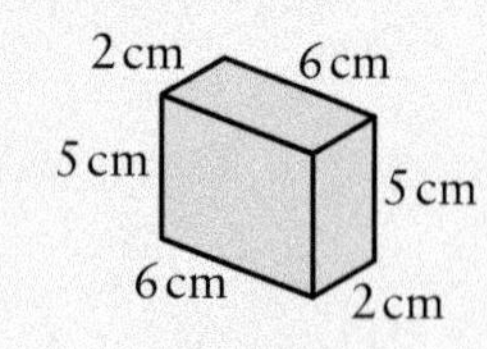

3 An open cuboid has no top face.
This open cuboid-shaped box has height 15 cm, width 30 cm and length 40 cm.

a Draw a net for this open box.

b Calculate the area of card used to make this box.

4 Calculate the surface area of each cuboid.

a
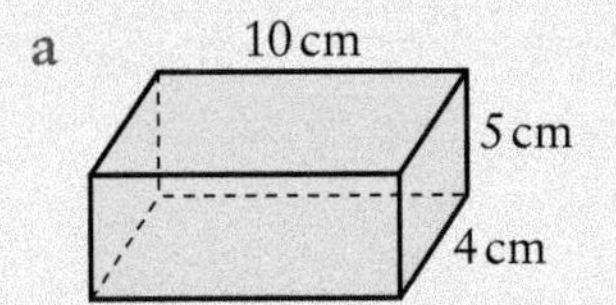

b
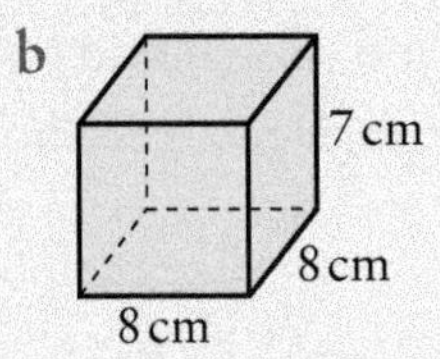

c
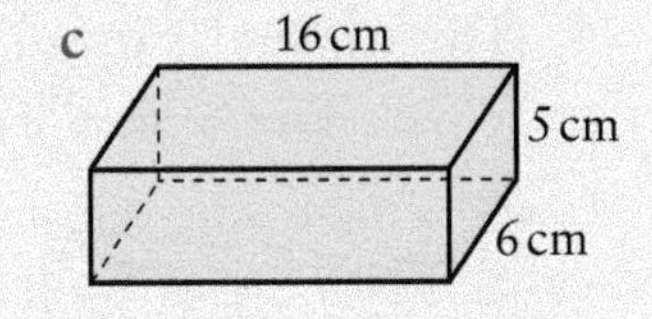

5 Calculate the surface area of this cuboid.

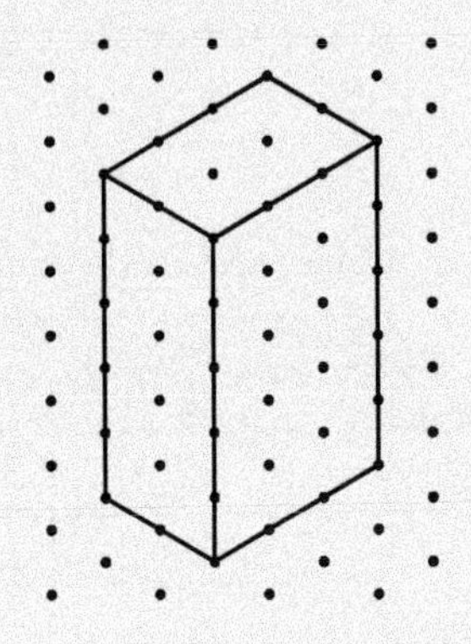

6 Calculate the surface area of each prism.

a
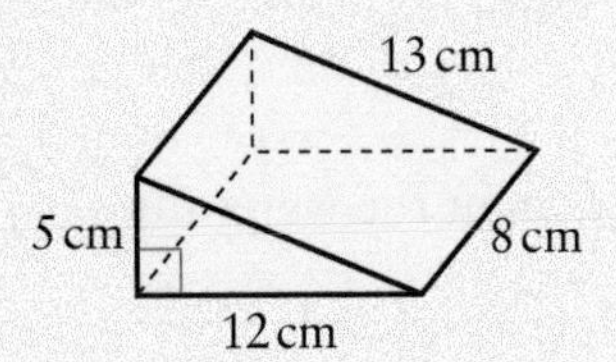

b
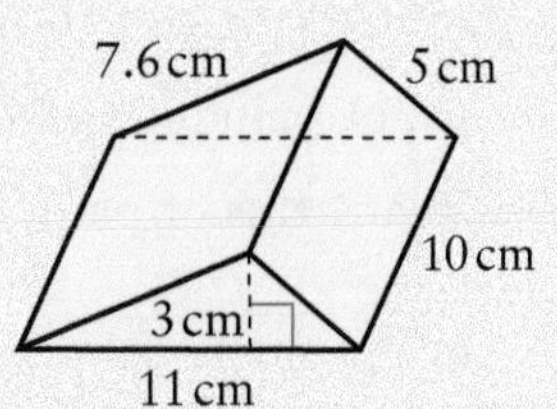

c
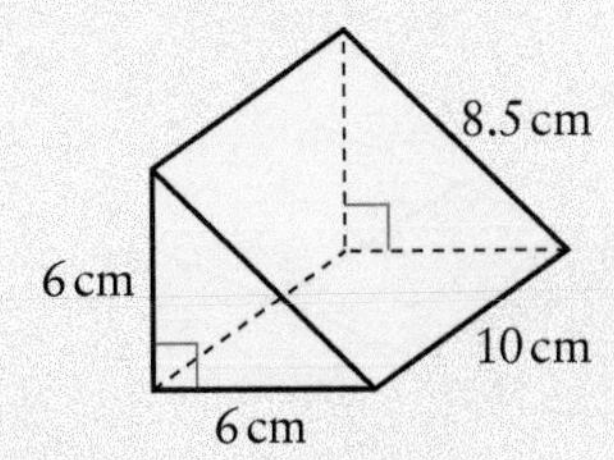

d
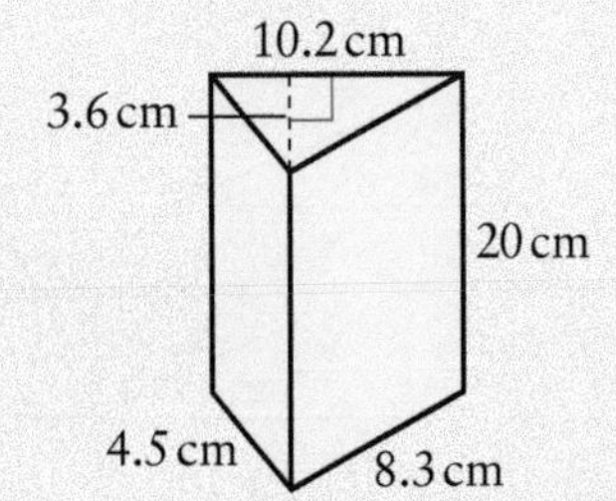

7 This wedge-shaped cushion has the measurements shown.
Calculate the area of fabric needed to make a cover for this cushion.

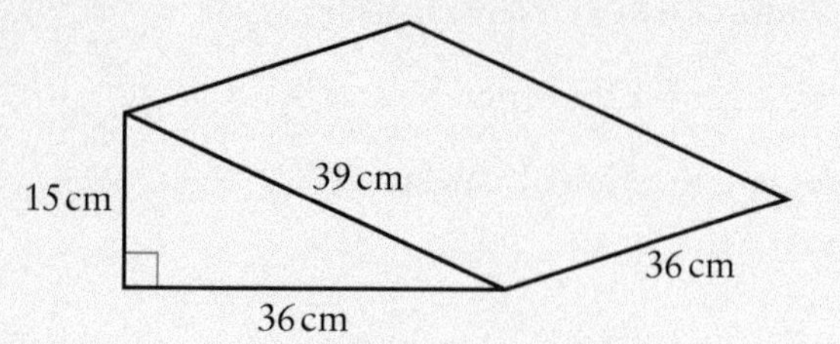

8 This triangular prism and this cuboid have the same volume.
Which has the smallest surface area? Show working to explain.

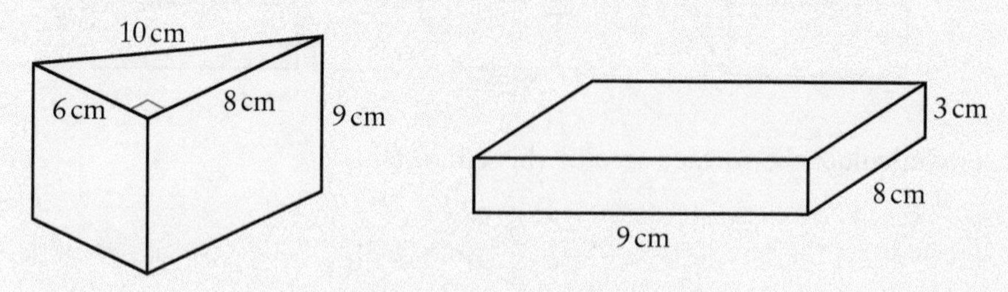

9 The diagram shows a water tank in the shape of an open cuboid.

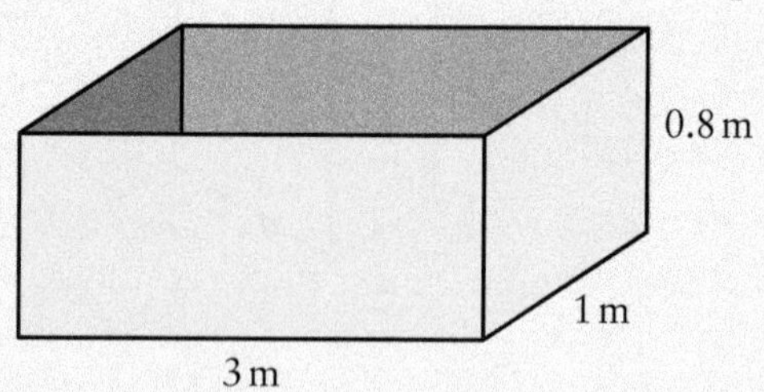

Each outside face of the tank is to be painted.

One tin of paint covers 4 m^2.

How many tins of paint are needed to paint the water tank?

10 For the 3D shape below, calculate:

a the volume

b the surface area.

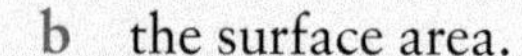

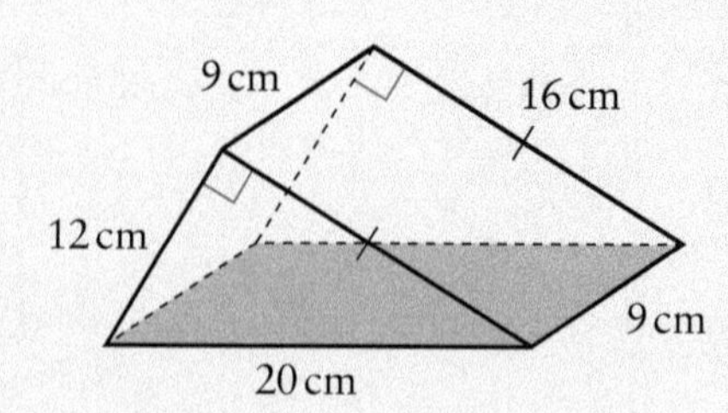

11 Write a formula for the surface area of a cube of side l.

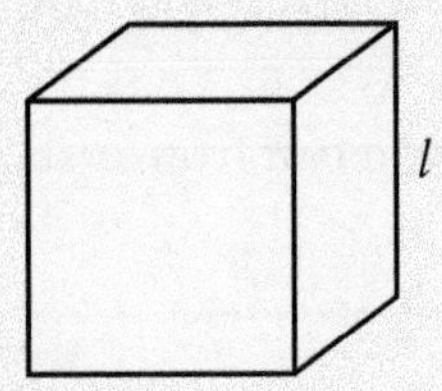

12 Five 2 cm cubes are placed in a row to make a cuboid shape:

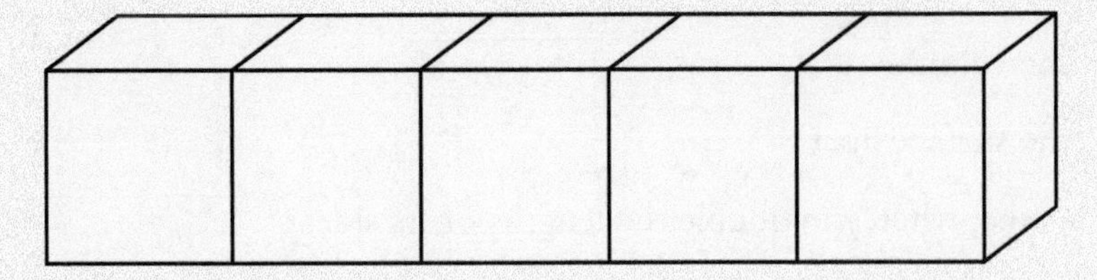

Work out:

a the overall volume

b the surface area.

13 A sculptor makes the model shown in the picture on the right for a larger bronze sculpture.

The sculptor wants to calculate the volume of bronze needed for the sculpture. What measurements and information does the sculptor need in order to calculate the volume of bronze?

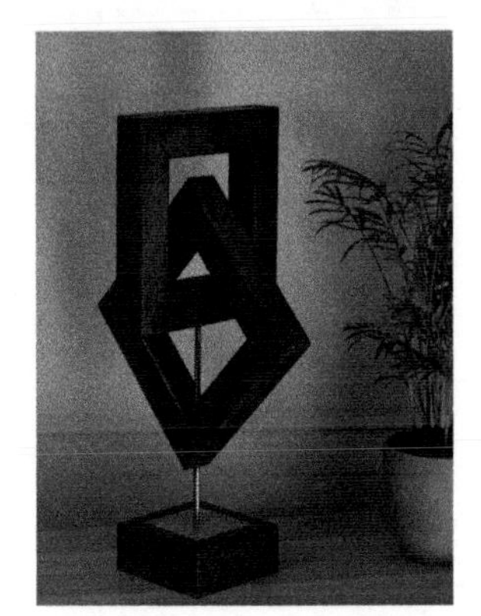

14 A cube has volume 343 cm³. Calculate its surface area.

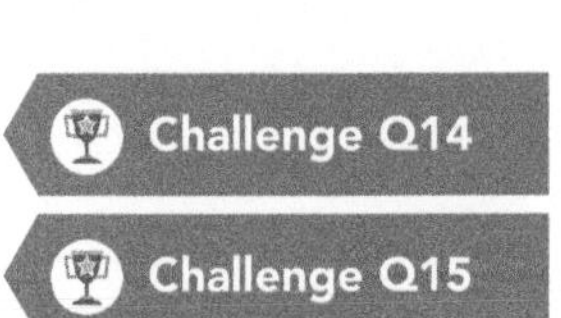

15 a Make an accurate drawing of the net of this prism on squared paper.

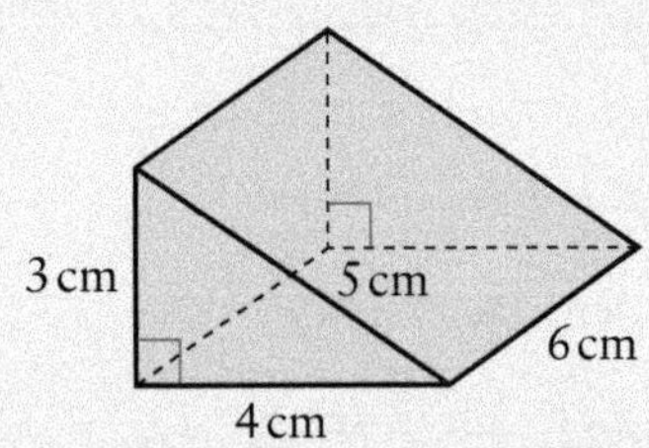

b Calculate the volume and surface area of the prism.

 Challenge Q16

16 Design a spreadsheet that calculates the volume and surface area of a cuboid when you input the length, width and height.

 Challenge Q17

17 Make the largest cube you can from a sheet of A4 paper, and calculate its surface area.

Investigation 9.2

 Self-management skills

What is the effect on:

1 the volume

2 the surface area

of a cube, when you double the lengths of its sides?

What is the effect of halving the lengths?

What is the effect of multiplying the lengths by n?

9.4 Units of area, volume and capacity

The volume of a 3D shape is the amount of space inside the shape.

The **capacity** of a 3D shape is the amount of liquid it can hold.

Volume of a liquid and capacity are usually measured in millilitres or litres.

You can convert between millilitres (ml), litres and cubic centimetres (cm^3)

$1\,\text{cm}^3 = 1\,\text{ml}$

$1000\,\text{cm}^3 = 1000\,\text{ml} = 1$ litre

Explore 9.6

 Hint

$1\,\text{cm} = 10\,\text{mm}$

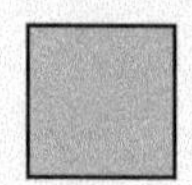

Work out:

a how many mm^2 there are in $1\,\text{cm}^2$

b how many cm^2 there are in $1\,\text{m}^2$

c how many mm^3 there are in $1\,\text{cm}^3$

d how many cm^3 there are in $1\,\text{m}^3$.

Worked example 9.5

A fish tank is in the shape of a cuboid with base 30 cm by 45 cm and height 35 cm.

The water level in the tank is 8 cm below the top of the tank.

Ignoring any fish and accessories, calculate the amount of water in the tank. Give your answer in litres.

Solution

Sketch the tank and label the measurements. Draw in the water level.

Calculate the volume of the 'cuboid' of water in cm^3.

Then convert to litres.

Sketch the tank and hence work out the depth of water in the tank:

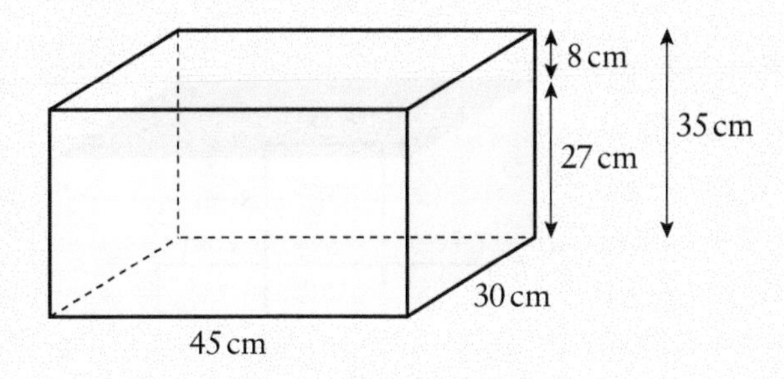

Calculate the volume of water in the tank:

volume of water $= 45 \times 30 \times 27 = 36\,450$ cm^3

Convert to ml and litres:

$36\,450$ cm^3 $= 36\,450$ ml $= 36.45$ litres

Reflect

How could you work out the capacity of the fish tank in litres?

Worked example 9.6

Tilly makes a cuboid with base 6 cm by 5 cm from cardboard.

They use 258 cm^2 of cardboard to make the cuboid.

Calculate the volume of the cuboid.

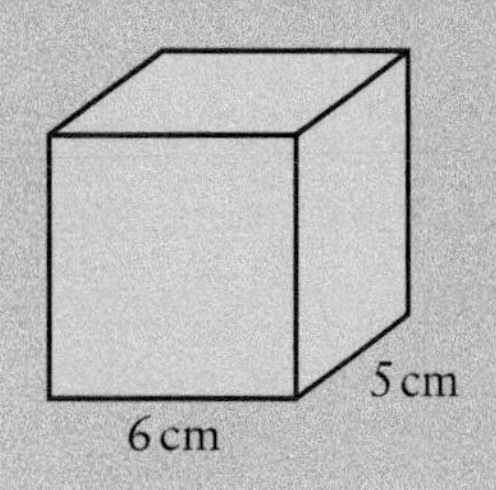

Solution

Label the height of the cuboid x.

Write and solve an equation to find x.

Sketch the cuboid:

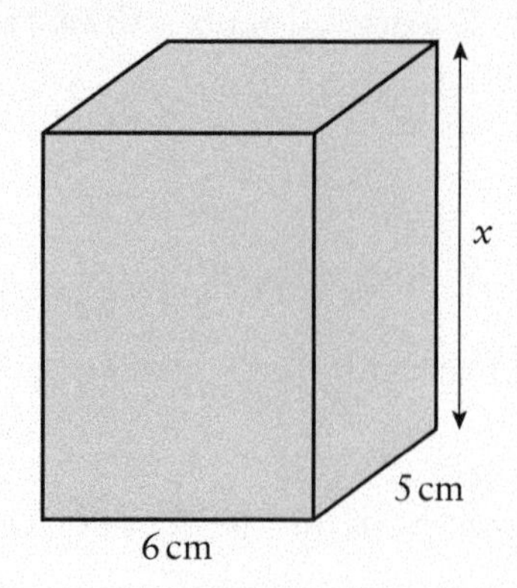

Write an expression for the area of each face of the cuboid:

6 cm
30
5 cm
6 cm 5 cm 5 cm
x $6x$ $5x$ $6x$ $5x$ x
30

$6x + 5x + 6x + 5x + 30 + 30$

Collect like terms:

$22x + 60$

Write an equation for the surface area of the cuboid:

$22x + 60 = 258$

Solve for x:

$22x = 198$

$x = 9\text{cm}$

Calculate the volume of the cuboid:

volume $= 5 \times 6 \times 9 = 270\text{cm}^3$

Reflect

Does it matter which letter you use to represent the missing length?

Do you need to sketch a net of the cuboid to help you write the equation for the surface area?

Practice questions 9.4

1 Convert each area measurement to the units shown:

a $15\,cm^2 =$ ______ mm^2

b $380\,mm^2 =$ ______ cm^2

c $3\,m^2 =$ ______ cm^2

d $45\,000\,cm^2 =$ ______ m^2

e $5\,m^2 =$ ______ mm^2

f $9000\,mm^2 =$ ______ cm^2

2 Convert each volume measurement to the units shown:

a $7\,cm^3 =$ ______ mm^3

b $23\,cm^3 =$ ______ mm^3

c $6\,m^3 =$ ______ cm^3

d $850\,000\,cm^3 =$ ______ m^3

e $1.4\,m^3 =$ ______ cm^3

f $2700\,mm^3 =$ ______ cm^3

3 Convert each measurement to the units shown:

a $12\,cm^3 =$ ______ ml

b $5600\,mm^3 =$ ______ ml

c $600\,cm^3 =$ ______ litres

d 45 litres = ______ cm^3

e $1\,m^3 =$ ______ litres

f 5470 litres = ______ m^3

4 Calculate the capacity of this cuboid shaped juice carton, in millilitres.

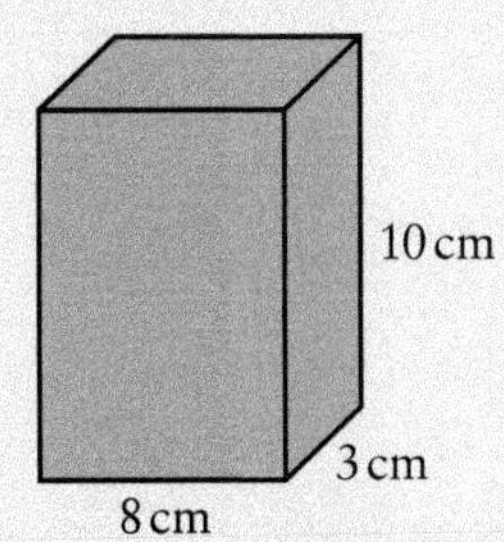

5 The diagram shows the dimensions of a tent.

Work out the area of fabric needed to make this tent, including the floor.

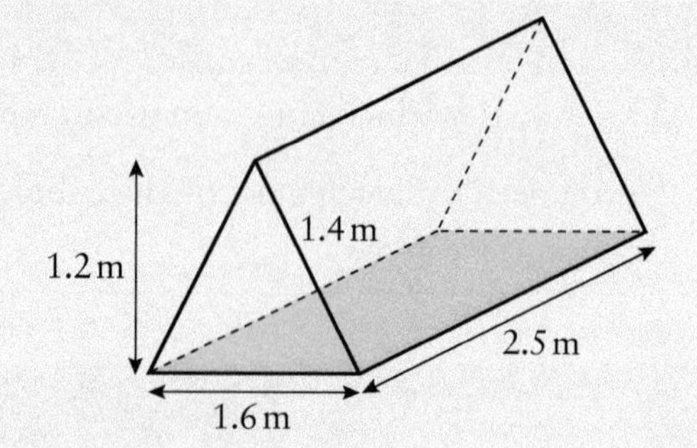

6 Jem builds this raised bed for planting vegetables.

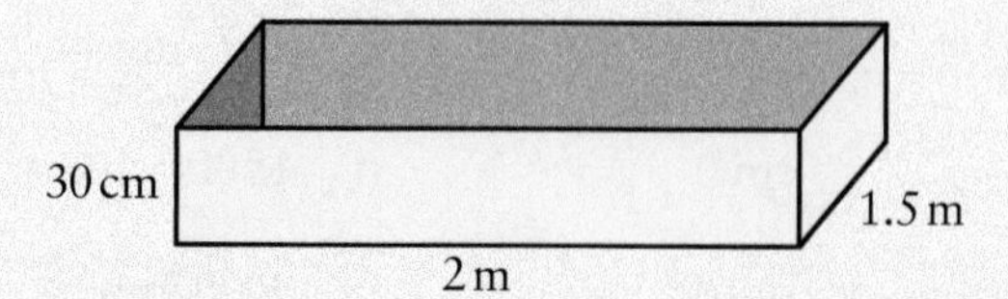

a Work out the volume of soil that Jem needs to fill this raised bed.

Soil is sold in 50-litre sacks.

b How many sacks of soil are needed to fill the raised bed?

7 The diagram shows the plans for a shed. All measurements are in metres.

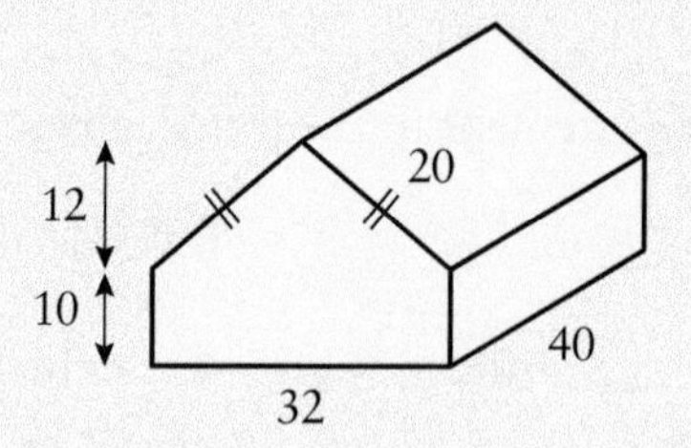

The sides and roof of the shed are made from corrugated iron.

Calculate the area of corrugated iron needed to make the shed.

8 The diagram shows the measurements of a swimming pool.

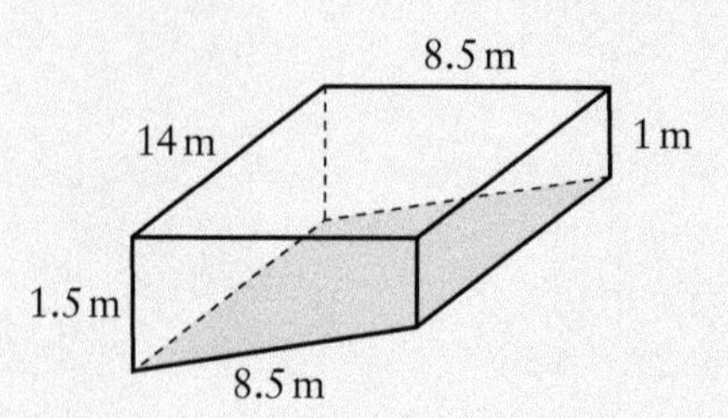

a The walls and the base of the swimming pool are to be tiled. Work out the area of tiles needed, to the nearest square metre.

b How many litres of water will it take to fill the swimming pool?

9 An ingot of copper is a trapezoidal prism as shown:

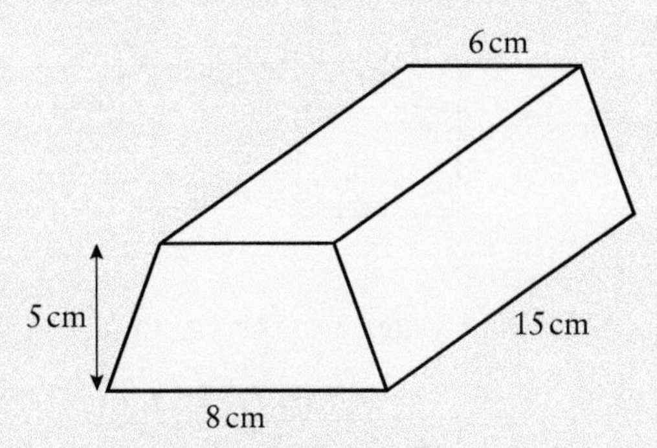

The ingot of copper is melted down to make cubes of side 8 mm.

a How many of these cubes can be made from the ingot?

b How many ml of liquid copper is used to make each cube?

10 The volume of this prism is $150\,\text{cm}^3$. Find its surface area.

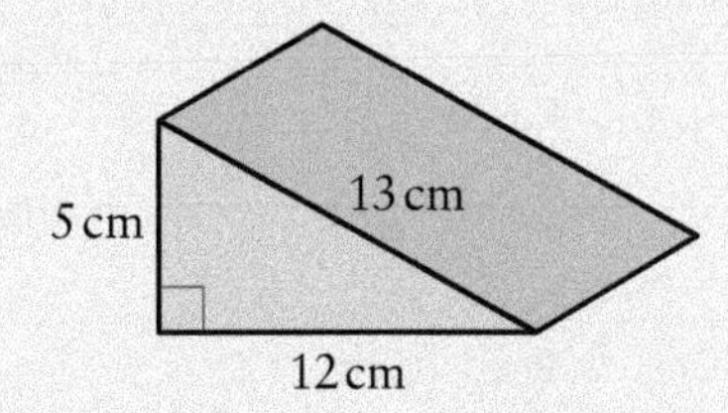

11 A drinking trough for animals has a trapezium cross-section and is 1.3 m long.

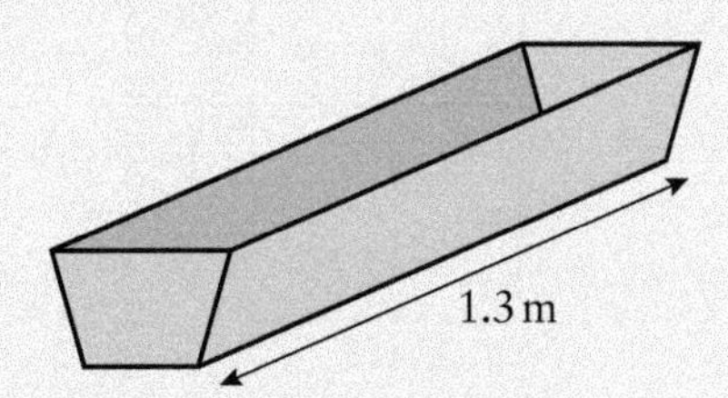

Here are the dimensions of the cross-section:

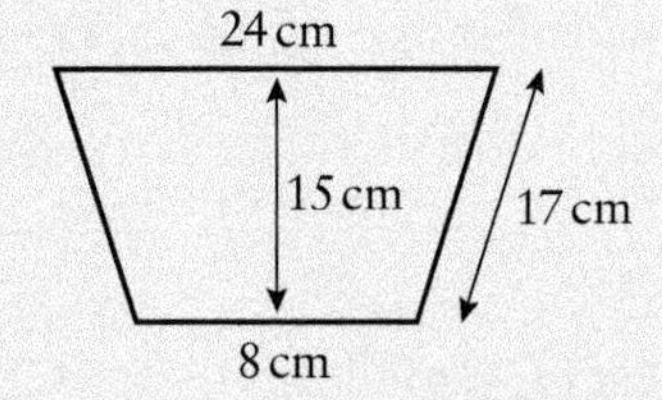

The drinking trough is made from thin steel sheet.

Calculate:

a the area of steel sheet needed to make the trough

b the capacity of the trough, in litres.

Fact

An ingot is a block of metal, usually with trapezium cross-section.

12 The surface area of this cuboid is $232\,cm^2$. Calculate its volume.

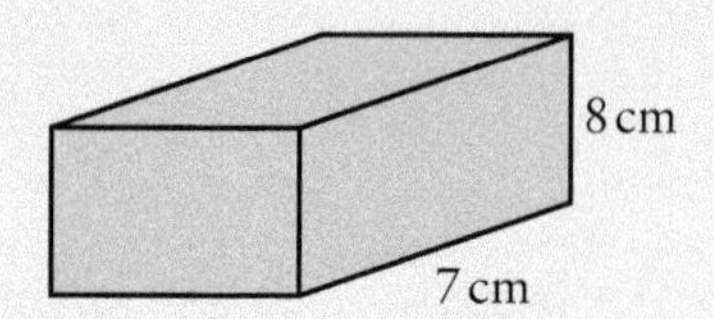

13 Lucy pours 12 litres of water into this container:

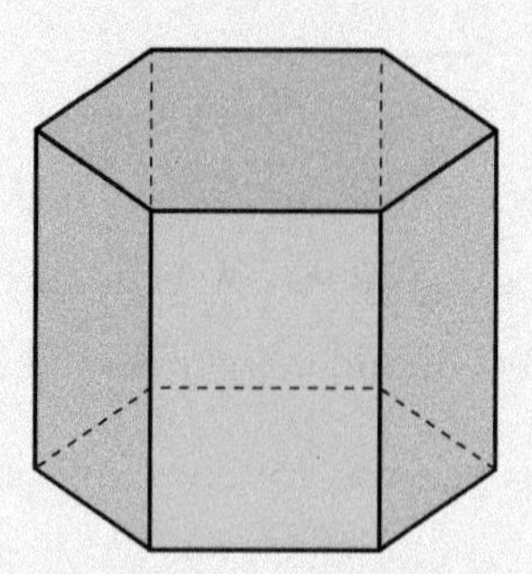

The base of the tank is a hexagon of area $260\,cm^2$.

How deep is the water in the tank, to the nearest millimetre?

Challenge Q14

14 This carton holds 1 litre of milk.

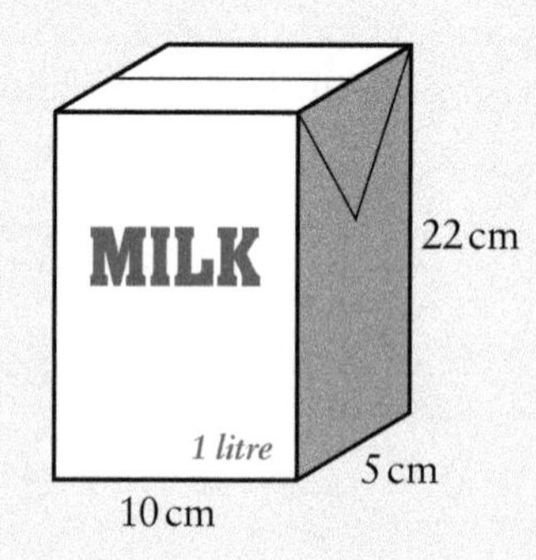

a Calculate the depth of milk in the carton.

The carton is turned over so it stands on the shaded face.

b What is the depth of milk in the carton in this position?

Challenge Q15

15 Water flows into this swimming pool at a rate of 0.8 litres per second.

How long will it take to fill the pool?

Give your answer in hours and minutes.

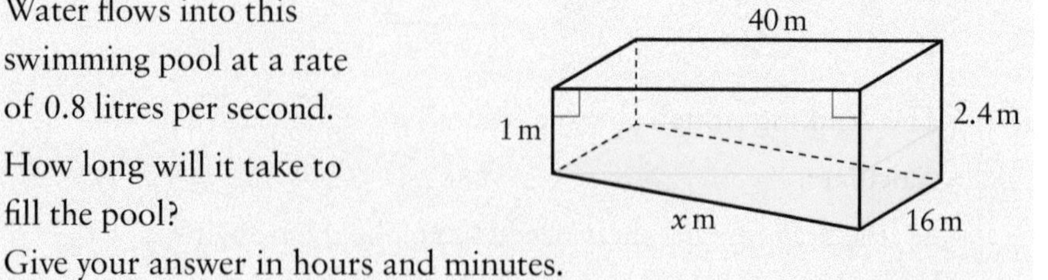

16 On a building site, the contractors have to remove 50 cm of top soil before they start building houses.

The building site is a rectangle with area 4 hectares.

1 hectare is the area of a square 100 m by 100 m.

Calculate the volume of top soil the contractors have to remove, in cubic metres.

17 Sketch and label the dimensions of two cuboids with capacity 500 ml. Which of the two cuboids has the smaller surface area?

Challenge Q17

Self-assessment

- I can find the formula for the area of a triangle.
- I can find the formula for the area of a trapezium.
- I can calculate the area of a trapezium.
- I can work out an unknown length on a trapezium when I know its area and the other lengths.
- I can calculate the area of rhombuses and kites.
- I can describe a 3D shape by describing its faces, edges and vertices.
- I can identify a 3D shape from a description of its faces.
- I can find the relationship between the numbers of faces, edges and vertices of a 3D shape with flat surfaces.
- I can name prisms and other 3D shapes.
- I can draw prisms on isometric paper.
- I can interpret diagrams drawn on isometric paper.
- I can calculate the volume of a prism.
- I can calculate the volume of a composite shape made from prisms.
- I can work out a missing length on a prism when I know its volume and the other lengths.
- I can sketch a net of a 3D shape.
- I can identify nets of a cube.
- I can find the formula for surface area of a cuboid.
- I can calculate the surface area of prisms.
- I can convert between metric units of volume.
- I can convert between cm^3, m^3 and litres.
- I can calculate the capacity of a 3D object in ml or litres.
- I can work out a missing length on a prism when I know its capacity and the other length.
- I can calculate the volume of a prism, given its surface area and two dimensions.
- I can calculate the surface area of a prism, given its volume and two dimensions.

? Check your knowledge questions

1 Calculate the area of each trapezium.

a

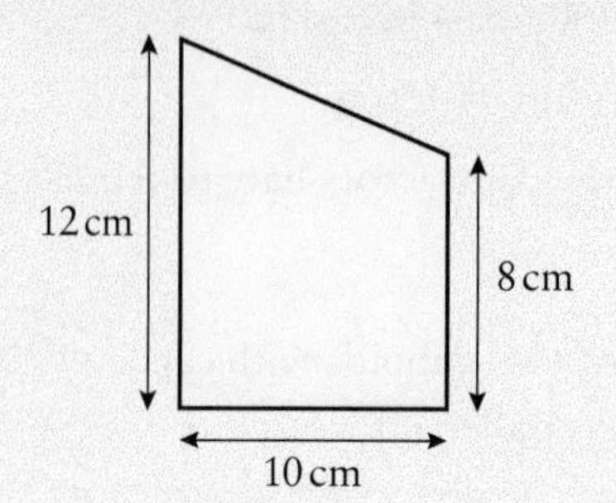

b

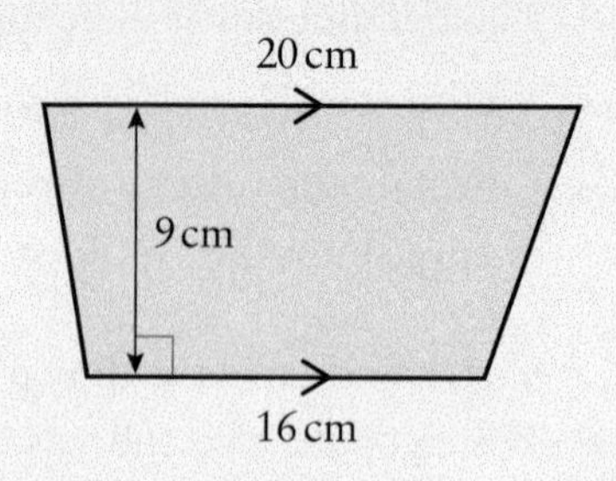

2 This trapezium has area $50\,cm^2$. Calculate the length of its base.

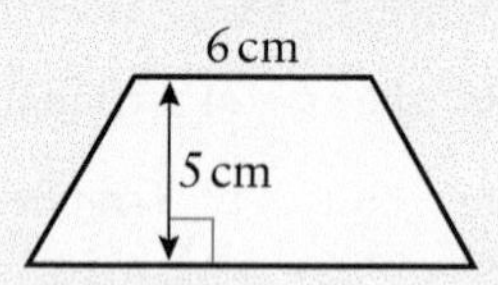

3 Calculate the area of each shape:

a

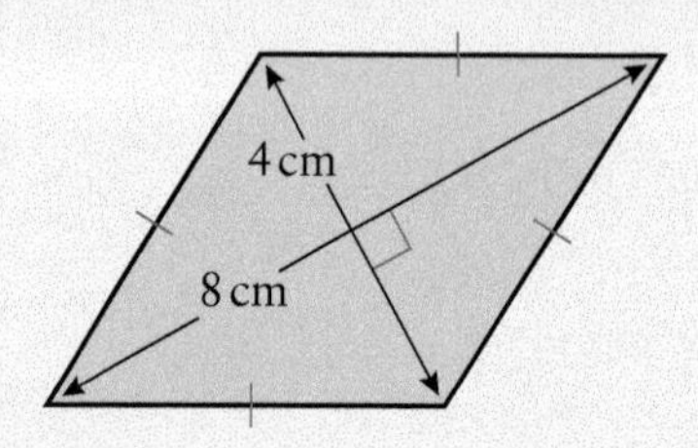

b

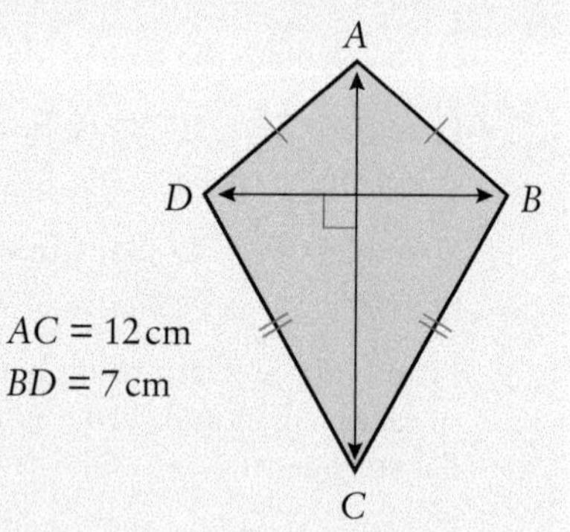

4 Work out the area of this shape:

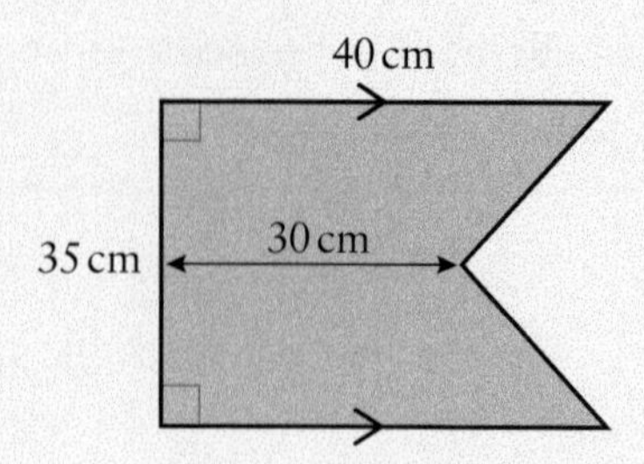

5 A prism has eight faces. Six of the faces are congruent rectangles. What shape are the other two faces?

6 a Draw this 3D shape on isometric paper.

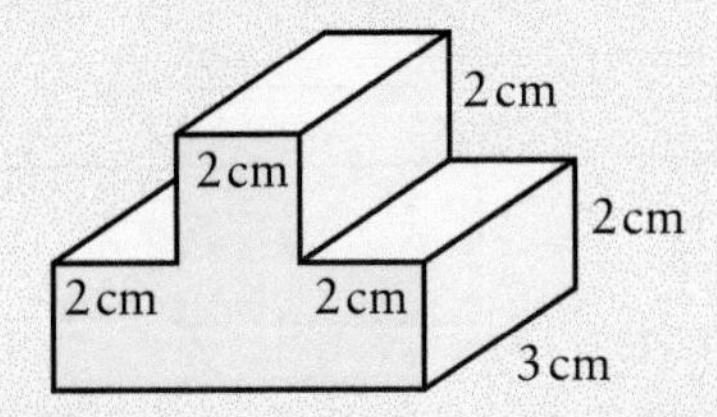

b Work out the volume of the 3D shape.

7 Calculate the volume of each 3D shape.

a

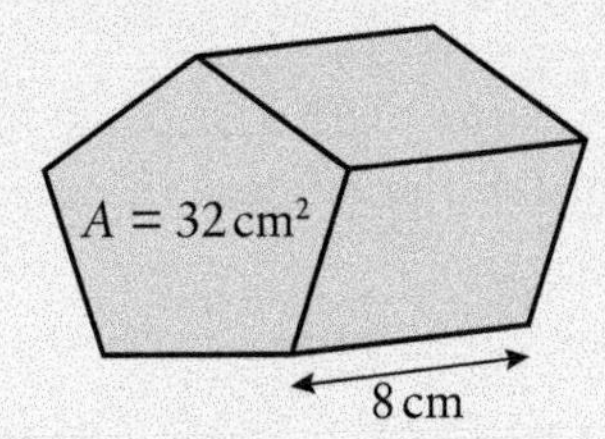

b

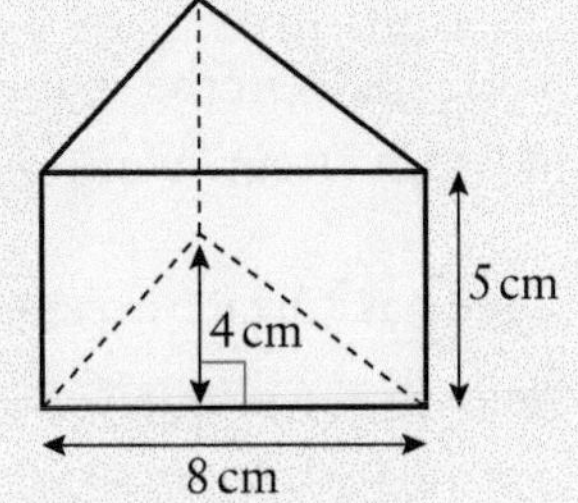

c

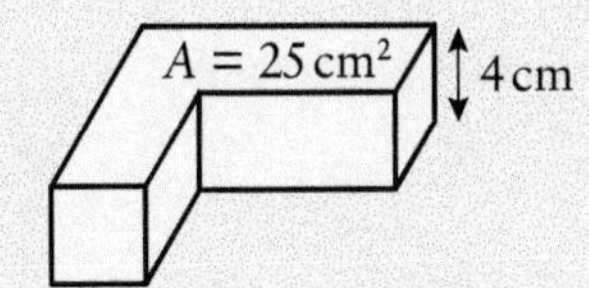

8 This triangular prism has volume $36\,\text{cm}^3$. Work out its height.

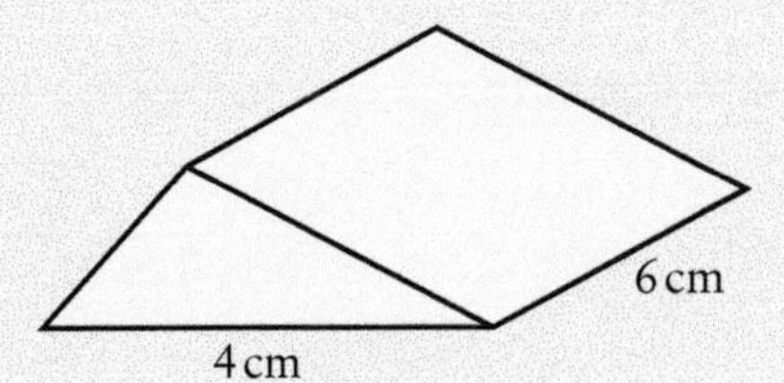

9 Work out the surface area and volume of this cuboid.

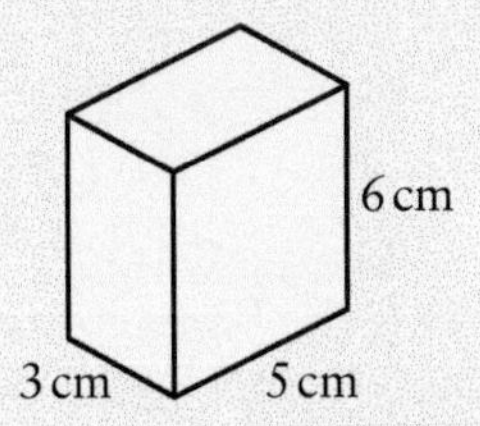

10 Calculate the volume and surface area of this cuboid.

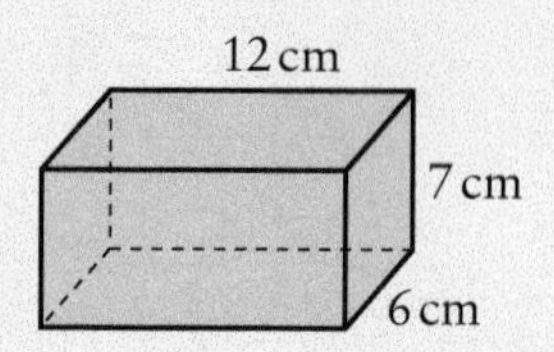

11 Convert these measures into the units shown.

a 5 litres = ____ cm^3

b 54 mm^2 = ____ cm^2

c 24 m^2 = ____ cm^2

d 370 cm^3 = ____ litres

12 An oil tank is in the shape of a cuboid with length 1.2m, height 1 m and width 70 cm. Work out the capacity of the oil tank in litres.

13 A cube has surface area 486 cm^2.

Calculate its volume.

14 A cuboid shaped fish tank contains 32 litres of water.

The depth of the water is $\frac{2}{3}$ of the height of the fish tank.

The base of the tank has area 2400 cm^2.

Find the height of the fish tank.

10 Statistics

10 Statistics

KEY CONCEPT

Relationships

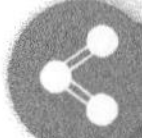

RELATED CONCEPTS

Models, Representation

GLOBAL CONTEXT

Fairness and development

Statement of inquiry

Representing relationships with a fair model helps to improve decision making.

Factual

- How do we measure central tendency?
- How do we measure dispersion?

Conceptual

- Why do we have more than one measure of average?

Debatable

- When assessing student performance, should we look at the mean of a student's scores or the median of the scores?

Do you recall?

1 You wish to find the most preferred colour among the students in your school, so you collect data from the students in your mathematics class. Have you conducted a census of the population or a survey of a sample of the population? Is the data you collect qualitative or quantitative?

2 Make a bar graph and pie chart for the following data:

Grade, x	Frequency, f
3	2
4	5
5	10
6	8
7	3

3 Which scatter graph suggests that the higher the temperature, the higher the ice cream sales? What does the other graph suggest?

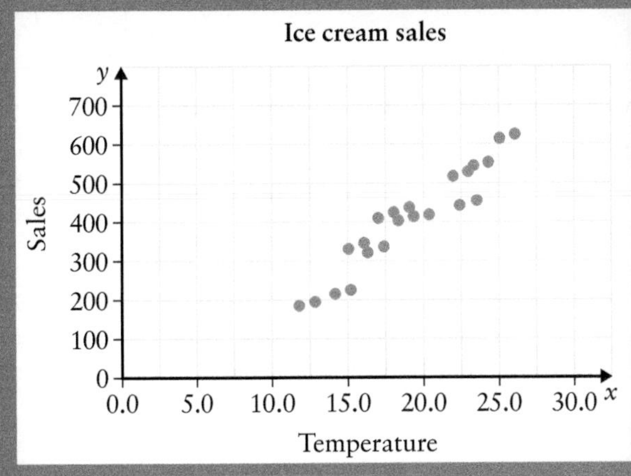

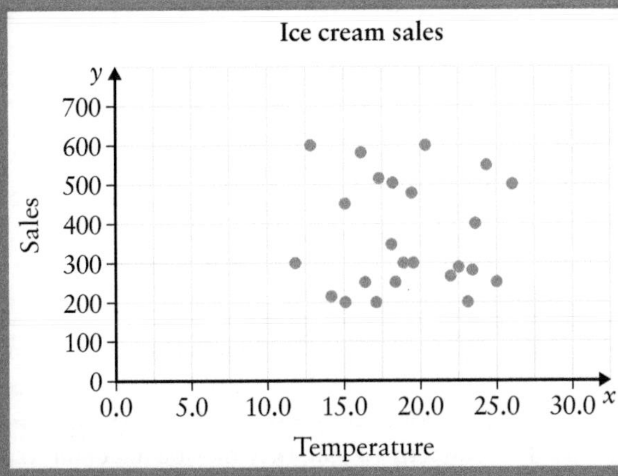

It is not a surprise to say that 'the most important commodity in the world today is information'. Much of the information we are faced with is numerical. There are not many of us who will be able to go through life without having to look at some numerical data and then draw conclusions from the data.

10.1 Collecting data

10.1.1 Surveys and questionnaires

In MYP Year 1, you created questionnaires and surveys to collect data for analysis.

Explore 10.1

Collect data on students in your class.

Include: age, height, handspan, favourite colour.

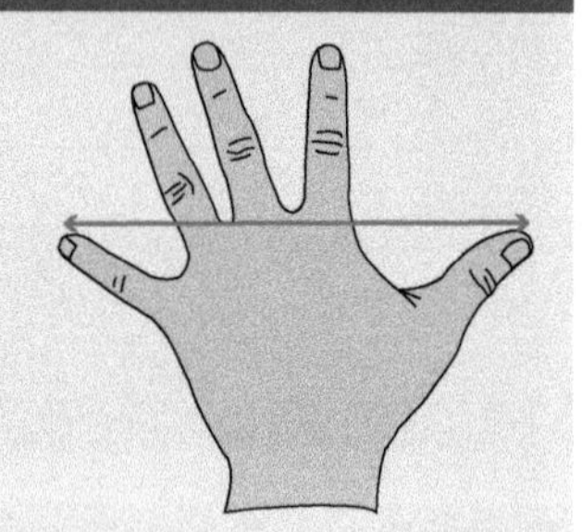

How did you collect the data? Compare with other students in your group. Which ways of collecting and recording the data do you think are best?

You can collect data by carrying out a survey. A survey usually asks every person for the same information. You can ask each person the survey questions and record the answers, or you can give each person a questionnaire.

Name ____________

Age: years ☐ months ☐

Do you have a pet? Yes ☐ No ☐

Distance travelled to school:

less than 5 km ☐

5–10 km ☐

more than 10 km ☐

Amount of pocket money each week ________

Favourite film ________

A name can help ensure you do not count multiple responses from the same person, but will people feel comfortable giving you the information you want if it is not anonymous?

Boxes to fill in make sure that everyone gives the same amount of information.

Tick boxes make it easy to answer a question.

Make sure the answer options do not 'overlap' but cover all possibilities.

Try to avoid asking personal questions like this. Even if you are not asking for names, if respondents are simply handing their surveys directly to you, they may be reluctant to give honest answers to such questions.

If you give a questionnaire to a large group of people asking 'What is your favourite film?', each person could choose a different film, so you would get a lot of different answers.This is called an open question. If you want to know what type of film to show at the end of term, you could change it to a closed question, and ask instead:

Which is your favourite type of film?

Comedy ☐ Romance ☐ Action ☐ Science fiction ☐ Other ______

This narrows the choice and will show you the most popular film types, but it could also lead to people not thinking of other genres.

You can design and use a data collection sheet to record the data you collect.

One way of recording people's survey responses is in a tally chart:

Number of brothers	Tally	Frequency
0	IIII	4
1	~~IIII~~ ~~IIII~~ I	11
2	~~IIII~~ II	7
3	II	2
4		0
5		0
More than 5	I	1

Tally marks are grouped in fives, so they are easy to count.

This has the highest frequency. It is the most common answer, or the mode.

Include all possible answers in this column. 'More than 5' saves you having to write rows for 6, 7, 8, 9, 10, etc.

When you have collected all the data, write the totals in the Frequency column.

The **mode** is the data item that appears most often. In a frequency table, the mode is the data item with the highest frequency.

Here is a grouped frequency table. The data is grouped into **classes**.

Number of days	Frequency
0–5	2
6–10	5
11–15	8
16–20	6

This class has the highest frequency. It is the most common answer, or the modal class.

In a grouped frequency table, the **modal class** is the class with the highest frequency.

When you write your own survey questions or design a data collection sheet, try it out with a few people before you do the whole survey. This will help you to identify and correct any problems with the questions and answer options, or with how you collect or record the answers.

Worked example 10.1

Noah writes this question for a survey:

How often do you go to the cinema each month?

Never ☐ 1–3 times ☐ 3–5 times ☐ 5–8 times ☐

a Explain two things that are wrong with this question.

b Write a better version of this question.

Solution

a The answer options overlap.
3 times is included in 1–3 times ☐ 3–5 times ☐.

The same is true for 5 times. Which box would you tick for '3 times', or '5 times'?

The answer options do not cover all possibilities. If you go more than 8 times you cannot record your answer.

b Correct the errors you described in part **a**.

How often do you go to the cinema each month?

Never ☐ 1–3 times ☐ 4–5 times ☐ 6–8 times ☐

more than 8 times ☐

Worked example 10.2

The frequency table shows the times, to the nearest second, that students took to complete a puzzle.

Time (seconds)	Frequency
41–45	2
46–50	5
51–55	8
56–60	6
61–65	3
66–70	1

a How many students:

i took between 45.5 and 50.5 seconds to complete the puzzle

ii took more than 60 seconds to complete the puzzle

iii completed the puzzle?

b Another student took 45.3 seconds to complete the puzzle. In which group would you record this student's time?

c What is the modal class?

Solution

Time (seconds)	Frequency
41–45	2
46–50	5
51–55	8
56–60	6
61–65	3
66–70	1

Five students took 46–50 seconds.

The highest frequency is 8
This class has the highest frequency.

3 + 1 students took more than 60 seconds.

a i Five students.

ii Four students took more than 60 seconds.

iii 2 + 5 + 8 + 6 + 3 + 1 = 25 students completed the puzzle.

Add the frequencies to find the total frequency: the total number of students.

b 41–45 seconds.

45.3 seconds rounds to 45 seconds to the nearest second.

c 51–55 seconds.

Reflect

How could you use the answer to parts **a ii** and **iii** to work out the number of students that took a minute or less to complete the puzzle?

What type of graph can be used to represent the information?

10.1.2 Sampling methods

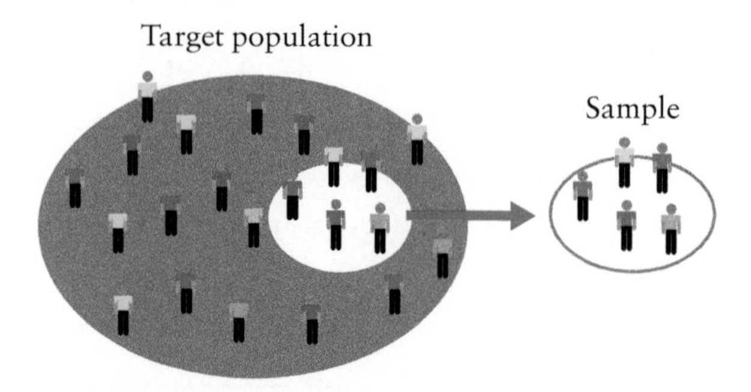

Fact

A sample refers to a smaller, more manageable version of a larger group. It is a subset containing the characteristics of a larger population. Samples are used when population sizes are too large to include all possible members or observations. A sample should represent the population as a whole and not reflect any bias toward a specific attribute.

Using samples allows us to conduct our studies more easily and in a timely fashion.

To achieve an unbiased sample, the selection should be random so everyone from the population has an equal and likely chance of being added to the sample group.

If you want to work out the average height of everyone in your class, you can do so easily. But what if you want to work out the average height of all MYP Year 2 students in the world during the 2020–2021 school year? Why would this be a problem? What would you do?

Recall from MYP Year 1 that, when you gather information from the entire **population**, you are conducting a **census**, and when you gather information from a **sample** of the population only, you are conducting a **survey**.

If you want to know the average height of a student in your mathematics class, then you can work out the population average by conducting a census of your class.

To find the average height of all MYP Year 2 students in the world, you could collect height data from a sample of students by means of a survey and then calculate an estimate of the world average by working out the average height of this sample of MYP Year 2 students. The sample average can be used as an estimate of the population average.

Explore 10.2

Go back to collecting data about all MYP Year 2 students' heights in the world. Can you find/estimate the height of a typical MYP Year 2 student? If yes, suggest how this might be done. If not, give reasons for your response.

Sampling considerations

You are an environmentalist, and you want to estimate the number of fish in a lake. Do you think you can catch all the fish in the lake in order to count them? You are a government official and you would like to know how the attitudes of people in your country changed over the last five years concerning the 35-hour work week. Will you take a census?

These two cases are examples where collecting information from the whole population is not possible or practical and that is why you will need to take samples.

In both cases, the sample you choose should be **representative** of the whole population. In the first case you may choose a spot in the lake at **random** and you make a catch there. Just to be sure it is representative; you may choose more than one spot and you may repeat on other days.

Hint

Random means that any spot in the lake has a chance of being chosen. Samples should closely resemble the population. All the participants in the sample should share the same characteristics and qualities. So, if the study is about male college first-year students, the sample should be a small percentage of males that fit this description. Similarly, a study on the sleep patterns of single women over 50, the sample should only include women within this demographic.

In addition to the above, an important consideration when choosing a sample is **size**. The larger the sample, the greater the proportion of the population you are using in your calculations, which should lead to a more accurate estimate. Mathematical tools are used to determine the minimum sample size for reliable estimates. The larger does not mean the better all the time though. Sampling can be compared to a cook tasting food – a teaspoon may be too small; a kitchen scoop is too large and does not add any extra taste. A balance is given by a tablespoon.

How will you ensure that individuals or values will be chosen randomly?

Technology can be used to generate a random sample. Your teacher may share with you some tools that you can use with available technology: graphic display calculators (GDC) or spreadsheets.

Explore 10.3

Can you identify the potential bias in each of the following approaches to collecting data and how to eliminate this?

a Trying to find out the average height of women around the world by randomly choosing 10 women from different countries.

Hint

A sample is biased if individuals or groups from the population are not represented in the sample.

b Trying to find out the most common way students travel to and from school by standing outside the school cafeteria after lunch and surveying the first 20 students that exit.

c Trying to determine the happiness of people in your region by conducting a survey outside the unemployment office.

d Trying to determine the views of people in your region on a political issue by conducting a survey:

i outside a local place of worship

ii on the campus of a university.

To be truly random when choosing subjects for a survey, you need to have a list of them in advance and can use technology to generate a random sample of them.

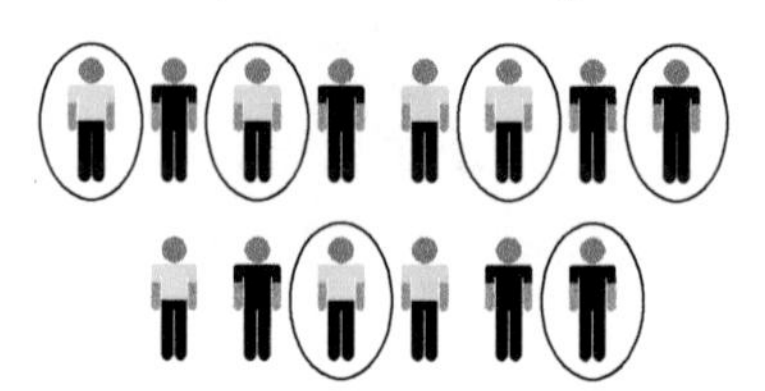

If the population in question is relatively small, then we use what is called **simple random sampling**. A simple random sample is a subset of a population in which each member of the subset has an equal chance of being selected. A simple random sample is meant to be an unbiased representation of a group.

An example of a simple random sample would be the names of 25 employees being chosen out of a hat from a company of 250 employees. In this case, the population is all 250 employees, and the sample is random because each employee has an equal chance of being chosen.

If the population is large and is made up of groups of similar subjects, called **strata**, then we use **stratified random sampling** is used when the population has several important and **distinct groups** that should be fairly or **proportionately** represented in the analysis.

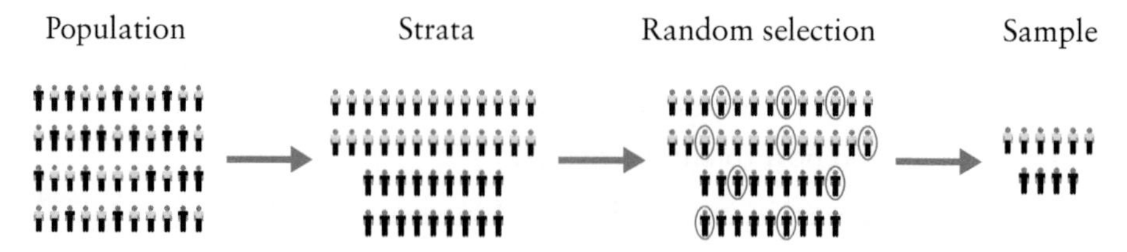

Stratified sampling

For example, a population of 20 000 has 30% with university education, 50% with only high school education and the rest did not finish their high school. If we want to survey a sample of 800 people, then we randomly select 30%, or $0.30 \times 800 = 240$ people from among those with university education, 50%, or $0.50 \times 800 = 400$ people from those with high school education and 160 from the rest of the population.

Another method is **systematic** sampling.

Systematic sampling is a type of random sampling method in which sample members from a population are selected according to a random starting point but with a fixed, periodic interval. This interval, called the sampling interval, is calculated by dividing the population size by the desired sample size.

Systematic sampling

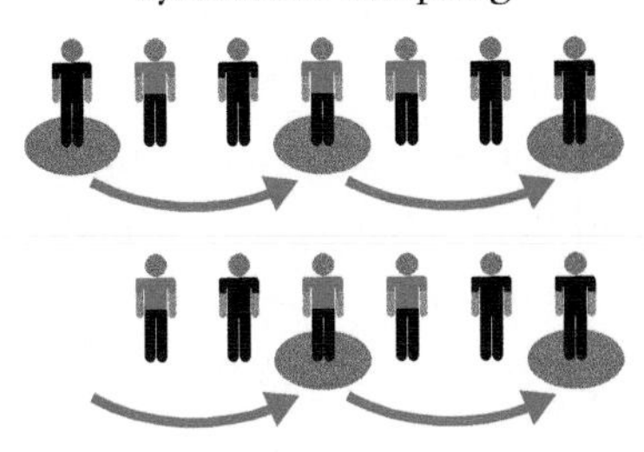

For example, if you wanted to select a random group of 1000 people from a population of 50 000 using systematic sampling, all the potential participants must be placed in a list and a starting point would be selected. Once the list is formed, every 50th person on the list (starting the count at the selected starting point) would be chosen as a participant, since $\frac{50\,000}{1000} = 50$

For example, if the selected starting point were 20, the 70th person on the list would be chosen followed by the 120th, and so on.

Explore 10.4

Which of the sampling methods would you use in each of the following situations?

a To find the average per person income of countries around the world.

b Your school needs to select a representative group of 16 teachers to take part in an upcoming convention. They want to do that by choosing members at random. The school has 30 male teachers and 50 female teachers.

c You need to collect data about your city's population preference for longer store opening hours.

Practice questions 10.1

1 The MYP class at Oakville School is conducting surveys to collect data.

a State whether the data collected in each survey is **qualitative** or **quantitative** and **discrete** or **continuous**.

i Number of pets per family

ii 100 metre sprint time

iii Number of basketball free throws made in 60 seconds

iv Favourite music genre.

b State whether the data in each of the following surveys could be collected from the entire group (population) or just a part of the group (sample).

If a sample is sufficient, which sampling technique would you use (simple random, systematic, stratified)? If stratified, identify suitable strata to survey.

i 100 metre sprint time of MYP Year 2 students in your school

ii Number of words correctly recalled by students in your school from a list of 20 words after looking at them for 60 seconds

iii Time to run a half marathon (13.1 miles) taken by an adult male or female

iv Preferred mobile device of secondary students in your school.

2 The number of students in a school in MYP Year 1 is 75, in MYP Year 2 is 60 and in MYP Year 3 is 87. If you want to survey the students using stratified sampling, how many students should you survey in each year group if your sample size is:

a 10 students b 30 students.

3 Write these numbers of tally marks in numerals.

a 𝍸𝍸𝍸 b 𝍸|||| c 𝍸|

4 Draw tally marks to represent:

a 5 b 8 c 21

5 Rob asked students at his school for the number of minutes it takes them to walk to school. He listed the results in the table below.

a Copy this tally chart and complete the frequency column.

Number of minutes	Tally	Frequency
15	III	
16	~~IIII~~ II	
17	~~IIII~~ ~~IIII~~ I	
18		
19	~~IIII~~ ~~IIII~~ IIII	
20	~~IIII~~ ~~IIII~~ ~~IIII~~ ~~IIII~~ III	

b What is the modal number of minutes?

Hint Q5b

The modal number of minutes is the most common number of minutes.

6 Here is a list of the fruit that a group of students chose for lunch:

orange, banana, apple, banana, orange, pear, pear, grapes, orange, orange, banana, grapes, banana, banana, grapes, grapes, pear, orange, apple, banana

a Draw a frequency table for this data with the following columns:

Type of fruit	Tally	Frequency

b Which fruit is the mode?

c Calculate the total frequency. How can you use this to check that you have included all the data items?

7 Here are the heights of 20 people, in metres:

1.45	1.63	1.72	1.51	1.63
1.67	1.54	1.75	1.68	1.60
1.73	1.69	1.42	1.77	1.64
1.49	1.46	1.58	1.62	1.72

a Copy and complete this frequency table for the data.

Height (m)	Frequency
1.41–1.50	
1.51–1.60	

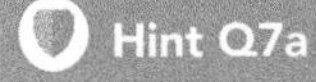
Hint Q7a

Follow the pattern for the height classes, so that each class is the same width.

b How many people are over 1.60 m tall?

c What is the modal class?

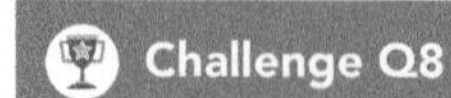

8 Ben designs this data collection sheet to record students' marks in a test. The test has a total of 50 marks.

Mark	Frequency
0–10	
10–20	
20–30	
30–50	

a Why can't you record a mark of 20?

b Draw an improved version of their data collection sheet.

10.2 Analysing data

10.2.1 Averages or measures of central tendency

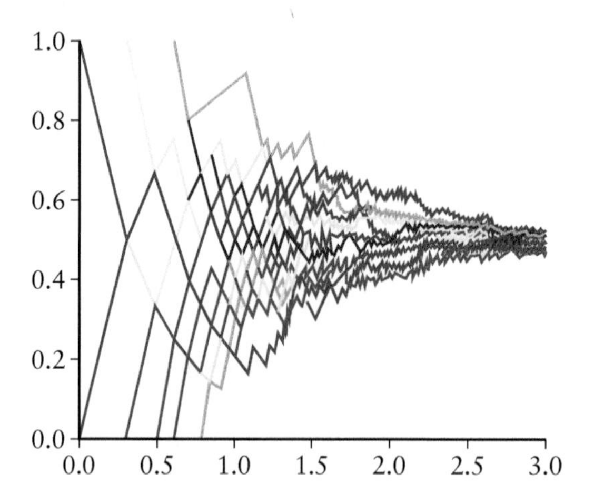

When we are given a data set, our first question must be: 'What does it tell me?'

If you want to get a quick, rough idea of how a group of students has performed on a task, how much it costs to live in a certain neighbourhood, or what the average salary paid by an IT company is, then looking at every student's score, the value of every house, or every employee's salary will be too time-consuming and, perhaps, not possible.

A measure of central tendency is a single value that attempts to describe a set of data by identifying the central position within that set of data. As such, measures of central tendency are sometimes called measures of central location.

The **mean** (average) of a data set is found by adding all numbers in the data set **and** then dividing by the number of values in the set. The **median** is the middle value when a data set is ordered from smallest to largest. The **mode** is the number that occurs most often in a data set.

Explore 10.5

Decide on one value which is useful for describing the whole group. Write a sentence or two justifying this choice (and possibly from whose perspective it is useful).

Test scores (out of 100) of a class:

87, 65, 72, 91, 83, 77, 89, 90, 74, 86, 79, 92, 89

Monthly salary (in dollars) of teachers in a department:

9600, 8900, 6200, 5600, 4900, 4200, 4100, 4100, 4000

Explore 10.6

In many restaurants, serving staff split the tips they receive evenly. Tip sharing encourages them to work together and support each other and leads to a better experience for the customers.

At the end of the evening, Aimee, Barbora, Cristian, Devon and Eunseo pool their tips. Individually, they received 49, 62, 55, 44 and 67 pounds, respectively.

Can you find out how much each should receive after sharing the tips?

In the previous two Explore activities, you probably mentioned some of the averages we will discuss here (which are not the only ones!).

The **mode** is the most frequent value.

The mode of the data set 15, 13, 12, 19, 12, 11, 12 is 12 since this number occurs more frequently than the others (three times, in this case).

The **median** is the middle value when the data is arranged in ascending (or descending) order.

Look at the data set 15, 13, 12, 19, 12

First, put it in order, from least to greatest:

12, 12, 13, 15, 19

median

The median is the number in the middle, 13

Fact

A set of numbers can have more than one mode. The data set 2, 2, 7, 10, 10, 12 has two numbers which appear more frequently than the others: 2 and 10. This set is **bimodal.**

Fact

When there is an even number of data values, there are two middle values. The median is halfway between them, such as here:

2, 2, 7, 10, 10, 12

median = 8.5

Hint

You can use your GDC or software in finding central values of data.

In many real-life situations, we talk about the **average**, such as average income, average height, average length, average speed, and so on. In fact, we mean the 'arithmetic average' better known as the **mean**.

It is straightforward to calculate: add up all the numbers, then divide by how many numbers there are. In other words, it is the sum divided by the count.

Look at the data set 15, 13, 12, 19, 12

The mean is $\dfrac{15 + 13 + 12 + 19 + 12}{5} = \dfrac{71}{5} = 14.2$

Reflect

Which of these three averages is best to use and when?

Worked example 10.3

Find the mode, median and mean of each of these data sets.

a 37, 42, 25, 42, 42, 32, 41, 51, 32

b −4, 7, 0, 2, −3, 4, 1, −2

Solution

To find the median, arrange the numbers in ascending order. This will also help you to see the mode(s) more easily. You can then sum the values and divide by the number of values to find the mean.

Hint

Here is an output of a GDC showing the answers

a

```
Math Rad Fix6   ab/c a+bi
Mean(list 1)
                    38.222222
Median(list 1)
                    41.000000
```

Putting the set in order gives 25, 32, 32, 37, 41, 42, 42, 42, 51

Both 32 and 42 occur more than once, but 42 occurs most frequently, so, the mode is 42

There are nine values, so the 5th value is the middle one. Hence, the median is 41. Notice that there are four numbers before 41 and four after.

The mean is $\dfrac{25 + 32 + 32 + 37 + 41 + 42 + 42 + 42 + 51}{9} = 38\dfrac{2}{9} \approx 38.2$

b

```
Mean(list 2)
                    0.625000
Median(list 2)
                    0.500000
```

Putting the set in order gives –4, –3, –2, 0, 1, 2, 4, 7

All values appear the same number of times, so there is no mode.

There are eight values, which can be split into two groups of four. So, the median is between the fourth and fifth values of 0 and 1

Hence, the median is $\frac{0+1}{2} = \frac{1}{2}$

The mean is $\frac{-4 + (-3) + (-2) + 0 + 1 + 2 + 4 + 7}{8} = \frac{5}{8} = 0.625$

Reflect

Which averages can be found when the data is qualitative?

Practice questions 10.2.1

1 Calculate the mode, median and mean of each set of numbers.

a 4, 6, 7, 10, 7, 3, 8, 2, 2, 7, 4

b 4, 6, 7, 10, 7, 3, 8, 2, 2, 7, 4, 4

c 4, 6, 7, 10, 7, 3, 8, 2, 2, 7, 4, 4, 4

d 4, 6, 7, 10, 7, 3, 8, 2, 2, 7, 4, 4, 4, 1

2 Consider these sets of numbers:

$A = \{12, 14, 17, 17, 20, 25, 25, 25, 26, 29\}$
$B = \{42, 44, 47, 47, 50, 55, 55, 55, 56, 59\}$
$C = \{-8, -6, -3, -3, 0, 5, 5, 5, 6, 9\}$
$D = \{24, 28, 34, 34, 40, 50, 50, 50, 52, 58\}$.

a Calculate the mode, median and mean of A.

b Calculate the mode, median and mean of B. Compare these with your results for set A.

c What do you think the mode, median and mean of C will be? Calculate them to check.

d What do you think the mode, median and mean of D will be? Calculate them to check.

3 The mean of a set of 20 numbers is 38. What is the sum of the numbers?

4 a The median of 83, 92, 101, x, 87 is 91. Find x.

b The mean of 83, 92, 101, x, 87 is 91. Find x.

c The median of 83, 92, 101, x, 87, 95 is 93. Find x.

d The mean of 83, 92, 101, x, 87, 95 is 93. Find x.

e The median of 83, 92, 101, x, 87, 95 is 93.5. Find x.

5 Write down a set of three numbers with:

a a mode of 7 and mean of 8

b a median of 1 and mean of 2 which has no mode.

6 Write down a set of four numbers with:

a a median of 9 and a mode of 10

b a median of 6 and which is bimodal.

10.2.2 Mean, median and mode from a frequency table

Frequency tables

Explore 10.7

The grades for a class of IBDP students at a large international school, on their final mathematics exams, are listed in the frequency table below:

Grade, x	Frequency, f
2	2
3	7
4	17
5	28
6	19
7	8
Total	**81**

a Can you find the mode?

b Can you find the median?

A frequency table is an efficient way of arranging your data, enabling you to easily spot patterns in the data.

Mean

As you may recall from the previous section, to find the mean you add up all the numbers, then divide by how many numbers there are.

Using the data from the Explore activity, since there are two students with 2, we add 2 two times and similarly we add 3 seven times, and 4 17 times, and so on.

Instead of adding all the values one by one, you can use multiplication to add repeated values. The frequency tells you how many students scored each grade, so if you multiply each grade (x) by its frequency (f) and then add these results (fx), you get the total of all the students' grades, 403

Grade, x	Frequency, f	fx
2	2	$2 \times 2 = 4$
3	7	$3 \times 7 = 21$
4	17	68
5	28	140
6	19	114
7	8	56
Total	**81**	**403**

Dividing this by the total number of students, 81, gives the mean 4.975 or approximately 5

Fact

The mean of a sample of x-values is represented by the symbol $\bar{x}$, which is called x bar, and is used on GDCs.

Your GDC, or other suitable software, can give you the results without the need for creating the fx column. You just enter your data in two columns: one for the score and one for the frequency. Here is the output of a GDC.

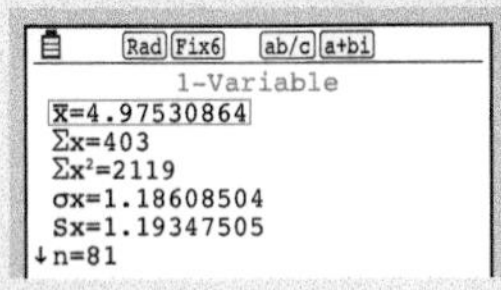

Rad Fix6 ab/c a+bi
1-Variable
Med=5
Q3=6
maxX=7
Mod=5
Mod:n=1
Mod:F=28

Worked example 10.4

The owner of a bookshop in a small town is interested in knowing the average number of visitors per day. She recorded the number of visitors over 20 days. The results are given in the table.

Work out the mode, median and mean of the number of visitors.

Visitors	Frequency
5	1
6	5
7	5
8	4
9	3
10	2
Total	**20**

Solution

The mode

Clearly, the highest frequency is 5, thus, the modes are 6 and 7 visitors per day. This is a bimodal distribution.

The median

To work out the median we can use the frequency column. There are 20 scores (an even number) so we want the average of the 10th and 11th observations. We can find out these two observations by adding the cumulative frequency (Cu. F) column to the table.

Visitors	Frequency	Cu. F
5	1	1
6	5	6
7	5	11
8	4	15
9	3	18
10	2	20
Total	**20**	

We progressively add down the frequency column until we reach the 10th and 11th scores.

We can see that there were six scores of 6 or less, and 11 scores of 7 or less. This means that both the 10th and 11th scores must have been 7

Of course the average of two 7s is 7

Therefore, the **median** is 7 visitors per day.

The mean

To work out the mean, we need to create the fx column as on the previous page, then we divide the sum of that column by the total number, 20

Visitors	Frequency	fx
5	1	1
6	5	30
7	5	35
8	4	32
9	3	27
10	2	20
Total	**20**	**145**

$$\text{mean} = \frac{\text{sum of } fx \text{ column}}{\text{total observations}}$$

$$= \frac{145}{20} = 7.25$$

This is approximately 7 visitors per day.

Hint

When you are not required to 'work out' the calculations, available technology will give you the same results. The important thing here is how you enter the data into your GDC or software – you need to indicate clearly which column is the frequency column. Here is a GDC output for the Worked example 10.4

```
Math Rad Fix6 ab/c a+bi
Mean(list 1,list 2)
                 7.450000
Median(list 1,list 2)
                 7.000000
Mod
     {6.000000,7.000000}
```

Grouped data

To draw a frequency distribution table where there are many different numerical outcomes, we usually group the outcomes into classes (intervals). Take for example the scores of 56 students in a spelling test listed in the following table:

Scores	Frequency
16–20	3
21–25	6
26–30	9
31–35	9
36–40	17
41–45	10
46–50	2

Fact

16–20 means all scores more than or equal to 16, but less than or equal to 20

If we try to work out the central values in a manner similar to the previous work, we are faced with a problem: When the class is 16–20 for example, what do we consider x to be?

A more convenient table could be constructed using class centres (also called mid-class values). Each class is represented by its median or mean. Here is an adjusted table:

Class	Class centre	Frequency	Cumulative frequency	fx
16–20	18	3	3	54
21–25	23	6	9	138
26–30	28	9	18	252
31–35	33	9	27	297
36–40	38	17	44	646
41–45	43	10	54	430
46–50	48	2	56	96
Totals			**56**	**1913**

Hint

Here is a GDC output for the test scores:

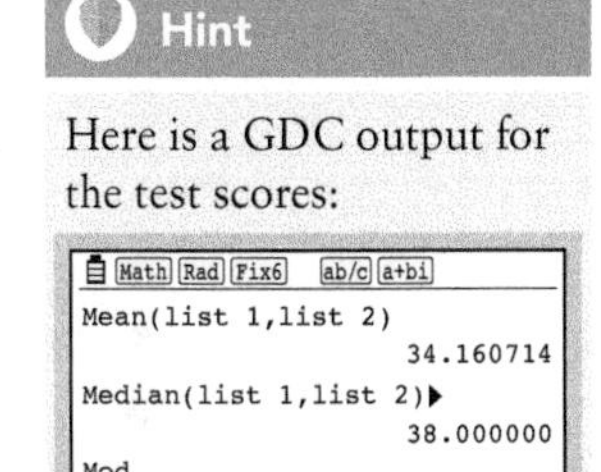

Note that the GDC lists the median and mode as 38 which is the class centre for class 36–40.

The rest of the work would be similar to the work in the previous example. One important difference is that our calculations now will give us approximate answers rather than exact ones. The class centre 18, of the 16-20 class assumes all numbers in that class to be 18 for example.

The mode according to the adjusted table is 38 since it has the highest frequency. We say that 36–40 is the **modal class.**

Since we have 56 observations, the median is between the 28th and 29th observations. Looking at the cumulative frequency column, the median must be in the class 36–40. We cannot locate the exact value of the median.

There are ways of estimating what the median is, but at the moment leave it for your GDC or other suitable software.

The mean is calculated in the same manner as before:

$$\text{mean} = \frac{\text{sum of } fx \text{ column}}{\text{total observations}} = \frac{1913}{56} = 34.16 \approx 34$$

Practice questions 10.2.2

1 Make a frequency table to match the following bar graph, and work out the mean, median and mode of the data presented.

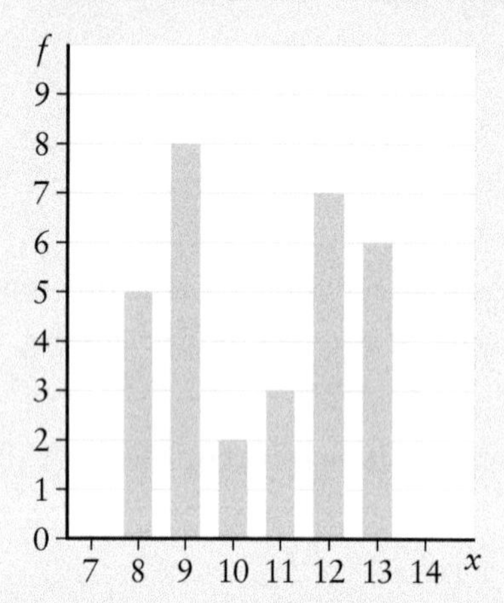

2 Complete the grouped frequency table.

Classes	Frequency, f	Mid-class values, x	fx
8–11	9	9.5	
12–15	8	13.5	
16–19	11	17.5	
20–23	7	21.5	
	Total =		Total =

a What is the modal class?

b In which group is the median value?

c Estimate the mean.

3 a Record the following data in a grouped frequency table with eight class intervals of equal width.

36	60	51	55	35	59	74	70	72	50
70	36	58	51	29	52	27	52	64	48
59	76	76	63	39	35	72	71	24	74
28	79	30	55	31	71	66	33	26	55

Classes	Frequency, f	Mid-class values, x	fx
	Total =		Total =

b Work out an estimate of the mean of the data by completing the last two columns of the table.

c The true mean of the 40 values is 52.8. What is the percentage error of your estimate?

d In which class interval does the median value lie? How do you know?

e What is the modal class?

f Draw a bar chart to represent the data in the frequency table.

4 The frequency table shows the times, to the nearest second, that students took to complete a puzzle.

Time (seconds)	Frequency
41–45	2
46–50	5
51–55	8
56–60	6
61–65	3
66–70	1

Add two columns to the table that will help you calculate an estimate of the mean time to complete the puzzle.

5 Conduct a survey for which you expect to get a wide range of values. Collect at least 40 values. Create a grouped frequency table of your results which has 5 to 10 class intervals. Using mid-class values, find an estimate of the mean from the table. How does this compare with the true mean?

6 What are the class intervals in the frequency table if the mid-class values are:

a 22, 25, 28, 31, 34 and 37

b 160, 165, 170, 175 and 180?

10.2.3 Deciding which average to use

Explore 10.8

Calculate the mean and median of the following sets of data:

a 1, 3, 5, 7, 9 b 1, 3, 5, 7, 19 c 1, 3, 5, 7, 90 d 1, 3, 5, 7, 900

What do you see? Write a short paragraph with your observation.

Explore 10.9

For each of the following three situations, decide on one number to represent the entire set, to help the people in each situation.

a David's family can afford to buy a house for €230 000 and are looking at house prices in different neighbourhoods. A housing development they like has one house valued at €215 000, one at €220 000, three at €225 000 each, two at €235 000 each, three at €240 000 each and one mansion worth €1 550 000, and they want to know if the neighbourhood is within their means.

b A clothing boutique has sold eight pairs of jeans of size 32, one size 34 and one size 36 this week, and the owner wants to know what size a typical customer wears.

c A class had results of 97%, 97%, 96%, 87%, 80% and 11% in a recent test, and the teacher wants to understand how well the students did and whether a retest is necessary.

Write a sentence describing the conditions when you think the mode is the best measure to use and when you think the median is the best measure to use.

Worked example 10.5

Which average was used in each of the following situations?
Explain why the average used to describe the set could be misleading.
State which average would be best to use and why.

a SAT mathematics grades: 350, 350, 500, 640, 700, 710, 750, 790, 800
average grade = 350

b Electric car prices ($): 37 000, 39 000, 44 500, 50 000, 141 000
average price = $62,300

c Location of dart throws: 1, 5, 5, 5, 5, 13, 17, 17, 17, 17
average location = 9

Solution

a 350 is the mode, which also happens to be the lowest score. Seven out of nine students scored better than this, so this is not representative of the group. The mean is 620 (to the nearest 10), which six students outperformed. This is better and would often be used to report the performance of the group, but perhaps it is not the best since most students scored highly, and the two very low scores skew the results. The median of 700 probably best represents how the students in this group did on the SAT.

b $62 300 is not the median, so it must be the mean. It is greater than all but one of the prices in the list, so it does not give an accurate idea of how much an electric car is likely to cost (and it is also less than half the price of the most expensive car!). The median price of $44 500 gives the consumer a better idea of how much they can expect to pay.

c The mean is $\frac{102}{10} = 10.2$ and the median is between the 5th and 6th numbers, 5 and 13, so 9 is the median $\left(\frac{5 + 13}{2}\right)$. To say that a score was 9 on average would be fine, but here we are interested in the location of the dart throw. The 9 sector is on the opposite half of the board from where every throw landed! In this particular situation, it would probably be best to say that, most of the time, 5s and 17s are thrown, which are the two modes.

Fact

To skew something means to change its direction or meaning, or to pull it further in a certain direction.

Investigation 10.1

You apply for a job at a small IT company with 20 employees because they advertise that their average salary is €80 000, and you read in the news that the CEO has a salary of €1 000 000. You earned €60 000 in your present job and are looking to earn more in your new job. Might this company be what you are looking for?

Consider the different amounts that the company's employees might earn if the reported average of €80 000 is

1 the mean salary

2 the median salary

3 the modal salary.

Fact

An outlier (extreme value) in a data set is an observation that is very far from the centre. The mean of a data set is usually 'pulled' towards outliers and thus, may not be very useful in representing such data sets.

Explore 10.10

Most US universities require you to submit scores from a standardised test, such as the ACT or SAT, when applying. Your top university choice in the US says that the median mathematics score of students they accepted last year was 650. Your score is 600, so you are feeling down and decide not to apply. Is this the right decision?

Approximately what proportion of students that the university accepted scored higher than 650?

Investigation 10.2

Different countries have different scales for clothing sizes, and you may have found that an article of clothing produced by one brand fits differently to the same size article of another brand.

Imagine that you are starting a company to produce and sell clothing.

In how many sizes will you offer your clothes? What will you call the sizes? What will be the dimensions of those sizes? How will you determine which sizes to offer? Will you produce different quantities of different sizes or the same quantity of each? Do these answers differ depending on the item of clothing?

Decide on a particular article of clothing your company will produce. Describe the process you would go through to determine the answers to some or all of the above questions before you would be able to start designing and producing garments (i.e. what information you would collect, which measurements/calculations you would make, etc). Provide a sketch of the garment with dimensions of the different sizes. Use appropriate mathematical terminology in your description.

Practice questions 10.2.3

1 Calculate the mean, median and mode of each data set below. State which average best represents the group for each set, giving a reason for your answer.

a Ages of the teachers in the mathematics department (years):
58, 52, 47, 61, 24

b Ages of people in a tour group (years):
45, 45, 33, 18, 13, 43, 31, 42

c Prices of e-bikes in a shop (euros):
10 500, 9500, 8300, 7500, 6300, 5200, 4900, 4500, 4400, 4300

2 Which average(s) of the data set 25, 39, 40, 41, 41, 41, 42, 42, 42, 42, 43, 45 is (are) most appropriate if these numbers represent:

a the prices of different cars, in thousands of dollars, and you are trying to decide how much money you will need to save to purchase one

b the sizes of shoes sold today in a shop, and the shop owner needs to decide which sizes need immediate restocking

c the grades on a mathematics contest with a maximum of 45 marks, and the teacher wants to see how the class performed?

3 What does it mean to say: 'The average person in my class has 1.5 pets at home'? How could this have been calculated? Which average do you think is best to use in this situation, and why? Do you think you should report the average as 1.5 or as something else?

10.2.4 Comparing distributions using measures of central tendency and spread

In several situations, looking at only one number to represent a data set may not give clear indication of the nature of this data.

Explore 10.11

After studying a unit in your science class, an assessment of three groups in your class gave the following scores (out of 100):

Group A: 42, 50, 90, 98

Group B: 60, 70, 70, 80

Group C: 62, 67, 68, 83

Can you help analyse which group's performance indicates good group work in learning the unit?

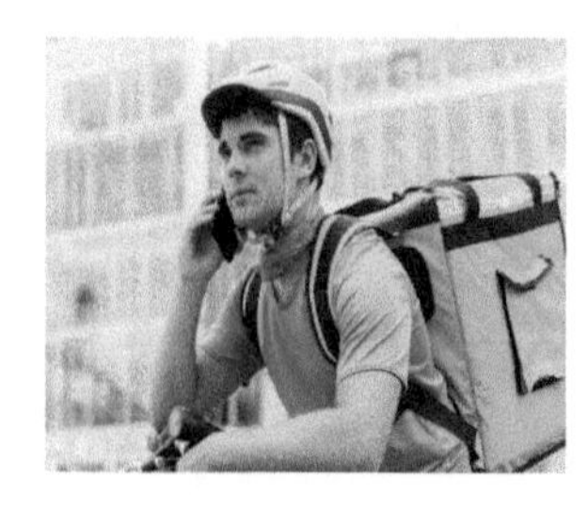

Worked example 10.6

Teachers at school often buy lunch from the nearby Vegan Cafe. They phone one of two delivery companies at the start of their 45-minute lunch break to bring their order. Cheetah Bike Couriers (C) delivered the lunch on eight occasions this year with delivery times of 20, 18, 34, 24, 30, 20, 26 and 28 minutes and Tortoise Tricycle Deliveries (T) brought the lunch seven times in 26, 27, 22, 28, 27, 25 and 26 minutes.

One teacher calculated the mean, and another calculated the median for delivery times of each company. Their numbers were not much different. Can you help them decide which service is better for their purpose?

Solution

You need to work out which company, C or T, has the faster service. If the central values are close, we must look at the service that gives teachers more time to eat their lunch.

First, you need to calculate the mean and median for each company's delivery times. Then we look at which company possibly gives more 'free' time at lunch.

We work out the mean either with a GDC or manually.

The mean for C is $\frac{20 + 18 + 34 + 24 + 30 + 20 + 26 + 28}{8} = \frac{200}{8}$

$= 25$ minutes.

The median can be found after arranging the times in ascending order:

18, 20, 20, 24, 26, 28, 30, 34

Since there are eight numbers, we choose the average of the 4th and 5th observations, i.e. $\frac{24 + 26}{2} = 25$ minutes.

Similarly, the mean for T is 26 minutes $\frac{22 + 25 + 26 + 26 + 27 + 27 + 28}{7}$

$= \frac{181}{7} = 25.86 \approx 26$, and the median is the 4th observation after arranging the numbers in ascending order:

22, 25, 26, 26, 27, 27, 28, that is 26 minutes.

Looking at how much time this gives the teachers to eat their lunch, on average:

with company C, they have $45 - 25 = 20$ minutes to eat, with T they have $45 - 26 = 19$ minutes, so C seems better but only very slightly.

Hint

Entering the data into a spreadsheet or GDC will help you find the statistics you need and allow you to focus on analysing the data. Here is a spreadsheet output:

	C	T
	20	26
	18	27
	34	22
	24	28
	30	27
	20	25
	26	26
	28	
Mean	25	25.8571
Median	25	26

and here is a GDC output:

```
Math Rad Fix6  ab/c a+bi
Mean(list 1)
                25.000000
Median(list 1)
                25.000000
```

```
Math Rad Fix6  ab/c a+bi
Mean(list 2)
                25.857143
Median(list 2)
                26.000000
```

However, looking at the times for C you can see that their fastest time is 18 while their slowest time is 34 minutes which, in this case, leaves teachers only 11 minutes to have lunch. Looking at the times for T, the fastest is 22 minutes and the slowest is 28, which still leaves teachers 17 minutes to have their lunch. This means that, while C has delivered the lunch to school very quickly on at least one occasion, it has also taken a very long time, so C is not very **reliable**, as their delivery times can vary quite significantly (their longest time was almost double their fastest).

T is more **consistent** with their deliveries, so even though they are slightly slower on average than C, the teachers can be fairly sure that they will have enough time to eat their lunch each day if they stick with T. While they could have a long, relaxed lunch break with C, they also risk having to eat their food very quickly with little more than 10 minutes before their next lesson starts!

The discussion above points to the fact that the spread of C's times is:

$34 - 18 = 16$ minutes

while the spread of time for T is

$28 - 22 = 6$ minutes

This spread is called the **range**.

Thinking skills

Communication skills

Fact

The **range** gives a rough idea of how the scores are spread. It is defined as the difference between the highest and lowest values.

range = highest value – lowest value

Practice questions 10.2.4

1 Two sports cars are driven around a racing circuit several times. Car A has a mean lap time of 37.2 seconds and a range of 13 seconds, whereas car B has a mean lap time of 40.0 seconds and a range of 4.1 seconds.

 a Which car is more consistent? How do you decide?

 b Which car is faster? How do you decide?

 c Can you decide which car is better using the given statistics?

2 a Calculate the mean, median, mode and range of these test percentages in mathematics classes A and B:

 $A = \{50, 51, 51, 52, 56, 100\}$ and $B = \{36, 51, 51, 52, 84, 86\}$

b According to the statistics you have calculated, you would expect these classes to be quite similar. Do you think they are? Why or why not? (Remember to think of them in context: these are the scores on a test by students in two classes.)

3 A hockey team has two strikers who are being considered for the Best Striker Award. In 10 matches, Su-Bin and Laura had the following number of goals and assists in each match:

Fact

You make an 'assist' when you pass to a player, and then, without losing possession, they score a goal.

Match	Su-Bin		Laura	
	Goals	Assists	Goals	Assists
1	1	1	0	0
2	1	2	4	1
3	1	1	1	0
4	2	1	0	1
5	1	2	5	0
6	1	1	0	1
7	2	2	0	2
8	1	1	4	0
9	1	1	0	0
10	1	1	1	1

a Who is the best striker, and what is your criteria for determining this? Justify your decision with reference to calculations of an average and/or range.

Anji is also on the team and scored 14 goals in the 10 matches, including an incredible four hat-tricks (matches when she scored three goals), and assisted once in each of eight matches.

b Would your decision for award recipient change if Anji were also being considered? Refer to the average and range in your justification.

Self-assessment

- I can look for bias in survey questions.
- I can decide on an appropriate sampling method for conducting a survey.
- I can find the mode, median and mean of a set of values presented as a list.
- I can find the mode, median and mean of a set of values presented in a frequency table.
- I can choose appropriate classes and create a grouped frequency table for a set of values.
- I can find the modal class from a grouped frequency table.
- I can find which class the median value is in for a grouped frequency table.
- I can use mid-class values to help calculate an estimate of the mean from a grouped frequency table.
- I can decide which average is best to represent a data set in a given context.
- I can find the range of a data set.
- I can use the range along with an average to compare two data sets.

Check your knowledge questions

1 Sasha wants to find out about people's reading habits. Here are her questions:

1. What is your favourite type of book? __________

2. How many times last week did you read a book for 15 minutes or more?

1–3 times ☐ 3–5 times ☐ 5–7 times ☐ more than 7 times ☐

Explain what is wrong with each question and write a better version.

2 Farid wants to ban homework and wants to find out what other students' views are on this issue. Here is his questionnaire:

Name __________ Year group _______

Do you agree that homework should be banned? Yes ☐ No ☐

Why?

Does not help learning ☐ Takes too much time ☐

For which year groups should homework be banned? (Tick one.)

All ☐ Primary ☐ MYP 1–3 ☐

MYP 4–5 ☐ Diploma Programme ☐

a Do you think Farid needs to know the students' names? Explain your answer.

b What is one disadvantage of asking students to give their names?

c There is a problem with each of the three questions or the choices they provide. What are they?

d Carefully consider how to improve this questionnaire, and design a data collection sheet for it.

e Try asking your classmates your questions and recording their answers.

3 For his mathematics assignment, a boy on the high school basketball team creates a survey to find out students' favourite break time activity.

a Identify the potential bias if he surveys only:

i his teammates

ii other boys in his year group

iii his year group

iv primary school students.

b If you were to carry out this survey in your school using stratified sampling, which groups of students would you collect data from in representative proportions?

4 Find the mode, median, mean and range of the following sets of values, to an appropriate degree of accuracy.

a 11, 20, 21, 28, 18, 22

b 14 cm, 25 cm, 32 cm, 35 cm, 29 cm

c 75 kg, 82 kg, 56 kg, 92 kg, 86 kg, 90 kg, 70 kg

d $5.40, $3.20, $6.95, $9, $6.75, $7.55, $10.25, $9.50

5 The school handball team has scored the following number of goals in each of their first five matches: 6, 7, 4, 10, 12. How many goals must they score in their sixth match to have a mean of eight goals per match?

6 Write two different sets of eight numbers each that have a mean of 10 and a range of 12

7 The bar chart shows the number of people living in each house on Shelley Street.

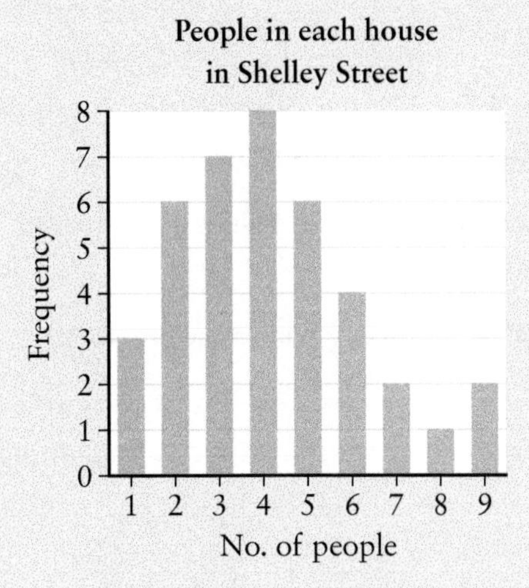

a Make a frequency table for the data shown in the bar graph.

b Use your table to work out the mean, median, mode and range of the data.

8 Here are the heights (in metres) of 20 people:

1.45	1.63	1.72	1.51	1.63
1.67	1.54	1.75	1.68	1.60
1.73	1.69	1.42	1.77	1.64
1.49	1.46	1.58	1.62	1.72

a Complete the table, which shows how to write class intervals using inequalities, and find an estimate of the mean value from the table.

Classes	Frequency, f	Mid-class values, x	fx
$1.40 \leqslant h < 1.50$		1.45	
$1.50 \leqslant h < 1.60$		1.55	
$1.60 \leqslant h < 1.70$			
$1.70 \leqslant h < 1.80$			
	Total =		Total =

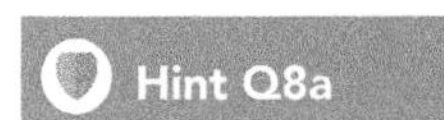
Hint Q8a

Be careful which class you put 1.60 m into.

b Find the exact mean of the heights.

c Find the percentage error of the estimate in part **a**.

9 The following are the lap times, in seconds, of two racing car drivers on the same track, in order from the beginning of the season.

Driver A: 11.25, 11.31, 11.22, 11.14, 11.07, 10.97

Driver B: 11.14, 10.92, 11.21, 11.30, 11.23

a Which driver achieved the best lap time this season?

b Which driver performed more consistently? Justify your answer using mathematical terminology.

c Which driver had a better season? Justify your answer using mathematical terminology.

d Which average, the mean or the median, is a better average to use in this situation? Why?

10 The girls and boys in a class each took 10 penalty shots at a football goal and 10 foul shots at a basketball hoop. These are their individual results:

Girls	Football	Basketball
A	2	3
B	4	3
C	3	3
D	5	6
E	2	5
F	2	1
G	3	0

Boys	Football	Basketball
Z	5	3
Y	6	3
X	4	0
W	2	0
V	4	3

Justify your answers to the following questions. Use the mean as your average.

a Who are better at taking football penalties in this class: girls or boys?

b Who are better at taking basketball foul shots: girls or boys?

c Based on this data, who are better overall: girls or boys?

d Is the class better at taking football penalties or basketball foul shots?

e In which sport:

i is the class more consistent

ii are girls more consistent

iii are boys more consistent?

11 Draw a bar graph to represent this data on average life expectancies in different regions of the world.

Region	Average life expectancy (years)
Europe	78
Western Pacific	77
Americas	77
South East Asia	70
Eastern Mediterranean	69
Africa	61

a What is the problem with finding the mean of these life expectancies to give an average life expectancy for the world?

b Can you find a way of calculating a reasonable estimate for the average life expectancy for the world using these values?

12 Write two sets of data values (five data points in each set) that:

a both have the same mean, but different ranges

b both have the same range, but different medians

c both have the same median, but different means

d both have the same median and mean, but different ranges.

13 The table shows the times taken by two athletes to run 100 m.

	Mean	Range
Athlete A	12.4 seconds	2.3 seconds
Athlete B	10.7 seconds	5.1 seconds

Which athlete would you pick to run 100 m for your athletics team? Justify your answer.

14 Here are the ages of six people waiting at a bus stop:

12, 15, 16, 20, 23, 24

a Find the mean and median ages.

A person aged 70 comes to wait at the bus stop.

b What effect does this have on

i the mean age

ii the median age?

15 The hourly wages of five people in a company are:

\$17, \$20, \$18.50, \$15.75, \$21

If another employee is hired at a rate of \$6 an hour, which is affected more, the mean or the median? Explain and justify your answer.

16 The pie chart shows students' grades on a recent test.

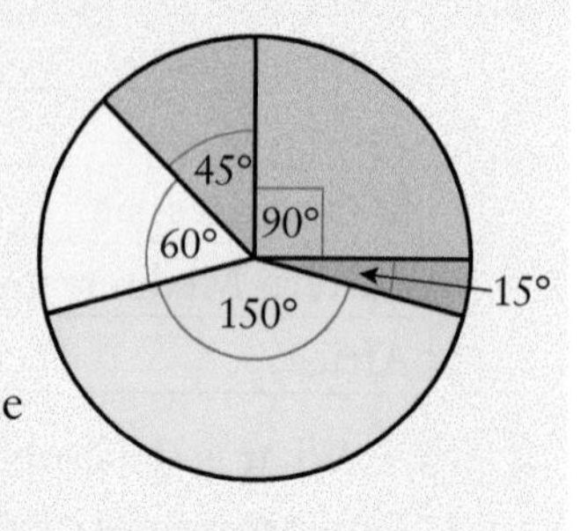

The mode was a 4, a quarter of the students scored the highest grade of 7, 5 was the median grade, the fewest number of students scored the lowest grade of 3, and 12.5% of the year group, or nine students, scored grade 6

a Draw a frequency table to show the grades of the class.

b Calculate the mean grade on the test.

17 This graph shows the scores of a class on a maths test and an English test, both of which were out of 20 marks.

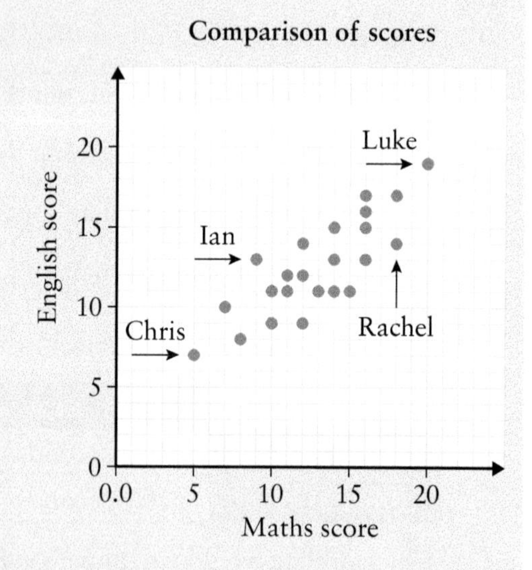

a What was Rachael's score on the English test?

b What was the range of scores on the maths test?

c What were the modal English and maths scores?

d Make two frequency tables: one for the English scores and one for the maths scores.

e 'In English, Ian's score was exactly in the middle of Luke's score, which was the highest, and Chris' score, which was the lowest, so his is the median score.' Do you agree with this statement? Why or why not?

f What were the median and mean English and maths scores?

g Did the class do better in English or maths? Justify your answer by referring to an average and the range.

11 Probability

11 Probability

KEY CONCEPT

Logic

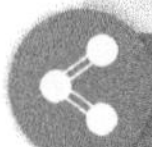

RELATED CONCEPTS

Quantity, Representation, Systems

GLOBAL CONTEXT

Fairness and development

Statement of inquiry

Using a logical system of measure to quantify chance helps us to make fair decisions.

Factual

- What are simple events?

Conceptual

- What does the probability of an event represent?

Debatable

- How accurate are the probabilities we calculate at representing real life?

Do you recall?

1 How to convert between decimals and percentages?

a 0.4   b 2.3%   c 0.07

2 How to convert between fractions and decimals?

a $\frac{2}{5}$ b 0.3   c $0.\dot{3}$

3 How to simplify fractions?

a $\frac{2}{12}$ b $\frac{6}{21}$ c $\frac{12}{54}$

4 How to multiply fractions by whole numbers?

a $\frac{3}{4} \times 12$ b $\frac{1}{3} \times 15$ c $\frac{4}{5} \times 2$

11.1 The basics of probability

11.1.1 The language of probability

Fact

A **probability experiment** is a repeatable procedure with a set of possible results. A coin flip is a probability experiment.

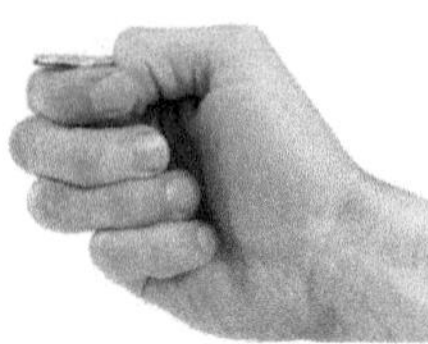

'Probability is nothing but common sense reduced to calculation'

Pierre-Simon Laplace.

Many events cannot be predicted with absolute certainty. The best we can say is how **likely** they are to happen, using the ideas of probability.

For example, when a coin is flipped, there are 2 possible outcomes: heads or tails, which are usually denoted as H or T. We say that the probability of the coin showing H is 0.5, or 50%.

Probability does not tell us precisely what will happen, it is just a guiding light so that we have some idea of what to be expecting.

In a 'flipping the coin' experiment, if we flip the coin 50 times, how many heads will come up?

Since the probability of heads is 50%, we can expect 25 heads. But when we actually try it we might get 23 heads or 28 heads ... or another number of heads.

Explore 11.1

Can you describe the likelihood of:

a being late for school tomorrow

b rolling an even number on a regular six-sided dice?

What is different about these likelihoods?

Worked example 11.1

You are picking one marble from a bag containing three blue marbles and one red marble. What is the probability that you pick a blue marble?

Solution

The number of ways this can happen is three since there are three blue marbles. The total number of outcomes is four, since there are four marbles altogether.

So, probability of picking a blue marble $= \frac{3}{4} = 0.75 = 75\%$

We can show probability on a sliding scale:

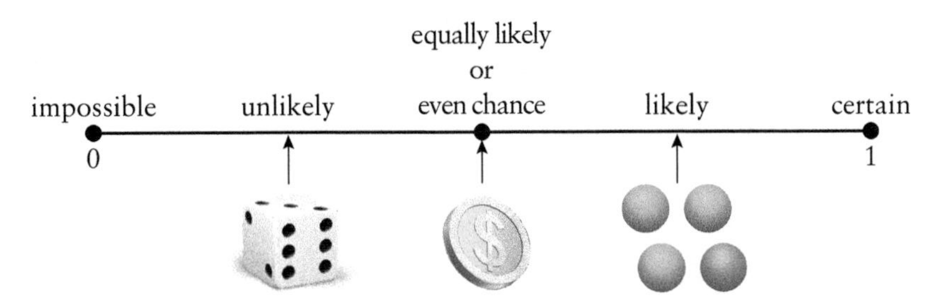

While this type of scale can be used to describe, for example, the likelihood of being late for school tomorrow, it is useful to also be able to quantify the numerical likelihood of an event occurring.

For example, an impossible event means there is no chance of it happening, meaning it would have a probability equal to 0

On the other hand, a certain event has a 100% chance of happening.

The likelihood of an event happening, therefore, lies between 0 and 100% (inclusive).

While percentages are used to measure probability, it is also common to use fractions or decimals. Thus, we can say that the likelihood of an event happening lies between 0 and 1

An **event** consists of one or more outcomes of an experiment.

Reminder

Recall, from your work on percentage conversion, that an event with a 100% likelihood can also be said to have a probability of 1

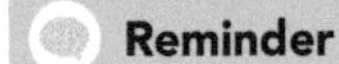

Recall that $50\% = \frac{1}{2} = 0.5$

Worked example 11.2

Use the terms impossible, unlikely, equally likely, likely or certain to classify the likelihood of the following events:

a a coin toss landing on heads

b rolling a 2 on a regular six-sided dice

c the Sun rising tomorrow morning.

Solution

a Each side of a coin is equally likely to come up.

b As there are six possible outcomes 1, 2, 3, 4, 5 or 6, rolling specifically a 2 on any given roll is unlikely, but definitely not impossible.

c It is certain that the Sun will rise tomorrow (or at least as certain as something can be!).

A sample space is the name we give to all the possible outcomes of an experiment. For example, in the figure on the right, the event of rolling a standard dice, has the sample space $\{1, 2, 3, 4, 5, 6\}$. Rolling a 4 is one of the possible outcomes. Such outcomes are called **sample points.** In fact, we can say that the sample space is made up of sample points.

Fact

An unbiased coin or dice is one where the outcomes are equally likely. Sometimes we also encounter biased coins or dice in probability questions, where one face is more likely to appear than the other.

When rolling a dice, landing on any one face is a simple event, equally likely as other events in this experiment. This means that rolling a 3 is just as likely as rolling any other number on the dice. Each face of the dice has a 1 in 6 probability of appearing on any given roll. Another way of writing this is

$P(3) = \frac{1}{6}$ where P(3) stands for the probability of rolling a 3.

We can generalise this observation to say that:

$$P(\text{of an event}) = \frac{\text{number of ways an event can happen}}{\text{total number of possible outcomes}}$$

Practice questions 11.1.1

1 What is the numerical likelihood of an event which:

a is impossible b has an even chance c is certain.

2 Decide whether the follow are outcomes or not:

a a coin showing tails
b rolling a dice
c picking a colour at random
d selecting a spade from a pack of playing card
e choosing a name out of a hat.

For the events in Questions 3–5, classify them as: impossible, unlikely, equally likely, likely, or certain.

3 A snowy day in March in Sweden.

4 It is literally raining cats and dogs.

5 What is the probability that the spinner will stop on red?

a
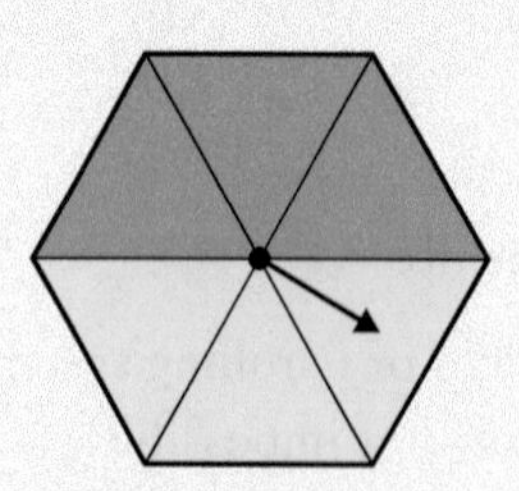

b
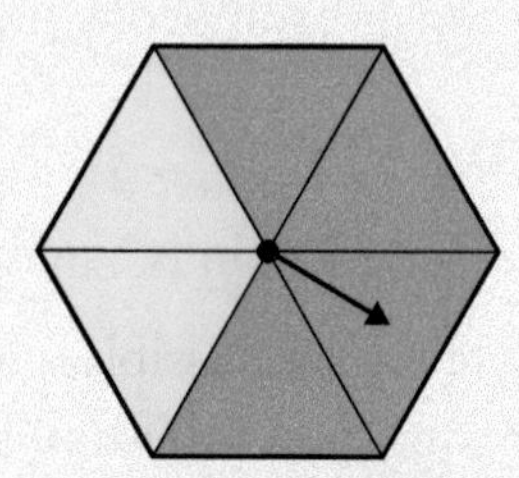

c
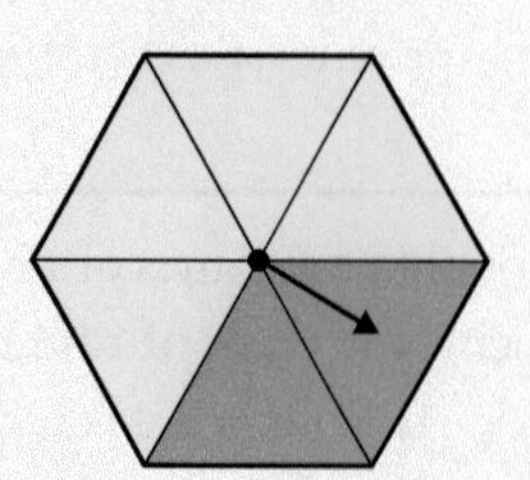

d
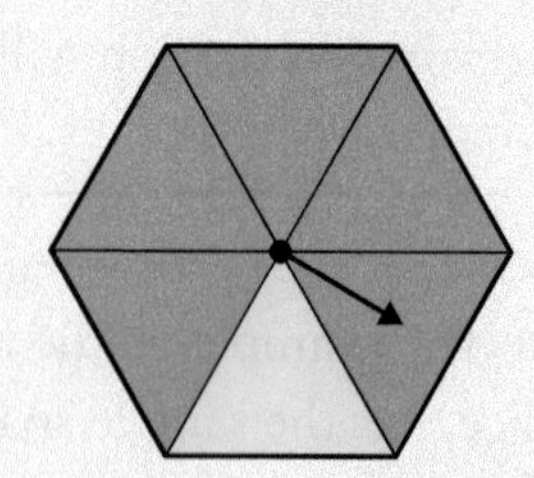

6 a What is the probability the spinner will stop on:

i yellow ii blue iii red?

b What answer do you get if you add all three probabilities together? Can you think of a reason to justify your answer?

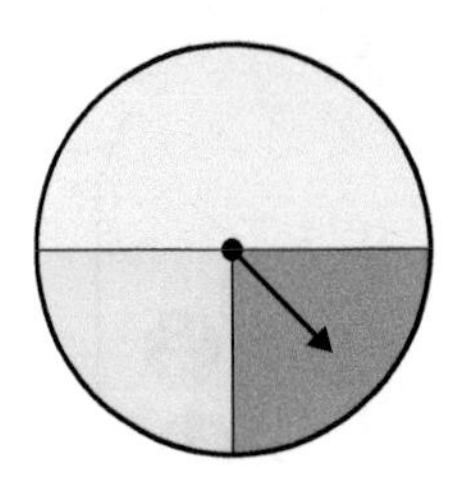

7 You roll a dice. Write, as a fraction, the probability that you will throw:

a a 2 b a zero

c an odd number d either a 1 or a 6

e a number less than 4 f a number less than 7?

11.1.2 Probability experiments

As you saw in the last Section, you can use probability to help you anticipate the outcomes of certain events.

In this section you will learn how to calculate the probability of certain events occurring.

Explore 11.2

On a fair, six-sided dice can you find the probability of rolling:

a an even number b a square number.

Fact

Probability questions will often state that a coin is fair to certify that it is not biased.

A general way to find the probability of any specific event of any probability experiment is:

$$\text{P(event } X) = \frac{\text{total number of ways event } X \text{ can occur}}{\text{total number of all possible outcomes}}.$$

We can shorten this using the notation:

$$\text{P(X)} = \frac{n(\text{X})}{n(\text{S})}$$

where $n(\ )$ replaces 'number of'.

Reminder

In the last Section we learnt that a list of all possible outcomes to an event is called the sample space, which can be abbreviated as S.

Many situations used to clarify probability ideas involve flipping a coin or rolling a dice.

Flipping coins

Generally this has only two outcomes: 'heads' or 'tails'. You are already familiar with the probability generated by flipping a coin from the last sub-section.

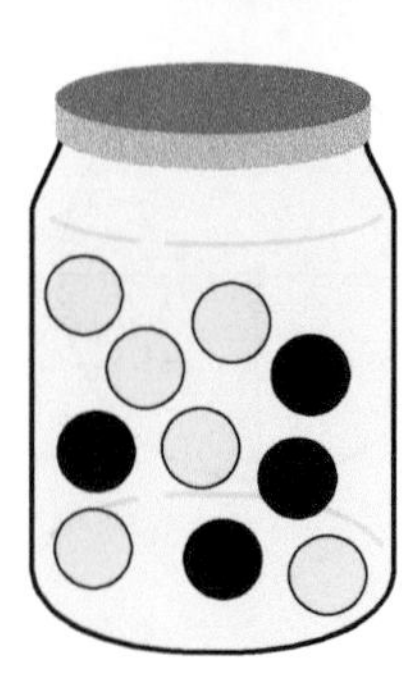

Worked example 11.3

A jar contains four black counters and six yellow counters.

A counter is chosen at random.

Find the probability of choosing:

a a black counter b a yellow counter.

Solution

a There are four black counters and 10 counters in total:

$$P(\text{black}) = \frac{4}{10} = \frac{2}{5}$$

b There are six yellow counters and 10 counters in total:

$$P(\text{yellow}) = \frac{6}{10} = \frac{3}{5}$$

Rolling dice

This is classically a dice with six sides, however there are also instances where a dice might have 4, 8, 12 or 20 sides. Each of these dice has the numbers one, 2, 3, … all the way to the number of sides on the dice. For example, a 12-sided dice will have the numbers 1–12. Since there is only one of each side, landing on any given side has a probability of $\frac{1}{\text{number of sides}}$.

Worked example 11.4

If you roll a fair six-sided dice, what is the likelihood of rolling:

a the number 5 b a prime number c a 9?

Solution

a Since there are six sides (possible outcomes) only one of which is the number 5:

$$P(\text{rolling } 5) = \frac{\text{total number of ways to roll 5}}{\text{total number of possible outcomes}} \qquad P(5) = \frac{1}{6}$$

b Since there are three prime numbers on a six-sided dice: 2, 3 and 5

$$P(\text{prime}) = \frac{1}{2}$$

It is good practice to simplify the fraction if possible.

c There is no 9 on a six-sided dice, making it impossible:

$$P(9) = \frac{0}{6} = 0$$

Fact

Events like rolling a 5 are *simple* events. A simple event is one that can only happen in one way – in other words, it has a single outcome. On the other hand, a *compound* event involves the probability of more than one outcome. Rolling a prime number, is a **compound** event because it can happen when you roll a 2, 3 or 5.

Other probability experiments involve drawing playing cards, therefore it is useful to know the 'make up' of a standard pack of cards.

Drawing playing cards

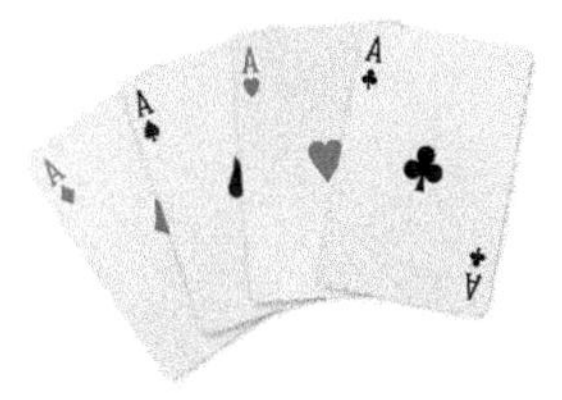

Traditionally, playing cards come in four suits: diamonds, spades, hearts and clubs, as shown opposite.

Note that two suits are black and two are red, this splits the deck evenly in two. Finally, each suit has a set of number and picture cards. The numbers go from 1 to 10 and the picture cards include King, Queen and Jack (although these take other names in different cultures). The card with number 1 is sometimes also called the Ace. A standard pack of playing cards has 52 cards. Each suit is comprised of 10 number cards and three picture cards.

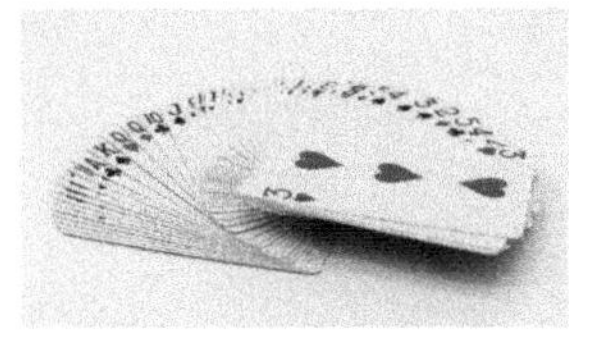

Worked example 11.5

When randomly selecting a card from a standard deck, what is the likelihood of drawing:

a a 7

b a picture card

c a heart

d the Ace of spades?

Hint

Drawing a 7 is simple, drawing a face card or a heart is compound.

Solution

a There are four cards with a 7 on them in a deck of 52 cards

$P(7) = \frac{4}{52} = \frac{1}{13}$

As always, simplify if possible.

b There are three types of picture card (J, Q, K) in each suit and four suits:

$P(\text{picture}) = \frac{3 \times 4}{52} = \frac{12}{52} = \frac{3}{13}$

c Each suit has 13 cards:

$P(\text{heart}) = \frac{13}{52} = \frac{1}{4}$

d There is only one Ace of spades in the pack of 52 cards:

$P(A\spadesuit) = \frac{1}{52}$

Practice questions 11.1.2

1 Write out the sample space of:

a drawing a card from a deck of 52 playing cards and looking for a suit

b an eight-sided dice numbered from 1 to 8

c the coloured tokens in the jar from Worked example 11.3

2 When rolling a standard six-sided dice, calculate the likelihood of rolling:

a the number 4

b an even number

c a multiple of 3

d the number 10

3 When rolling a 12-sided dice, numbered 1 to 12 calculate the likelihood of rolling:

a the number 7

b an odd number

c a multiple of 4

d a number less than 3

e a prime number.

4 When rolling a 20-sided dice numbered 1 to 20, what is the likelihood of rolling:

a the number 18

b a multiple of 5

c a prime number

d a number greater than or equal to 9.

5 You draw a card at random from a standard deck of cards. Calculate the probability that the card is:

a red

b an Ace

c a spade

d a number (excluding Aces)

e the 7 of diamonds.

6 A jar has 10 counters inside: five are green, two are red and the rest are blue. When drawing one at random, find the likelihood the counter is:

a green

b red

c blue.

7 Yolanda has a bowl of sweets on her desk. Eight are strawberry flavoured, four are peach and six are grape. She chooses one at random to eat. What is the likelihood she chooses the flavour:

a strawberry
b peach
c grape?

8 A jar contains 16 counters that are either blue or green. Given that the likelihood of drawing a blue counter is $\frac{3}{4}$, how many green counters are there?

Challenge Q8

9 A box has 18 chocolates that are either dark chocolate or hazelnut. Given that the likelihood of drawing a hazelnut is $\frac{2}{3}$, how many dark chocolates are there?

Challenge Q9

10 Throw two regular unbiased dice.

Challenge Q10

a How many possible outcomes are there? List five of them.

b What is the probability that the two numbers on the dice are the same?

11.2 Applying probability

11.2.1 Complementary events

Explore 11.3

Sometimes a probability event has a direct opposite, called a complement. That is, the complement of an event consists of all outcomes that are not the event.

Can you tell what the complement of each of these events is? Give reasons for your decision:

a flipping a coin and it landing on heads

b rolling an even number on a dice

c the weather being sunny tomorrow

d drawing a card from a standard deck and getting spades.

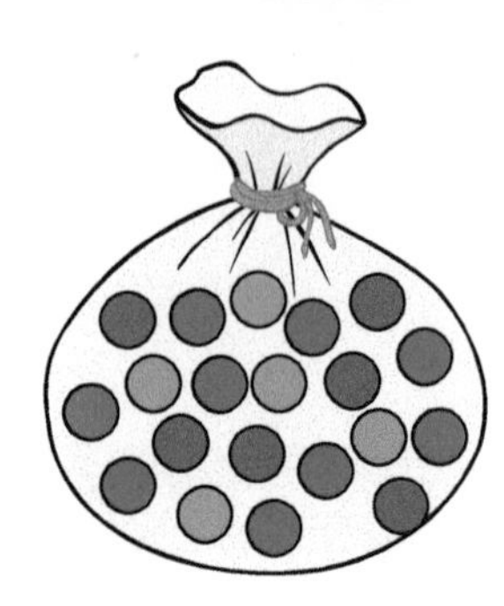

Let us consider the following activity.

You have a bag with 20 marbles – nine red, six blue and five green. You choose one marble at random without looking. What is the probability the chosen marble will be:

a blue **b** not blue **c** red

d not red **e** green **f** not green?

Now look at the probabilities you found and answer the following questions by filling in the blanks:

1 $P(\text{blue}) + P(\text{not blue}) = \frac{6}{[\ ]} + \frac{14}{[\ ]} = [\ ]$

2 $P(\text{red}) + P(\text{not red}) = \frac{[\ ]}{[\ ]} + \frac{[\ ]}{[\ ]} = [\ ]$

3 $P(\text{green}) + P(\text{not green}) = \frac{[\ ]}{[\ ]} + \frac{[\ ]}{[\ ]} = [\ ]$

As you see, the pairs of events are complementary, P(E) and P(not E), and the sum of their probabilities is 1.

Fact

If E is an event, then E′ denotes the complement of the event and $P(E) + P(E') = 1$ and $P(E') = 1 - P(E')$ or $P(E) = 1 - P(E')$.

Worked example 11.6

When you throw two dice, the probability of throwing a sum of 8 is $\frac{5}{36}$. What is the probability of not throwing a sum of 8?

Solution

You are not required to find the probability of throwing all the other sums because it would require a lot of work from you. You just find the probability of the complement, that is

$$P(\text{not throwing a sum of 8}) = 1 - P(\text{throwing a sum of 8})$$
$$= 1 - \frac{5}{36} = \frac{31}{36}.$$

Why is the complement useful?

It is sometimes easier to work out the complement first. Consider the following example.

Worked example 11.7

When you throw two dice, what is the probability that the two scores are different?

Solution

Different scores are, for example, getting a 4 and a 6, or getting a 3 and a 1. If you try and list them all it is quite a long list:

$A = \{(1, 2), (1, 3), \dots, (6, 5)\}$

The complement, that is when the scores are not different, i.e. the same, has only six possible outcomes:

$A' = \{(1, 1), (2, 2), (3, 3), (4, 4), (5, 5), (6, 6)\}$

Hence, $P(A) = 1 - P(A') = 1 - \frac{6}{36} = \frac{30}{36} = \frac{5}{6}$

Hint

Remember that the sample space of throwing two regular dice has 36 outcomes.

11.2.2 Experimental probability

Up until now we have been learning about theoretical probabilities. That is, we have been learning how to calculate values that are true in a perfect, theoretical world. But real life is a bit different. For example, weather forecasters may anticipate rain on Monday, but Monday may turn out to be a sunny day. However, even if our predictions are not perfect, probability theory can give us a reasonable idea of how reality may turn out.

Explore 11.4

Thinking skills

Imagine rolling a standard six-sided dice 12 times.

a How many times would you expect to get the number 5?

b How many times would you expect to not get 5? Explain your reasoning.

Now *roll* a six-sided dice 12 separate times.

Record the number of times you roll a 5.

c Look back at your predictions from part **b**.

Are you surprised by your results? Explain your reasoning.

d How might this change if you were to repeat the experiment 100 more times?

We use probability to help predict the most likely outcomes of future events.

- Every person should try to calculate the chance of success before beginning a new venture so that failure can be avoided or plans changed so that the chance of success can be increased.
- The captains of sporting teams (e.g. football) will assess the probabilities of each possible action when making decisions. e.g. Should we attack or defend? What strategy is likely to give us our best chance of winning?
- Insurance companies work out the likelihood of death or accident to calculate the cost of policies.

Even though calculating probabilities will help us in making decisions, it does not really foretell what will actually happen. For example, an airline claims that there is only 1% chance of losing luggage while travelling with them. You travel with them and your luggage gets lost! Does that mean that they are misleading you?

The answer is no. If you travel with them a large number of times, on average they may lose your luggage 1% of the time.

Connections

The sciences often require some understanding of probability to make predictions about the future. An obvious example is the study of weather patterns, called meteorology, which is responsible for weather forecasts. However, many other fields in engineering and science require predictive elements.

For the Explore activity, this means that for 12 dice rolls we would expect

$E(X) = \frac{1}{6} \times 12 = 2$

11.2.3 Modelling probability from data

When a meteorologist tries to predict the weather or an insurance company tries to predict the likelihood of a certain incident, they need to use real-time and historical data to make the most accurate prediction they can.

The most straightforward method of doing this is looking at the data you have and assuming the trend you observe will stay the same or similar in the future.

Investigation 11.1

Research skills

Do April showers bring May flowers?

April 2020 in Vienna, Austria, saw 15 rainy days.

What is the chance of rain on any given day in April 2020?

Could we use this to predict rain in subsequent years? Explain your reasoning.

When looking into historical data for the last 10 years, on average there have been only 13 rainy days in April.

What is the chance of a rainy April day in Vienna, Austria over the last 10 years?

Is the 10-year average more or less accurate at predicting weather in future years? Explain your reasoning.

What other data might help make a more accurate prediction about the weather on a given day in April 2025?

Apply this type of thinking to your home city.

How many rainy days are there in April in your own region?

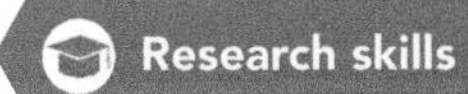

Worked example 11.8

The school newspaper is conducting a poll to predict the student elections. During lunch, the student reporters collect data from 42 students: 18 plan to vote for Rania, 16 plan to vote for Jonas, while the rest are undecided.

Based on these results, find the probability of:

a a student voting for Rania

b a student being undecided.

Solution

a Since 18 of the 42 students chose Rania, it is their ratio:

$$\text{P(Rania)} = \frac{18}{42}$$

Simplify to give the final answer

$$= \frac{3}{7}$$

b First calculate the number of undecided students:

$$n(\text{undecided}) = 42 - (18 + 16) = 8$$

$$\text{P(undecided)} = \frac{8}{42} = \frac{4}{21}$$

Practice questions 11.2

1 Write the complement of the following events:

a rolling an odd number

b drawing a picture card

c drawing a diamond

d a coin landing on tails.

2 Use appropriate notation to write out the probability of the complement of these events:

a P(Even) = 0.6

b P(Rain) = 0.35

c P(Faulty) = 0.05

d P(True) = 0.99

3 Taylor has missed the school bus eight times in the first 60 days of school.

a What is the probability Taylor makes the school bus, on any given day?

b How many times should he expect to miss the bus in the next 25 days?

4 Jan has planted 82 lemon seeds and has had 24 successfully germinate (grow successfully).

a From this experience, what is the probability the next seed fails to grow?

b If Jan plants 10 more seeds, how many can be expected to germinate?

5 A company producing microchips has two machines making products. In the last month, Machine A has product 1550 microchips with 80 having deficiencies. Machine B has produces 850 microchips with 60 having deficiencies. Using this data, estimate the probability of:

a machine A producing a working microchip

b machine B producing a working microchip

c the next microchip being defective.

Hint Q5c

Consider the total production and deficiencies of both machines.

6 A pollster investigating the upcoming election found that 2509 respondents of their survey supported the Green Party, 3150 supported the Red Party, 2117 the Blue Party and 1540 abstained from answering. Find the probability that a randomly selected individual supports:

a the Green Party   b the Red Party   c the Blue Party.

7 An insurance provider for smart devices has data that indicates that from 500 smart-phone purchases, 170 will need screen replacements.

a What is the probability a customer with a smartphone will need a screen replacement on their device?

Challenge Q7b

b If replacing a screen costs €50, what would be a profitable price for the company to sell screen-protection insurance for, based on the information available?

8 In a standard pack of cards there are two red suits (hearts and diamonds) and two black suits (clubs and spades). When a card is chosen at random, what is the probability of selecting:

a a black card

b a card that is not red

c either a black card or a red card

d neither a black card nor a red card?

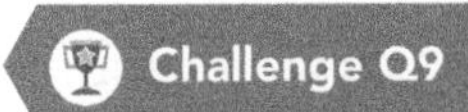

9 A stamp is to be selected at random from a collection. We know some probabilities already.

$$P(\text{Australian}) = \frac{3}{10},\ P(\text{Polish}) = \frac{1}{17},\ P(\text{French}) = \frac{37}{170}$$

Find the probability that the stamp selected is:

a not Polish   b not Australia   c not French

d none of the mentioned nationalities.

Challenge Q10

10 From a standard pack of 52 cards, one card is taken at random. Find the probability that the card will be:

a a diamond   b not a diamond   c a 4

d not a 4   e a picture card   f not a picture card

g a 3, 4, 5, 6 or 7   h not a 3, 4, 5, 6 or 7   i either red or black.

Reflect

What is the main assumption we make when calculating probability from data?

In what situations is this assumption likely to be correct?

What could make the predictions more accurate?

Self-assessment

- I can classify outcomes as impossible, unlikely, equally likely, likely or certain.
- I can identify the possible outcomes of simple events.
- I can calculate the probability of simple events.
- I can list sample spaces.
- I can understand standard probability experiments, like flipping coins, drawing cards or rolling dice.
- I can interpret probability symbols and notation.
- I can identify the complement of an event.
- I can calculate the probability of an event's complement.
- I can predict basic probabilities from data.

? Check your knowledge questions

1 Classify the following outcomes as impossible, unlikely, equally likely, likely or certain:

a drawing a red card from a standard deck

b a coin landing on tails three times in a row

c picking a chocolate cookie from jar with only chocolate cookies.

2 Decide whether or not you can calculate the probability of the following events. If you can, write down the probability of the event:

a rolling a 7 on an eight-sided dice

b choosing your name out of a hat with 19 other names in it

c your friend picking blue as a random colour.

3 A jar has 15 counters inside: six are red, five are black and the rest are white. When drawing one at random, find the likelihood the counter is:

a red b black c white.

4 You roll a 12-sided dice.

What is the likelihood of rolling:

a the number 3

b a multiple of 6

c a prime number

d a number less than 5

e a number with the digit 1?

5 You draw a card at random from a standard deck of cards, replace it, and reshuffle.

Calculate the probability that the card is:

a a 5

b not a 5

c a heart

d not a heart

e a picture card

f not a picture card.

6 Write out the following in words:

a P(odd) = 0.5

b $P(A') = \frac{1}{3}$

c P(fail) = 0

7 Write out the following using appropriate notation:

a The probability of rain is 70%

b The probability of drawing an even card is 0.4

c The probability of it not being a picture card is $\frac{3}{4}$

8 Diego wants to raise awareness about poverty in his neighbourhood. His research discovers that of the 25785 residents living in his area, approximately 5435 rely on government programs for food security. He also finds that 7820 are listed below the poverty line.

a What is the probability that a person in Diego's area:

i relies on the government for food

ii lives below the poverty line.

b How can Diego use these statistics to help his community?

 Challenge Q9

9 A jar with blue, yellow and red tokens has a total of 45 tokens. The chance of drawing a blue token is $\frac{1}{3}$ and the chance of drawing a yellow token is $\frac{2}{5}$

How many red tokens are in the jar?

10 A national airline claims that loss of luggage while travelling with them can happen only 3% of the time.

a What is the probability that luggage will not be lost traveling with this airline?

b In 2019, a passenger travelled with this airline 12 times and her luggage was lost on three occasions. What probability does this reflect?

c Is there a contradiction between your answers to parts **a** and **b**?

11 I toss a fair coin three times and record the results in order: (H, H, H), (H, H, T), ...

a List all possible outcomes.

b What is the probability of getting

i two heads

ii one head

iii no heads

iv at least one head?

Chapter 1 answers

Do you recall?

1 Natural numbers, $\mathbb{N}$; Integers, $\mathbb{Z}$; Rational numbers, $\mathbb{Q}$; Real numbers, $\mathbb{R}$

2 14

3 1, 2, 3, 4, 6, 12

4 7, 14, 21, 28

5 $\frac{3}{10}$; 0.3

6 The first angle is acute; the second angle is obtuse; the third angle is right.

7 The first triangle is equilateral; the second triangle is isosceles; the third triangle is scalene.

8 2

9 Perimeter = 40 cm; area = 43 cm^2

10 180 cm^3

11 Quantitative: number of marks scored in a test; time taken to run 100 m; height
Qualitative: eye colour; favourite food

12 Examples might include: trial and improvement, make a list, make a table, eliminate possibilities, make a diagram, look for a pattern, work backward, simplify the problem, use an equation

Practice questions 1.1

1 a 38 452 b 508 405 c 6 407 040

2 a 90 000 + 2000 + 600 + 70 + 3
b 400 000 + 50 000 + 8000 + 60 + 2
c 50 000 000 + 400 000 + 30 000 + 2000 + 100 + 6

3 a 346 b −4520
c 6896 d −85

4 74

5 −13

6

	$\mathbb{N}$	$\mathbb{Z}$	$\mathbb{R}$
−11		✓	✓
$\frac{3}{7}$			✓
23	✓	✓	✓
−6.25			✓
$-\frac{8}{2}$		✓	✓
3.6			✓
0	✓	✓	✓
7821	✓	✓	✓

7 a Any real number between 3 and 4 e.g. 3.5
b Any real number between 12.6 and 12.7 e.g. 12.65
c Any real number between −62 and −61 e.g. −61.5
d Any real number between 0.06 and 0.062 e.g. 0.061
e Any real number between $\frac{4}{9}$ and $\frac{5}{9}$ e.g. $\frac{9}{18}$

8 a 8063 b 227 000
c 53 004 d 5 000 009

9 a four thousand eight hundred and twelve
b thirteen thousand nine hundred and one
c five hundred thousand seven hundred and eighty
d seven million eighty two thousand eight hundred and seven

10 a tens b tenths
c thousands d hundreds
e thousandths

11 a 13286 b 132 063 c 7 050 705

12 a 61 093, 62 213, 62 341, 63 812, 63 976
b 78 012, 78 302, 78 365, 78 514, 78 546
c 302 645, 302 709, 302 739, 302 963, 302 972

13 a 0 b 13 c 5 d 99
e 7 f 63 g 98 h 8

14 50

Practice questions 1.2

1 a 403 b 0 c 97
d 81 e 0 f 42

2 a 0 b 653 c 217
d any number, e.g. 5
e 7 f 5 g 0 h 0

3

Number	Divisible by 2	Divisible by 3	Divisible by 4	Divisible by 5
1004	✓		✓	
1485		✓		✓
1020	✓	✓	✓	✓
13 476	✓	✓	✓	
91 800	✓	✓	✓	✓

Number	Divisible by 6	Divisible by 8	Divisible by 9	Divisible by 10
1004				
1485			✓	
1020	✓			✓
13 476	✓			
91 800	✓	✓	✓	✓

4 a 1, 2, 3, 4, 6, 8, 12, 24
b 1, 2, 3, 5, 6, 9, 10, 15, 18, 30, 45, 90
c 1, 2, 4, 29, 58, 116
d 1, 3, 5, 7, 15, 21, 35, 105

5 a 15 b 51 c 22 d 1

6 a 5, 10, 15, 20 b 9, 18, 27, 36
c 12, 24, 36, 48 d 30, 60, 90, 120

7 a 40 b 45 c 120 d 140

8 a 2, 3, 5, 7, 11, 13, 17, 19, 23, 29
b 32, 33, 34, 35, 36, 38, 39

9 a $2 \times 2 \times 2 \times 3$ or $2^3 \times 3$
b $2 \times 2 \times 2 \times 2 \times 3 \times 3$ or $2^4 \times 3^2$
c $5 \times 5 \times 3 \times 3$ or $5^2 \times 3^2$
d $2 \times 2 \times 3 \times 3 \times 7$ or $2^2 \times 3^2 \times 7$

10 a $4^2 = 4 \times 4 = 16$
b $10^3 = 10 \times 10 \times 10 = 1000$
c $9^2 = 9 \times 9 = 81$
d $2^4 = 2 \times 2 \times 2 \times 2 = 16$

11 8

12 10:00

13 $400 = 2^4 \times 5^2$ and $1080 = 2^3 \times 3^3 \times 5$
a HCF is $40 = 2^3 \times 5$
b LCM is $10\,800 = 2^4 \times 3^3 \times 5^2$

Practice questions 1.3

1 a −3 b 15 c −8 d −14
e 19 f 5 g −11 h −2
i 8 j −1 k −23 l 16

2 a −2 b −15 c 42 d 0
e −42 f −64 g 16 h 48
i 6 j −11 k −4 l 7
m −6 n 7 o −11 p 26

3 a 14 b −28 c 29 d −14

4 a Kiev, Ukraine
b Istanbul, Turkey
c Kiev, Copenhagen, Belgrade, Amsterdam, Hamburg, Istanbul

5 −4 °C

6 24

7 Kathmandu is 10:45 hours ahead New York

8 a −6 and −5 b −6 and 7

Practice questions 1.4

1 a $\frac{6}{10} = \frac{3}{5}$ b $\frac{43}{100}$
c $\frac{802}{1000} = \frac{401}{500}$ d $\frac{43}{10}$

2 a 0.6, 0.607, 0.61, 0.619
b 0.04, 0.044, 0.4, 0.44
c 0.085, 0.09, 0.1, 0.8
d 0.07, 0.077, 0.707, 0.77

3 a 21.66 b 2.94 c 21.902 d 0.363

4 a 0.15 b 543.1 c 9.26 d 172.8

e 2.198 f 52.32

5 a 10.35 b 0.03075 c 0.07402

d 43.65 e 4 f 31

6 a 5 b 57 c 33 d 15

7 a 7.4 b 4.1 c 0.1 d 20.0

8 a 24.77 b 5.90 c 0.08 d 3.99

9 €50.78

10 991.76 cents

11 $9.27

Practice questions 1.5

1 a $3\frac{2}{3}$ b $1\frac{3}{5}$ c 11 d $34\frac{1}{10}$

2 a $\frac{5}{2}$ b $\frac{18}{5}$ c $\frac{92}{9}$ d $\frac{17}{3}$

3 a 7 b 12 c 4 d 12

4 a $\frac{3}{4} < \frac{4}{5}$ b $\frac{3}{8} = \frac{9}{24}$

c $\frac{5}{6} > \frac{7}{9}$ d $\frac{1}{7} > \frac{14}{100}$

5 a $\frac{2}{3}$ b $\frac{1}{4}$ c $\frac{4}{3}$ d $\frac{13}{10}$

6 a $\frac{13}{10}$ b $\frac{7}{24}$ c $\frac{19}{24}$ d $\frac{7}{36}$

7 a $11\frac{1}{4}$ b $4\frac{19}{20}$ c $4\frac{19}{40}$ d $4\frac{5}{6}$

8 a $\frac{3}{8}$ b $\frac{27}{40}$ c $1\frac{1}{4}$ d $18\frac{2}{5}$

9 a $1\frac{1}{4}$ b $1\frac{1}{20}$ c $1\frac{1}{2}$ d $1\frac{69}{76}$

10 a 80 cm b 80 mins

c 375 ml d 2700 g

11 a $\frac{1}{5}$ b $\frac{1}{5}$ c $\frac{1}{24}$ d $\frac{1}{60}$

12 160 mins

13 $\frac{3}{8}$

14 1 hour 20 minutes

15 180

Practice questions 1.6

1 a $\frac{1}{100}$ b $\frac{1}{4}$ c $\frac{57}{50}$ d $\frac{19}{400}$

2 a 17% b 15% c 262.5% d 14%

3 a 0.14 b 0.85 c 1.26 d 0.099

4 a 42% b 30% c 9% d 205%

5 a €54.60

b 7.8 km

c 3.6 mins or 3 mins and 36 s

d 84 g

6 a 80% b 5% c 10% d 70%

7 a $13.20 b $52.80 c $47.52

8 a 10% b 30% c $33\frac{1}{3}\%$

d $26\frac{2}{3}\%$ e $73\frac{1}{3}\%$ f 60%

Practice questions 1.7

1 a Student's own answer

b certain

c Student's own answer

d even chance

e impossible

f Student's own answer

2 Answer will depend on student's answers to question 1

3 a $\frac{1}{5}$ b $\frac{2}{5}$ c $\frac{7}{10}$ d $\frac{2}{5}$

4 a $\frac{1}{2}$ b $\frac{1}{3}$ c $\frac{1}{6}$

d $\frac{2}{3}$ e $\frac{1}{3}$

Practice questions 1.8

1 a straight b obtuse c right d reflex e acute

2 a complementary angles b supplementary angles c vertically opposite angles
d corresponding angles e alternate angles f co-interior angles

3 a 32° b 133° c 240° d 105° e 73°
f 60° g 68° h 122°

4 a pentagon b ellipse c trapezium d rhombus

5 a

Polygon	Number of sides	Sum of the interior angles
Triangle	3	180°
Quadrilateral	4	360°
Pentagon	5	540°
Hexagon	6	720°
Heptagon	7	900°
Octagon	8	1080°

b $n - 2$

6 a 70° b 74° c 90° d 120°

7 a 60°, acute-angled, equilateral b 40°, acute-angled, isosceles
c 30°, obtuse-angled, scalene d 35°, right-angled, scalene
e 45°, right-angled, isosceles f 40°, obtuse-angled, isosceles
g 65°, acute-angled, scalene h 50°, acute-angled, isosceles

8

Shape number	Name	Prism	Pyramid	Curved surface	Number of faces	Number of vertices	Number of edges
1	Cube	Yes	No	No	6	8	12
2	Rectangular prism	Yes	No	No	6	8	12
3	Triangular prism	Yes	No	No	5	6	9
4	Cylinder	Yes	No	Yes	3	0	2
5	Cone	No	Yes	Yes	2	1	1
6	Square-based pyramid	No	Yes	No	5	5	8
7	Rectangular-based pyramid	No	Yes	No	5	5	8
8	Triangular-based pyramid	No	Yes	No	4	4	6
9	Sphere	No	No	Yes	1	0	0

9 a rectangular prism b triangular prism c square-based pyramid

Practice questions 1.9

1 a 08:15 b 19:40 c 03:52 d 22:37

2 199 mins or 3 hrs and 19 mins

3 5 hrs 10 mins

4 1 hr 35 mins

5 a 80 cm b 63.45 m
c 43 000 mm d 362 000 cm

6 a 720 cm or 7.2 m
b 643 mm or 64.3 cm
c 10 832 m or 10.832 km
d 930 cm or 9.3 m

7 a 15 cm
b 360 cm or 3.6 m
c 28 mm

8 a 13.5 cm^2
b 4500 cm^2 or 0.45 m^2
c 49 mm^2

9 a 24 m^2 b 17.5 cm^2 c 30 cm^2

10 a 210 m^3 b 84 m^3 c 45 m^3
d 1980 m^3 e 2520 m^3

Practice questions 1.10

1 a

s	1	2	3	4	5
m	4	7	10	13	16

b The number of matchsticks is 3 times the number of squares plus one.
c $m = 3 \times s + 1$
d i 61 ii 91 iii 301

2 a $7k$ b $3 + 4z$ c $\frac{6}{d}$
d xyz e $\frac{6h}{z}$ f $9(f + 4)$

3 a $2 \times b - 7$ b $7 \times r - s \div 3$
c $8 \times (a + 5)$ d $(p + 4) \div 5$

4 a 10 b 7 c 19 d 17
e −2 f 4 g 40 h 4

5 a D b E c C d A
e F f B

6 a $8xy$ b $15d$ c $6ab$
d $27g$ e $-12p$ f $-32k$
g $14mn$ h $5e - 4f$

7 a $4c$ b $-7h$ c $\frac{-3t}{u}$
d $\frac{3a}{b}$ e $-5x$ f $-4a$
g $\frac{3r}{2s}$ h $\frac{3k}{2l}$

8 a $4v$ b $2n$ c $18a + 4b$
d $10k + 10l$ e $12f - 6e$ f $p + q$
g $3t^2 - 4t$ h $3w^2 - 2w$

9 a $6x + 2$ b $18q - 30$
c $3f + f^2$ d $4r^2 - 32r$
e $28a + 21b$ f $18g^2 - 6gh$
g $12n + 14$ h $4t - 4$

10 a p^3 b m^3n^2 c $30a^2$
d $15g^3$ e $140x^2y^2$ f $15r^3s$

11 a $h = 3$ b $d = 7$ c $r = 1$
d $w = 10$ e $y = 6$ f $z = 5$
g $s = 3$ h $q = 15$

12 a $x + (x - 6) = 78$
Sam is 36 and Bob is 42
b $x + 3x = 56$
Tanya has 14 books.
c $3x - 8 = -14$
The number is −2
d $7(x + 3) + 4(x) = 54$
Children's tickets cost €3 and adult tickets cost €6.

Practice questions 1.11

1 a discrete b continuous
c discrete d continuous
e continuous f discrete

2 a

Outcome	Frequency
1	6
2	6
3	3
4	9
5	5

b 29 c 4

d

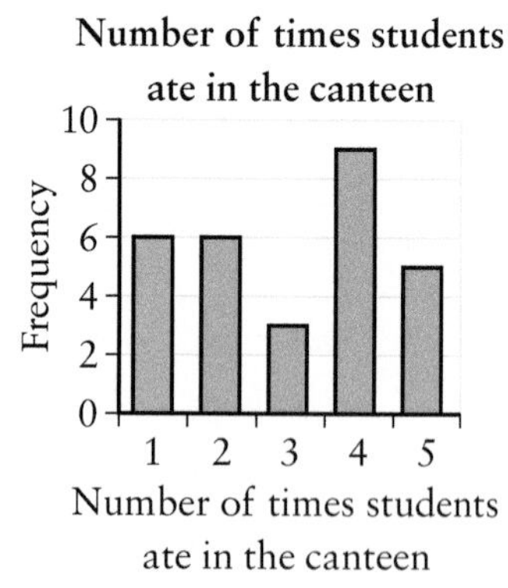

3

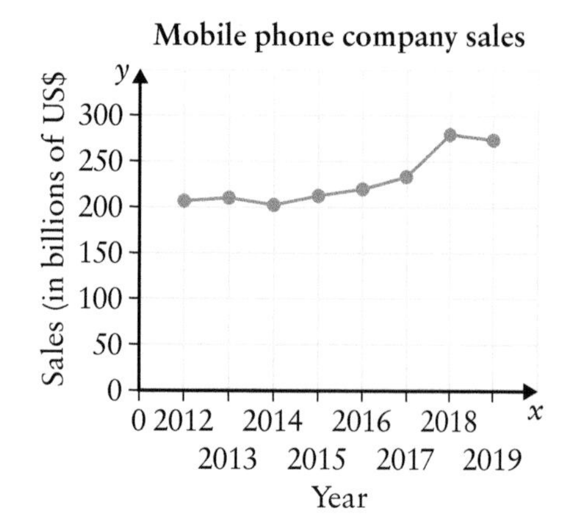

4

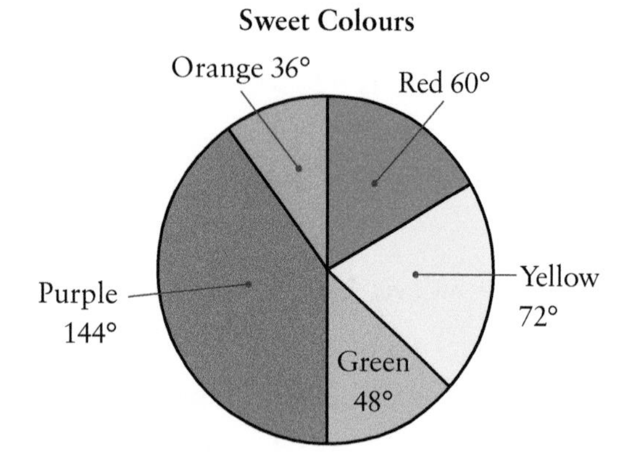

5 a 7 b 8 and 9 c 6.9 d 7.5

Practice questions 1.12

1 Three legged stools: 17
Four legged stools: 9

2 12 ways (MJCP, MJPC, MACP, MAPC, JMCP, JMPC, JACP, JAPC, AMCP, AMPC, AJCP, AJPC)

3 26 children in grade 1 can swim.

	Grade 1	Grade 2	Grade 3	Total
Can swim	26	21	12	59
Cannot swim	11	8	18	37
Total	37	29	30	96

4 Sarah Lee (attends Averly International), Becky Taylor (attends Rice High School), Karen Foster (attends Oaks Hill School)

5 The ladder has 23 rungs.

6 i 31 ii 51 iii 151

7 Matthew is 47.

8 12 hours

9 a 24 ways
b 34
c 4 and 9
d 20 games
e Florian is 23.
f 12 chickens
g 1
h 24 unique outfits
i Mr French teaches Craft, Mr English teaches French, Mr Kraft teaches Music and Mr Musik teaches English.

Practice questions 1.13

1 a 6 b 18 c 24

2 5 students study both French and Spanish.

3 a 23 b 12 c 65

4 48 students passed both.

5 a 4 b 2 c 21
d 28 e 8

6 a 17 b 23 c 27
d 6 e 3 f 5
g 36 h 20 i 65

Check your knowledge questions

1

	$\mathbb{N}$	$\mathbb{Z}$	$\mathbb{R}$
−9		✓	✓
15	✓	✓	✓
7.5			✓
$-\frac{3}{2}$			✓

2 a 144 b 29 c 0
d 67 e 4 f 26

3 16

4 $3 \times 3 \times 3 \times 5 \times 5$ or $3^3 \times 5^2$

5 72

6 a −15 b −13 c 1 d −30
e −28 f 45 g −10 h 6

7 −8 and 6

8 a 23.352 b 48.93
c 71.06 d 1.038

9 a 3.0 b 142.1
c 53.4 d 64.0

10 a $5\frac{9}{10}$ b $3\frac{3}{7}$ c $1\frac{19}{20}$ d $\frac{2}{5}$

11 375 m

12 $\frac{1}{12}$

13 a $15
b 12 minutes
c 180 ml

14 75%

15 a $\frac{1}{6}$ b $\frac{1}{12}$ c $\frac{1}{60}$ d $\frac{11}{15}$

16 a 140° b 20° c 80° d 135°
e 55° f 110° g 70° h 95°

17 a 06:23 b 23:07 c 09:41 d 16:35

18 a Saturday b Tuesday c Friday

19 a 52 cm b 15.7 cm

20 a 147 cm^2 b 9 cm^2

21 a 183.6 cm^3 b 308 mm^3

22 a 1 b 24 c −1 d −2

23 a $4a + 3b$ b $-5x - 6y$
c $2r^2 - 3r$ d $-3k + 4l$

24 a $5x - 15$ b $6 - 12d$
c $2y^2 - 4y$ d $10c^2 + 15bc$

25 a $m = 8$ b $x = 13$
c $w = 12$ d $m = 8$

26 a $x = 3$ b $m = 3$
c $a = 3$ d $p = 3$

27 a

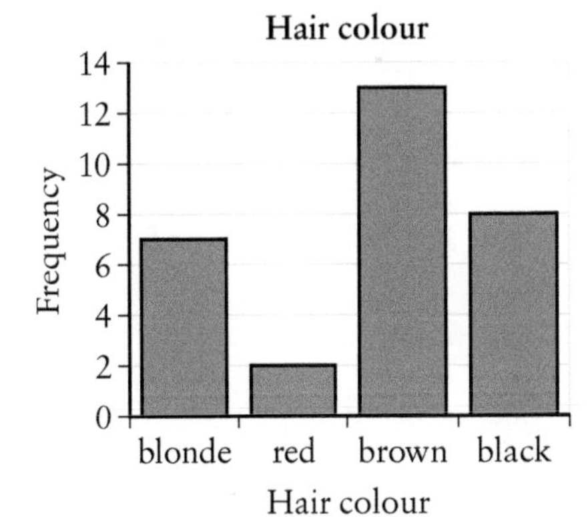

b

Hair Colours
Black 96°
Blonde 84°
Red 24°
Brown 156°

28 a 2 b 3.5 c 4 d 5

29 a 12 adult tickets
b 21 handshakes
c 24 sweets

30 a 25 b 36

31 10 hours 20 minutes (= 8 × 1 hr 10 mins + 1 hr)

Chapter 2 answers

Do you recall?

1 0.4

2 $\frac{7}{20}$

3 0.23

4 $45

Practice questions 2.1.1

1 a 1% b 27% c 30%
d 55% e 62.5% f $208.\dot{3}\%$
g 28.6% h $183.\dot{3}\%$

2 a 5% b 66% c 10%
d 145% e 1.5% f 237.5%

3 a $\frac{1}{5}$, 0.5, 75% b 0.63, 72%, $\frac{3}{4}$
c 18%, $1\frac{1}{8}$, 1.8, d 65%, $\frac{2}{3}$, 0.68

Practice questions 2.1.2

1 a $\frac{9}{100}$ b $\frac{49}{100}$ c $\frac{3}{5}$
d $\frac{8}{25}$ e $1\frac{3}{20}$ f $3\frac{1}{4}$

2 a 0.14 b 0.03 c 0.4
d 5 e 0.1224 f 0.155

Practice questions 2.1.3

1 a 80 g b 33.2 m c $120
d €8.76 e 6.4 g f 2.1 ml

2 a 1.2 kg b 60 mm c 840 litres
d €2300 e 145.5 cm f 40 g

3 30% of 30

4 £412

5 70 million tons

6 21 million tons

7 2020

8 €7650

Practice questions 2.1.4

1 a 60% b 75% c 25% d 62.5%

2 a 10% b $16\frac{2}{3}\%$ c 400%
d $66\frac{2}{3}\%$ e 70% f $333\frac{1}{3}\%$
g $27\frac{1}{2}\%$ h 0.125%

3 Lauren scores 60% and Phillip scores 75%. Phillip scores the higher percentage.

4 $10\frac{1}{4}\%$

5 $22\frac{1}{2}\%$

6 80%

7 25

Practice questions 2.2

1 a C b B c F d H
e G f A g E h D

2 a €880 b 4.48 litres
c 111 g d 144 min

3 a $64 b 1.95 kg
c 576 mm d 58.8 litres

4 11.2 billion people

5 €322

6 $738

7 €43 260

8 $6930

Practice questions 2.3

1 a 400 cm b 700 t
c €1350 d 700 litres

2 $3900

3 40 000 million tons

4 €860

Practice questions 2.4

1 No because we don't know the initial price of the skateboards in either shop.

2 €14

3 $100.80

4 The basketball costs €14.22 after the discount at Sports Life and it costs €13.26 after the discount at Sportland. The basketball from Sportland is cheaper by €0.96.

5 a $196 b 30%

6 a 5% b 20%

Practice questions 2.5

1 a 6 : 5 b 6 : 1
c 5 : 1 d 5 : 1 : 6

2 a 2 : 1 b 1 : 2 c No

3 a $\frac{2}{7}$ b $\frac{5}{7}$

4 a 15 : 12 b 12 : 15 c 15 : 27
d $\frac{15}{27}$ e 12 : 27 f $\frac{12}{27}$

5 a Black b $\frac{7}{12}$ c 1 : 12

6 a No it is not sweeter. For example, if you are making 100 ml of the drink at a ratio of 1 : 3 you would use 25 ml of juice and 75 ml of water. If you are making 100 ml of the drink at a ratio of 1 : 4 you would use 20 ml of juice and 80 ml of water. So with the ratio 1 : 4 there is less juice and more water hence the drink is not sweeter than recommended.

b Yes it is sweeter. For example, if you are making 60 ml of the drink at a ratio of 1 : 3 you would use 15 ml of juice and 45 ml of water. If you are making 60 ml of the drink at a ratio of 1 : 2 you would use 20 ml of juice and 40 ml of water. So with the ratio 1 : 2 there is more juice and less water hence the drink is sweeter than recommended.

Practice questions 2.6

1 a 4 b 25 c 3 d 12

2 a 3 : 2 b 6 : 1 c 3 : 2
d 30 : 1 e 7 : 12 f 2 : 5
g 8 : 1 h 8 : 1 : 9

3 a 7 : 5 b 4 : 1
c 1 : 5 d 17 : 12 : 5

4 a 9 : 4 b 4 : 9
c 9 : 13 d 9 : 4 : 13

5 1 : 1000

6 a 7 : 5 b 5 : 6 c 7 : 6
d 7 : 5 : 6 e 7 : 18

Practice questions 2.7

1 a 15 : 9 b 1638 : 1092
c 65 : 195 : 325

2 385

3 €964

4 25 kg tin, 5 kg lead

5 480

6 16 kg gravel, 24 kg sand, 32 kg cement

7 4.2 cm, 8.4 cm, 12.6 cm

8 a i J ii H iii D iv I v C

b 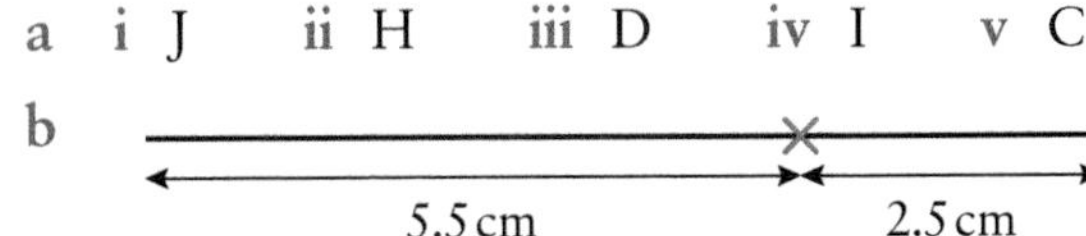

9 Town A = 18 030, Town B = 48 080, Town C = 19 232

Practice questions 2.8

1 a 2 km/h b 4 kg/$ c 5 c/g
d 28 students/teacher
e 2.5 kg/m^2 f 48 litres/h

2 a 192 b 300 c 3000
d 120 e 240 f 2000

3 52 km/h

Practice questions 2.9

1 a 240 km/h b 144 litres/day
c 0.04 litres/min d 18 km/h

2 a Low Fat Supermilk 1.75 litres
b 500 g of tomatoes

3 525 km

4 Car A is more economical because Car A consumes 0.07 litres/km and Car B consumes 0.08 litres/km.

5 3 hours and 45 mins

6 90

7 ¥31 800

8 2 mins 15 s

9 $748

10 288

11 a 18 000 000 km
b 8 min 16 s

12 A$81.50

13 €1 = R19.69

14 2 min 24 s

Practice questions 2.10

1 a 500 cm b 10 000 cm c 20 000 m

2 a 1 : 1000 b 1 : 1500
c 1 : 20 000 d 1 : 20 000

3 a 15 m b 26 m
c 1 m d 17.5 m

4 a 5 mm or 0.5cm b 35 mm or 3.5 cm
c 72 cm d 75 mm or 7.5 cm

5 Student's own answer, for example, using a rectangular page with dimensions 21 cm by 30 cm:
a 1 : 2000
b 1 : 100
c 1 : 25 000

6 3.9 m by 5.6 m

7 e.g. 1 : 400

8 a 80 km b 96 km
c 220 km d 104 km

Check your knowledge questions

1 a 33% b 12% c $41\frac{2}{3}\%$
d $177\frac{7}{9}\%$ e $216\frac{2}{3}\%$ f $46\frac{2}{3}\%$

2 a 9% b 72% c 30%
d 267% e 4.52% f 431.5%

3 a $\frac{1}{10}$ b $\frac{6}{5}$ c $\frac{16}{25}$ d $\frac{9}{400}$

4 a 0.23 b 0.05 c 0.6 d 1.75

5 a 3150 g b 34.1 cm c 240 kg
d 86.4 min e 3.84 ml f 1.8 g

6 $78.30

7 a $1\frac{2}{3}\%$ b 25% c $33\frac{1}{3}\%$ d 80%

8 $291.20

9 448 cm

10 €416

11 1296

12 a 800 g b 190 mm
c $104 d 800 ml

13 250 000

14 a 20% b 10%

15 a 30 : 1 b 1 : 20
c 100 : 1 d 6 : 25

16 18

17 324

18 Albert: $173 715, Bob: $289 525, Charles: $57 905

19 4 ml

20 250 g of chocolate for €4.50

21 a 420 km b 52 litres

22 54 km/h

23 16.7 m/s

24 €15.95

25 €135.20

26 500 revs/min

27 a 400 cm or 4 m b 620 cm or 6.2 m
 c 30 cm d 370 cm or 3.7 m

28 a 100 cm or 1 m b 190 cm or 1.9 m
 c 3.03 cm d 3.76 cm

Chapter 3 answers

Do you recall?

1 a −3 b 5 c −13 d −12
 e 4

2 a −27 b 40 c −28 d −24

3 a 5^2 b 2^4 c $(-4)^3$

4 A **variable** is a letter or symbol used in mathematics to represent a number, EX: x or ab

5 An **expression** is a collection of numbers and/or variables, for example '$2x + 3y$'.

6 **Terms** are the individual elements of an expression, usually separated by addition or subtraction. For example, the expression '$2x + 3y$' has two terms: $2x$ and $3y$.

7 **Coefficients** are numbers in front of variables, for example 3 is the coefficient of $3y$. Remember, the operation happening between a variable and its coefficient is multiplication, so we could also rewrite $3y$ as $3 \times y$.

8 **Like terms** are terms with the same variable, for example $2x$ and $5x$. In contrast $2x$ and $3x^2$ are not like terms because x and x^2 are different.

Practice questions 3.1.1

1 a ×2 b 24, 48, 96 c geometric

2 a +4 b 13, 17, 21 c arithmetic

3 a +10 b 40, 50, 60 c arithmetic

4 a ÷3 b $3, 1, \frac{1}{3}$ c geometric

5 a −9 b −19, −28, −37 c arithmetic

6 a ×3 b 9, 27, 81 c geometric

7 a +12 b 40, 52, 64 c arithmetic

8 a ×−2 b 8, −16, 32 c geometric

9 a −7 b −26, −33, −40 c arithmetic

10 a Add the amount of digits in the next term in the units place.
 b 1234, 12345
 c It is neither arithmetic or geometric.

11 a First letter of the next day in the week
 b S, S
 c Monday, Tuesday, Wednesday, Thursday, Friday correspond to the days of the week.

12 a Square root of the total number of terms
 b $\sqrt{6}, \sqrt{7}$
 c $\sqrt{1} = 1$ and $\sqrt{4} = 2$

13 a Subtract the number of terms currently present from the previous term.
 b −15, −21
 c This is the negative version of triangular numbers, with the addition of zero.

14 a First letter of number corresponding to total term numbers
 b S, S
 c One, Two, Three, Four, Five match the pattern.

15 a Multiply by the number of terms.
 b 120, 720
 c $1 \times 2 = 2, 2 \times 3 = 6, 6 \times 4 = 24$

16 a The next score up in tennis
 b 40, game
 c Love is the starting score (zero) in a game of tennis, followed by 15, then 30

Practice questions 3.1.2

1 $y = 3x - 1$

a

x	1	2	3	4	5
y	2	5	8	11	14

b +3

c The coefficient of x

2 $y = 4 - 2x$

a

x	1	2	3	4	5
y	2	0	−2	−4	−6

b −2

c The coefficient of x

3 $y = 5x + 3$

a

x	1	2	3	4	5
y	8	13	18	23	28

b +5

c The coefficient of x

4 $y = -1 - x$

a

x	1	2	3	4	5
y	−2	−3	−4	−5	−6

b −1

c The coefficient of x

5 $y = -10 + x$

a

x	1	2	3	4	5
y	−9	−8	−7	−6	−5

b +1

c The coefficient of x

6 a i $b = 1 - a^2$

a	1	2	3	4
b	0	−3	−8	−15

ii $b = a^2 + a$

a	1	2	3	4
b	2	6	12	20

b The patterns formed by b are not simple arithmetic patterns, and instead are changing with each new term. For example in the first pattern 0, −3, −8, −15 the pattern is −3, −5, −7 showing that it is changing.

7 a +5 b −2 c +1

8 a 45 b 34 c −6

d 27 e 37

9 a 14 b −5 c −18

d −16 e 12 f 20

g −38 h 20 i 6

Practice questions 3.1.3

1 a Two more dots drawn vertically on the left hand side.

b Two more each time

c $C = 2F + 1$ where C is number of circles and F is figure number so for $F = 6$, $C = 13$

d

Figure 5

Figure 6

2 a +2

b $T = 2F + 1$ where T is number of toothpicks and F is figure number. So for $F = 10$, $T = 21$

c You can see if it works for the figures you have.

d e.g. Both lead to a formula of the type $y = 2x + 1$, i.e an arithmetic sequence based on adding 2 each time.

3 a 15

b 6

c e.g. It increases by one more than the version number.

d The terms in the pattern do not form an arithmetic sequence. The difference in successive terms increases by one each time.

Practice questions 3.2.1

1 a $3a, 2a$ b $2x^2, x^2$
c $3fg, -fg$ d $-2ab, ba$

2 a $7q - 2$ b $2 + p$
c $7 - 2d$ d $7c - 5$
e $7u + 4$ f $4 - a$

3 a $4 + 3b$ b $9z + 2y$
c $c - 3d$ d $3a + 2b$
e $10t - 1$ f $6 - 3x + 5h$
g $11k - 5j + 4$

4 a $-a^2 + a + 11$ b $7g^2 + 2g - 3$
c $3r^4 + 14r$ d $-3k^3 + 12k$
e $-3z^2 + 4z$ f $5u^3 + 3u + 2$
g $7y^2 - 7y + 5$

5 a $4pqr + 4p^2 + 2pq$ b $-h^2 - h + 9$
c $4m + 2n$ d $a - 6b + 5ab$
e $2z^3 - z^2 - z$ f $8x^2 + y$
g $13f + g^2$

6 a $m^2 + 6m$ b $5wx - 2x$
c $a + 8b + 3ab$ d $e^2 - 4ef - f$

Practice questions 3.2.2

1 a $6x$ b $18h^3$ c $5ac$
d $30p$ e $10u^4$ f $72mn$
g $56ab$ h $36fgh$ i $4.8s^2t$

2 a $-3a$ b $-16d$ c $36m$
d $-28w$ e $-35fg$ f $-55x$
g $27jk$ h $30np^2$ i $-36v^2w$

3 a $8pq$ b abc c $6xy$
d $-12hj$ e $-40z$ f $42ab$
g $-12fg$ h $75w^2yz$ i $-5.6km^3n^2$

4 a $5x$ b $16m$ c $4u$
d $4b$ e $20st$ f $-3r$
g $43h$ h $10q$ i $-3uv$

5 a -18 b 27 c 33

Practice questions 3.2.3

1 a a^2 b d^4 c r^5
d m^7 e $-e^3$ f $5j^3$

2 a $z \times z \times z$
b $r \times r \times r \times r \times r$
c $t \times t$
d $-p \times -p \times -p \times -p \times -p$

3 a $10a^3$ b $14w^2$ c $12f^3$
d $12k^3$ e $9h^4$ f $12r^5$
g $56m^6$

4 a $3mn$ b $6h^2k$ c $24bc$
d $28st^3$ e $5mn^5$ f $14g^2h^3$

5 a $-32k^4$ b $42d^6$ c $-54b^5$
d $5v^3$ e $8p^5$

6 a $25h^2$ b $9a^2$ c $36j^4$
d $16k^6$ e $4n^4$

7 a $-48jk^2$ b $-2m^3n^3$ c $30p^3q^3$
d $-56a^2b^3$ e $63x^2y^4$

Practice questions 3.2.4

1 a $\frac{f}{2}$ b $\frac{3}{b}$ c $\frac{3x}{4}$ d $\frac{5}{3h}$

2 a $\frac{2}{5}$ b $\frac{4}{3}$ c $\frac{5}{3}$ d 3

3 a $\frac{k^2}{5}$ b $\frac{2}{3t^3}$ c $\frac{4w}{7}$ d $\frac{2}{5z^3}$

4 a $\frac{j}{k^3}$ b $\frac{mn}{6}$ c $\frac{3y}{4z}$ d $\frac{3a}{2b^3}$

Practice questions 3.2.5

1 a 21 b $7a^2b$ c 21

2 a -4 b ac c -4

3 a -4 b $4a$ c -4

4 a $20xy$ b $\frac{m}{2}$ c $-47jk$
d $\frac{11f}{4}$ e $\frac{f}{4}$

5 a $7st$ b $\frac{d^2}{2c}$ c $\frac{2a^2 + 7b}{7a}$

d $7uw$ e $-\frac{3h^2}{2}$

Practice questions 3.3.1

1 a $15 + 9x$ b $20f - 10$

c $h^2 + h$ d $7t^2 + 9t$

e $2b - 2b^2$ f $56g^3 + 72g^2$

2 a $-w - 2$ b $-30 + 6x$

c $6p - 3q$ d $-8t + t^2$

e $-8c^2 + 14cd$ f $4h^2 + 5h^3$

3 a $8x + 11$ b $10 - 10a$

c $3jk + 8j + 3k^2$ d $5z + 62z^2 - 9z^3$

e $35m - 56 - 3m^2$ f $4r^3 - r^2 + r - 2$

4 a $5p + 10q - 15$ b $9u - 63v - 72$

c $-6 - pq + 9p$ d $6xz - 6z^2 + 3z$

e $-8x - 4x^2 + 32$ f $-2j^2 + 2jk - 14j$

5 a $54f^2 - 2fg + 27f + 8g^2 - 40g$

b $-54k + 26k^2 - 7$

Practice questions 3.3.2

1 a Factorised completely

b $t^2(14t + 13)$

c $-3(x + 3y)$

d Factorised completely

e $6m(2 - m)$

f $-3k(2j + 5)$

2 a $2(w - 2z)$

b $5u(1 + 3u)$

c $-(8f + 7g)$

d $3x(3 + x)$

e $5b(a - 4)$

f $4mn(1 + 2n)$

g $7r^2(2r - 5)$

h $3q^2(pq - 2)$

3 a $3j(1 + 6k)$

b $6m(-3 + 4n)$ or $-6m(3 - 4n)$

c $14q(3q + 2r)$

d $8h(9 - 2h^2)$

e $7w(3x - 2)$

f $4s(-9t + 7s)$ or $-4s(9t - 7s)$

4 a $-(7j + 2k)$

b $-2(xy + 3z)$

c $-n(m + 3)$

d $-g(7 + g)$

e $-4q(2p + 3)$

f $-8s^2t(5 + 3st)$

5 a $j(2 - 3k + j)$

b $3(2h^2 + 7h - 12)$

c $-(3 + 2r + 5s)$

d $3d(3 + 5c + d)$

e $7m(2m^2 - 4 - n)$

f $-u(5v + 6u^2 + 7uv^2)$

Practice questions 3.3.3

1 a £1.90c + £8.50p

b £1.90(8) + £8.50(4) = £49.20

2 a $42n + 21w$

b $42(2) + 21(11) =$ €315

3 a $30b^2$ b $30b$

4 a $49y^2$ b $28y$

5 a $5j^2 + 3j$ b $5j + 3 + 6k$

6 a

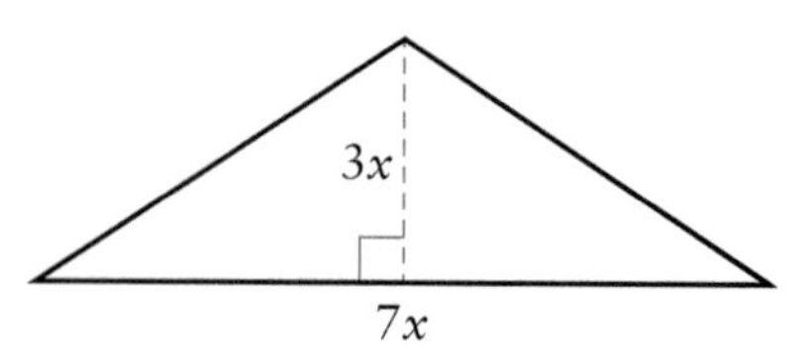

b $\frac{21x^2}{2}$

7 a 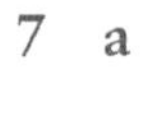

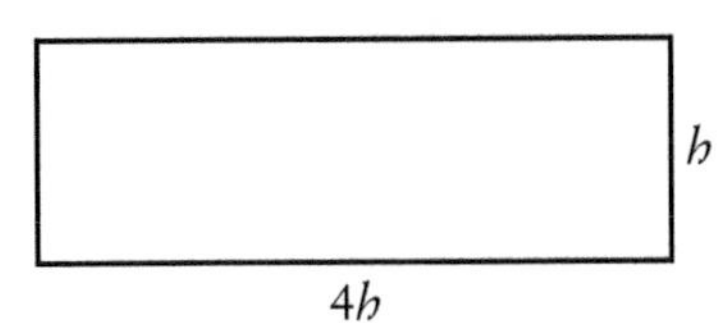

b $4h^2$

8 a

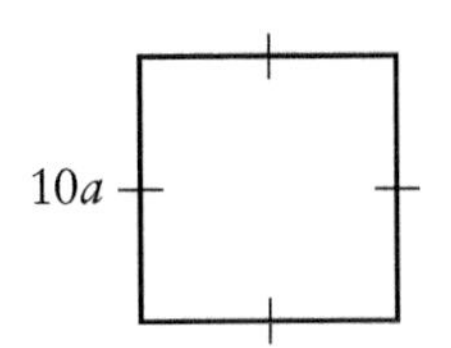

b $100a^2$

9 a

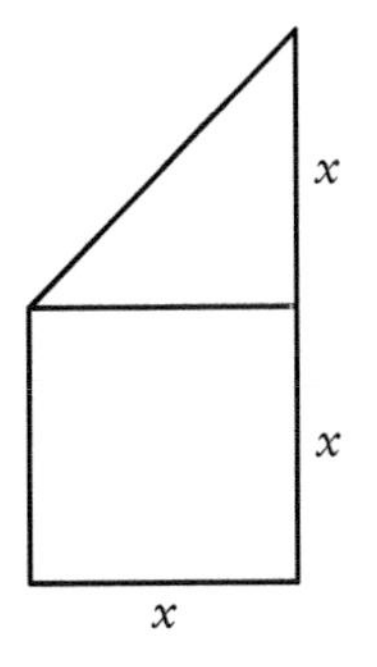

b $\frac{3x^2}{2}$

Check your knowledge questions

1 a Divide by 2 each time

b $1, \frac{1}{2}, \frac{1}{4}$

c Geometric

2 a

x	1	2	3	4	5
y	6	10	14	18	22

b Add 4 each time c Arithmetic

d $4x$ e 26, 30, 34

3 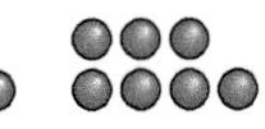 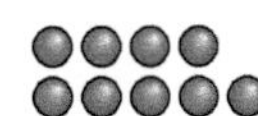 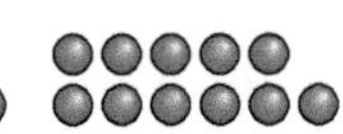

Figure 1 Figure 2 Figure 3 Figure 4

4 a 0 b -8 c 16

d -8 e 108

5 a $a^2 + 3a - 2$ b $3pqr + 6qr$

c $49h^4$ d $-72m^5$

e $25z^2$ f $3f$

g $13x$ h $\frac{2jk^2}{3}$

i $3w^2$ j $\frac{2s^2}{5t^3}$

k $\frac{y}{8}$

6 a $3p + 6$ b $8x^2 - 8x$

c $2z^3 - 2z^2 + 6z$ d $-b^2 + 7b - 6$

e $9a + a^2$

7 a $2b(5b - 7a)$ b $3t(2s^2 + 5s - t^2)$

c $-6z(5z^2 + 8 + 3z)$

8 a

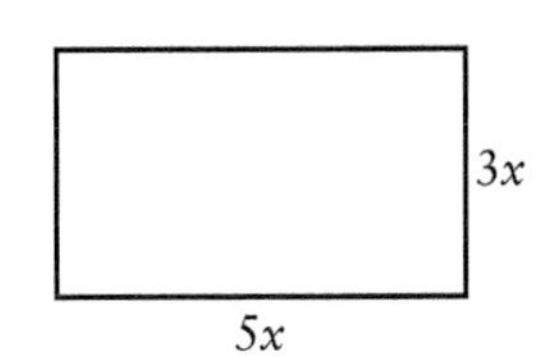

b $15x^2$ c $16x$

Chapter 4 answers

Do you recall?

1 An equation is a mathematical statement consisting of an 'equals' symbol. You are required to 'solve' an equation. For example: Solve the equation $3x + 5 = 14$, Answer $x = 3$

2 10

3 9

4 6

5 a $2 \times 2 \times 2$

b $-1 \times -1 \times -1 \times -1 \times -1$

c 7×7

6 a $8 + 3^2 = 8 + 9 = 17$

b $5 - 2 \times 6 = 5 - 12 = -7$

c $(3 - 7) \div (-2) = (-4) \div (-2) = 2$

Practice questions 4.1.1

1 a Adding 2 b Dividing by 7
c Multiplying by 3 d Subtracting 15
e Square rooting a number

2 a $+c$ b $\times 7$ c $+3$ d $\div -8$
e $+u$ f $\times 9$ g $\div f$

3 a i -2 ii $\div 3$
b i $+2$ ii $\div 7$
c i $+6$ ii $\times 2$
d i -8 ii $\times -1$ (or $\div -1$)
e i -1 ii $\times 4$
f i $\div 2$ ii -1
g i -3 ii $\div -2$
h i $\times 3$ ii -4
i i -5 ii $\times -6$
j i $\div -3$ ii $+5$
k i $\times 7$ ii $+8$

4 a $-2, \div 3$ b $3f$

5 a $\div 4, +1$ b $m - 1$

6 a $-8, \times -1$ b $-j$

7 a $-1, \times 2$ b $\frac{k}{2}$

8 a $-3, \div -5$ b $-5n$

9 a $\div 4$ b -1
c $+1, \div 2$ d $\times 2$
e $\times -3$ f $+9, \times -1$
g $-7, \div -2$ h $\div 8, +1$
i $\times 5, -2, \times -1$ j $\sqrt{\ }, -6, \div -2$

Practice questions 4.1.2

1 $p = -11 \Rightarrow 2 + (-11) = -9$ ✓
2 $q = -2 \Rightarrow 7 + 3(-2) = 1$ ✓
3 $r = -4 \Rightarrow 9 - (-4) = 13$ ✓
4 $s = -4 \Rightarrow (-4) - 6 = -10$ ✓
5 $t = 6 \Rightarrow 15 - 2(6) = 3$ ✓

6 a $a = 1$ b $b = -4$ c $c = 7$
d $d = 8$ e $e = 21$ f $f = 51$
g $g = -3$ h $h = -12$

7 a $u = 2$ b $v = -2$ c $w = 10$
d $x = -3$ e $y = 9$ f $z = 12$
g $j = 7$ h $k = -5$ i $m = -14$
j $n = -4$ k $s = -12$
l $r = 3$ or -7

Practice questions 4.1.3

1 $x = -1$
2 $x = 2$
3 $x = 3$
4 $x = -1$
5 $x = \frac{1}{2}$
6 a 3 b -1 c 4
d -1 e 6
7 $x = -2$
8 7 mm
9 £8
10 9
11 6 cm
12 17 students
13 14 years old
14 12 dice

Practice questions 4.2.1

1 a False b True c True
d True e False f True
g False h True

2 a iv $x < 3$
b iii $x \neq 3$
c ii $x \geqslant 3$
d v $x \leqslant 3$
e i $x > 3$

3 a y, number line: open circle at 4, arrows both directions (4)

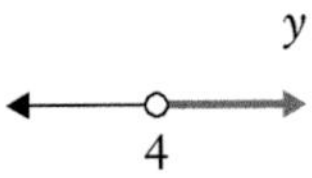

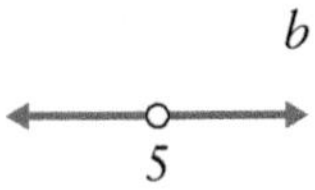

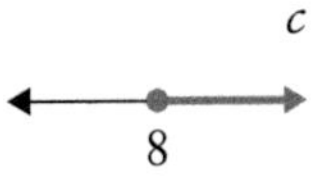

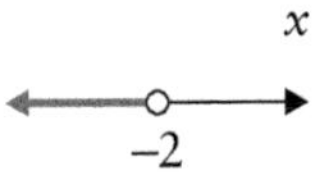

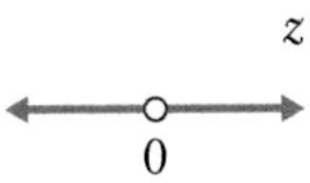

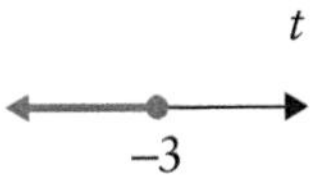

b b, number line: open circle at 5 (5)

c c, number line: closed circle at 8 (8)

d x, number line: open circle at −2 (−2)

e z, number line: open circle at 0 (0)

f t, number line: closed circle at −3 (−3)

g a, number line: open circle at 1 (1)

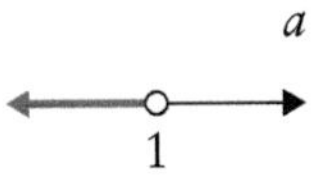

Practice questions 4.2.2

1 $x > 3$

2 $g \neq 2$

3 $b \geqslant -5$

4 $x < -7$

5 $a \leqslant -2$

6 a $k \neq 7$ b $f > 7$ c $b \leqslant 5$

d $m < -1$ e $j \geqslant -4$

7 a $n > 2$ b $f \leqslant 4$ c $r \neq -2$

d $x < -6$ e $a \geqslant -5$

8 a $a > 2$ b $b \leqslant -2$ c $c \neq 6$

d $d < 10$ e $e \leqslant -3$ f $f > 12$

g $g \neq 13$ h $h \leqslant 5$ i $i < -14$

j $j \neq -4$ k $k < 0$

Practice questions 4.3.1

1 a 26 cm b 14 m c 3.8 cm

2 a 3 cm b 1 km c 1.4 m

3 a 8 cm b 2 m c 1.3 m

4 a $x = a - 3$ b $x = \frac{b}{2}$

c $x = \frac{c + 1}{3}$ d $x = 4d$

e $x = 9 - e$ f $x = 5f + 2$

g $x = \frac{g - 2y}{2}$ h $x = \frac{h}{3} + 1$ or $\frac{h + 3}{3}$

i $x = -2(i - 1)$ or $-2i + 2$ or $x = 2(1 - i)$ or $x = 2 - 2i$

j $x = \frac{j + 1}{3y}$ k $x = \pm\sqrt{k}$

5 a $m = 1$ b $m = 7$ c $m = -2$

d $m = 2$ e $m = -4$

6 a $y = 4$ b $y = 3$ c $y = 1$

d $y = -2$ e $y = 2$ f $y = 1$

g $y = -6$ h $y = 9$ i $y = 4$

Practice questions 4.3.2

1 a $A = \frac{1}{2}bh$ b $A = x^2$ c $A = rt$

2 a (triangle with sides a, a, b)

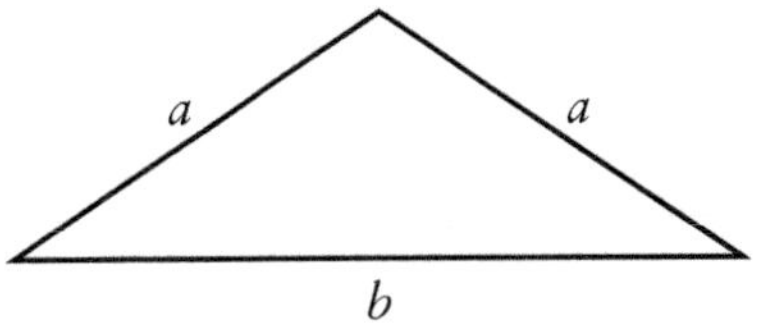

b $P = 2a + b$ c 31 cm

3 a (hexagon with all sides x)

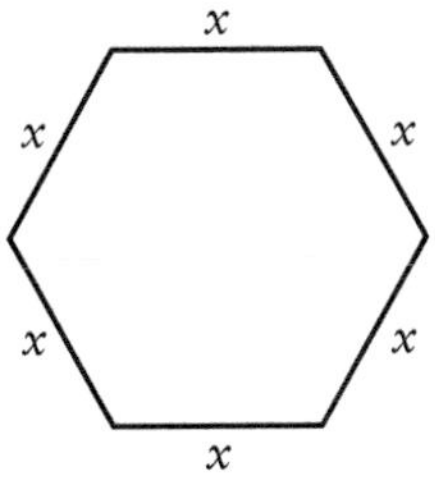

b $P = 6x$ c 54 cm d 1.1 m

4 a $x = \frac{a + b}{2}$ b 7 c 20.5

d $a = 2x - b$ e 30

5 a $T = c + 2b$ b €52

c i $c = T - 2b$ ii 27 cookies

6 a $C = 0.89s + 1.49a$ b £7.14

c £12.50 d $s = \frac{C - 1.49a}{0.89}$

e 2 sparkling waters

f Since you cannot buy 2.8 sparkling waters, you would need to round down for this problem, as you do not have enough money for 3 full bottles.

Check your knowledge questions

1 a Multiply by 4 b Add 7
 c Divide by 3 d Subtract 15

2 a ÷5 b +6
 c ×3 d −7, ×−1
 e −1, ÷2 f ÷3, +5 or +15, ÷3
 g × 7, − 2

3 a $a = 16$ b $b = 2$
 c $c = -2$ d $d = 12$
 e $e = -4$ f $f = 12$
 g $g = 3$

4 a $3n = n + 8$, $n = 4$
 b $2p + 6 = 20$, Tyler has 13
 c $6l = 12$, 4 metres.

5 a
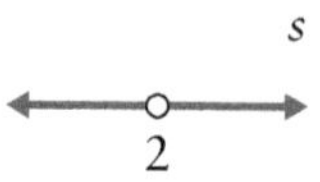

 b
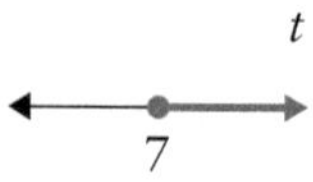

 c
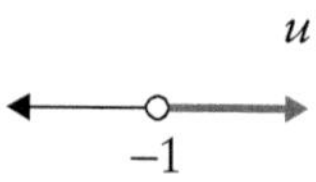

 d
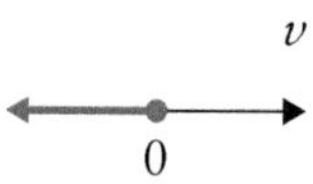

 e
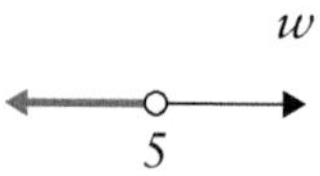

6 a $x > -9$ b $d \neq 13$ c $f \leqslant -4$
 d $y > 10$ e $m > 30$ f $s \leqslant -9$

7 a 50 kg m/s^2 b 1800 kg m/s^2
 c 4.62 kg m/s^2

8 a $u = v - at$ b $a = \dfrac{v - u}{t}$
 c $t = \dfrac{v - u}{a}$

9 a
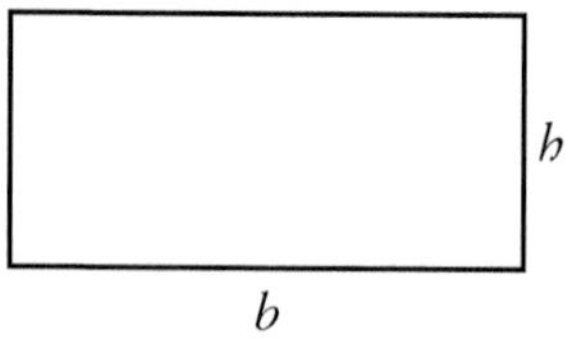

 b $P = 2b + 2h$ c $A = bh$
 d i 1.8 m ii 1.44 m^2

10 a $C = 8.9s + 12.8g + 14.5v$
 b i €108.60 ii €181 iii €362
 c 7 scarves

Chapter 5 answers

Do you recall?

1 $x = 4$

2
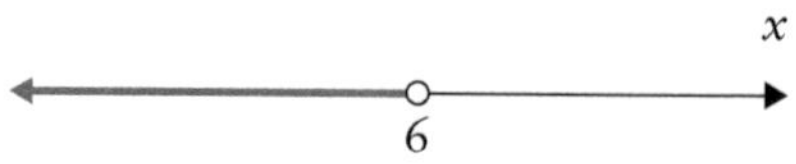

3 −9.8

Practice questions 5.1

1 a E2 b C7 c B1 d G8

2 a i Mt Scollen
 ii point where Matthew's River meets Stella Stream
 b 295505, 288482
 c Camp Michael to Mt Scollen, Phillipa's Campsite to Mt Scollen, Camp Michael to Phillipa's Campsite

3 a i San Salvador ii Maiz Islands
 b i 12°N 86°W ii 10°N 84°W

4 a
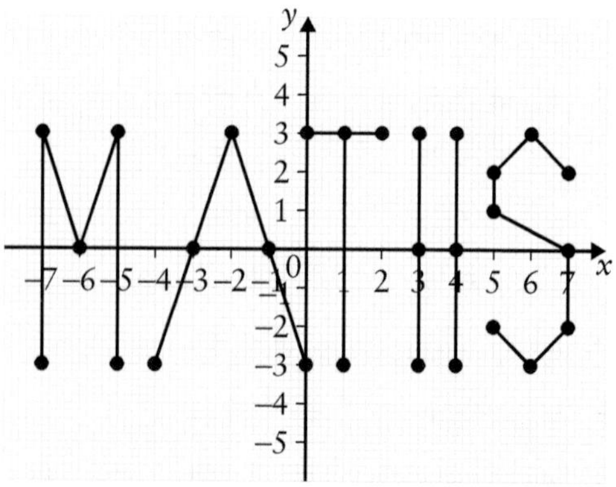

 b (−7, −3) to (−7, 3); (−7, 3) to (−6, 0); (−6, 0) to (−5, 3); (−5, 3) to (−5, −3)

5 Students play the game.

6 a i isosceles triangle

ii right(-angled) triangle

iii rectangle

b (4, −1) c (−2, 1), (−2, −3)

d y-axis

7 a III b D, E

c I and IV d I and III

8 a −2 or 2

b −1.5

Practice questions 5.2

1 a

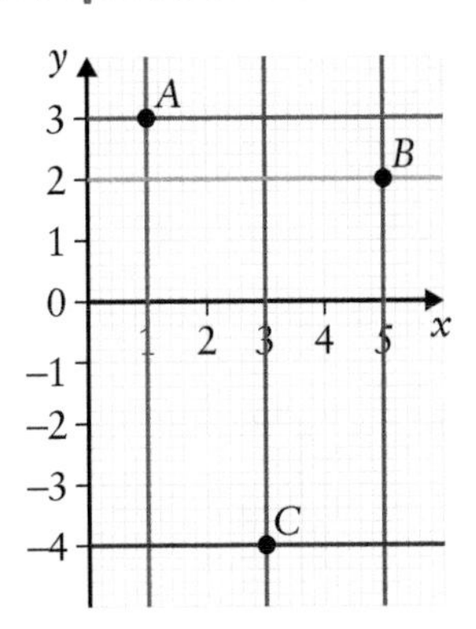

b $x = 1, x = 5, x = 3$

c $y = 3, y = 2, y = -4$

2

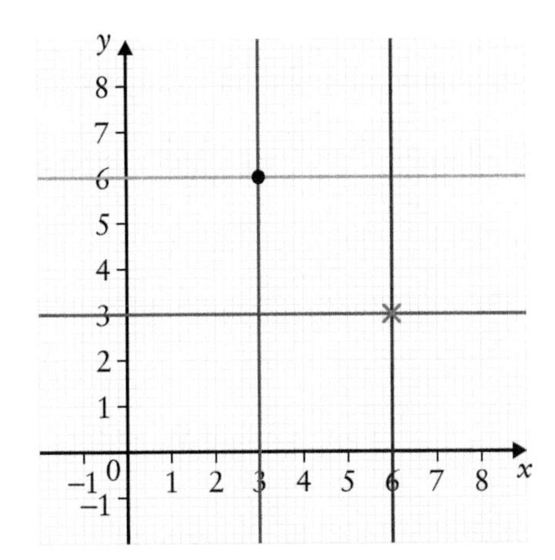

3 Any three points in form (____, 9)

4 a C b $y = -2$ c $x = 1$

d $y = 3$ e $x = 3$ f $x = 0$

g (3, 3)

5 a–d

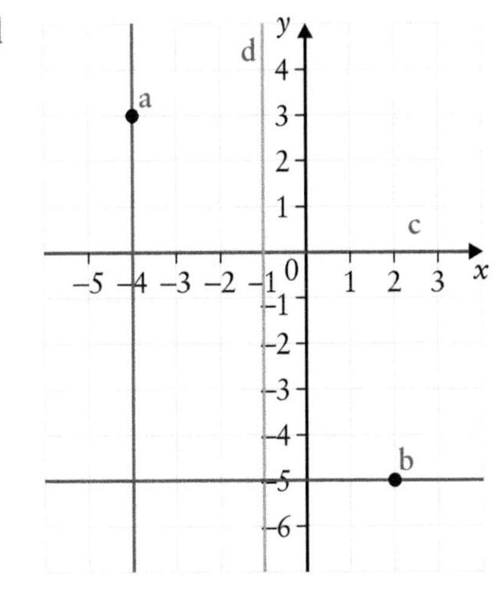

e 15 square units f (−1, 0)

6 a horizontal (12 vs 9) b $y = -2$

c $y = 4$ d A, D, E

e $(-1, 4); x = -1$ f $(-3, 2.5); y = 2.5$

Practice questions 5.3

1 a Rajiv, John

b Rajiv

c Indu

2 a B; line goes up by more than the others; 40%

b C; grades vary the least

c A, because after the third test A had achieved an average mark of close to 90%

3 a i In 2014, adults in the US watched TV for an average of 4 hours 20 minutes each day.

ii In 2019, adults in the US spent an average of 3 hours and 35 minutes a day on their TVs.
Or In 2019, adults in the US spent an average of 3 hours and 43 minutes a day on their mobile devices.

iii In 2015, adults in the US spent an average of 6 hours and 59 minutes each day with their TVs and mobile devices turned on.

iv The total screen time per day on mobile devices for adults in the US is predicted to increase significantly from 2014 to 2021.
Or The total screen time per day for adults in the US is predicted to increase slightly from 2014 to 2021.

b These points are predictions based on trends in the data and business projections in the industry.

4 a They finished together.

b Alena

c 5 min

d Alena seems to have backtracked or run in the wrong direction about 1.2 km into the race, because she moved closer to the start between 13:20 and 13:30.

e Where the graphs cross, their straight-line distance from the start is the same at that moment, however they may be in different locations (e.g. if the start is the centre of a circle, they could be standing at different points on circumference of the circle).

f 6 km in 1 h 20 min is 4.5 km/h.

Practice questions 5.4

1 Any three of (−9, −1), (−6, 0), (−3, 1), (0, 2), (3, 3).

Clearly there are an infinite number of solutions if you choose decimal values.

2

x	y
4	1
6	4
2	−2

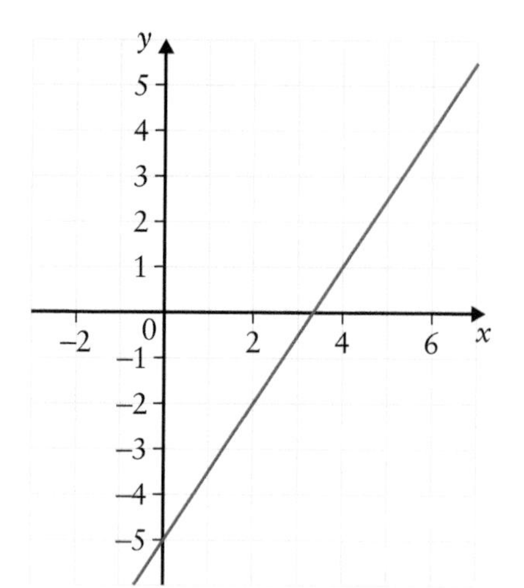

3

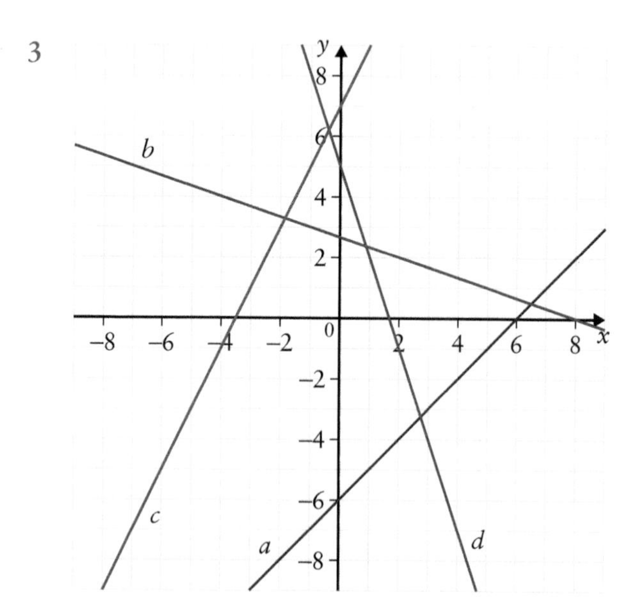

4 a b

b $x + y = 2$

c (3, −1)

d (0, 2)

e $2x - 6y = 12$, since line c is the only one with both coordinates positive to the right of the graph.

f 10 square units

5 a

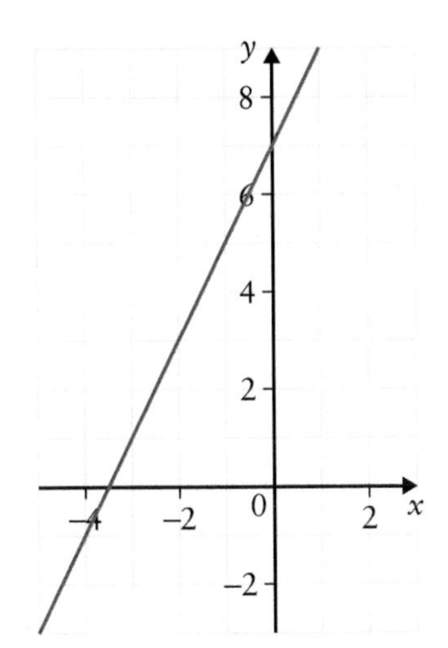

b c

c $y = -\frac{2}{3}x + 2$

6 The left side of each equation is the same, but the constant on the right is the opposite sign.

The graphs have the same steepness but the y-intercept is on the opposite side of the origin.

Practice questions 5.5

1 a (12, 0), (0, 12)

b (1, 0), (0, 1)

c (−3, 0), (0, −3)

d (8, 0), (0, 8)

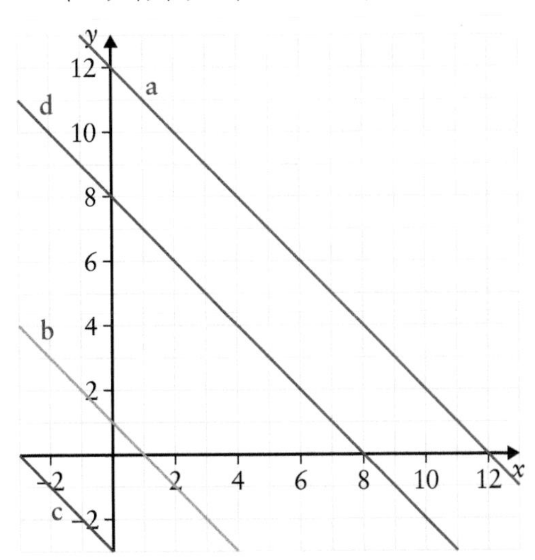

2 b, e and f

3 a $x + y = -4$

b $x + y = -2$

c $x + y = 0$

d $x + y = 1$

e $x + y = 3.5$

Check your knowledge questions

1 FLAT MAPS; The Earth's surface is curved, so flat maps distort the true distances and areas.

2 a D, H

b C, G

c $B; y = 3$

d $G; x = 5$

e A, L

f $x + y = 7$

g F, H

h $B; x + y = 1$

i $x + 2y = 4$

3 a The price of the car is approximately 40% of its original value, and it still has approximately 40% of its lifespan remaining, so it is a fair price to buy the car at.

b When the dotted blue line is below the orange line, you get a greater percentage of life out of the car than you pay for, which indicates good value for your money. The greatest difference seems to be for a car between one and three years old, so this is the best value.

c i 12.5% ii 6.25%

4 $y = 4$; horizontal

5 a They intersect at the point (4, 2).

b

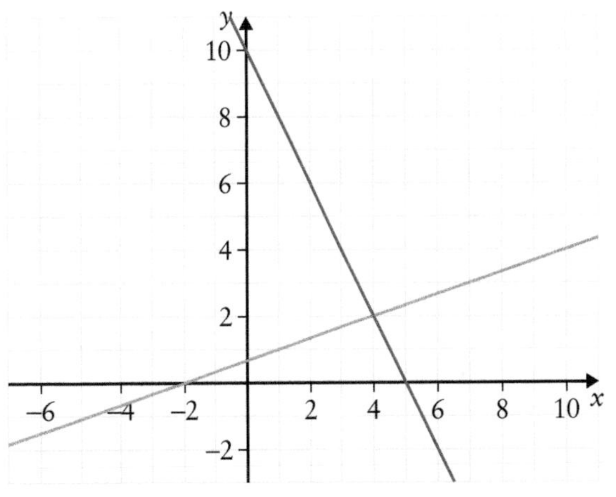

c They are all parallel, and they cross the y-axis at the value indicated by the constant on the right side of the equation.

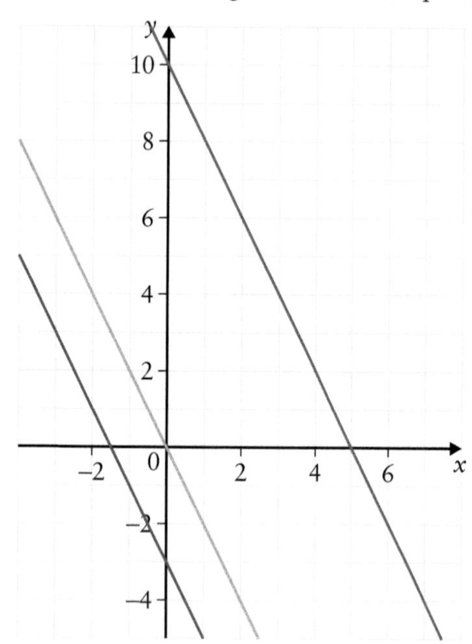

6

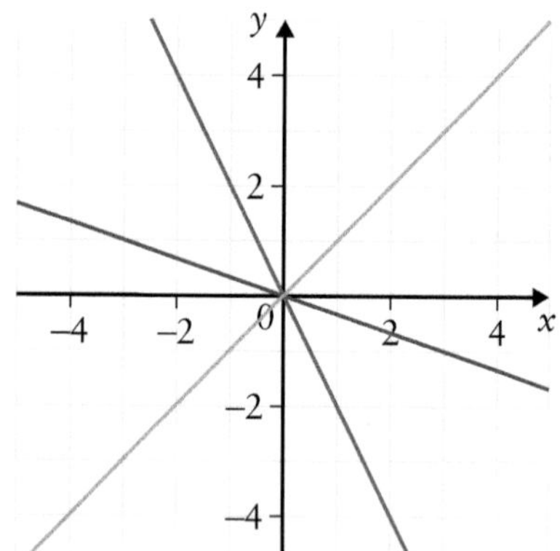

(0, 0), because when we substitute $x = 0$ into the equation, we get $By = 0$, and dividing by B gives $y = 0$.

7 a

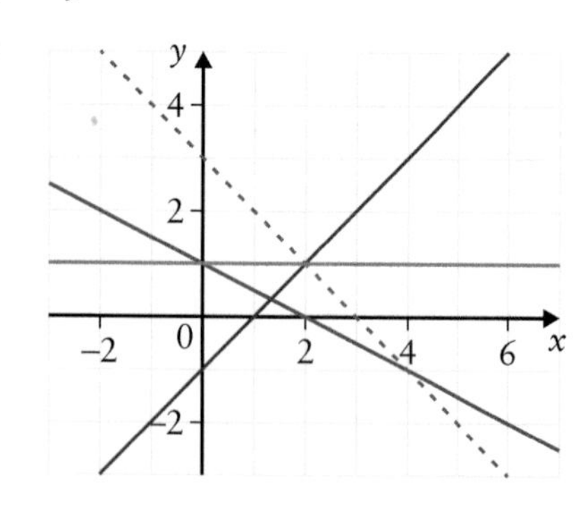

b For example: $x + 2y = 2$ is the odd one out, because it is the only one not to pass through (2, 1).

$y = 1$ is the odd one out, because it is the only one that does not have an x-intercept.

$x - y = 1$ is the odd one out, because it is the only one that does not have a positive y-intercept. (Saying that it is the only horizontal one would not be a good reason, because it does not say what the others have in common.)

$x + y = 3$ is the odd one out, because it is the only one that has a y-intercept that is not one unit away from the origin.

Chapter 6 answers

Do you recall?

1 90°

2 180°

3 360°

4 Complementary angles add to 90°. Supplementary angles add to 180°.

5 Transversal

6 An equilateral triangle has three equal sides and three equal angles (all 60°).

An isosceles triangle has two equal sides and two equal angles.

A scalene triangle has no equal sides and no equal angles.

Practice questions 6.1.1

1 a $\alpha = 47°, \alpha = 12° + 35°$

b $\beta = 34°, 75° = 41° + \beta$

c $\theta = 128°, \theta = 73° + 55°$

d $\gamma = 102°, 160° = 43° + \gamma + 15°$

e $\alpha = 252°, \alpha = 66° + 66° + 120°$

f $\theta = 13°, 270° = \theta + 85° + 155° + 17°$

2 a $\alpha = 55°$, angles in a right-angle sum to 90°.
b $\beta = 25°$, angles in a right-angle sum to 90°.
c $\gamma = 64°$, angles in a right-angle sum to 90°.
d $\theta = 19°$, angles in a right-angle sum to 90°.
e $\alpha = 12°$, angles in a right-angle sum to 90°.
f $\beta = 25°$, angles in a right-angle sum to 90°.
g $\gamma = 66°$, angles in a right-angle sum to 90°.
h $\theta = 32°$, angles in a right-angle sum to 90°.
i $\alpha = 19°$, angles in a right-angle sum to 90°.

3 a $\alpha = 125°$, angles on a straight-line sum to 180°.
b $\beta = 45°$, angles on a straight-line sum to 180°.
c $\gamma = 39°$, angles on a straight-line sum to 180°.
d $\theta = 106°$, angles on a straight-line sum to 180°.
e $\alpha = 70°$, angles on a straight-line sum to 180°.
f $\beta = 151°$, angles on a straight-line sum to 180°.
g $\gamma = 100°$, angles on a straight-line sum to 180°.
h $\theta = 36°$, angles on a straight-line sum to 180°.
i $\alpha = 117°$, angles on a straight-line sum to 180°.

4 a $\alpha = 20°$, angles in a right-angle sum to 90°.
b $\beta = 15°$, angles on a straight-line sum to 180°.
c $\theta = 34°$, angles in a right-angle sum to 90°.
d $\gamma = 18°$, angles on a straight-line sum to 180°.
e $\alpha = 29°$, angles on a straight-line sum to 180°.

5 a $\alpha = 30°$
b $\beta = 36°$
c $2\gamma = 36°, 3\gamma = 54°$
d $5\theta = 100°, 4\theta = 80°$
e $\alpha = 30°, 2\alpha = 60°, 3\alpha = 90°$
f $\beta = 18°, 3\beta = 54°, 6\beta = 108°$

6 $3\alpha + 40° = 180°, 3\alpha = 140°, \alpha = 46.7°$ (1 d.p.)
Not all the angles are the same.

7 $4\alpha = 180°, \alpha = 45°$
$2\theta + 100° = 180°, \theta = 40°$
Therefore, α is greater.

8 Any value of α and β such that $\alpha + \beta = 90°$, where $\alpha \neq \beta$ such as $\alpha = 40°$ and $\beta = 50°$

9 $2\theta + 90° = 180°, \theta = 45°$ so Fahad is correct.

10 $\theta = \frac{2}{3}\alpha, \alpha + \frac{2}{3}\alpha = 90°$ or $\frac{5}{3}\alpha = 90°, \alpha = 54°$ and $\theta = 36°$

11 $\alpha + \beta = 180°, 180° \div 10 = 18°, \alpha = 126°$ and $\beta = 54°$

12 $\alpha + \beta = \theta = 180°, \alpha = 172.8°$ and $\beta = 7.2°$

Practice questions 6.1.2

1 a $\alpha + 130° + 80° = 360°, \alpha = 150°$
b $\beta + 150° + 105° + 90° = 360°, \beta = 15°$
c $\gamma + 68° + 72° + 72° + 68° = 360°, \gamma = 80°$
d $\theta + 162° + 90° + 90° = 360°, \theta = 18°$
e $\theta + 162° + 90° + 90° = 360°, \theta = 18°$

2 a $\alpha = 85°$, vertically opposite angles are equal.
b $\beta = 166°$, vertically opposite angles are equal.
c $\theta = 67°$, vertically opposite angles are equal.
d $\alpha = 49°$, vertically opposite angles are equal.
$\beta = 41°$, vertically opposite angles are equal.

e $\gamma = 107°$, vertically opposite angles are equal.

f $\alpha = 66°$, vertically opposite angles are equal.
$\beta = 109°$, vertically opposite angles are equal.
$\theta = 137°$, vertically opposite angles are equal.

3 a $2\alpha + 166° + 90° + 34° = 360°, \alpha = 35°$

b $\beta = 154°$, vertically opposite angles are equal.
$\gamma + 154° = 180°, \gamma = 26°$

c $2\theta + 90° + 36° + 144° = 360°, \theta = 45°$

d $\theta + 59° = 360°, \theta = 301°$

e $\alpha = 25°$, vertically opposite angles are equal.
$\beta + 25° = 180°, \beta = 155°$
$\gamma + 90° + 25° + 155° + 25° = 360°, \gamma = 65°$

f $\gamma + 129° = 180°, \gamma = 51°$
$2\gamma + \beta = 129°$, vertically opposite angles are equal.
$102° + \beta = 129°, \beta = 27°$

g $\alpha + 43° + 51° + 52° = 180°, \alpha = 34°$

h $3\beta = 66°$, vertically opposite angles are equal. $\beta = 22°$

i $\alpha = 132°$, vertically opposite angles are equal.
$\beta + 132° = 180°, \beta = 48°, \gamma = 48°$

4 a $\alpha + 111° = 180°, \alpha = 69°$
$\beta = 111°$ as vertically opposite angles are equal.
$\gamma = 69°$ as vertically opposite angles are equal.

b $8\theta = 360°, \theta = 45°$

c $\alpha + 120° = 70° + 80°$ as vertically opposite angles are equal. $\alpha = 30°$

d $\beta + 155° = 180°, \beta = 25°$
$\gamma = 155°$ as vertically opposite angles are equal.
$\theta = 25°$ as vertically opposite angles are equal.

5 a $5\alpha = 360°, \alpha = 72°$

b $12\beta = 360°, \beta = 30°$

c $22\gamma = 360°, \gamma = 16.36°$

d $12\theta = 360°, \theta = 30°$

e $15\alpha = 360°, \alpha = 24°$

f $5\beta + 90° = 180°, \beta = 18°$
$3\beta = \theta$ as vertically opposite angles are equal, $\theta = 54°$
$\gamma + \theta + 2\beta + 90° + 3\beta = 360°, \gamma + 54° + 36° + 90° + 54° = 360°, \gamma = 126°$

6 $\alpha + \beta = 360°$
$360° \div 20 = 18°$
$\alpha = 126°$ and $\beta = 234°$

7 No, Anya is not correct. $\theta + 130° + 170° = 360°, \theta = 60°$

8 No, Dara is not correct. $10\theta + 30° = 360°, \theta = 33°$

9 $\alpha + \beta + \gamma + 20° = 360°, \alpha + \beta + \gamma + = 340°$
$340° \div 17 = 20°$
$\alpha = 40°, \beta = 160°$ and $\gamma = 140°$

10 $360° \div 12 = 30°$
The obtuse angle $= 5 \times 30° = 150°$ and the reflex angle $=360° - 150° = 210°$.

11 $\alpha + \beta + 3° = 360°$
$\alpha : \beta = 100 : 110$
$357° \div 210 = 1.7°$
$\alpha = 170°$ and $\beta = 187°$

12 120°

Practice questions 6.2

1 a

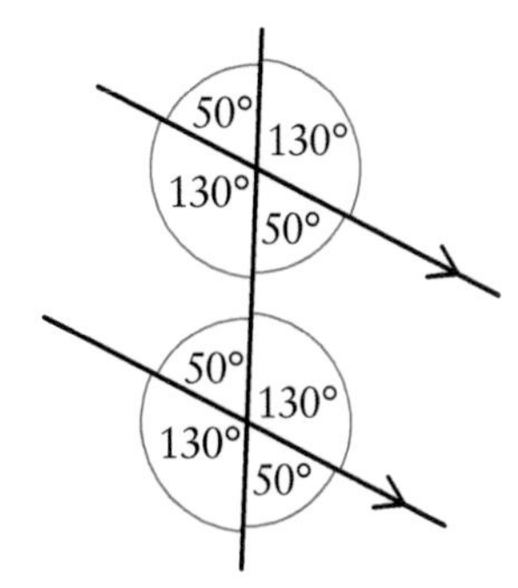

b

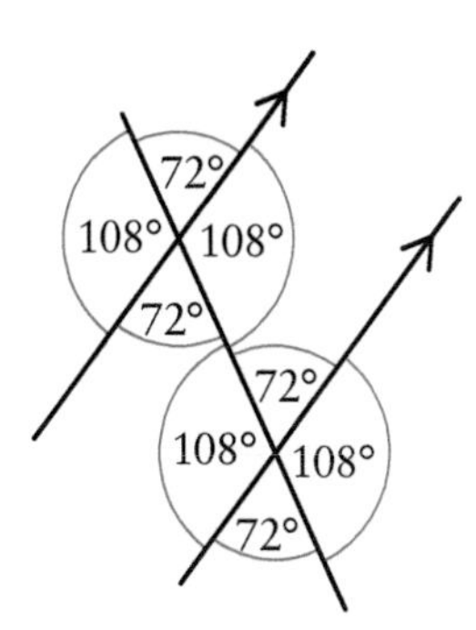

c

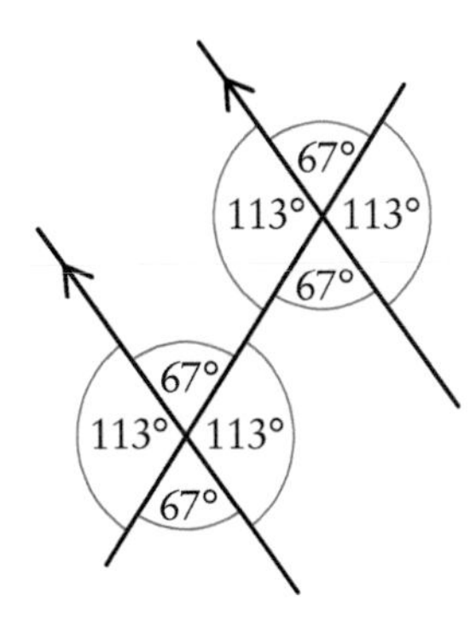

2 a

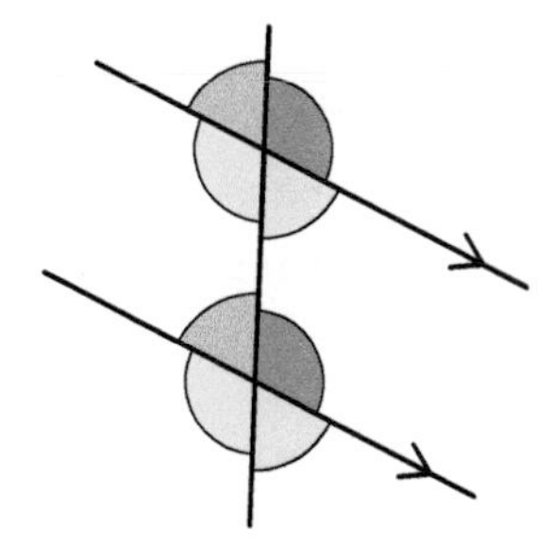

b

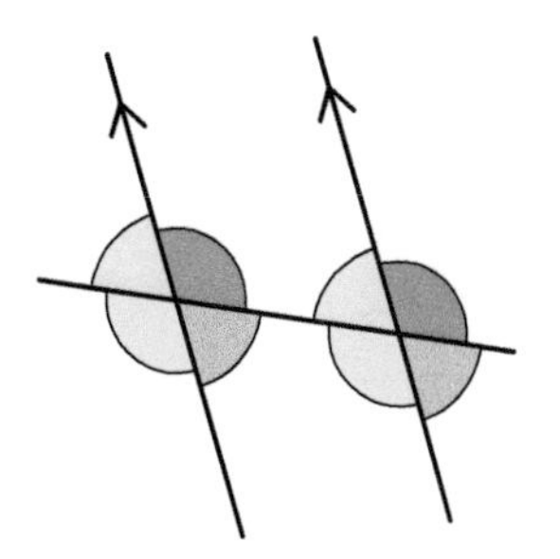

c

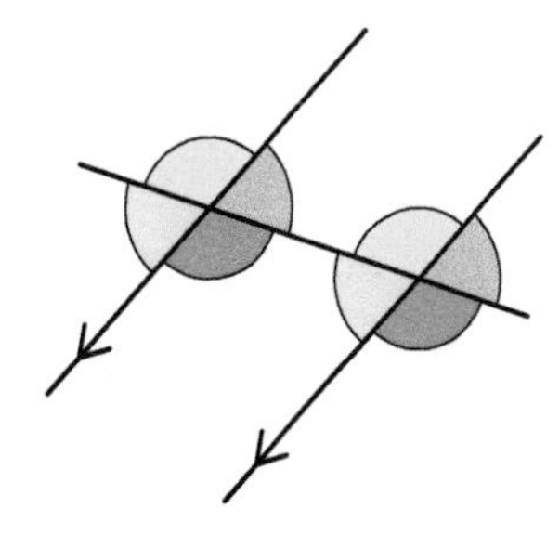

3 a $\alpha = 110°$, corresponding angles are equal.

b $\beta = 57°$, corresponding angles are equal.

c $\gamma = 72°$, corresponding angles are equal.

d $\theta = 123°$, corresponding angles are equal.

e $\alpha = 107°$, corresponding angles are equal.
$\beta = 107°$, corresponding angles are equal.

4 a

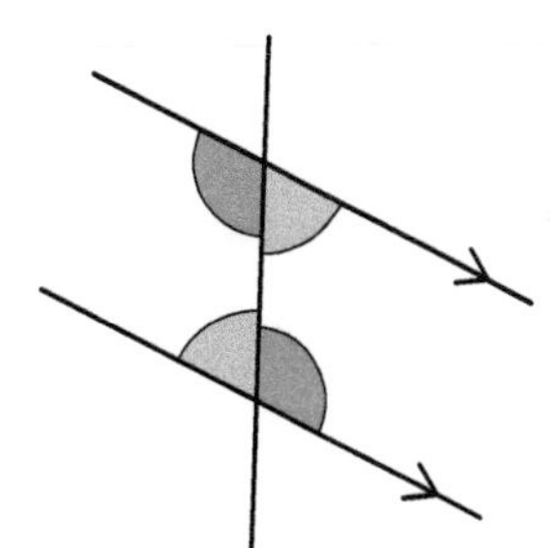

b

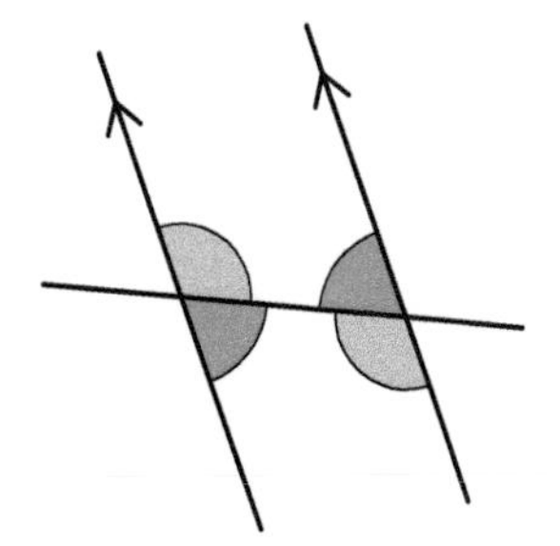

c

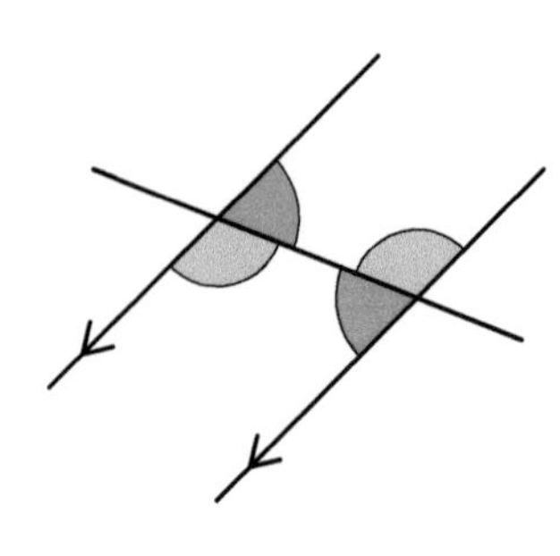

5 a $\alpha = 135°$, alternate angles are equal.
b $\beta = 63°$, alternate angles are equal.
c $\gamma = 75°$, alternate angles are equal.
d $\theta = 154°$, alternate angles are equal.
e $\gamma = 72°$, alternate angles are equal.
$\beta = 84°$, alternate angles are equal.
f $\alpha = 83°$, alternate angles are equal.
$\beta = 83°$, alternate angles are equal.
g $\gamma = 79°$, alternate angles are equal.
$\theta = 144°$, alternate angles are equal.

6 a

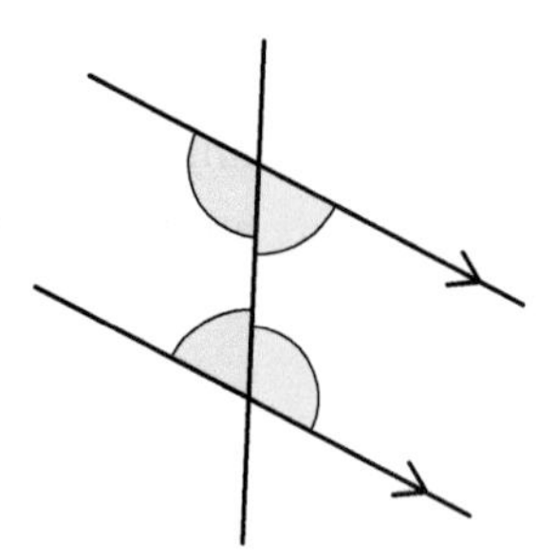

b

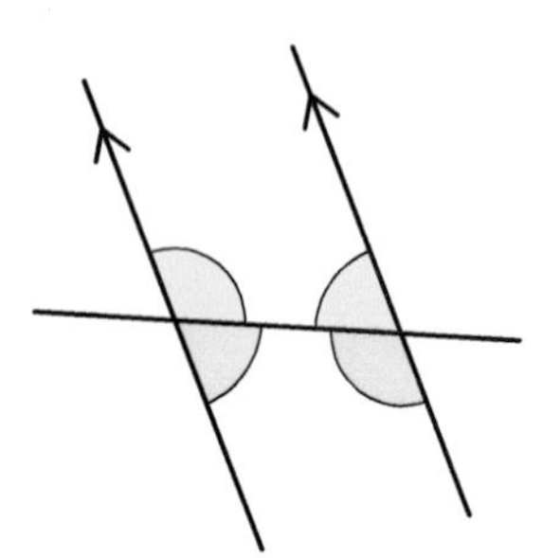

c

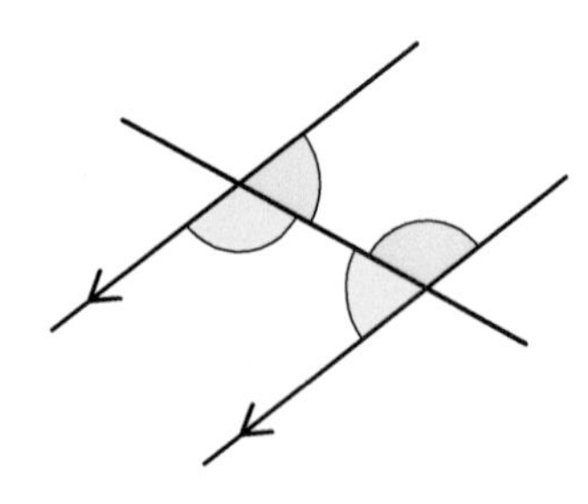

7 a $\alpha = 33°$, co-interior angles sum to 180°.
b $\beta = 96°$, co-interior angles sum to 180°.
c $\theta = 68°$, co-interior angles sum to 180°.
$\gamma = 60°$, co-interior angles sum to 180°.
d $\beta = 29°$, co-interior angles sum to 180°.
$\alpha = 151°$, co-interior angles sum to 180°.
e $\gamma = 103°$, co-interior angles sum to 180°.
$\theta = 55°$, co-interior angles sum to 180°.

8 a $\alpha = 104°$, corresponding angles are equal.
b $\beta = 24°$, co-interior angles sum to 180°.
c $\gamma = 68°$, corresponding angles are equal.
d $\theta = 79°$, alternate angles are equal.
e $\alpha = 166°$, corresponding angles are equal.
$\beta = \alpha = 166°$, alternate angles are equal.
f $\gamma = 126°$, co-interior angles sum to 180°.
$\theta = 136°$, alternate angles are equal.
g $\alpha = 73°$, corresponding angles are equal.
$\beta = 107°$, co-interior angles sum to 180°.
h $\theta = 73°$, corresponding angles are equal.
$\gamma = 107°$, co-interior angles sum to 180°.
i $\alpha = 63°$, alternate angles are equal.
$\beta = 47°$, co-interior angles sum to 180°.

9 a $\alpha = 54°$, corresponding angles are equal.
$\beta = 126°$, angles on a straight-line sum to 180°.
b $\gamma = 39°$, vertically opposite angles are equal.
$\theta = 39°$, corresponding angles are equal.
c $\alpha = 74°$, co-interior angles sum to 180°.
$\beta = \alpha = 74°$, alternate angles are equal.
d $\alpha = 29$, vertically opposite angles are equal.
$\beta = 29°$, corresponding angles are equal.
$\gamma = 151°$, co-interior angles sum to 180°.
$\theta = \gamma = 151°$, vertically opposite angles are equal.

e $\gamma = 44°$, vertically opposite angles are equal.
$\alpha = 44°$, corresponding angles are equal.
$\beta = \alpha = 44°$, alternate angles are equal.
$\theta = 136°$, co-interior angles sum to 180°.

f $\alpha = 130°$, angles at a point sum to 360°.
$\beta = 60°$, co-interior angles sum to 180°.
$\gamma = 130°$, corresponding angles are equal.

10 b and h, c and e, j and p, k and m.

11

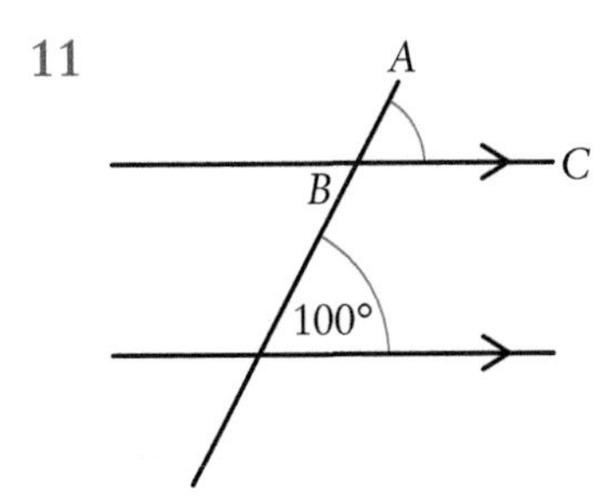

12 a and e, b and f, c and g, d and h, i and m, j and n, k and o, l and p.

13 Students write their own question, for example:
Find the size of angle ABC.

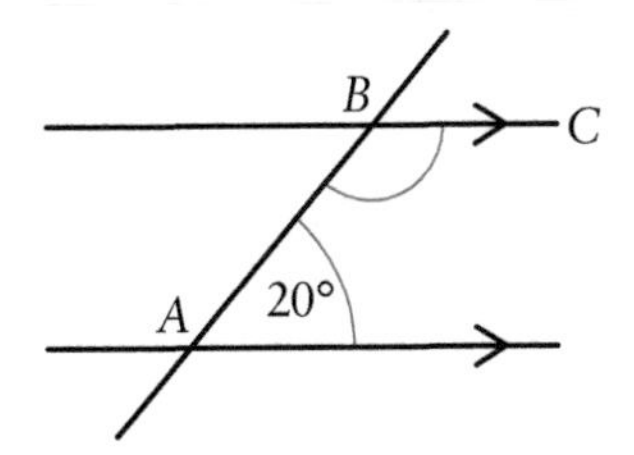

14 No, Max is not correct. Angle BEF is co-interior with angle EBC.

15 No, Carlos is incorrect. Angle $DEF = 64°$ as they are corresponding angles not alternate angles.

16 Zahra is correct. Angles $LNO + ONQ = 180°$ as angles on a straight-line sum to 180°, angle $ONQ = 91°$. If lines MO and PR are parallel then angle ONQ = angle RQS as corresponding angles are equal. As angles ONQ and RQS both equal 91° the two lines must be parallel.

17 Students write their own five questions with labelled diagrams.

Practice questions 6.3

1 a $\alpha = 70°$, angles in a triangle sum to 180°.

b $\beta = 70°$, angles in a triangle sum to 180°.

c $\gamma = 50°$, angles in a triangle sum to 180°.

d $\theta = 15°$, angles in a triangle sum to 180°.

e $\alpha = 71°$, angles in a triangle sum to 180°.

f $\alpha = 29°$, angles in a triangle sum to 180°.

g $\beta = 123°$, the exterior angle of a triangle is equal to the sum of the two interior angles at the other two vertices of the triangle.

h $\gamma = 155°$, the exterior angle of a triangle is equal to the sum of the two interior angles at the other two vertices of the triangle.

i $\theta = 108°$, the exterior angle of a triangle is equal to the sum of the two interior angles at the other two vertices of the triangle.

2 a $\alpha = 18°$, $3\alpha + 95° + 31° = 180°$ as angles in a triangle sum to 180°.

b $\beta = 27°$, $4\beta + 72° = 180°$ as angles in a triangle sum to 180°.

c $\gamma = 20°$, $9\gamma = 180°$ as angles in a triangle sum to 180°.

d $\theta = 53°$, $\theta + 90° = 143°$ as the exterior angle of a triangle is equal to the sum of the two interior angles at the other two vertices of the triangle.

e $\beta = 10°$, $9\beta = 90°$ as the exterior angle of a triangle is equal to the sum of the two interior angles at the other two vertices of the triangle.

f $\alpha = 12°$, $7\alpha = 84°$ as the exterior angle of a triangle is equal to the sum of the two interior angles at the other two vertices of the triangle.

3 a $10\alpha = 180°$ as angles in a triangle sum to $180°$, $2\alpha = 36°$, $3\alpha = 54°$ and $5\alpha = 90°$

b $3\beta = 180°$ as angles in an equilateral triangle sum to $180°$, $\beta = 60°$, $\beta = 60°$, and $\beta = 60°$

c $12\gamma = 180°$ as angles in an isosceles triangle sum to $180°$, $2\gamma = 30°$, $5\gamma = 75°$ and $5\gamma = 75°$

d $5\theta = 150°$ as the exterior angle of a triangle is equal to the sum of the two interior angles at the other two vertices of the triangle, $2\theta = 60°$ and $3\theta = 90°$

e The interior angles of an equilateral triangle are $60°$. As the exterior angle of a triangle is equal to the sum of the two interior angles at the other two vertices of the triangle, $\beta = 120°$

f The interior base angles of the isosceles triangle are $80°$ each. $\beta + 160° = 180°$ as angles in a triangle sum to $180°$, $\beta = 20°$

4 a $\theta = 122°$, $\theta + 24° + 34° = 180°$ as angles in a triangle sum to $180°$

b $\alpha = 70°$, $\angle ACB = 40°$ as corresponding angles are equal. $\alpha + 70° + 40° = 180°$ as angles in a triangle sum to $180°$

c $\beta = 50°$, $\beta + 75° + 55° = 180°$ as angles in a triangle sum to $180°$

$\gamma = 40°$, $\gamma + 90° + \beta° = 180°$ as angles in a triangle sum to $180°$

d $\theta = 55°$, $\angle GHJ + 70° + 70° = 180°$ as angles in an isosceles triangle sum to $180°$, $\angle GHJ = 40°$. $\theta + 70° + 55° = 180°$ as angles in a triangle sum to $180°$

e $\beta = 79°$, corresponding angles are equal.

$\gamma = 42°$, $\gamma + 59° + \beta = 180°$ as angles in a triangle sum to $180°$

5 a $\alpha = 108°$, $\alpha + 72° = 180°$ as angles on a straight line sum to $180°$

$\beta = 89°$, $\angle BDE + \alpha + 20° = 180°$ as angles in a triangle sum to $180°$, $\angle BDE = 52°$ and $\beta + 52° + 39° = 180°$ as angles in a triangle sum to $180°$

b $\alpha = 100°$, $\alpha + 80° = 180°$ as co-interior angles sum to $180°$

$\beta = 24°$, $\angle MUS = 56°$ as vertically opposite angles are equal and $\beta + \angle MUS + \alpha = 180°$ as angles in a triangle sum to $180°$

c $\alpha = 74°$, $\alpha + 57° + 49° = 180°$ as angles in a triangle sum to $180°$

$\beta = 57°$ as alternate angles are equal.

$\gamma = 49°$ as alternate angles are equal.

d $\alpha = 53°$, $\angle BCD + 90° + 53° = 180°$ as angles in a triangle sum to $180°$, $\angle BCD = 37°$ and $\angle BCD + 90° + \alpha = 180°$ as angles in a triangle sum to $180°$

$\beta = 37°$, $\alpha + \beta = 90°$ as angles in a right-angle sum to $180°$

e $\gamma = 48°$, $\gamma + 97° + 35° = 180°$ as angles in a triangle sum to $180°$

$\theta = 99°$, $\theta + \gamma = 147°$ as the exterior angle of a triangle is equal to the sum of the two interior angles at the other two vertices of the triangle.

f $\alpha = 107°$, $\alpha + 49° = 156°$ as the exterior angle of a triangle is equal to the sum of the two interior angles at the other two vertices of the triangle.

$\gamma = 49°$ as alternate angles are equal.

$\beta = 24°$, $\angle CBD + 156° = 180°$ as angles on a straight line sum to $180°$, $\angle CBD = 24°$ and alternate angles are equal.

6 a False, $40° + 100° + 50° = 190°$ not $180°$.
b False, $60° + 30° + 40° = 130°$ not $180°$.
c False, $180° \div 3 = 60°$ not $50°$.
d True, $50° + 50° + 80° = 180°$.
e True, $70° + 80° + 30° = 180°$.
f False, $80° + 80° + 50° = 210°$ not $180°$.
g True, $24° + 48° + 108° = 180°$.
h True, $3 \times 60° = 180°$.

7 a $\alpha = 72°$ as alternate angles are equal.
$\beta = 93°$, $\alpha + \beta + 15° = 180°$ as angles in a triangle sum to $180°$

b $\alpha = 95°$ as alternate angles are equal.
$\beta = 40°$, $\beta + 140° = 180°$ as angles on a straight-line sum to $180°$
$\gamma = 45°$, $\alpha + \beta + \gamma = 180°$ as angles in a triangle sum to $180°$

c $\alpha = 53°$ as vertically opposite angles are equal.
$\gamma = 45°$ as alternate angles are equal.
$\beta = 82°$, $\alpha + \beta + \gamma = 180°$ as angles in a triangle sum to $180°$

8

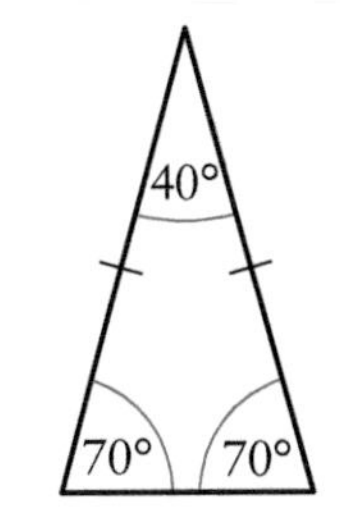

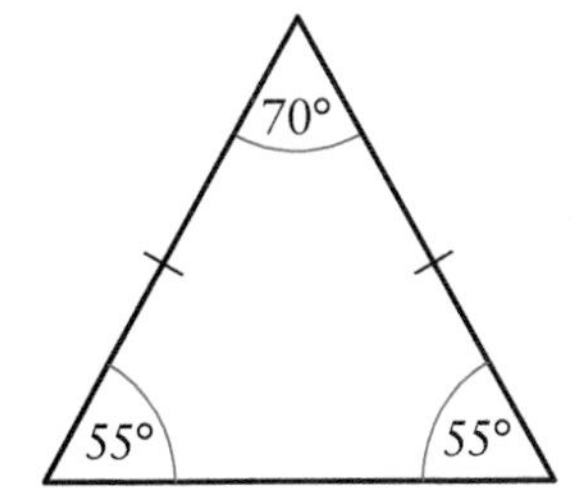

9 $\alpha = 50°$ and $\beta = 60°$
$\alpha + \beta + 70° = 180°$ as angles in a triangle sum to $180°$, $\alpha + \beta = 110°$, $110° \div 11 = 10°$

10 $\alpha = 20°$, $\beta = 40°$ and $\gamma = 120°$
$\beta = 2\alpha$ and $\gamma = 3\beta = 6\alpha$, $\alpha + 2\alpha + 6\alpha = 180°$ as angles in a triangle sum to $180°$, $9\alpha = 180°$

11 a True, vertically opposite angles are equal.
b True, angles in a triangle sum to $180°$.
c False.
d False.
e True, alternate angles are equal.
f True, corresponding angles are equal.
g True, alternate angles are equal.
h True, vertically opposite angles are equal.
i True, $a + b + c = 180°$ and $k + l = 180°$ as both are angles on a straight line and sum to $180°$.

12 Jamal is incorrect, $\alpha + 72° = 180°$ as co-interior angles sum to $180°$ so $\alpha = 108°$.
$\theta + 63° = 180°$ as co-interior angles sum to $180°$ so $\theta = 117°$.

13 $\theta = 35°$, $\angle ADB + 90° + 20° = 180°$ as angles in a triangle sum to $180°$ so $\angle ADB = 70°$ and $\theta + \theta = \angle ADB$ as the exterior angle of a triangle is equal to the sum of the two interior angles at the other two vertices of the triangle.

14 $\alpha = 100°$, $\beta = 50°$ and $\gamma = 30°$
$\alpha + \beta + \gamma = 180°$ as angles in a triangle sum to $180°$.
The ratio of $\alpha : \beta : \gamma = 100 : 50 : 30$,
$180° \div 180 = 1°$

Practice questions 6.4

1 a $\alpha = 85°$, $\alpha + 80° + 95° + 100° = 360°$
b $\beta = 47°$, $\beta + 53° + 125° + 135° = 360°$
c $\gamma = 63°$, $\gamma + 90° + 90° + 117° = 360°$
d $\alpha = 75°$, $\alpha + 61° + 108° + 116° = 360°$
e $\theta = 227°$, $\theta + 33° + 43° + 57° = 360°$
f $\alpha = 64°$, $\alpha + 116° + 64° + 116° = 360°$

2 a $\alpha = 15°$, $5\alpha + 90° + 135° + 60° = 360°$
b $\beta = 9°$, $11\beta + 106° + 39° + 116° = 360°$
c $\gamma = 100°$, $2\gamma + 81° + 79° = 360°$
d $\theta = 67°$, $4\theta + 61° + 31° = 360°$

3 a $\alpha = 40°$, $2\alpha = 60°$, $3\alpha = 90°$, $9\alpha = 360°$
b $\beta = 36°$, $2\beta = 72°$, $3\beta = 108°$, $4\beta = 144°$, $10\beta = 360°$
c $\gamma = 30°$, $3\gamma = 90°$, $5\gamma = 150°$, $12\gamma = 360°$
d $\theta = 24°$, $5\theta = 120°$, $8\theta = 192°$, $15\theta = 360°$

4 a $\alpha = 27°, \alpha + 90° + 63° = 180°$
$\beta = 30°, \beta + 123° + 90° + 90° + 27° = 360°$

b $\gamma = 21°, \gamma + 84° + 75° = 180°$
$\theta = 62°, \theta + 72° + 121° + 105° = 360°$

c $\alpha = 240°, \alpha + 120° = 360°$
$\beta = 46°, \beta + 240° + 25° + 49° = 360°$

d $\theta = 99°, \theta + 81° = 180°$
$\gamma = 226°, \gamma + 22° + 13° + 99° = 360°$

e $\alpha = 106°, \alpha + 135° + 60° + 59° = 360°$
$\beta = 74°, \beta + 106° = 180°$

f $\gamma = 60°, \gamma + 35° + 85° = 180°$
$\theta = 153°, \theta + 40° + 72° + 95° = 360°$

5 a $\theta = 150°, 360° - 26° - 34° - 90° = 210°,$
$\theta + 210° = 360°$

b $\alpha = 105°, \alpha + 75° = 180°$
$\beta = 65°, \gamma = 10°, \gamma + 65° + 105° = 180°$

c $\alpha = 35°, \alpha + 115° + 30° = 180°$

d $\theta = 65°, \theta + 55° + 60° = 180°$

e $\theta = 130°, \theta + 50° = 180°$

f $\beta = 52°, \beta + 65° + 63° = 180°$

6 $360° \div 36 = 10°$, angles are 40°, 50°, 120°, 150°

7 Yes Arlo is correct as co-interior angles sum to 180°

8 No, two of the angles are formed by joining two of the angles in the equilateral triangles meaning they are 120° each, the other two angles are 60° each.

9 $\alpha = 76°, \alpha + 93° + 11° = 180°$
$\theta = 59°, 2\theta + 138° + 104° = 360°$

10 Yes, Anika is correct. $360° \div 10 = 36°$.
Largest angle = 144°, smallest angle = 36°,
$144° - 36° = 108°$.

Check your knowledge questions

1 a $\theta = 145°, \theta = 78° + 67°$

b $x = 52°$, angles in a right-angle sum to 90°.

c $\theta = 69°$, angles on a straight line sum to 180°.

d $\beta = 45°$, angles at a point sum to 360°.

e $\alpha = 152°$, vertically opposite angles are equal.

f $\theta = 130°$, vertically opposite angles are equal.
$\alpha = 130°$, corresponding angles are equal.
$\beta = 50°$, angles on a straight line sum to 180°.

g $a = 50°$, angles on a straight line total 180°, the base angles in the isosceles triangle are equal to 65°, and the angles in a triangle sum to 180°.

h $a = 48°$, the angles in a quadrilateral sum to 360°.

i $x = 122°$, co-interior angles sum to 180°.
$y = 122°$, vertically opposite angles are equal.

2 a $8\alpha = 180°, \alpha = 22.5°$

b $5\alpha = 140°, \alpha = 28°$

c $18\beta = 360°, \beta = 20°$

3 $\alpha = 1.5\theta$, $1.5\theta + \theta = 90°$, angles in a right-angle sum to 90°, $\theta = 36°$, $\alpha = 54°$

4 $360° \div 24 = 15°$, $\alpha = 285°$, $\beta = 75°$

5 For example,

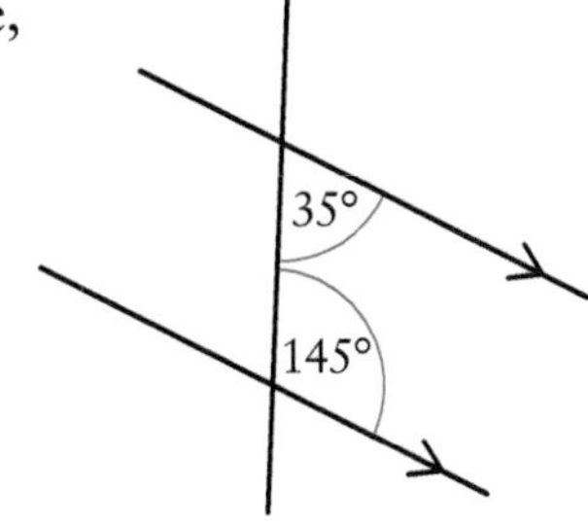

6 a $\theta = 60°$, the angles in a triangle sum to 180° and the angles in an equilateral triangle are equal.

b $\alpha = 65°$, corresponding angles are equal and the angles in a triangle sum to 180°
$\beta = 95°$, angles on a straight line sum to 180° and alternate angles are equal.

c $\gamma = 36°$, angles in a quadrilateral sum to 360°.

7 $\gamma = 44°$, vertically opposite angles are equal.
$\alpha = 44°$, corresponding angles are equal.
$\beta = 44°$, corresponding angles are equal.
(γ and β are corresponding angles.)
$\theta = 136°$, angles on a straight line sum to 180°.

8 $\alpha = 35°$, $\alpha + 80° + 65° = 180°$ as angles on a straight line sum to 180°.
$\theta = 80°$ as alternate angles are equal.

Chapter 7 answers

Do you recall?

1 Student's own answer

2

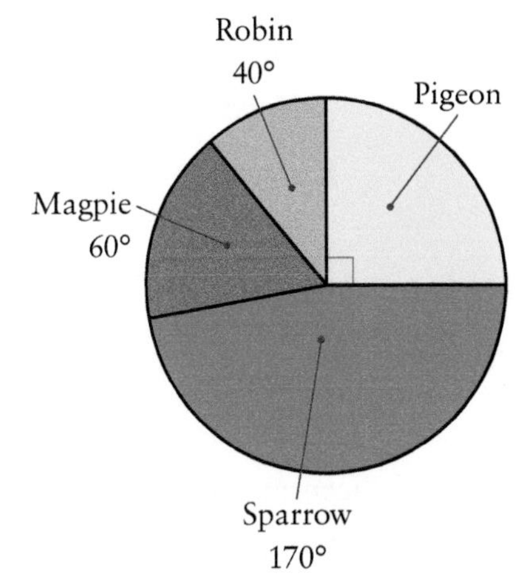

3 85°

4 a isosceles b equilateral

c right-angled

5 a 160 b 81 c 100

6 a 3 b 5 (and −5)

7 a 18 b 2.6 (1 d.p.)

8 a 15 cm^2 b 10 cm^2 c 16 cm^2

Practice questions 7.1

1 a (minor) segment b radius

c (major) arc d diameter

e (minor) sector f chord

2 a tangent b sector

c semicircle (or arc)

3 a diameter b radius

c chord d tangent

4 No. In a pie chart, sets of data are represented in **sectors** of a circle.

5 Student's own accurate drawing of circle radius 7 cm

6 Student's own accurate drawing of circle radius 5.5 cm

7 a 40°

b Isosceles, has two equal sides (the radii).

8 Student's own accurate drawing of inscribed regular octagon

9 a Student's own accurate drawing

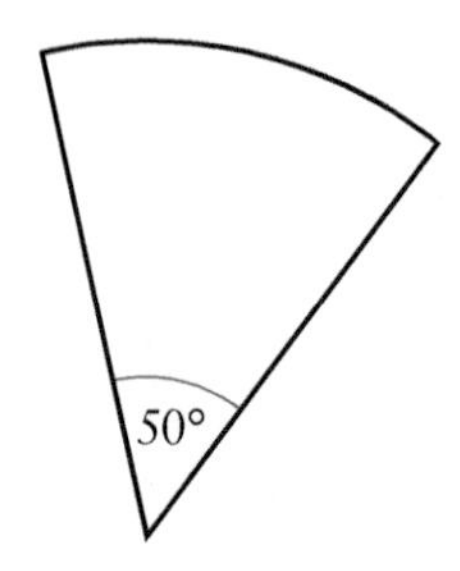

b Student's own accurate drawing

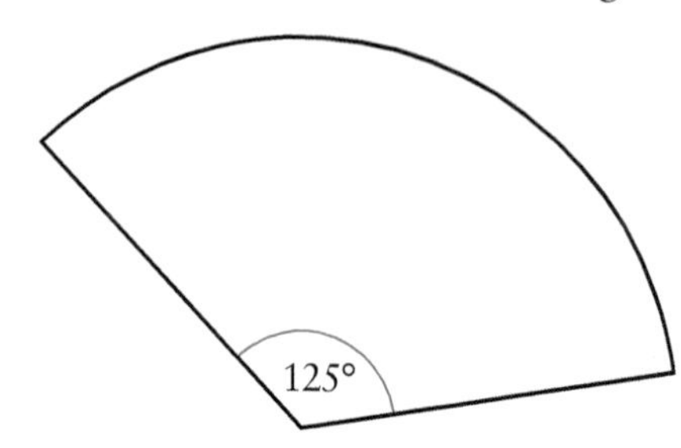

10 Student's own accurate drawing (square and semicircle)

11 Student's own accurate drawing (square and quarter circles)

12 Student's own accurate drawing

13 Student's own accurate drawing of inscribed hexagon, divided into 6 equilateral triangles.

Explanation that each triangle is isosceles (radii) and angle, therefore base angles of each triangle are $\frac{1}{2}(180 - 60) = 60$.

Therefore triangle is equilateral.

Practice questions 7.2

1 a 25.1 mm b 44.0 cm

c 62.8 mm d 66.0 cm

e 29.5 mm f 38.0 mm

2 a 15 cm b 18 cm

3 a 23.6 cm b 16.7 cm

4 3.77 m (2 d.p.)

5 395.84 m6 455

7 64.8 cm

8 7.6 cm

9 22.28 cm

10 159 mm

11 28.3 cm

12 4.7 cm (1 d.p.)

13 No, if the perimeter of the square is an integer, then the circumference of the circle is an integer. Hence its diameter, calculated by circumference divided by π will be an irrational number, as π is an irrational number. You cannot draw a circle with an irrational diameter (you can get close but never exact!).

14 a 40 030 km

b 41 287 km (nearest km)

Practice questions 7.3

1 a 153.9 cm^2 b 3.1 cm^2

c 50.3 m^2 d 78.5 m^2

e 38.5 mm^2 f 22.9 cm^2

g 366.4 cm^2 h 95.0 cm^2

2 a 33.18 cm^2 b 6.16 cm^2

3 12 m^2 (by rounding all values to nearest whole number)

4 a 4.52 m^2 b 7.54 m

5 $12.6\ m^2$

6 $707\ m^2$

7 3.1 cm

8 23.3 cm

9 1.95 m

10 No, max diameter that will fit is 3.5 m.
Area of circle with diameter 3.5 m is $9.62\ m^2$

11 a $50.3\ cm^2$ b $30.9\ cm^2$

12 12.6 cm (1 d.p.)

13 A3 paper dimensions: 29.7 cm by 42 cm
Therefore you can cut $5 \times 8 = 40$ circles of diameter 5 cm from the paper.

14 Doubling the radius doubles the circumference and multiplies the area by 4.

Practice questions 7.4

1 a $157.1\ mm^2$ (1 d.p.) b $50.3\ cm^2$ (1 d.p.)
c $127.2\ cm^2$ (1 d.p.)

2 a 22.0 cm (1 d.p.) b 11.8 cm (1 d.p.)
c 18.8 cm (1 d.p.)

3 a 30.85 cm (2 d.p.) b 43.19 cm (2 d.p.)
c 15.00 cm (2 d.p.)

4 $1.27\ m^2$ (2 d.p.)

5 a $41.1\ cm^2$ (1 d.p.) b 24.6 cm

6 Perimeter = 36.6 cm, Area = $89.1\ cm^2$

7 $2241\ cm^2$

8 11.3% (1 d.p.)

9 $7.97 (nearest cent)

10 a 6.113 km (to the nearest metre)
b 73 km

11 a 26.7 cm (1 d.p.) b 96.1 cm (1 d.p.)

12 Area = $235.6\ cm^2$, Perimeter = 62.8 cm (1 d.p.)

13 40.0 cm (1 d.p.)

14 $70.9\ cm^2$ (1 d.p.)

15 $3.39\ m^2$ (2 d.p.)

16 Both areas are $30.9\ cm^2$ (1 d.p.) or $144 - 36\pi$

17 31.4 cm

18 a Inside edge is shorter.
e.g. for outer lane, outside edge = $70 + 10\pi = 101.4$ m
Inside edge = $70 + 8\pi = 95.13$ m
b 6.28 m

Check your knowledge questions

1 a diameter b radius
c (minor) arc d semicircle
e (minor) sector

2–5 Student's own diagrams

6 a 25.1 cm (1 d.p.) b 56.5 cm (1 d.p.)

7 a $28.3\ cm^2$ (1 d.p.) b $153.9\ cm^2$ (1 d.p.)

8 Area $\approx 48\ cm^2$, Circumference ≈ 24 cm

9 10.3 cm

10 5.6 cm

11 a 25.7 cm b 27.4 cm
c 25.0 cm d 24.6 cm

12 a $39.27\ cm^2$ b $21.21\ cm^2$
c $49.13\ cm^2$ d $7.73\ cm^2$

13 15.9 cm

Chapter 8 answers

Do you recall?

1 a QR b RS c ∠QRS
d ∠QRP = 54°, ∠QSP = 90°, ∠RQS = 36°

2 a 5 cm b 6 cm c C and D
d Draw a circle of radius 4 cm from A and a circle 7 cm from B
The points of intersection of the two circles give the required points.

Practice questions 8.1

1 a i $\angle MNO$ ii $\angle QRS$ iii $\angle RST$

b i KT ii MN iii RS

c She is correct. If the shapes are congruent then they are identical in shape and size, therefore, the area of the two shapes is the same.

2 a Any rotation of:

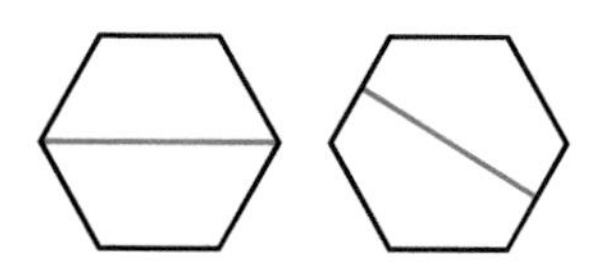

b Any rotation of:

 or 

c Any rotation or reflection of:

 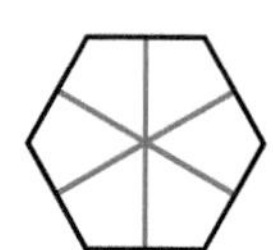 

d Any rotation of:

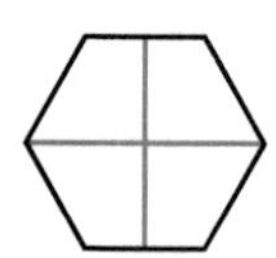

e See above for some alternative solutions.

3 a Any rotation of:

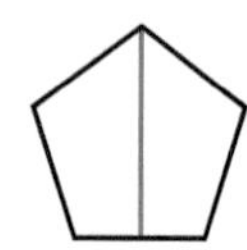

b

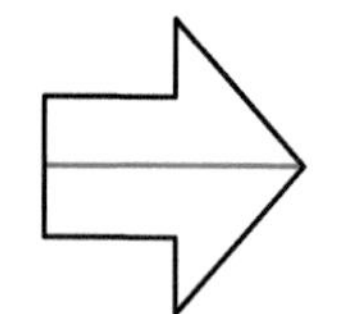

c

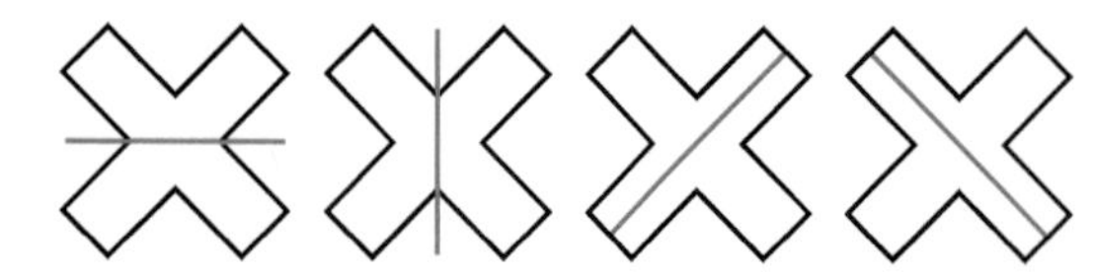

d

e

f

g

h

4 a Congruent b Congruent

c Not congruent d Congruent

e Congruent f Not congruent

g Not congruent h Congruent

5 a All have an area of 4 square units.

b A and C.

c A and B have the same area (4 square units) and perimeter (10 units).

d A and B are not congruent. They are not identical in shape.

6 a Any one of:

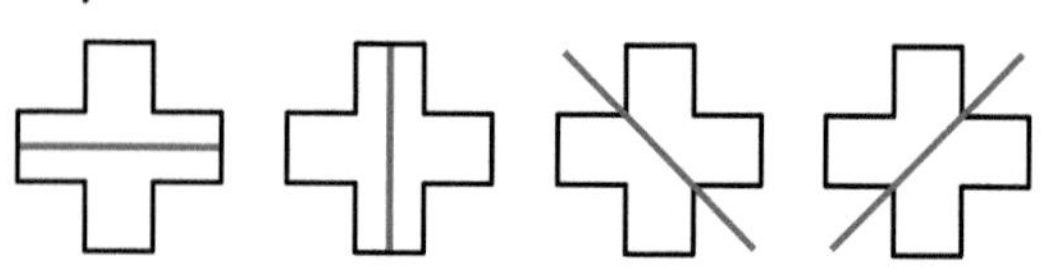

b Any one of:

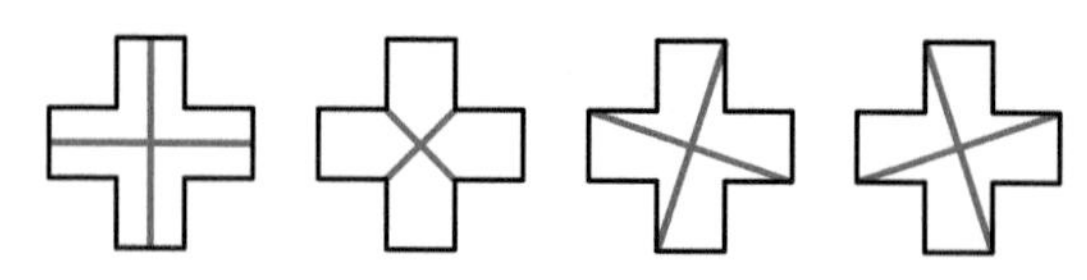

c

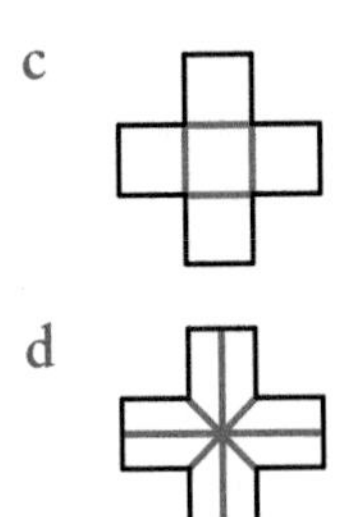

d

e See above for some alternative solutions.

7 a i $\angle CDE$ ii $\angle CED$ iii $\angle DCE$

b i DE ii EC iii DC

8 a No they are not congruent. The radius of one is larger than the other.

b They must have the same radius.

9 Kiara is correct. The two shapes are not congruent. The second shape is larger than the first shape .

10 Opel is not correct. For two squares to be congruent their sides have to be the same length.

11 a ABD and ABC, ACD and BDC, AED and BEC

b ABD and CDB, ABC and CDA, AED and CEB, ABE and CDE

Practice questions 8.2.1

Diagrams may vary depending on the side used as the base of the triangle. However, all triangles should be congruent to the ones given and should be constructed using the method shown.

1 a

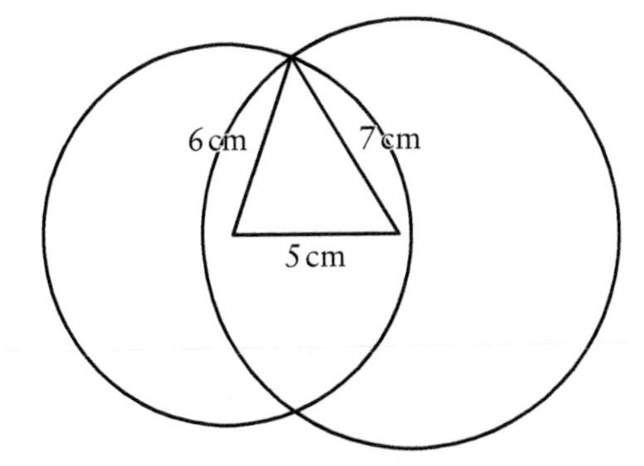

b Students construct a triangle with sides of length 8 cm, 5 cm and 8 cm using the same method as in Q1a.

c Students construct a triangle with sides of length 9 cm, 5 cm and 7 cm using the same method as in Q1a.

2 a Students construct a triangle with sides of 7 cm, 8 cm and 9 cm using the same method as in Q1a.

b Students construct a triangle with sides of 3.5 cm, 4 cm and 4.5 cm using the same method as in Q1a.

3 Matching sides: AB and PQ; BC and QR; CA and RP

Matching angles: $\angle BAC$ and $\angle QPR$; $\angle ABC$ and $\angle PQR$; $\angle ACB$ and $\angle PRQ$

4 a Students construct an equilateral triangle with sides of length 4 cm using the same method as in Q1a.

b Students construct a triangle with sides of length 5 cm, 3 cm and 3 cm using the same method as in Q1a.

5 Students construct an equilateral triangle with sides of length 6 cm using the same method as in Q1a.

6

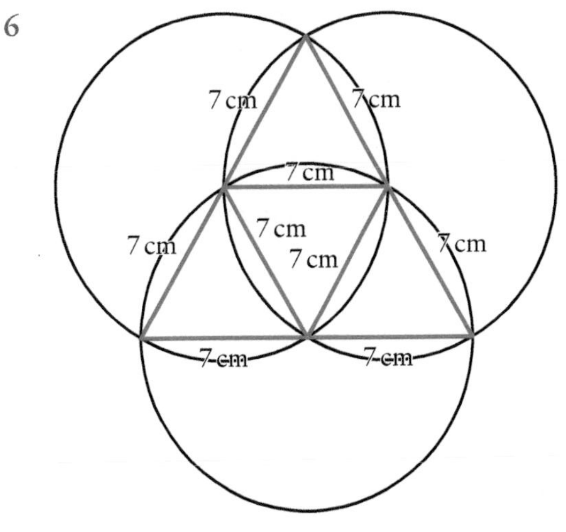

7 In order to construct a triangle the total of the two shorter sides must be greater than the longer side.

8 Students construct a triangle with sides of length 5 cm, 7 cm and 7 cm and a triangle with sides of length 5 cm, 5 cm and 7 cm using the same method as in Q1a.

9

10 Proceed as if you are drawing an equilateral triangle. (Pick your own radius for the circle)

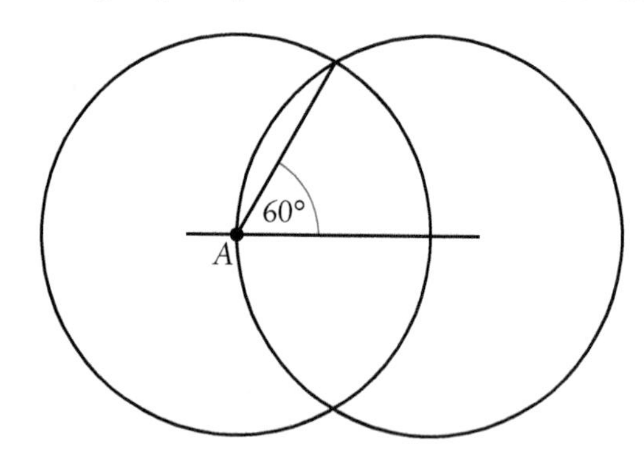

Practice questions 8.2.2

1 Students draw two different acute angles and bisect each of them using this method:

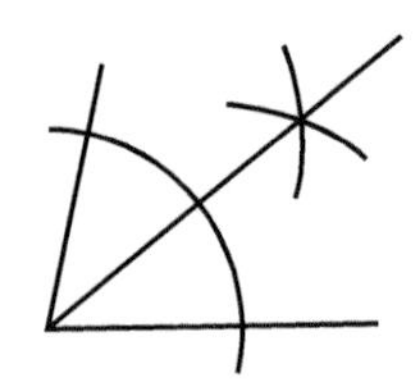

2 Students draw two different obtuse angles and bisect each of them using the method in Q1.

3 a Students draw an angle measuring 50° and bisect it using the method in Q1.

b Students draw an angle measuring 85° and bisect it using the method in Q1.

c Students draw an angle measuring 110° and bisect it using the method in Q1.

d Students draw an angle measuring 162° and bisect it using the method in Q1.

4 Students draw a scalene triangle with one obtuse angle and bisect the obtuse angle using the method in Q1.

5 a Students construct a triangle with sides of length 8 cm, 4 cm and 7 cm using the same method as in Q1 from Practice questions 8.2.1.

b Students bisect angle *ABC* using the method in Q1.

6 a Students construct an equilateral triangle with sides of length 7 cm using the same method as in Q1 from Practice questions 8.2.1.

b Students bisect an angle in their triangle using the method in Q1.

c 30°

Practice questions 8.2.3

1

2 a Students draw a line segment 10 cm long and then construct its perpendicular bisector using the same method as in Q1.

b Students draw a line segment 7 cm long and then construct its perpendicular bisector using the same method as in Q1.

c Students draw a line segment 12 cm long and then construct its perpendicular bisector using the same method as in Q1.

d Students draw a line segment 6 cm long and then construct its perpendicular bisector using the same method as in Q1.

3 Students draw a scalene triangle and then construct a perpendicular bisector to one side of the triangle using the same method as in Q1.

4 a Students construct a triangle with sides of length 7 cm, 6 cm and 5 cm using the same method as in Q1 from Practice questions 8.2.1.

b Students construct the perpendicular bisector to side *AB* using the same method as in Q1.

5 a Students draw a line segment 10 cm long and then construct its perpendicular bisector using the same method as in Q1.

b Students construct the angle bisector of one of the right angles using the method used in Practice questions 8.2.2. Q1.

c 45°

6

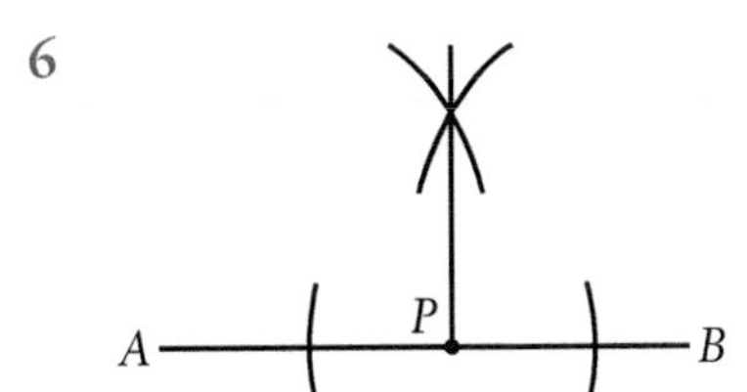

7 a Students construct a perpendicular line to the line *AB* from point *P* using the method used in Q6.

b Students construct a perpendicular line to the line *AB* from point P using the method used in Q6.

8

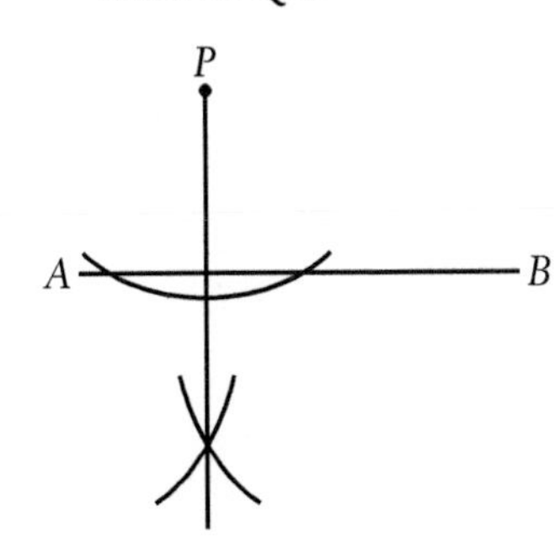

9 a Students construct a perpendicular line to the line *AB* from point P using the method used in Q8.

b Students construct a perpendicular line to the line *AB* from point P using the method used in Q8.

10

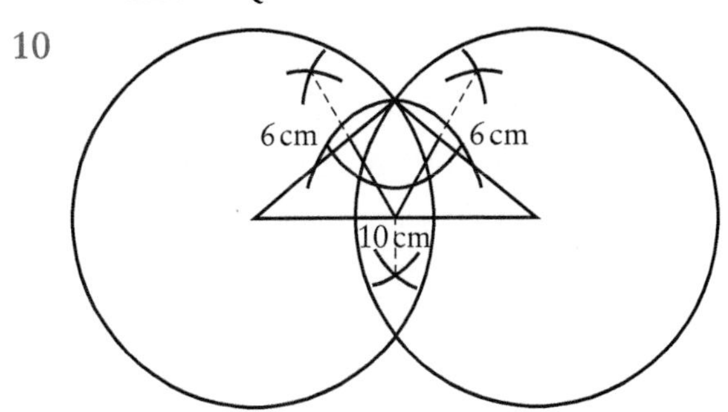

11

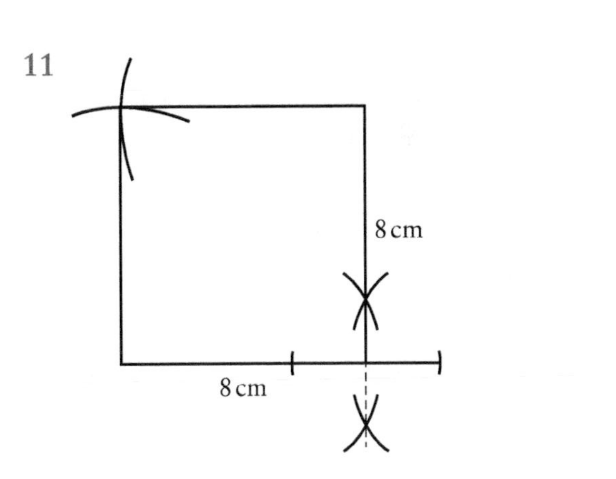

Check your knowledge questions

1 There is more than one answer for each question part, here is an example for each:

a

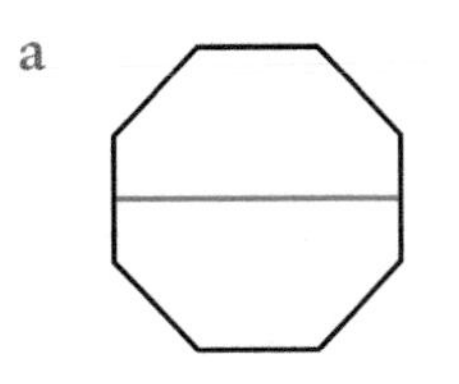

b

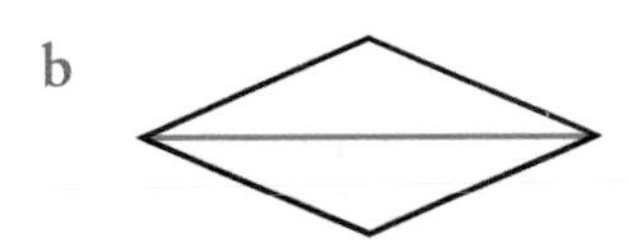

c

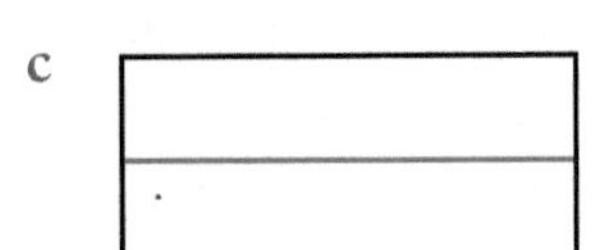

2 a i $\angle CED$ ii $\angle CDE$ iii $\angle ECD$

b i ED ii CD iii EC

3 Students construct a triangle with an area of 20 cm², for example, a right angled triangle with base 8 cm and height 5 cm or an isosceles triangle with base 10 cm and height 4 cm.

4 Students construct an equilateral triangle with sides of length 4 cm using the same method as in Q1 from Practice questions 8.2.1.

5 Students construct a triangle with sides of length 4 cm, 3 cm and 5 cm using the same method as in Q1 from Practice questions 8.2.1.

6 Students construct a triangle with sides of length 6 cm, 6 cm and 9 cm and a triangle with sides of length 6 cm, 9 cm and 9 cm using the same method as in Q1 from Practice questions 8.2.1.

7 Students draw an obtuse angle and bisect it using the method in Q1 from Practice questions 8.2.2.

8 Students construct an isosceles triangle with one obtuse angle using the same method as Q1 from Practice questions 8.2.1. and bisect the obtuse angle using the method in Q1 from Practice questions 8.2.2.

9 Jay has drawn his arcs from the end of each line. He should have first drawn an arc from the vertex that crosses both lines.

10 Students construct a perpendicular bisector on one of the sides of their answer from Q4. If correct it would bisect the opposite angle (if extended).

11 a Students construct a perpendicular line to the line AB from point P using the method used in Q6 in Practice questions 8.2.3.

b Students construct a perpendicular line to the line AB from point P using the method used in Q8 in Practice questions 8.2.3.

Chapter 9 answers

Do you recall?

1 a 30 cm^2 b 36 cm^2 c 48 cm^2

2 a triangular prism

b cuboid

c cube

d right-angled triangular prism

e square-based pyramid

3 a 180 cm^3 b 125 cm^3

4 a $x = 11$ b $x = 18$

Practice questions 9.1

1 a 60 cm^2 b 27 cm^2 c 51 cm^2

2 4875 cm^2

3 a 24 cm^2 b 80 cm^2 c 104 cm^3

4 40 cm^2

5 White 882 cm^2, Green 1029 cm^2, Red 1029 cm^2, Black 588 cm^2

6 Area 40.255 m^2; 14 tins

7 650 m^2

8 a 9 cm b 57.8 mm (1 d.p.)

c 10 cm d 19 cm

9 top 18 cm, bottom 22 cm

10 11 cm and 22 cm

11 $a = 3$, $b = 3$

12 630 cm^2

13 a 310.1 cm^2 (1 d.p.) b 49 cm^2

14 69.2 cm^2

15 54.5% (1 d.p.)

16 a 4 cm b 24 cm^2

c 48 cm^2 d Same area, 48 cm^2

e Area $= \frac{1}{2}xy$

17 a 25 cm^2 b Area $= \frac{1}{2}xy$

18 a 42 cm^2 b 31.5 cm^2

19 30 cm

20 Student's own answers

Practice questions 9.2

1 a, b, f and g are prisms

2 a

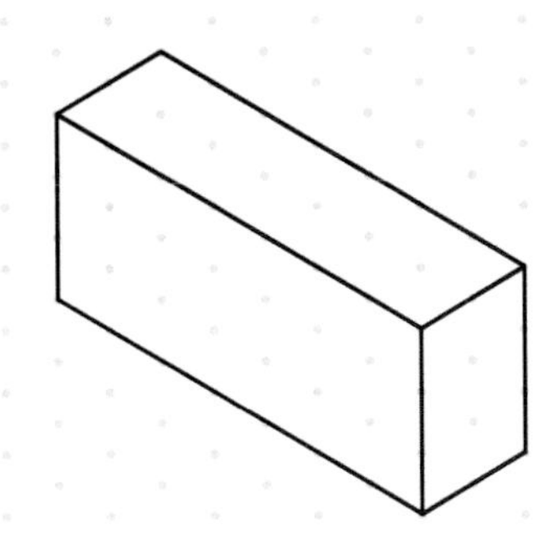

b

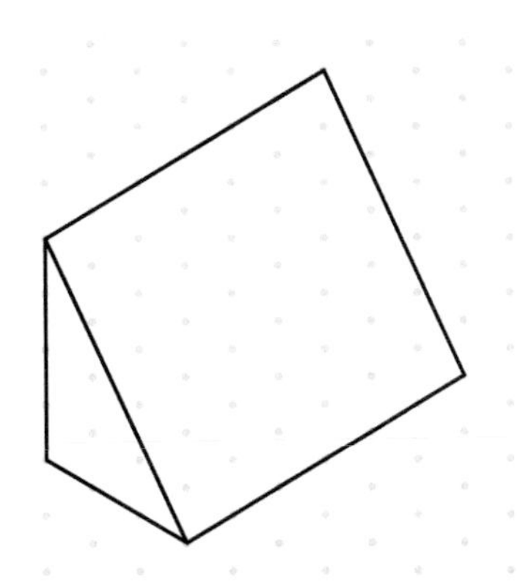

3

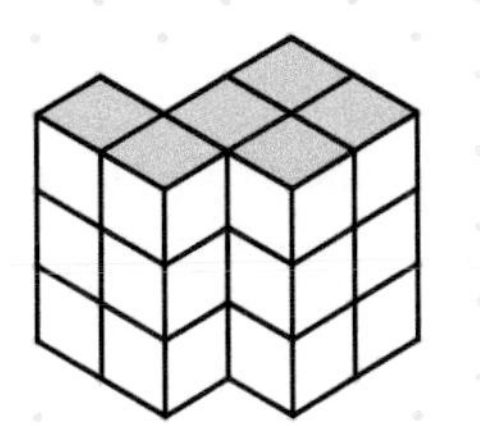

4

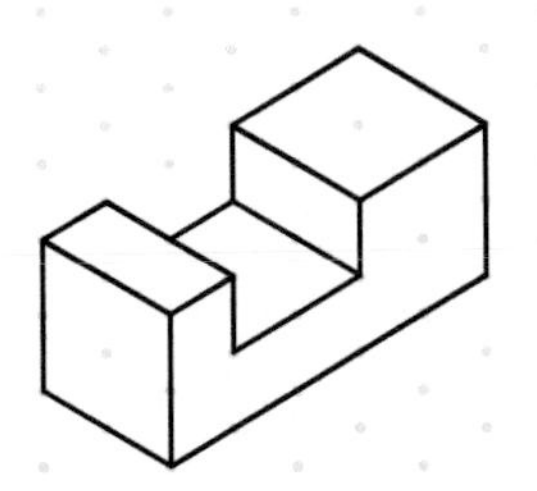

5 a Triangular prism

b i Two pentagon faces, five rectangular faces, 15 edges, 10 vertices

ii Two octagon faces, eight rectangular faces, 24 edges, 16 vertices

6 a 42 cm^3 b 315 cm^3

c 75 cm^3

7 a 54 cm^3 b 175 cm^3

c 320 cm^3 d 378 cm^3

8 a 66 cm^3 b 140 cm^3

9 860 cm^3

10 a 480 cm^3 b 315 cm^3 c 80 cm^3

11 337.5 m^3

12 1.68 m^3

13 a 8 cm b 5 cm

14 Prism A: Prism A 240 cm^3, Prism B 128 cm^3

15 244 cm^3

16 18.9 cm^3

17 31.4 cm^3

18 6 cm

19 Student's own answers

Practice questions 9.3

1 a

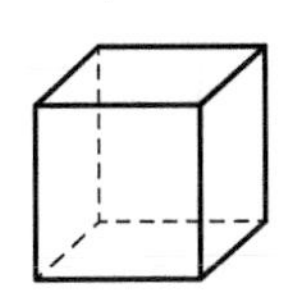

b

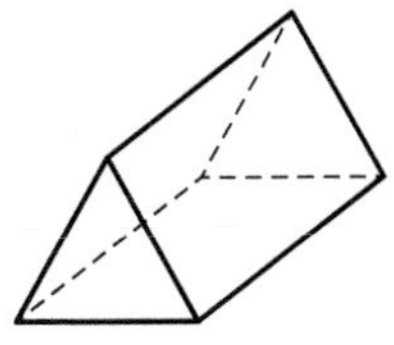

c

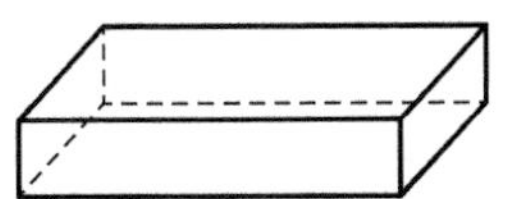

d

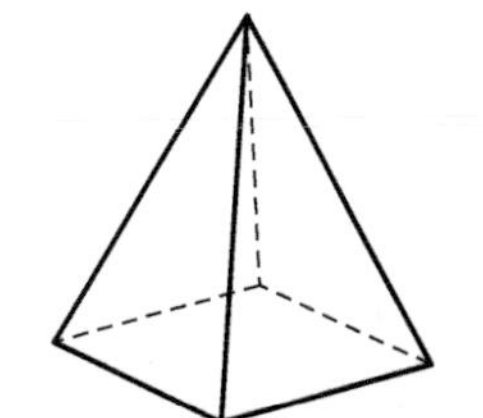

e

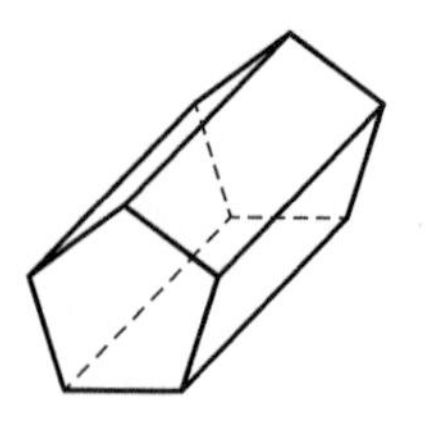

f

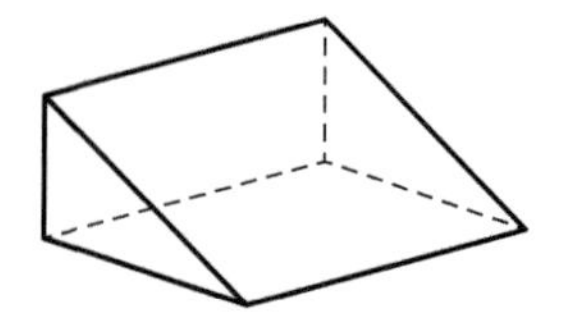

2 a

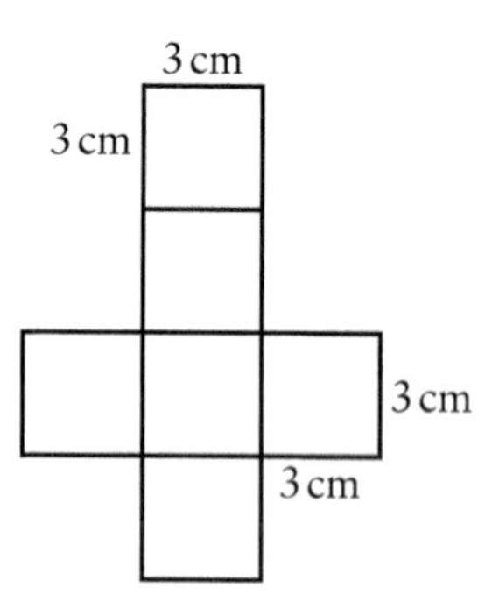

b

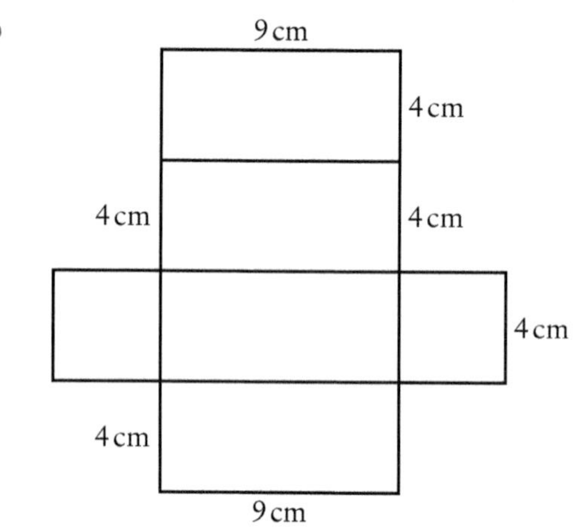

c

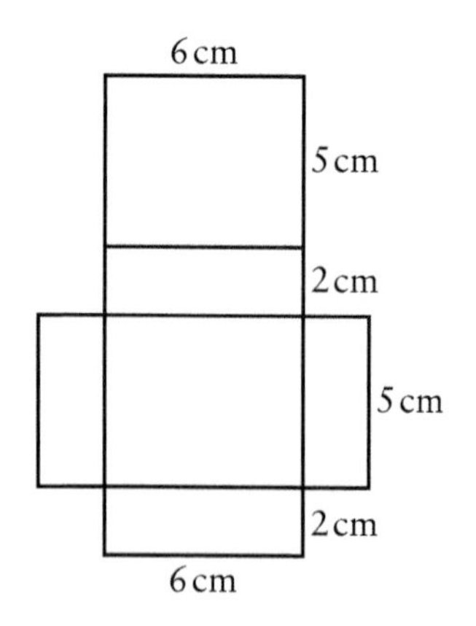

3 a

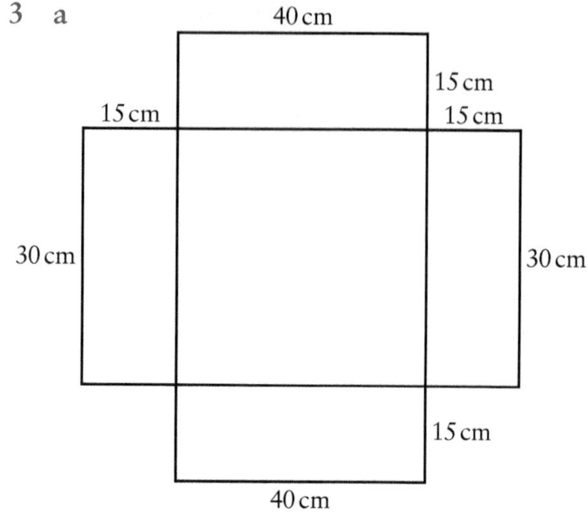

b $3300\ \text{cm}^2$

4 a $220\ \text{cm}^2$ b $352\ \text{cm}^2$ c $412\ \text{cm}^2$

5 $62\ \text{cm}^2$

6 a $300\ \text{cm}^2$ b $269\ \text{cm}^2$ c $241\ \text{cm}^2$
d $496.72\ \text{cm}^2$

7 $3780\ \text{cm}^2$

8 Cuboid. Its surface area is $246\ \text{cm}^2$, while that of the of the prism is $264\ \text{cm}^2$.

9 4 tins (surface area is $12.4\ \text{m}^2$)

10 a $864\ \text{cm}^3$ b $624\ \text{cm}^2$

11 Surface area $= 6l^2$

12 a $40\ \text{cm}^3$ b $88\ \text{cm}^2$

13 Length, width and height of the outer square and inner square. Scale factor of enlargement.

14 $294\ \text{cm}^2$

15 a Student's accurate drawing
b Volume $36\ \text{cm}^3$, surface area $84\ \text{cm}^2$

16 Student's spreadsheet

17 Student's own answer

Practice questions 9.4

1 a $1500\ \text{mm}^2$ b $3.8\ \text{cm}^2$
c $30\,000\ \text{cm}^2$ d $4.5\ \text{m}^2$
e $5\,000\,000\ \text{mm}^2$ f $90\ \text{cm}^2$

2 a $7000\ mm^3$ b $23\,000\ mm^3$

c $6\,000\,000\ cm^3$ d $0.85\ m^3$

e $1\,400\,000\ cm^3$ f $2.7\ cm^3$

3 a 12 ml b 5.6 ml

c 0.6 litres d $45\,000\ cm^3$

e 1000 litres f $5.47\ m^3$

4 240 ml

5 $12.92\ m^2$

6 a $0.9\ m^3$ b 900 litres = 18 sacks

7 $3424\ m^2$

8 a $177\ m^2$ (to the nearest square metre)

b 148 750 litres

9 a 1025 b 0.512 ml

10 $210\ cm^2$

11 a $5940\ cm^2$ b 31.2 litres

12 $224\ cm^3$

13 46.2 cm

14 a 20 cm b 9.1 cm (1 d.p.)

15 377 hours 47 minutes (to the nearest minute)

16 $20\,000\ m^3$

17 Student's own answers

Check your knowledge questions

1 a $100\ cm^2$ b $162\ cm^2$

2 14 cm

3 a $16\ cm^2$ b $42\ cm^2$

4 $1225\ cm^2$

5 Regular hexagons

6 a

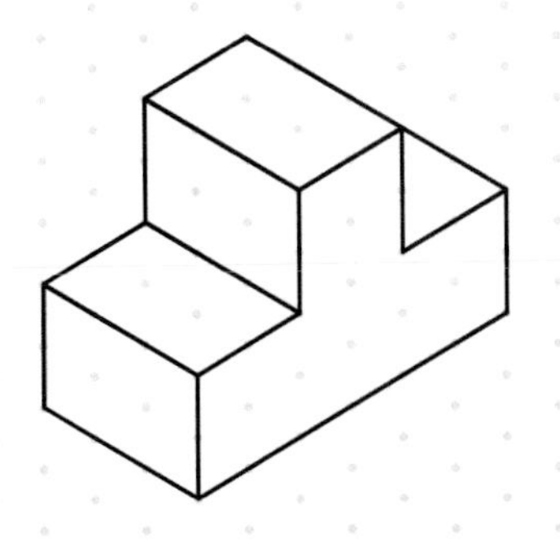

b $48\ cm^3$

7 Surface area $126\ cm^2$, volume $90\ cm^3$

8 a $256\ cm^3$ b $80\ cm^3$ c $100\ cm^3$

9 Volume $504\ cm^3$, surface area $396\ cm^2$

10 3 cm

11 a $5000\ cm^3$ b $0.54\ cm^2$

c $240\,000\ cm^2$ d 0.37 litres

12 840 litres

13 $729\ cm^3$

14 20 cm

Chapter 10 answers

Do you recall?

1 Survey of a sample; Quantitative.

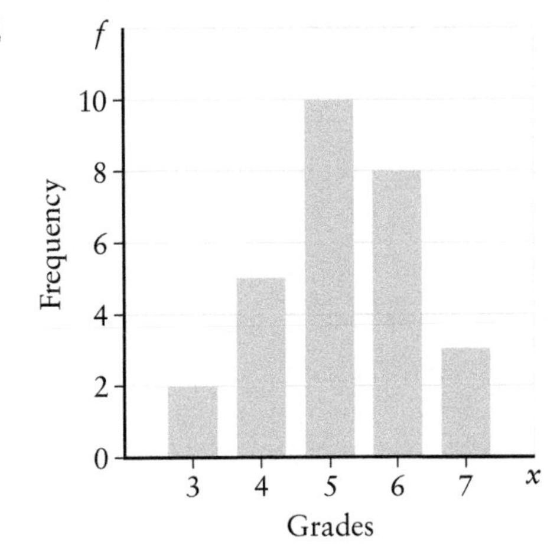

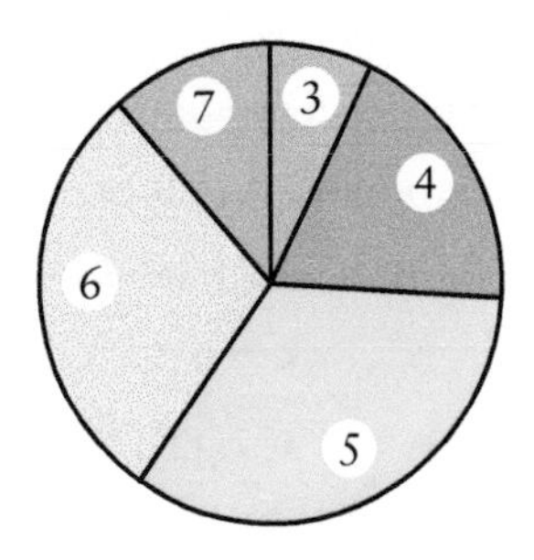

3 The top graph suggests that the higher the temperature, the higher the sales.
The lower graph suggests that ice cream sales do not depend on the temperature.

Practice questions 10.1

1 a i Quantitative and discrete

ii Quantitative and continuous

iii Quantitative and discrete

iv Qualitative

b i If the population (year group) is small then collect the data of all the population. If the population is large it would depend on the make-up of the population. If a single gender school you might want to use simple random sampling. If a mixed gender school you might want to use stratified sampling where the genders are the two strata.

ii If the population (school size) is small you might want to collect the data of all the population. If the population is large then you might want to use stratified sampling where the strata are the different year groups.

iii Break the population (people running the marathon) into male and female runners and then use systematic sampling within these two groups to establish a suitable sample size.

iv Similar arguments to part ii above.

2 a MYP Year 1 = 3 students. MYP Year 2 = 3 students, MYP Year 3 = 4 students

b MYP Year 1 = 10 students. MYP Year 2 = 8 students, MYP Year 3 = 12 students

3 a 15 b 9 c 6

4 a 卌 b 卌 ||| c 卌 卌 卌 卌 |

5 a frequencies: 3, 7, 11, 0, 14, 23

b 20 minutes

6 a

Type of fruit	Tally	Frequency
Apple	||	2
Banana	卌 |	6
Grapes	||||	4
Orange	卌	5
Pear	|||	3

b banana

c 20; The number of items in the list should match this value.

7 a

Height (m)	Frequency
1.41–1.50	4
1.51–1.60	4
1.61–1.70	7
1.71–1.80	5

b 12 people

c 1.61–1.70m

8 a It is unclear which class it belongs to.

b Mark: 0–10, 11–20, 21–30, 31–40, 41–50

Practice questions 10.2.1

1 a mode = 7, median = 6, mean = 5.45 (2 d.p.)

b mode = 4 and 7, median = 5, mean = 5.33 (2 d.p.)

c mode = 4, median = 4, mean = 5.23 (2 d.p.)

d mode = 4, median = 4, mean = 4.93 (2 d.p.)

2 a Set A: mode = 25, median = 22.5, mean = 21

b Set B: mode = 55, median = 52.5, mean = 51

As all numbers are increased by 30, all values of, mode, median and mean will increase by 30.

c Set C: mode = 5, median = 2.5, mean = 1 (all decrease by 20 from set A to set C)

d Set D: mode = 50, median = 45, mean = 42 (all doubled from set A to set D)

3 760

4 a $x = 91$ b $x = 92$ c $x = 94$

d $x = 100$ e $x = 95$

5 a 7, 7, 10

b There are many possibilities here, for example, 0, 1, 5 or −1, 1, 6 or anything of form a, 1, b ,where the numbers are in order and $a + b = 5$

6 a There are many possibilities here, for example, 7, 8, 10, 10 or 6, 8, 10, 10 or anything of form x, 8, 10, 10 where $x \leqslant 7$

b There are many possibilities, for example, 5, 5, 7, 7 or anything of the form x, x, y, y where $x + y = 12$

Practice questions 10.2.2

1

x	Frequency, f
8	5
9	8
10	2
11	3
12	7
13	6

Mean = 10.55 (2 d.p.), median = 11, mode = 9

2 Total $f = 35$; fx: 85.5, 108, 192.5, 150.5; Total $fx = 536.5$

a 16–19

b 16–19

c 15.33 (2 d.p.)

3 a

Classes	Frequency, f	Mid-class values, x	fx
24–30	6	27	162
31–37	6	34	204
38–44	1	41	41
45–51	4	48	192
52–58	6	55	330
59–65	5	62	310
66–72	7	69	483
73–79	5	76	380
Total	40		2102

b Estimate of mean = 2102/40 = 52.55

c Percentage error
= Error/Exact value × 100%
= (52.8 – 52.55)/52.8 × 100%
= 0.47% (2 d.p.)

d As the total frequency is 40, the median interval is the interval which contains the 20th and 21st data point. Both lie in the interval 52–58, hence this is median interval.

e Modal class = 66–72

f

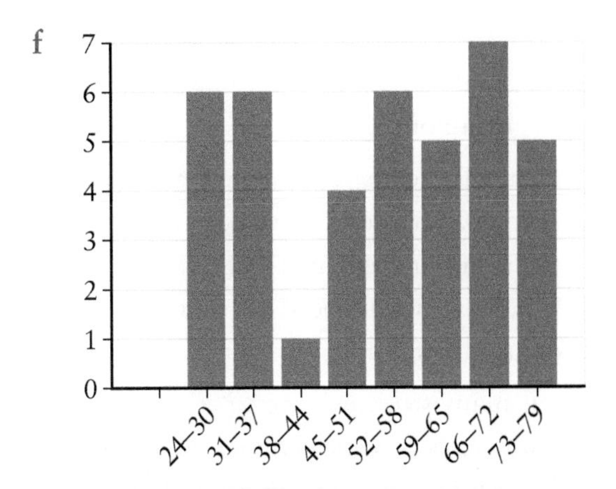

4

Time (secs)	Frequency, f	Mid-class values, x	fx
41–45	2	43	86
46–50	5	48	240
51–55	8	53	424
56–60	6	58	348
61–65	3	63	189
66–70	1	68	68
Total	25		1355

Estimate of mean = 1355/25 =54.2

5 Student's own results

6 a 21–23, 24–26, 27–29, 30–32, 33–35, 36–38

b 158–162, 163–167, 168–172, 173–177, 178–182

Practice questions 10.2.3

1 a Mean = 48.4, Median = 52, No mode. Median is better than mean, since 24 skews the mean.

b Mean = 33.75, Median = 37.5, Mode = 45. Median is better than mean, since half the group are in their 40s.

c Mean = 6540, Median = 5750, No mode. Median is better since values in lower half are fairly close, while values in upper half are spread out widely.

2 a median

b mode

c mean or median

3 It means they have between 1 and 2 pets on average. This could have been calculated using the mean or median. To decide which average is best, we would need to be familiar with the dataset. For example,

– if many students have 1 pet, and a few have multiple pets, which caused the mean to be 1.5, perhaps the median or mode would be best to use;

– if half the students have 1 pet and half the students have 2 pets, which caused both the mean and median to be 1.5, perhaps mentioning both modes would be best.

– if there are more modes, and neither the mean nor median are integer values, perhaps it makes the most sense to say the average is 'between 1 and 2 pets', since we are looking at discrete data.

Practice questions 10.2.4

1 a Car B, because the range is smaller.

b Car B *could* have had the overall fastest lap, but it seems more likely that car A had both the fastest and slowest laps overall. Car A is clearly faster on average, since it has a lower mean time.

c It would probably be helpful to have more information, such as the median times and the minimum times.

2 a both A and B: mean = 60, median = 51.5, mode = 51, range = 50

b The statistics calculated in part **a** would initially have us believe that the two classes are similar. However on closer inspection the performance of pupils in class A are very similar with one outlier of 100, whereas in class B there is a bigger spread of ability.

3 a Su-Bin goals: mean = 1.2, median = 1, mode = 1, range = 1

Su-Bin assists: mean = 1.3, median = 1, mode = 1, range = 1

Laura goals: mean = 1.5, median = 0.5, mode = 0, range = 5

Laura assists: mean = 0.6, median = 0.5, mode = 0, range = 2

b Su-Bin is the better striker as she is simply more consistent. She has scored and assisted in every game.

Anji goals: mean = 1.4, median = 1, mode = 0 or 0 & 3, range = 3

Anji assists: mean = 0.8, median = 1, mode = 1, range = 1

For similar reasons of consistency, Su-Bin is better striker (other arguments could be put forward depending on criteria for *Best Striker.* i.e does assisting count? Its goals that matter!)

Check your knowledge questions

1 Question 1 needs to have categories/genres to pick from e.g. fiction, thriller, mystery, biography, comic strip, etc. Alternatively pick three from a long list.

In question 2 there is an overlap in the boxes e.g. 3 appears in two of the boxes.

Better categories would be:

0–2 times ☐ 3–5 times ☐

6–8 times ☐ more than 8 times ☐

2 a It may help Farid to know their names, simply from an organisational point of view, to ensure that he does not give multiple questionnaires to the one person.

b People are sometimes not completely honest when a survey is not anonymous.

c Q1: leading question – it seems the person giving the survey wants you to say 'yes'.

Q2: only provides reasons to support a previous 'yes' response and does not allow for other reasons if you have said 'no'.

Q3: this question should allow for multiple answers.

d Student's own answer

e Students ask their classmates their questions and record their answers.

3 a i Likely to all give the same response.

ii Likely not to give sports or activities played only or primarily by girls.

iii Other year groups may have different interests than students in his year group.

iv Primary students may have different interests than secondary students.

b Stratify by both year group and gender

4 a mode = none, median = 20.5, mean = 20, range = 17

b mode = none, median = 29, mean = 27, range = 21

c mode = none, median = 82, mean = 78.7, range = 36

d mode = none, median = 7.25, mean = 7.33, range = 7.05

5 9 goals

6 Examples 4, 10, 10, 10, 10, 10, 10, 16 and 3, 10, 10, 10, 10, 10, 12, 15

7 a

No. people, x	Frequency, f	fx
1	3	3
2	6	12
3	7	21
4	8	32
5	6	30
6	4	24
7	2	14
8	1	8
9	2	18
	$\Sigma f = 39$	$\Sigma fx = 162$

b mean = 4.15, median= 4, mode= 4, range = 8

8 a missing f: 4, 3, 8, 5, Total = 20

missing x: 1.65, 1.75

missing fx: 5.8, 4.65, 13.2, 8.75, Total = 32.4

mean estimate = 1.62

b 1.615

c 0.310% (3 s.f.)

9 a B

b A, since A's range is smaller, but only slightly.

c B had the fastest time, whereas A had the lower median time. They both had the same mean time.

A was slightly more consistent. Overall, A had the better season but not much in it!

d median

10 a mean of boys = 4.2, mean of girls = 3 hence boys are better at taking penalties.

b mean of boys = 1.8, mean of girls = 3 hence girls are better at taking foul shots.

c Combining both penalty shots and foul shots, the percentage success rate for both boys and girls is the same. Hence no difference.

(Percentage girls = 42/140 × 100% = 30%, Percentage boys =30/100 × 100% = 30%)

d Simply by combining totals, the class is better at taking penalties than taking foul shots.

(42 penalties against 30 foul shots)

e i The class has a lower range in successful penalties, hence more consistent at taking penalties,

ii Girls have a smaller range in successful penalties, hence more consistent at taking penalties,

iii Boys have a smaller range in successful foul shots, hence more consistent at taking foul shots.

11 a Each region has a different population, so averages for smaller regions will skew the mean in its direction. Also, we should check that every country in the world has been included in one of these regions.

b If we can find the population of each region, we can use those values as a frequency column and calculate the mean as we do from a table.

12 One example of each is given

a Set A: 11, 12, 13, 14, 15
Set B: 9, 11, 13, 15, 17

b Set A: 18, 19, 20, 21, 22
Set B: 19, 20, 21, 22, 23

c Set A: 13, 14, 15, 16, 17
Set B: 12, 13, 15, 18, 19

d Same answer as part **a**.

13 Given that the values in the table are based on a significant number of runs and that the World Record is close to 10 seconds, pick B as he/she has a much lower mean. The fact that the range is high for this person could be accounted by the result from one bad run e.g. pulled up!

14 a mean = 18.3 (3 s.f.), median = 18

b i The mean increases to 25.7 i.e. an increase of roughly 40%.

ii The median increases to 20 i.e. an increase of roughly 11%.

15 The mean is affected twice as much as the median, as it decreases by $2.07 (11%), and the median decreases by $0.75 (4%).

16 a

Grade	Frequency
3	3
4	30
5	12
6	9
7	18

b 5.125

17 a 14

b 15

c English: 11, Maths: 16

d

Score	5	6	7	8	9	10	11	12
Maths	1	0	1	1	1	2	2	3
English	0	0	1	1	2	1	5	2

Score	13	14	15	16	17	18	19	20
Maths	1	3	1	4	0	2	0	1
English	3	2	2	1	2	0	1	0

e The median is not necessarily exactly between the minimum and maximum, it is the middle value when all the values are put in order (or exactly in between the two middle values, if there are an even number of values in the list). It must take account of the frequencies.

f English: median = 12, mean = 12.5 (3 s.f.);
Maths: median = 13, mean = 12.9 (3 s.f.)

g The class performed slightly better in maths when comparing both mean and median scores, however the scores for maths were also more spread out, as evidenced by the ranges, and the lowest score overall was in maths.

Chapter 11 answers

Do you recall?

1 a 40% b 0.023 c 7%

2 a 0.4 b $\frac{3}{10}$ c $\frac{1}{3}$

3 a $\frac{1}{6}$ b $\frac{2}{7}$ c $\frac{2}{9}$

4 a 9 b 5 c 1.6

Practice questions 11.1.1

1 a zero b 0.5 c 1 or 100%

2 a outcome

b not an outcome – a probability experiment

c not an outcome – a probability experiment

d outcome

e not an outcome – a probability experiment

3 Unlikely – Snow days are reasonably rare in Sweden in March. But they do occur.

4 Impossible

5 a $\frac{1}{2}$ b $\frac{2}{3}$ c $\frac{1}{3}$ d $\frac{5}{6}$

6 a i $\frac{1}{2}$ ii $\frac{1}{4}$ iii $\frac{1}{4}$

b 1, the three probabilities account for all possibilities.

7 a $\frac{1}{6}$ b 0 c $\frac{1}{2}$

d $\frac{1}{3}$ e $\frac{1}{2}$ f 1

Practice questions 11.1.2

1 a {hearts, diamonds, clubs, spades}

b {1, 2, 3, 4, 5, 6, 7, 8}

c {black, yellow}

2 a $\frac{1}{6}$ b $\frac{1}{2}$ c $\frac{1}{3}$ d 0

3 a $\frac{1}{12}$ b $\frac{1}{2}$ c $\frac{3}{12} = \frac{1}{4}$

d $\frac{2}{12} = \frac{1}{6}$ e $\frac{5}{12}$

4 a $\frac{1}{20}$ b $\frac{4}{20} = \frac{1}{5}$ c $\frac{8}{20} = \frac{2}{5}$

d $\frac{12}{20} = \frac{3}{5}$

5 a $\frac{1}{2}$ b $\frac{1}{13}$ c $\frac{1}{4}$

d $\frac{36}{52} = \frac{9}{13}$ e $\frac{1}{52}$

6 a $\frac{1}{2}$ b $\frac{2}{10} = \frac{1}{5}$ c $\frac{3}{10}$

7 a $\frac{8}{18} = \frac{4}{9}$ b $\frac{4}{18} = \frac{2}{9}$ c $\frac{6}{18} = \frac{1}{3}$

8 4 green counters

9 6 dark chocolates

10 a 36. Examples are (1, 5), (1, 6), (3, 1), (3, 3), (6, 4)

b $\frac{1}{6}$

Practice questions 11.2

1 a Rolling an even number

b Drawing a number or ace

c Not drawing a diamond

d A coin landing on heads

2 a P(odd) = 0.4

b P(no rain) = 0.65

c P(not faulty) = 0.95

d P(False) = 0.01

3 a $\frac{52}{60} = \frac{13}{15}$ b $\frac{2}{15} \times 25 = \frac{10}{3} \approx 3.33$

4 a $\frac{58}{82} = \frac{29}{41}$ b $\frac{24}{82} \times 10 \approx 2.93$

5 a $\frac{1470}{1550} = \frac{147}{155}$ b $\frac{790}{850} = \frac{79}{85}$

c $\frac{140}{2400} = \frac{7}{120}$

6 a $\frac{2509}{9316} \approx 0.269$ b $\frac{3150}{9316} \approx 0.338$

c $\frac{2117}{9316} \approx 0.227$

7 a $\frac{170}{500} = 0.34$

b Anything above €17 could potentially be profitable, however it still would depend since this probability is an estimate.

8 a $\frac{1}{2}$ b $\frac{1}{2}$ c 1 d 0

9 a $\frac{16}{17}$ b $\frac{7}{10}$ c $\frac{133}{170}$ d $\frac{72}{170} = \frac{36}{85}$

10 a $\frac{1}{4}$ b $\frac{3}{4}$ c $\frac{1}{13}$

d $\frac{12}{13}$ e $\frac{3}{13}$ f $\frac{10}{13}$

g $\frac{5}{13}$ h $\frac{8}{13}$ i 1

Check your knowledge questions

1 a Equally likely b Unlikely

c Certain

2 a Yes: $\frac{1}{8}$ b Yes: $\frac{1}{20}$

c No: Cannot be calculated

3 a $\frac{6}{15} = \frac{2}{5}$ b $\frac{5}{15} = \frac{1}{3}$

c $\frac{4}{15}$

4 a $\frac{1}{12}$ b $\frac{2}{12} = \frac{1}{6}$

c $\frac{5}{12}$ d $\frac{4}{12} = \frac{1}{3}$

e $\frac{4}{12} = \frac{1}{3}$

5 a $\frac{1}{13}$ b $\frac{12}{13}$

c $\frac{1}{4}$ d $\frac{3}{4}$

e $\frac{3}{13}$ f $\frac{10}{13}$

6 a The probability of an odd number is half/a half/one-half.

b The probability of not A is one-third.

c The probability of failing is zero.

7 a P(rain) = 0.7 b P(even) = 0.4

c P(face) = $\frac{3}{4}$

8 a i $\frac{5435}{25785} \approx 0.211$ ii $\frac{7820}{25785} \approx 0.303$

b He could start a project or political campaign to raise awareness of these statistics or start a food drive to help those in need.

9 12

10 a 0.97 b $\frac{1}{4}$ c Yes

11 a (H, H, H), (H, H, T), (H, T, H), (H, T, T), (T, T, T), (T, T, H), (T, H, T), (T, H, H)

b i $\frac{3}{8}$ ii $\frac{3}{8}$ iii $\frac{1}{8}$ iv $\frac{7}{8}$